BUSINESS DATA COMMUNICATIONS

Business Data Communications

Raymond R. Panko
University of Hawaii at Manoa

PRENTICE HALL
Upper Saddle River, NJ 07458

Library of Congress Cataloging-in-Publication Data
Panko, R. R.
Business Data Communication/ Raymond R. Panko
p. cm.
Includes bibliograghical references and index.
ISBN 0-13-308164-8
1. Business—Data processing—Case studies. 2. Business-
Communication Systems—Data processing—Case studies. 3. Local
area networks (computer networks)—Case studies. 3. local area
networks (computer networks)—Case studies. 4. Data transmission
systems—Case studies. 4. Data HF5548.2.P2594 1997
005.7'1—dc20

96-7869
CIP

Director of Production and Manufacturing: *Joanne Jay*
Managing Editor: *Katherine Evancie*
Project Manager: *Susan Rifkin*
Manufacturing Buyer: *Alana Zdinak*
Editor-in-Chief: *Rich Wohl*
Acquisitions Editor: *Jo-Ann Deluca*
Design Director: *Patricia Wosczyk*
Art Director: *Sue Behnke*
Designer: *Maureen Eide*
Cover Designer: *Maureen Eide*

© 1997 by PRENTICE-HALL, INC.
Simon & Schuster/A Viacom Company
Upper Saddle River, NJ 07458

All rights reserved. No part of this book may be repro-
duced, in any form or by any means,
without permission in writing from the publisher.

Printed in the United States of America

10 9 8 7 6 5 4 3 2 1

ISBN 0-13-308164-8

Prentice-Hall International (UK) Limited, *London*
Prentice-Hall of Australia Pty. Limited, *Sydney*
Prentice-Hall Canada, Inc., *Toronto*
Prentice-Hall Hispanoamericana, S.A., *Mexico*
Prentice-Hall of India Private Limited, *New Delhi*
Prentice-Hall of Japan, Inc., *Tokyo*
Simon & Schuster Asia Pte. Ltd., *Singapore*
Editora Prentice-Hall do Brasil, Ltda., *Rio de Janeiro*

*I would like to dedicate this book
to Dr. Douglas C. Engelbart, who invented
hypertext, the mouse, groupware,
document-centric computing, and several more
important things we haven't come
to appreciate yet.*

About the Author

Ray Panko was introduced to electromagnetic waves as an undergraduate physics major. He worked on propagation at Boeing, but he also got his MBA, which introduced him to issues beyond technology.

He did his Ph.D. at Stanford, in Communications, in the 1970s. This allowed him to study both raw technology and real-world applications. He learned advanced programming from Don Knuth. Bruce Lusignan, who invented VSATs, hired him to work on one of the first VSAT projects. He designed a broadband cable LAN for the campus under Ed Parker. On that project, he worked with Paul Baran, the inventor of packet switching. He also worked on projects using cable television for interactive residential services.

His doctoral dissertation was funded by the Executive Office of the President of the United States. To do the work, he moved to SRI International. In the Telecommunications Sciences Center, he worked on videoconference, computer conferencing, and electronic mail. He also worked in the Augmentation Research Center, which was run by Doug Engelbart, who invented the mouse, hypertext, and many other advances. His team even built the first client/server computer system.

After SRI International, he moved to the University of Hawaii, where he is currently a full professor in the College of Business Administration. He also engages in external consulting and training to stay grounded in current issues (and to support his teaching habit). Among his clients have been AT&T, Xerox Palo Alto Research Center, British Telecoms, Bell Canada, Trans-Canada Telephone System, Fujitsu, Wang Laboratories, Hughes Aircraft Space Division, RCA, Royal Dutch Shell, The Council of European Post and Telecommunications Administrations, RCA, 20th Century-Fox, ABC, CBS, NBC, HBO, NASA, National Institutes of Standards and Technology, NSF, Congressional Office of Technology Assessment, Defense Advanced Research Projects Agency, U.S. Army Development and Readiness Command, U.S. Army Ballistics Research Laboratory, Naval Electronics Laboratory Command, A.C. Nielson DATAQUEST, and Auerbach Publishers.

Table of Contents

PREFACE xvii

1 INTRODUCTION 1

A TALE OF TWO COMPANIES 1
Verifone 1   NuPools 3   Perspectives 4   The Chapter 4
THE BASIC COMMUNICATION MODEL 5
A Framework for Mediated Communication 5   The User Layer 6   The Application Layer 7   The Computer (Transport) Layer 8   The Transmission Layer 8   Subdividing the Transmission Layer 9
STANDARDS 10
Standards Agencies and Standards Architectures 10   OSI 12   TCP/IP 14   IPX/SPX 17   Systems Network Architecture (SNA) 17   Setting Organizational Standards 17
THE TRANSMISSION LAYER 18
Transmission Speed 18   LANs, MANs, WANs, and Internets 19
CONCLUSION 21   CORE REVIEW QUESTIONS 22
DETAILED REVIEW QUESTIONS 23
THOUGHT QUESTIONS 23   PROJECTS 24
REFERENCES 24

2 BEYOND THE TRANSMISSION LAYER 25

INTRODUCTION 25
THE COMPUTER (TRANSPORT) LAYER 26
Terminal–Host Systems 27   Networked PCs 28   RISC Workstation Networks 31   Multiple Computer Platforms and Networks 33
THE APPLICATION LAYER 33
Standards 34   Client/Server Processing 34   Electronic Mail 37   The World Wide Web 39
THE USER LAYER 42
INTERNET ACCESS VIA CSLIP AND PPP 43
Devices 44   Layering 45   Message Transmissions 47

Reception of the PDUs 49   Interfaces 50   Flexibility 50
Internet Access at VeriFone and NuPools 51
CONCLUSION 51   CORE REVIEW QUESTIONS 54
DETAILED REVIEW QUESTIONS 54
THOUGHT QUESTIONS 55   PROJECTS 55
REFERENCES 55

3 TERMINAL–HOST COMMUNICATION 56

INTRODUCTION 56
TERMINAL EMULATION 58
VT100 Terminals 58   Other Terminals 61
TELEPHONE TRANSMISSION 62
Digital and Analog Transmission 63   Modems 64   Modem Standards 64   Limits of the Telephone System 71
ASYNCHRONOUS ASCII TRANSMISSION 72
OTHER VT100 TRANSMISSION SETTINGS 74
Half-Duplex and Full-Duplex Transmission 74   The Serial Plug (Connector) 74
THE ISDN 76
The Basic Rate Interface (BRI) 77   Cost 78
CONCLUSION 80   CORE REVIEW QUESTIONS 81
DETAILED REVIEW QUESTIONS 81
THOUGHT QUESTIONS 82   PROJECTS 83

4 A SIMPLE ETHERNET LAN 84

INTRODUCTION 84
IEEE LAN STANDARDS 84
The IEEE 85   Layering 85
NUPOOLS: BUILDING A LAN WITH ETHERNET 10BASE-T: THE PHYSICAL LAYER 88
Network Interface Cards (NICs) 89   Unshielded Twisted Pair (UTP) 89   Operation of 10Base-T Hubs 91   Electrical Signaling 92
THE MAC LAYER: CSMA/CD AND THE 802.3 MAC LAYER FRAME 94
Media Access Control 94   The MAC Layer Frame 96   The 802.2 LLC Layer Frame 99
LARGER NETWORKS 100
Building a Larger Network by Adding Hubs 100   Mixing Other 802.3 Physical-Layer Technologies 103   10Base-F 105

CONCLUSION 106 CORE REVIEW QUESTIONS 107
DETAILED REVIEW QUESTIONS 108
THOUGHT QUESTIONS 108 PROJECTS 109
REFERENCES 109

5 OTHER LAN TECHNOLOGIES 110

INTRODUCTION 110
THE 802.5 TOKEN-RING NETWORK 110

*Ring Topology 111   Shielded Twisted Pair (STP) Wiring 112
Access Units 112   Speed 113   Token Passing 113   Market
Status of 802.5 114*

100 Mbps LANS 115

*Problems with 10 Mbps LANs 115   FDDI 116
802.3 100Base-X 117   802.12: 100VG-AnyLAN 119
100 Mbps Shared Media LANs in Perspective 120*

ALL-SWITCHED LANS 120

*The Death of Shared Media LANs 120   Switching 121
Switched Network 122   Alternative Routing 123   Ethernet
Switches 124   ATM Switches 124   Switching and the OSI
Architecture 125*

RADIO LANS 125
CONCLUSION 126 CORE REVIEW QUESTIONS 127
DETAILED REVIEW QUESTIONS 128
THOUGHT QUESTIONS 128 PROJECTS 128
REFERENCES 128

6 LOCAL INTERNETS 129

INTRODUCTION 129

*The Need for Local Internets 129   Internetting versus Switched
LANs 132   Internetting Technologies 132   The Chapter 133*

BRIDGES 133

*Learning 133   Bridged Internets: No Loops Allowed! 136
Disadvantages of Bridges 136   Advantages of Bridges 137
Bridge Standards 138*

SWITCHED LOCAL INTERNETS 139

Switching 139   Switched Ethernet Internets 140

ROUTERS 141

Sophistication 141   Multiprotocol Routers 145

INTERNETTING DEVICES IN PERSPECTIVE 145

*Routers 145   Bridges 146   Mixing Bridges and Routers 146
Switched Internets 147*

BACKBONES FOR LOCAL INTERNETS   147
Distributed Backbone Connections 147   Collapsed Backbone Internets 148
CONCLUSION   150   CORE REVIEW QUESTIONS   150
DETAILED REVIEW QUESTIONS   151
THOUGHT QUESTIONS   152   PROJECTS   152

7 ENTERPRISE INTERNETS 153

INTRODUCTION   153
Tiers of Carriers 154
TRANSMISSION CARRIERS AND DEREGULATION   154
Local Transmission Service 156   Circuits 157
LEASED LINE TRANSMISSION LINKS   160
Advantage and Disadvantage 161   Networks of Leased Lines 161   Virtual Private Networks 163
SWITCHED DATA NETWORKS   171
Switched Data Networks: Elements and Market Conditions 172
COMMON SWITCHED DATA NETWORK STANDARDS   173
X.25 Data Networks 174   ISDN Service 175   Frame Relay 176   Asynchronous Transfer Mode (ATM) 177   Frame Relay and ATM Pricing 177
CONCLUSION   180   CORE REVIEW QUESTIONS   181
DETAILED REVIEW QUESTIONS   181
THOUGHT QUESTIONS   182   PROJECTS   183

8 PC NETWORKING 184

INTRODUCTION   184
ELEMENTS IN A PC NETWORK   184
Client PCs 185   Servers 186   Server Application Software 188   Operating Systems 188   The Network 191
COMMUNICATION BETWEEN CLIENT PCS AND SERVERS   192
Communication Between a Client PC and a Single Server 192
FILE SERVICE   194
File Service: The User's View 194   File Service for Data Files 196   Program File Service 199   Redirection 202
OTHER SERVICES   204
Print Service 205   Client/Server Processing 205   Communication Servers 210   Electronic Mail Servers 212   Network Management Servers 212
THE NUPOOLS SERVERS   212

CONCLUSION 213 CORE REVIEW QUESTIONS 214
DETAILED REVIEW QUESTIONS 215
THOUGHT QUESTIONS 215 PROJECTS 216

9 NETWORK MANAGEMENT 217

INTRODUCTION 217
NETWORK MANAGEMENT SYSTEMS 218
The Overall Network Management System 219 The Network Management Program 220 Managed Devices 222 Distributed Network Management 223 Network Management Messages 224
NETWORK MANAGEMENT STANDARDS 225
Needed Standards 225 Simple Network Management Protocol (SNMP) 227 CMIP/CMIS 229 NetView 230 Other Network Management Standards 230 Limits of Standards 232 Cutting Across Standards 233
FAULT MANAGEMENT 233
Recognizing the Existence of Faults 234 Diagnosing Faults 234
CONFIGURATION MANAGEMENT 235
Changing States 235 Electronic Software Distribution and Control 236 Current Network Map 236 Remote Management 236 Directory Servers 237
PERFORMANCE MANAGEMENT AND PLANNING 237
Vital Statistics 237 What-If Analysis in Performance Testing 237 Planning New Networks 240 Client/Server Performance Management 241
SECURITY MANAGEMENT 241
Encryption and Authentication 241 Access Control 243 Virus Control 245
ACCOUNTING MANAGEMENT 245
OUTSOURCING 245
CONCLUSION 246 CORE REVIEW QUESTIONS 246
DETAILED REVIEW QUESTIONS 247
THOUGHT QUESTIONS 248 REFERENCES 248

10 NETWORKED COMMUNICATION APPLICATIONS 249

INTRODUCTION 249
The Two Application Chapters 249 Standards 250
ELECTRONIC MAIL 251
Basic Features 252 Advanced Features 252 Electronic Mail

Standards 254   Mail Etiquette and Legality 259   E-mail at Verifone and NuPools 259
GROUPWARE 262
Categories of Group Processes 263
KEYBOARD-BASED GROUPWARE 264
Services 264   Lotus Notes 268   Internet Groupware 270
VIDEOCONFERENCING 270
Room-to-Room Videoconferencing 270   Desktop Conferencing 273   Telephony 274
ELECTRONIC MEETING ROOMS 274
PRODUCTION SYSTEMS 276
CONCLUSION 277   CORE REVIEW QUESTIONS 278
DETAILED REVIEW QUESTIONS 279
THOUGHT QUESTIONS 279   REFERENCES 279

11 NETWORKED DATABASE APPLICATIONS AND REENGINEERING 281

INTRODUCTION 281
DATA APPLICATIONS 281
Client/Server Databases 282   Software 285   Standards and Middleware 286   Data Warehouses for Decision Support 289   Three-Tier Processing for Production Data 290   Distributed Databases 291   Multimedia Databases 292   Commercial Online Services 293   Database Applications at VeriFone and NuPools 296
REENGINEERING 296
Obliterate 296   Consolidate 297   Teams 297   Interorganizational Systems 298   Outsourcing 298   New Channels 299   Telecommuting 299
CONCLUSION 299   CORE REVIEW QUESTIONS 300
DETAILED REVIEW QUESTIONS 300
THOUGHT QUESTIONS 301   PROJECTS 301
REFERENCES 301

A THE OSI ARCHITECTURE 303

INTRODUCTION 303
THE OSI ARCHITECTURE 303
The OSI Transmission Layers 304   OSI Internetting 309   OSI OSI Transport-Layer Standards 311   Application-Layer Protocols 312   Common Functions 312
CONCLUSION 315   CORE REVIEW QUESTIONS 315

DETAILED REVIEW QUESTIONS 316
THOUGHT QUESTIONS 316   REFERENCES 316

B THE INTERNET 317

INTRODUCTION 317
BASIC CONCEPTS 318
ORIGINS 319
*The ARPANET and Telnet 319   ARPANET: Electronic Mail 321
Connecting Networks 321   The Internet 322   Managing the
Internet 322*
HUMAN COMMUNICATION SERVICES 323
*LISTSERV 323   USENET 326   Real-Time Discussions 326
Newbies and FAQs 328   Videoconferencing 328   Internet
Telephone Service 329*
INFORMATION RETRIEVAL SERVICES 329
*Anonymous FTP 329   ARCHIE 333   Gopher 333
VERONICA 334   WAIS 334*
ACCESSING THE INTERNET 335
*Online Services 335   CSLIP and PPP Access 336   Access Via
the Corporate PC Network Internet Gateway Server 336
Firewalls 337*
TOMORROW'S INTERNET 337
Funding 337   Crime and Netiquette 337   Speed 338
CONCLUSION 338   CORE REVIEW QUESTIONS 339
DETAILED REVIEW QUESTIONS 339
THOUGHT QUESTIONS 340   PROJECTS 341
REFERENCES 341

C THE WORLD WIDE WEB AND BROWSERS 342

INTRODUCTION 342
THE WORLD WIDE WEB 342
*Hypertext 343   The World Wide Web 343   Surfing the Net,
Crawling the Web 346   Commercial Transactions 348
Speed 349*
HTML 349
*Creating HTML Pages 349   A Simple Page 350   Adding
Links 353   Adding Images 356   Other Markup Codes 359
Special Characters 360   Lists 360   Forms and CGI 361
Java 362*
BROWSERS 363

CONCLUSION 365 CORE REVIEW QUESTIONS 366
DETAILED REVIEW QUESTIONS 366
THOUGHT QUESTIONS 367 PROJECTS 367
APPENDIX TO MODULE C: HTML CODES 368

D ADVANCED TOPICS IN POINT-TO-POINT TRANSMISSION 373

INTRODUCTION 373
DIGITAL CODES AND ENCODING 373
*Pixel-Mapped Graphics Codes and Compression 374
Vector Graphics 376   Text and Graphics Documents 376*
BANDWIDTH, NOISE, AND SPEED 378
MODEMS 379
*Modem Transmission Speeds 379   Modem Error Correction 381
Modem Compression 382   Modem Intelligence 383   Modem Forms 384   Facsimile Modems 386
Other Types of Modems 388*
CORE REVIEW QUESTIONS 389 DETAILED REVIEW QUESTIONS 389 THOUGHT QUESTIONS 390
REFERENCES 390

E ADVANCED TOPICS IN V100 TERMINAL EMULATION 391

INTRODUCTION 391
VT100 TERMINAL EMULATION 391
VT100 Terminals 391   Setting Up the Communication Program 392   Host Setup 393   Saving Setup Information 394   File Transfer Protocols 394
CORE REVIEW QUESTIONS 395
THOUGHT QUESTIONS 395 PROJECTS 396

F ADVANCED TOPICS IN LAN TECHNOLOGY 397

INTRODUCTION 397
ADDITIONAL ETHERNET STANDARDS 397
*10Base5 398   802.3 Physical-Layer Standard 10Base2 401
Mixed 802.3 Networks 403*
THE 802.5 TOKEN-RING STANDARD 404
Differential Manchester Encoding 404   The 802.5 MAC-Layer Frame: Token Frame 406   The 802.5 MAC-Layer Frame: Full Frame 408   Error Handling 410

CORE REVIEW QUESTIONS 414 DETAILED REVIEW
QUESTIONS 414 THOUGHT QUESTIONS 415
PROJECTS 415 REFERENCES 415

G TELEPHONE SERVICE 416

POTS 416
THE TRADITIONAL TELEPHONE SYSTEM 417
A Telephone Call 417   Carriers 417   Regulation 419
CUSTOMER PREMISES EQUIPMENT 421
PBXs 421   PBX Networks 423   User Services 424
CARRIER SERVICES AND PRICING 425
Tariffs 426   Basic Voice Services 427   Advanced Services 428
CELLULAR TELEPHONES 429
Cellular Concepts 429   Analog Cellular Telephones: The First Generation 432   Digital Cellular: The Second Generation 434   Personal Communication Services: The Third Generation 434   LEO Services 436
CONCLUSION 436   CORE REVIEW QUESTIONS 436
DETAILED REVIEW QUESTIONS 437
THOUGHT QUESTIONS 438   PROJECTS 438
REFERENCES 439

H ADVANCED TOPICS IN INTERNETTING 440

BASIC TCP/IP CONCEPTS 440
*Layering 440   IP Numbers 441   Delivery 444
IP Addressing in More Detail 448   TCP and UDP 451
Supervisory Protocols 453   IP Version 6 459*
CORE REVIEW QUESTIONS 461   DETAILED REVIEW
QUESTIONS 462   THOUGHT QUESTIONS 463
REFERENCES 463

I MANAGING A PC NETWORK 464

INTRODUCTION 464
INSTALLING A FILE SERVER 464
Planning the Directory Structure 465   Selecting Server Hardware 468   Installing the Server Operating System 473   Installing Application Software 473
ADDING A USER 475

Installing Hardware and Software on a Client PC 475   Installing a User's Account on the File Server 481
ONGOING MANAGEMENT 482
Monitoring the Server 482   Protocol Analyzers 488   User Support 489
CONCLUSION 490   CORE REVIEW QUESTIONS 491
DETAILED REVIEW QUESTIONS 492   THOUGHT QUESTIONS 493   PROJECTS 493   REFERENCES 494

J ADVANCED TOPICS IN MANAGEMENT AND APPLICATIONS 495

INTRODUCTION 495
ENCRYPTION AND AUTHENTICATION 495
Encryption 495   Authentication 500
VIDEO COMPRESSION 502
DOCUMENT CONFERENCING 504
DATABASE MIDDLEWARE 505
CORE REVIEW QUESTIONS 507
DETAILED REVIEW QUESTIONS 507
THOUGHT QUESTIONS 508   REFERENCES 508

GLOSSARY 509

INDEX 540

Preface

THE BOOK

Ten years ago, MIS programs rarely offered data communications. Today, almost all MIS programs require at least one *datacoms* course. Many offer two or more. Networking is exploding in corporations as well. In the past, most MIS graduates were hired as programmers. Most firms today, however, are hiring more networking/user support specialists than programmers. Even if you want to be a programmer, furthermore, you will need to master networking. Programming today often requires you to write code that interacts with other programs across a network. Mainframes and minicomputers dominated the 1970s. PCs owned the '80s. The 1990s have become the *Network Era*.

A Modular Approach

When I started writing this book, I talked to a number of teachers. It was clear that there is no standard body of material that everybody covers. That's not surprising in such a new area, but it creates problems. Traditional rigidly structured text books make it difficult for teachers to select material to fit their courses. In addition, teachers often have to assign one or two supplementary books in additional areas they consider to be important.

To help teachers, this book has a *modular design*. The book begins with eleven *core chapters* that most teachers are likely to cover. Beyond that, teachers can enrich the material in core chapters by assigning specific material in *advanced modules*. This allows teachers to focus on transmission technology, on applications, on the Internet, or on whatever else they feel is most appropriate for their students. There is even enough material to use the book in a second datacoms course.

Market-Driven Content and Concrete Presentations

Benjamin Franklin once said that he wished to tell people "solid useful things." This book tries to do the same. For instance, Chapter 4 introduces LANs. Many texts limit themselves to abstract concepts, such as bus versus ring designs. This leaves students ill-prepared to install even a simple LAN that uses contemporary technology. In addition, many texts treat all LAN technologies equally, regardless of market acceptance. This leaves too little time to teach the information they will really need after graduation.

Chapter 4 of *Business Data Communications*, in contrast, focuses on 10Base-T LANs. This is the most popular LAN technology by far, and students need to know it well. In addition, new standards for 100 Mbps LANs and switched networks build on the structural design of 10Base-T.

The presentation of 10Base-T is very concrete. By the end of Chapter 4, a student should know how to specify, cost-out, and even build a 10Base-T LAN. Chapter 5 adds more LAN standards, but it too focuses. It emphasizes 100 Mbps LANs and switched LANs, which are beginning to spread rapidly. Chapter 5 downplays token-ring networks, but Module F offers material for teachers who wish to focus on that excellent but sales-challenged technology.

A TCP/IP Internet Focus

Traditionally, datacoms text books focused on *OSI* standards. However, above the lowest transmission layers, OSI has failed to make a big impact, while the TCP/IP architecture is growing explosively on the Internet, in corporate networks, and even in PC networks. In addition, OSI is more complex than other architectures at the upper layers, and its failure to integrate internetworking elegantly makes it a nightmare to cover properly.

Consistent with its market focus, the book is built around a Basic Communication Model (Figures 1–1 and 1–2) that is close in spirit to TCP/IP. In addition, there is a great deal of discussion of TCP/IP throughout the book, including Module H.

The *Internet*, of course, has become a major focus for networking specialists. Even within companies, *intranets* that use Internet technology internally are very hot. The first two chapters introduce the Internet and its most important services today—electronic mail and the World Wide Web. Chapter 2 specifically discusses how to link to the Internet. The Internet also figures prominently in most other chapters.

For teachers who wish to stress the Internet further, Modules B and C take the student more deeply into the Internet and its services. Module C focuses especially on the World Wide Web and Browsers. Module H, as noted above, focuses on TCP/IP standards. Module J provides more detailed information on electronic mail standards.

The Book's Website

Of course the book has a website. Please visit it at the following URL:

http://www.prenhall.com/panko

The website provides a number of *online exercises.* These exercises require students to follow links to other websites, write up their findings, and send their online homework to the teacher via electronic mail. For instance, one online exercise has students design a 10Base-T or 100Base-T LAN and then cost out the LAN using current vendor data.

Of course the website offers a lot more. Technology changes by the minute, so the website offers *technology and market updates.*

Links to other websites help students learn how to explore the Internet to find more information. There, you'll find links to *case studies.* Although the book follows a large firm (VeriFone) and a small firm (NuPools) through its core chapters, case material ages rapidly. By putting case material at the website, we get information to students while it is fresh and hot.

The website contains detailed *PowerPoint* presentations for each chapter and module—not just PowerPoint images of two or three figures from each chapter. For complex figures, you'll find a series of PowerPoint slides that build the figure gradually.

Applications

Not long ago, most datacoms texts focused on transmission issues. But datacoms professionals no longer have that luxury. They have to work with electronic mail, client/server computing, PC networking services, and other user-level services. In other words, they have to work with *all* the layers in the architecture.

The text introduces applications in Chapter 2, and it focuses on applications extensively at its end. Chapter 8 introduces PC networking services, and Module I looks at PC networking in more detail. Chapter 10 focuses on human communication services, while Chapter 11 focuses on networked database services. Modules I and J go into more detail on these topics. For telecommunications courses, Module G looks at telephone services.

Throughout the book, by the way, there is an emphasis on *client/server* processing. Chapter 2 introduces client/server computing and notes that it brings independence from computer platforms. Chapter 8 (PC networking) looks at client/server processing in more depth. Chapters 9 through 11 continue to explore client/server computing.

Study Aids

It's important to have study aids. In addition to the online exercises, each chapter and module has extensive review questions and projects. Core review questions cover the essence of the chapter or the module. Detailed review questions require a more in-depth understanding. Thought questions require students to integrate what they have learned and to develop insights. Projects involve hands-on work.

To make life a little easier for students, the full text of the review questions is available at the website. Students can download the questions, so that they only have to type the answers.

Feedback

If you have suggestions for additional material to put in the text book or website, or if you have complaints, please contact me directly. I'm *panko@hawaii.edu*. Sorry, but if you have a specific question, I won't have the time to answer it, unless you are the teacher in the course. You can also check out my home page at the following URL:

```
http://www.cba/hawaii.edu/panko
```

Acknowledgments

I would like to express my special thanks to Susan Rifkin, who herded this textbook through electronic production. There were so many places where the whole project could have fallen apart that the book's staying on schedule (despite my foul-ups) was a miracle. I would like to thank Chuck Gemby of University Graphics, who handled the typesetting of the manuscript; and I would like to thank Donna Mulder, who copyedited my manuscript, no doubt while holding her nose. She edited carefully, adding a great deal to the flow of the text. Yet she never succumbed to the strongest of all human needs—the desire to completely rewrite someone else's words.

I would like to thank two years of data communications students who struggled through drafts of this book and pointed out, sometimes with glee, its errors and shortcomings. You could have been much harder on me.

Finally, I would like to thank my family—my wife Romy, my first-born, Julia, and my son, David. Romy persevered through many of my long days. Julia gave me a lot of perspective on what is important in life. David was my fearless net surfer, whose questions constantly forced me to confront basic and advanced communication issues. He showed me that if you just jump in and try things, the computer and the network (usually) won't break.

CHAPTER 1

Introduction

A Tale of Two Companies

This book is about **data communications**—a broad term that encompasses *all forms of electronic communication involving computers.* To understand this abstract definition, we will look at two firms that use data communications extensively. We will be following these two firms throughout this book. For now, we will only look at them broadly.

VeriFone

Bennis and Slater [1968] coined the term **adhocracy** for a radically new form of organization. Instead of fixed departments, there would be constantly shifting project teams. These teams would form quickly as needs arose, and then disband to free people to work on the next project. Adhocracies could respond quickly to business problems and opportunities, out-innovating their slower monolithic competitors. More recently, a number of writers (for instance, Davidow and Malone, [1992]) have renewed this call for faster and more flexible organizational structures, usually under the label **virtual corporations.** In today's world, customers want their products and services immediately, and competitors who are too slow to respond will die.

While academics have been debating the merits of virtual corporations, a number of firms have been quietly transforming themselves into these new types of organizations. One of the most successful has been VeriFone. You may not be familiar with that name, but you are undoubtedly familiar with VeriFone's products. When you go into a retail store and hand the clerk your credit card, he or she will run it through a small box. Your credit approval will appear a few seconds later. VeriFone pioneered systems for **transaction automation,** and it still controls over half the market for these devices.

Why has VeriFone been so successful? One key has been its strong corporate culture of cooperation. All employees learn quickly that sharing information is critical. The company gives so much information to its employees that it has arranged for insider status with the Federal Trade Commission for over 10 percent of its employees.

The employees themselves are expected to share whatever they know with everyone else and to jump in whenever they are needed. In addition, many VeriFone employees constantly call on customers. This includes the firm's chief executive officer (CEO), Hatim Tyabi. Thanks to this intensive customer contact, Tyabi has been able to say [Galal, 1994, p. 7], "There's no opportunity we don't find out about before our competitors do." VeriFone learns about the market's needs early and responds quickly with new products.

Intensive customer contact is helped by the company's policy of radical geographical decentralization. Even when the company started, its five initial employees worked in Hong Kong, Washington, DC, Milwaukee, and two cities in Hawaii. Today, although the company is headquartered in Redwood City, California, fewer than 7 percent of its employees are located there. In many ways, VeriFone is like a doughnut. It has a middle, but there's nothing there physically. Instead, its employees are scattered across several major sites and many more minor sites around the world. This places its people near its customers, making constant contact almost automatic.

Geographical dispersion is fine for customer contact, but what about internal communication? Here the firm relies on data communications. Every new employee is issued a laptop computer before he or she gets a desk [Freedman, 1995], and every employee is expected to use the laptop constantly when away from the office. They are also expected to respond promptly when they get messages. Just as transaction automation is a real-time computer application, human communication at VeriFone has an equally strong sense of urgency [Freedman, 1995].

To give you a feeling for just how radically the company is structured, VeriFone's chief information officer (CIO), Will Pape, works out of his home in Santa Fe, New Mexico, far from the nearest VeriFone office. Yet Pape is a strong hands-on manager. Perhaps it is precisely because he is equidistant from everyone that he can work so easily with everyone in the firm, not merely those who are near.

In fact, the company uses time zones to its advantage. An employee in Redwood City may start a response to a customer. Then, at the end of the business day, he or she may pass the work in progress west to Hawaii, India, or one of VeriFone's other offices. The next morning the originator can see what others have added and can continue to work on that expanded base. We will see another example of this kind of round-robin project in the next chapter.

Although electronic communication is vital to the corporation's success, VeriFone's technology does not push the state of the art. One of VeriFone's main tools for human communication is simple electronic mail (email). The company's 2,100 employees send over a million messages a month (500 per employee) and many of these messages go to more than one person.

Even some corporate reports are distributed via email. Every day CIO Will Pape (who also has other broad responsibilities in the firm) summarizes key data and writes up a newspaper-like column. This column goes to almost every manager in the firm.

Of course, VeriFone realizes that email is not a very rich medium [Pape, 1995], so people travel a great deal and meet periodically face to face. The company is also moving toward richer communication media, such as videoconferencing. Yet VeriFone has no illusions that technology drives anything. The company's competitors often have better technology. VeriFone, however, knows how to *use* technology to support its radical and extremely effective organizational strategies.

NuPools

NuPools[1] is a much smaller company with only 15 employees. Based in Santa Barbara, California, NuPools serves people who own swimming pools or who want to own swimming pools. Clarence Whyte owns and manages the company.

NuPools has two operations in a single building. One is a retail store that sells pool supplies, such as chemicals, pool furniture, hot tubs, and almost everything else that pool owners could want. Haraj Sidhu is the retail manager. He has eight sales clerks and an inventory manager and purchasing agent, Kristi Stone.

In the back office are Whyte's desk and the desk of his controller, Larry Ochoa. This is also where the company handles the second part of its business, designing new pools and contracting for their construction. Luz Martinez is the contracting manager. She is also the engineer who designs new pools. The company does not do the actual construction. Instead, as a contractor, it hires construction firms, plumbers, and anyone else needed to do the work. Luz has two site managers, Kevin Brown and Todd Kimura, who travel to the various construction sites to oversee the work.

---

[1] The name and location of NuPools are fictitious, as are the names of NuPools' employees, in order to protect the privacy of the actual company. Some business details have also been distorted to disguise the actual firm.

Until recently, the company had stand-alone electronic cash registers and stand-alone PCs for everyone in the back office. Yet while the firm was computerized, these were individual islands of computing. The technology simply did not fit the way the firm's employees communicated intensively with one another. In addition, if someone on the sales floor needed to check on inventory, they had to ask Kristi Stone if they could use her PC, where the inventory data files were stored.

The first step in adding communication to computation was to install a small network to link the PCs in the back office. Once that was done, information sharing improved greatly. Even Clarence Whyte found himself checking inventory frequently because keeping adequate but not excessive inventory is critical to swimming pool retailing.

' Surprisingly, one of the firm's biggest uses is electronic mail. Although the people in the back office share a single working area, they are not always there at the same time. Instead of dropping notes on desks, NuPools employees write brief email messages to one another. In addition, Clarence often sends broadcast messages to his entire back office staff, to inform them about events or changes in policy. Fairly quickly, the firm added the equipment and software needed for the two site managers to communicate with the office while they were at a job site.

More recently, the firm has added a point-of-sale system to its local area network, linking its retail floor to the back office. Now a sales clerk can tell a customer if a product is in stock and, if not, when it is expected to arrive. In addition, Kristi Stone's inventory database is updated constantly, so she can keep stock lower yet reduce inventory outages at the same time.

Perspectives

Both VeriFone and NuPools began their plunges into data communications at almost the same time. This was no coincidence. Their moves into data communications were a natural outgrowth of the way that technology has matured.

During the 1980s, the personal computer spearheaded an explosion in computer use. By the late 1980s, 80 percent of all office workers were using computers, and about a third were using computers more than five hours a day [Steelcase, 1989].

By the beginning of the 1990s, there was a high enough density of computers in many businesses to justify communication-based applications, such as electronic mail and shared databases. At about the same time, networking costs fell enough to become attractive. Companies began to network their PCs within sites and eventually across sites as well.

The Chapter

In the next section, we will look at a basic communication model that divides data communications concerns into four layers. After we look at the model, we will discuss setting standards so that various transmission systems, computer platforms, applications, and users can work together. In the remainder of this chapter and in the next chapter, we will look at the four layers, working from the lowest layer to the highest.

THE BASIC COMMUNICATION MODEL

Companies like VeriFone and NuPools can only succeed if they have capable data communications staffs that can search the market for available technologies, implement the chosen technologies, and work with line managers to support business strategies. This is a complex job, at least in part because it requires skills in several different areas: transmission technologies, computer systems, application programs, and even work practices.

A Framework for Mediated Communication

Figure 1.1 provides a framework that we will use throughout this book to help us understand what a networking specialist must know to approach a situation. We will use the example of electronic mail, but the framework is useful for almost all other applications as well.

In many ways, Figure 1.1 is the core framework of this book. We will see it in almost every chapter, and the book generally looks at the elements in the figure from bottom to top as it progresses through its core chapters.

The figure shows two people engaging in electronic mail transmission. Person A is sending a message to Person B. Note that Person A does not talk to Person B directly. Rather, the process is complex, involving multiple layers of intermediary hardware, software, and transmission services. In general, we say that this process is **mediated** because the two people do not interact directly but only through the mediation of software, computers, and transmission technologies. In data communications, we always deal with mediated communication.

The key to mediated communication is **standards**—expectations of how people (and technology) are to behave when they communicate. By setting expectations, standards allow the receivers to react appropriately when someone (or some hardware or software processes) sends them a message. At the three technical layers, where there is no human intelligence at all, standards have to be very explicit.

In data communications, technical-layer standards have to be **multivendor standards** so that products from different vendors can work together. In Figure 1.1, Program 1 might come from Microsoft, while Program 2 might come from Novell. At the computer layer, in turn, Computer S is a Macintosh, while Computer T is an IBM computer. The two transmission-layer processes, X and

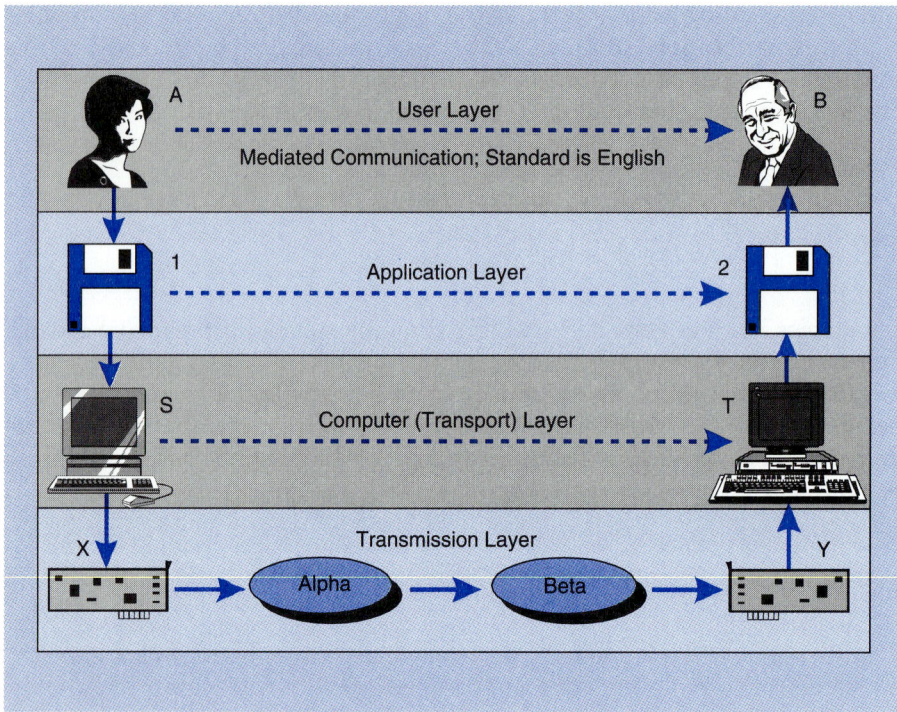

FIGURE 1.1
The Basic Communication Model

Person A sends a message to Person B. Person A uses electronic mail Program 1, while Person B uses email Program 2 from a different vendor. Program 1 works on Computer S, which is an Apple Macintosh personal computer. Program 2 works on Computer T, which is an IBM personal computer. Computer S has Transmission Process X, which attaches the computer to Network Alpha. In turn, Computer T attaches to Network Beta via Transmission Process Y. The transmission processes consist of hardware and associated software. Communication between Person A and Person B is mediated by the technology. There must be standards for communication at all four layers. For instance, at the user layer, English is the standard. Standards are also needed for the application layer, which governs communication between the application programs, the computer (transport) layer for communication between the different computers, and the transmission layer, for communication between each computer and the network and between networks.

Y, which are handled primarily in hardware, may also come from two different vendors. Even the two networks, Alpha and Beta, may come from different transmission vendors.

The User Layer

The communication process begins when the sender, Person A, has a message to send. Perhaps she needs to tell the receiver, Person B, that the planned release of a new product has been pushed up by a month. Person B will have to develop the initial distribution plan earlier than expected.

For this communication to take place, the two people will have to have certain things in common. Most obviously, there is the language used in the message. Unless both speak the same language, say English, they will not be able to communicate.

Their company will have to have established a broad set of business practices to make this communication work. For instance, suppose that Person A is a peer of Person B, that is, someone working at the same level in another department. Some companies allow and encourage this sort of direct and rapid information dissemination. Other firms require Person A to use the proper chain of command.

The importance of VeriFone's success in the transaction automation market has been due at least in part to its business practices. Other firms have VeriFone's electronic mail and other technologies, often in more sophisticated form. Yet VeriFone is able to move faster than its competitors because it has developed a corporate culture of information sharing, cooperation, and customer contact.

If we view Figure 1.1 as a set of layers, like the layers on a cake, we can call the layer containing Person A and Person B the **user layer.** This book will focus more on the lower layers of the framework because they require the special technical expertise of information systems (IS) professionals. But the user-layer strategies that companies create and implement bring the real gains from data communications technology. The job of lower layers is to provide enough functionality to support the user-layer strategies that firms wish to follow.

The Application Layer

In mediated communication, the two parties do not interact directly. Instead, when Person A wants to communicate, she types a message into her electronic mail program, Program 1.

Later when Person B logs into his email program, Program 2, the program tells him that new mail has arrived. Person B can bring up the message, read it, and take appropriate action.

The job of the **application layer** is simple—to ensure that two application programs will be able to work together. However, this layer is extremely complex in practice because there is an infinite number of applications. Different applications need different application-layer standards, of course. Electronic mail programs have entirely different coordination needs than database programs, for example. As a result, the application layer is open-ended, and new application-layer standards will always be needed as new applications emerge.

Typically there are competing standards at the application layer. For instance, in electronic mail, the most popular standards are SMTP, X.400, and MHS. As Chapter 10 discusses, these standards are mutually incompatible. So firms need to develop strategic plans for the application layer of our model.

They must either select a single corporatewide standard or find ways to translate between programs using different standards. This **competing standards problem** also occurs at lower layers.

The Computer (Transport) Layer

Email Program 1 resides on Computer S. Program 2 resides on Computer T. These two computers must be able to communicate in order for Programs 1 and 2 to be able to exchange their messages. For instance, Computer S will have to deliver Program 1's message within a message to Computer T. This message will have to say something like, "Here is a message for Program 2 that is running on you."

We call this layer of communication the **computer layer** or the **transport layer**.[2] The job of the computer/transport layer is to ensure successful interactions between computers despite the fact that they may come from different computer vendors.

The computer or transport layer is comparatively simple. There is nothing like the need for multiple application standards. Once two computers establish communications via a transport-layer standard, all of their application programs will be able to communicate with one another. While there are dozens of application-layer standards, there is only a handful of competing computer/transport-layer standards. We will look at this layer in Chapter 2 and in many later parts of the book.

The Transmission Layer

Much of the data communications professional's life will be spent working at the lowest layer, the **transmission layer.** The job of the transmission layer is to move streams of bits reliably between computers.

In Figure 1.1, Transmission Process X (a piece of hardware plus associated software) on Computer S first submits its message to Network Alpha. Network Alpha then transmits the message to Network Beta. At the other end, Network Beta delivers the message to Transmission Process Y on Computer T.

Although the transmission process has a simple basic job, technology is changing rapidly in this area. When new transmission technologies emerge, new standards are needed. As a result, there are literally hundreds of different transmission-layer standards. We will spend much of this book talking about transmission technologies. Chapters 2 through 7 will focus almost exclusively on the transmission layer.

---

[2]The term *transport layer* is a bit unfortunate because it sounds like something that happens at the transmission layer that lies below it. Unfortunately, this piece of terminology is well established.

CHAPTER 1   INTRODUCTION   9

Subdividing the Transmission Layer

Note in Figure 1.1 that transmission does not take place across a single network. Rather, the transmission must pass through two networks, Network Alpha and Network Beta. This means that we need *two layers* of transmission standards, as Figure 1.2 illustrates. (The figure adds yet a third network to emphasize the presence of multiple networks.)

Networks ■

These two layers rely on the concept of **network.** A network is an *any-to-any* transmission system. This means that any **station** can send a message to any other station simply by telling the network the **address** of the other station. The familiar telephone system is a network for voice communication because almost any telephone can reach almost any other, simply by dialing a number.

Subnet Layer ■

The lowest transmission sublayer in Figure 1.2 deals with transmission across a single network. In internetting terminology, a single network is called a **subnet.** Alpha and Beta are subnets in this figure. If you only have a single subnet, two stations communicate with each other using subnet (single-network) standards. NuPools, for instance, only has a single network.

FIGURE 1.2

The Subnet (Single-Network) and Internet Transmission Layers

In a network, any station can send a message to any other station simply by giving its address. When there are multiple transmission networks, there are two tasks, each requiring its own standards. The subnet (single-network) layer deals with standards for transmission among stations attached to a single subnet, such as Alpha or Beta. The internet layer, in turn, deals with the routing of messages across the internet. An internet is a collection of subnets that allows any-to-any communications among stations on different subnets.

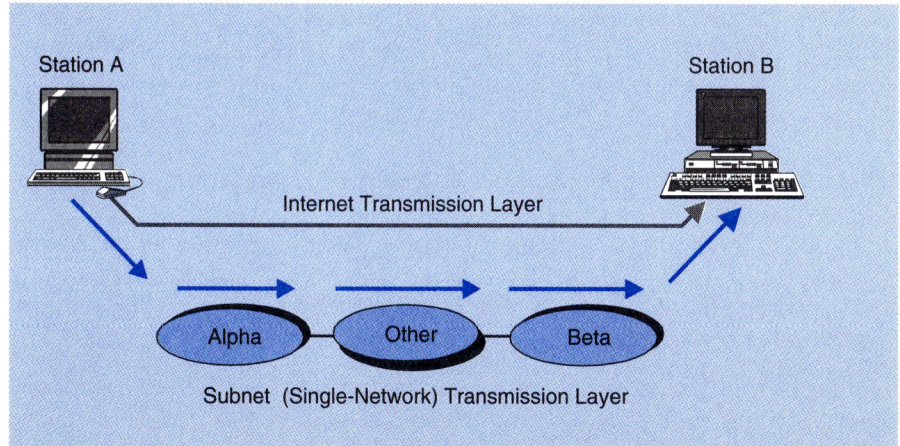

Internet Layer ■

In Figure 1.1, the message from Station S to Station T must pass through two interconnected subnets. We use the term **internet** for an interconnected set of subnets in which any station attached to any subnet can send a message to any other station on any of the internet's subnets simply by giving the internet address of the other station.

This means that Processes X and Y will have two addresses. Each will have an address on its own subnet, Alpha or Beta. Each will also have an internet address, as will every other station on the internet.

The term *internet* with a lowercase *i* refers to any internet—any interconnected set of networks with any-to-any connectivity. *For more on internets, see Chapters 6 and 7, and Module H.*

However, when people talk about the **Internet** with an uppercase *I*, they are referring to a specific internet. This is a worldwide internet that connects literally millions of computers and hundreds of millions of users around the world. The Internet is critical because it allows even home users to reach any of the millions of Internet **host computers** around the world that provide services. Thanks to such innovative applications as electronic mail and the **World Wide Web** (discussed in the next chapter), the Internet has become indispensable in many businesses. Every employee at VeriFone, for instance, has an Internet address, and even small companies like NuPools can create World Wide Web pages that act like vastly expanded Yellow Pages advertisements in the traditional telephone book. *For more on the Internet and the World Wide Web, see Modules B and C.*

STANDARDS

Now that we have looked at the basic elements needed for two (or more) people to work together via data communications, we will look at some of the concepts we have encountered in more depth.

In a sense, the most important concept we have encountered is *standards*. Without standards, there can be no meaningful communication. Recall that standards are expectations about how the communicating parties will act.

Standards Agencies and Standards Architectures

Standards do not appear by magic, of course. Their creation often takes years of effort.

Standards Agencies ■

It might be nice if there was only a single **standards agency,** that is, a single body that would create data communications standards. But this is not the case today. Figure 1.3 lists some of the major standards agencies that exist. There are also dozens of minor standards agencies, such as the Institute of Electrical and Electronics Engineers.

Some standards agencies are **de jure** standards agencies. This is from the Latin for *law*. These de jure standards agencies are somehow official. Most important, they are not controlled by a single vendor. We call their standards **open** because they are not under the closed control of a single vendor. Any vendor can build products according to these standards without fear of the standard being closed by a controlling vendor.

Other standards agencies build **proprietary** standards, meaning that the agency is a single company that builds and controls a set of standards. In some cases, these agencies publish the specifications to their standards and allow other vendors to build compatible products at reasonable royalty rates. Sometimes the developers are dominant vendors, such as IBM or Novell. In other cases, they may be small innovative firms. Proprietary standards are also called **ad hoc** standards.

Standards Architectures ■

Even in our simplified model in Figure 1.1, standards are needed at three technical layers. In real systems, standards must be set at seven or more layers for two application programs to communicate with one another.

If individual standards in different layers were designed in isolation, even

FIGURE 1.3

Standards Agencies and Architectures

There are several major standards agencies that produce technical standards, as well as many minor standards agencies, such as the Institute for Electrical and Electronics Engineers. Each major agency has an architecture or grand plan for setting standards. Within an architecture, standards are compatible. Standards from different architectures tend to be incompatible.

Architecture	Type	Responsible Agency
OSI (Reference Model of Open Systems Interconnection)	De jure	ISO (International Organization for Standardization)
		ITU-T(International Telecommunications Union-Telecommunications Standards Sector)
TCP/IP	De jure	IETF (Internet Society's Internet EngineeringTask Force)
IPX/SPX	Proprietary	Novell Corporation
SNA (Systems Network Architecture)	Proprietary	IBM Corporation

by a single agency, the result would be chaos. So each agency has a plan for developing standards. We call these plans **architectures.** The architecture of a house specifies the major elements to be created (rooms in a house) and how these elements will work together to create what is needed by the user—a livable house. Data communications architectures, in turn, specify what standards will be needed and how they will work together, so that Person A can send an email message to Person B.

In a house, once you define your architecture, you can begin designing individual rooms. Once you know that a room will be a bathroom, you can work on it while others work separately on the design and construction of other rooms. Similarly, in data communications architectures, once the architecture is set, specialists can work on the design of its individual standards.

In a house, if television technology changes, you can change your television room into a "multimedia room." In general, you do not have to change other rooms as well. In data communications architectures, the same thing happens. If a new transmission technology emerges, you can change a single subnet's technology without changing standards for internetworking, transport processes, or application programs.

OSI

The most widely taught architecture is **OSI,** the **reference model of Open Systems Interconnection.** *Reference model* means architecture. An *open system*, in turn, is one that is open to communicating with other systems using nonproprietary architectures. So OSI is an architecture for interconnecting systems that are open to communication via the nonproprietary standards created under the OSI architecture.

ISO

OSI is managed by two standards agencies. One is **ISO,** the **International Organization for Standardization.**[3] ISO sets computer standards. It also sets standards for motor oils, screws, and many other industrial products. ISO has a very broad charter.

Calling the architecture OSI was an unfortunate decision because it is all too easy to confuse *ISO the organization* with *OSI the architecture*.

ITU-T

While ISO has long been a key standards agency for computers, ITU-T is dominant in telecommunications[4] standards in telephony, television, radio, fac-

---

[3]While "International Standards Organization" would fit the initials better, "International Organization for Standardization" is ISO's standard English translation for its name.

[4]Telecommunications refers to all forms of electronic communications. Data communications is a subset of telecommunications involving computers.

simile, and other areas. **ITU-T**[5] stands for **International Telecommunications Union—Telecommunications Standards Sector.** As the name suggests, ITU-T is the standards-setting arm of International Telecommunications Union (ITU), which is itself an arm of the United Nations.

When it sets standards for telephony, television, radio, and facsimile, those standards usually dominate the marketplace, thanks to the cohesiveness of the world's telephone companies, which well understand the importance of worldwide integration.

Until recently, the ITU had two separate groups that set standards—the CCITT in wire-based communications and the CCIR in radio.[6] The ITU combined its standardization roles into ITU-T, recognizing the growing convergence of wire and radio transmission.

OSI ■

In 1978, ISO and ITU-T released the OSI framework to guide their future standards setting. OSI is the most open family of international standards. Standards have to go through long and open processes before they are accepted. OSI standards also tend to be very sophisticated because they are designed by large groups of professionals.

Although ISO and ITU-T manage the OSI standards-setting process, they do not produce all the standards themselves. For instance, in local area networks, the *Institute for Electrical and Electronics Engineers (IEEE)* creates most standards. In video, the *Motion Picture Experts Group (MPEG)* sets most standards. However, these other organizations develop their standards according to the OSI framework, and they submit their draft standards to ISO and ITU-T for acceptance. *For more on LAN standards, see Chapter 4, and Module F.*

Market Position ■

Open development and sophistication have many advantages, but speed and cost are not among them. Thanks to slow standards development, OSI standards tend to be late to come to market. When standards do appear, furthermore, it takes a long time for vendors to develop products meeting these sophisticated standards, and sophistication also makes products expensive. In marketing, "slow to market with high-cost products" is not a good recipe for success.

Once, it was widely believed that OSI would become the dominant standards architecture in data communications. In what we have called the *subnet layer*, that is certainly the case. For individual local area networks and wide area networks, OSI standards are completely dominant. Rather than fight this

---

[5]It would seem more logical to call it the ITU-TSS, but the organization uses one-letter abbreviations for all of its sectors.

[6]Consultative Committee for International Telephony and Telegraphy (CCITT) and Consultative Committee for International Radio (CCIR).

dominance, the other standards agencies specify OSI standards for their subnet layers.

In other layers, including the internetting, transport (computer), and application layers, OSI standards have not seen widespread acceptance, although some of its applications standards are widely used and its network management standards are somewhat accepted.

A key problem is that OSI developers were late to understand the importance of internetting (see Figure 1.2). As a result, companies had to turn to other standards architectures for their internetting standards. They tended to stay with these architectures when they selected their transport and application standards.

OSI Layers

Figure 1.4 compares the basic layers we have seen in Figures 1.1 and 1.2 to the way that OSI divides its standards into seven layers. Note that OSI divides the transmission layer of Figure 1.1 into three layers, the computer/transport layer into two, and the application layer into two. Note also that the OSI architecture has no clear internetting layer because internetting was not a major concern when OSI emerged. *For more on OSI layering, see Module A.*

TCP/IP

Table 1.1 shows that at the computer/transport layer the most widely used architecture is **TCP/IP,** which is named after two of its most popular standards, TCP and IP (Transmission Control Protocol and Internet Protocol). Today TCP/IP's market share at the transport layer is even larger.

Standards Setting

TCP/IP protocols were originally created for the Internet. Internet standards are set by the **Internet Engineering Tasks Force (IETF),** a collection of working groups building standards in key areas.

Watching the IETF work is watching rough democracy in action. IETF working groups work fast, using electronic mail to handle much of their communication. Working for a large vendor means very little, while technical expertise means almost everything. Discussions tend to be frank and sometimes brutal. IETF "standards" really consist of **requests for comments (RFCs),** some of which become widely accepted while others are ignored. In many ways, the IETF mirrors the chaos of the Internet, but it is successful nonetheless or perhaps precisely because of this "Let's do it, and let's do it right now" attitude.

Market Acceptance

While the process of setting TCP/IP standards is chaotic, it has resulted in widely accepted standards. One reason for this is TCP/IP's strength in inter-

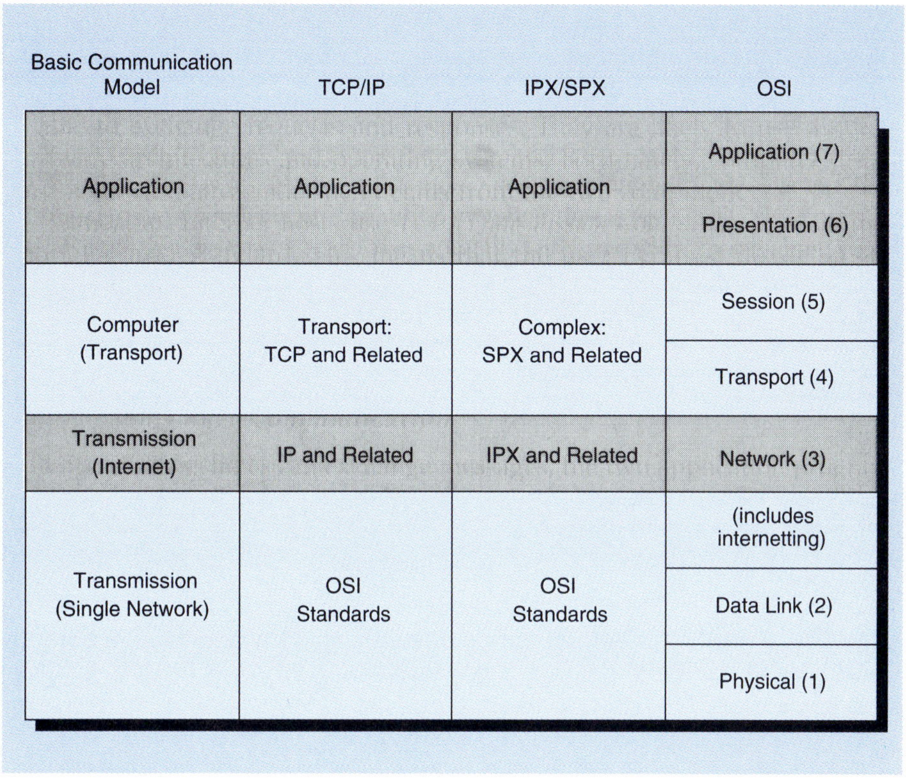

Basic Communication Model	TCP/IP	IPX/SPX	OSI
Application	Application	Application	Application (7)
			Presentation (6)
Computer (Transport)	Transport: TCP and Related	Complex: SPX and Related	Session (5)
			Transport (4)
Transmission (Internet)	IP and Related	IPX and Related	Network (3) (includes internetting)
Transmission (Single Network)	OSI Standards	OSI Standards	Data Link (2)
			Physical (1)

FIGURE 1.4

Layering in Major Data Communications Architectures

Each of the major architectures divides data communications into multiple layers. The TCP/IP architecture conforms closely to our three major layers and the two sublayers of the transmission layer. OSI has no distinct internetting layer and divides the application and transport layers in our Basic Communication Model into sublayers. SNA has a different philosophy than the other architectures, so the layer match-ups for SNA and our Basic Communication Model are not shown.

netting. The Internet was born by interconnecting many independent networks. The *Internet Protocol (IP)* standard was created in 1981. Later, as commercial organizations began to experience the need for internetting, they found IP readily available. *For more on the history of the Internet, see Module B. For more on internetting, see Chapter 6, Chapter 7, and Module H.*

Above the internetting layer, commercial organizations also found good transport-layer protocols, including **TCP (Transmission Control Protocol).** TCP is a simple protocol, so even PCs can run it without harming their performance. This made it inexpensive to implement, and companies that wished to implement it could find TCP (and IP) software for almost all of their com-

TABLE 1.1

Transport-Layer (Computer-Layer) Traffic

Agency	Architecture	Traffic Share	Special Market Strengths
Internet Society (IETF)	TCP/IP	38%	Internetting, transport
Novell Corp.	IPX/SPX	34%	PC networking
IBM Corp.	SNA	18%	Mainframe communication
ISO and ITU-T	OSI	—	Single networks
Other		10%	
Total		100%	

Notes: Traffic share is at the transport (computer–computer) layer
Abbreviations
  ISO: International Organization for Standardization
  ITU-T: International Telecommunications Union–Telecommunications Standards Sector
  OSI: Reference Model of Open Systems Interconnection
  SNA: Systems Network Architecture
Source for traffic share data: Molloy [1994].

puters. As a result, TCP/IP standards offered a ready way to handle both the transport and internetting layers. *For more on using TCP/IP with PCs, see Chapter 2. For more on TCP/IP layering, see Module H.*

Even better, in the eyes of some organizations, the IETF usually keeps its standards simple. In fact, many of its standards include the word *simple* in their names. As a result, the IETF develops its standards quickly, and after standards emerge, vendors develop simple TCP/IP products quickly. While many critical OSI standards were still "under construction," TCP/IP products were widely available. Perhaps best of all, simple standards mean inexpensive products. TCP/IP software is inexpensive for large computers, and for PCs, it is available through shareware.

At the application layer, IETF standards have had broad success on the Internet, with standards such as the Simple Message Transfer Protocol, Telnet, Gopher, and the World Wide Web. However, the application layer is open-ended, and IETF standards are not dominant for all applications. In network management, however, TCP/IP's Simple Network Management Protocol is the most widely used. *For more on Internet services, see Modules B and C.*

Layers ∎

Figure 1.4 shows that TCP/IP layering is similar to the layering that we presented in Figures 1.1 and 1.2. It specifies OSI standards for subnets, it has a distinct internetting layer, and it has single layers for transport and applications. *For more on TCP/IP layers, see Module H.*

IPX/SPX

Novell is dominant in PC networking today, thanks to its NetWare networking software (see Chapter 8). For its PC networking, Novell created a proprietary standards architecture, **IPX/SPX.** Figure 1.4 shows that IPX/SPX has a similar structure to TCP/IP. This is due at least in part to the fact that the two architectures have common roots.

Table 1.1 shows that IPX/SPX is the second most widely used architecture at the transport layer. However, while TCP/IP use is growing, IPX/SPX's share of the market is gradually shrinking. In fact, Novell itself has added TCP/IP options, although these are still not widely used. At least for the foreseeable future, IPX/SPX should continue to be very important, although this importance will reside in its niche market of PC networking and will depend entirely on Novell's market share.

Systems Network Architecture (SNA)

Four years before OSI, IBM introduced its proprietary **Systems Network Architecture (SNA).** SNA has become a de facto standard for mainframe communications.

Table 1.1 shows that SNA is the third most widely used architecture. However, this is due primarily to SNA's dominance in the mainframe market, which is no longer growing. IBM has attempted to extend SNA to other uses through its **Advanced Peer to Peer Networking (APPN)** extensions. But at least to date, these have not seen wide acceptance.

SNA has a profoundly different design philosophy than do other architectures, so we do not show its match-up with our Basic Communication Model. Different writers give different match-ups between SNA layering and other layered architectures such as OSI.

Setting Organizational Standards

While standards agencies offer many standards, individual firms must select **implementation profiles**—limited lists of the standards that they will implement internally. This allows them to build networks without having to deal with hundreds of different standards. Selecting the corporate implementation profile is one of the most critical tasks of data communications management.

For instance, VeriFone has a number of very firm corporate standards. For local area networks, it specifies 802.3 (Ethernet) technology. For file transfers, it specifies TCP/IP. In even requires that each employee have an Internet account and that this account have a directory called XFER as a place where other employees can send files. *For more on Ethernet LANs, see Chapter 4, and Module F.*

The Transmission Layer

Now that we have looked at the basic communication model and the issue of standards, we are ready to begin looking at the four layers individually. We will begin at the bottom layer, the transmission layer. This layer serves as a foundation for all other layers. In addition, it is at this layer that most data communications professionals do most of their professional work.

To connect our computers together, we typically use networks. As we saw earlier, a network has two main elements. First, there are the *stations* attached to the network. Second, there is the transmission network itself, which may consist of transmission lines, radio stations, switches, and other hardware and software needed to connect the stations attached to the network. Networking provides *any-to-any connectivity* among the stations.

Transmission Speed

Probably the most important characteristic of any network is its speed. We measure transmission speed in **bits per second (bps).** This assumes that the information is **binary,** consisting of 1s and 0s. So bits per second measures the number of 1s or 0s traveling through the network each second. If a network transmits 9,600 bits each second, this is a speed of 9,600 bps.

For higher speeds, we follow the metric notation, as Figure 1.5 illustrates. For thousands of bits per second, we use the metric symbol k for kilo. So we usually write 64,000 bps as 64 **kbps,** or we say 64 **kilobits per second.** *Note that the proper metric way to write kilo is with a lowercase k.* Computer scientists often use a capital K for RAM and disk capacity, but networking professionals usually follow the standard metric notation for kilo as a lowercase k.

For millions of bits per second, we write M for *mega.* So 10 million bits per second is 10 **megabits per second** or 10 **Mbps.** A thousand megabits per second is 1 **gigabit per second** or 1 **Gbps.** A thousand gigabits per second, in turn,

FIGURE 1.5

Notations for Transmission Speeds

We measure transmission speeds in bits per second (bps), kilobits per second (kbps), megabits per second (Mbps), gigabits per second (Gbps), and terabits per second (Tbps). Note the lowercase k in kilobits per second.

Prefix	Use	Metric Meaning	Example	Interpretation
kilo	kbps	1,000	64 kbps	64,000 bps
Mega	Mbps	1,000,000	1.544 Mbps	1,544,000 bps
Giga	Gbps	1000 million	2.4 Gbps	2,400 Mbps
Tera	Tbps	1 million million	2 Tbps	2,000 Gbps

Acronym	Name	Geographical Scope	Speed Range
LAN	Local Area Network	Single office Building Campus	10 Mbps 100 Mbps 1 Gbps (coming)
MAN	Metropolitan Area Network	Intracity and Contiguous cities	1 Mbps 100 Mbps
WAN	Wide Area Network	Intercity International	9,600 bps 64 kbps 1 Mbps (fairly common) 100 Mbps (emerging)

FIGURE 1.6

LANs, MANs, and WANs

is a **terabit per second (Tbps).** Most transmission today takes place in the kilobit and megabit ranges.[7]

Transmission speed actually refers to the duration of each symbol's appearance on the transmission line. If we transmit at only 10 bps, then each bit's signal is one tenth of a second long. But if we transmit at 100 Mbps, each bit has a duration of just one hundred millionth of a second (0.000,000,01 second). Obviously, as speed increases, it becomes harder to send and receive accurately. This means higher cost.

LANs, MANs, WANs, and Internets

Figure 1.6 illustrates that there are three main categories of networks according to geographical scope. Besides the distances each type of network spans, there are also differences in performance.

LANs ■

As their name implies, **local area networks (LANs)** are for short-distance communication. Figure 1.7 shows a simple LAN that serves a handful of PCs in a small office. A *wiring hub* provides the connection between the PCs. There is a single line between each PC and the hub. This line often consists of standard business telephone wiring. Inside each PC, furthermore, is a printed circuit expansion board called a network interface card (NIC). This simple arrangement costs less than $300 per PC, yet it provides good connectivity. *For more on LAN technology, see Chapters 4 and 5, and Module F. For more on wiring, see Chapters 4 and 5 and Module D.*

---

[7]A general rule is that there should be at least one nonzero digit to the left of the decimal point. So you would not write 0.3 kbps. You would write 300 bps.

Another general rule is that there should not be more than three digits to the left of the decimal point. So you would not write 1,200 kbps. You would write 1.2 Mbps.

The second rule is suspended for speeds under 10,000 bps. It is normal to write 9,600 bps.

FIGURE 1.7

PC Networking on a Small Local Area Network (LAN)

The PCs connect to a wiring hub. The connections often consist of standard business telephone wires. The hub manages the transmissions between stations. Inside each PC is a network interface card (NIC).

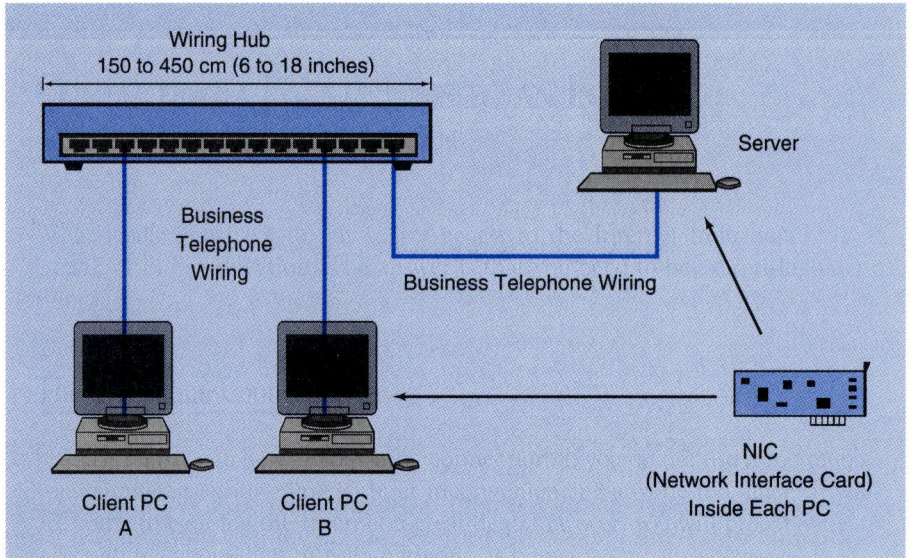

We will see in Chapter 6 that many LANs are a good deal larger than this. Some serve a single building. Others serve an entire site, such as a factory or university campus. In general, LANs can extend to roughly a kilometer and people will still consider them to be local area networks.

Figure 1.6 emphasizes that, whatever their scope, LANs have one thing in common. This is speed. Over short distances, high transmission speed is relatively inexpensive. Today most LANs operate at about 10 Mbps. Some companies are already upgrading their LANs to 100 Mbps. Higher speeds allow LANs to support applications that are rich in graphics or even full-motion video. They also allow each LAN to serve more users, just as highways with more lanes can carry more cars.

WANs ■

Sometimes networks have to reach beyond the property owned by the organization (the **customer premises**). **Wide area networks (WANs)** link distant sites. *For more on WANs, see Chapter 7.*

We noted previously that almost all LANs operate at megabit speeds today because high-speed transmission is not expensive over short distances. For long-distance transmission, however, speed is expensive. As a result, most WANs today operate at speeds of only 9,600 bps, 64 kbps, or about 1 Mbps. Speeds in the hundreds of megabits per second are now becoming available, but their high cost limits them to special uses.

MANs ■

Metropolitan area networks (MANs) handle communications within a city or metropolitan area consisting of several contiguous cities. Of course, companies can use WANs to link sites within a city. But as we just noted, WANs typically operate at rather low speeds. In contrast, many MANs operate at 100 Mbps. This sounds faster than LAN speeds, but the MAN's high-speed lines are shared by many users. Each gets only a fraction of this capacity. Effective throughput usually is lower than LAN speeds. MANs are still much faster than WANs, however.

Internets ■

As we saw earlier in this chapter, we often link our individual LANs, MANs, and WANs together into internets.

CONCLUSION

Data communications is a broad term that encompasses all forms of electronic communication involving computers.

By necessity, data communications involves *mediated communication* instead of direct face-to-face communication between people. Data communications professionals divide mediated communication into multiple *layers*. At each layer we need *standards*—expected ways of communicating—so that we can purchase layer processes from multiple vendors.

- The *user layer* governs interactions among people. Data communications professionals do not work primarily at this layer, but this is the critical layer for achieving benefits from technology.
- The *application layer* governs interactions among application programs. Because there are many types of application programs, there is a large number of application-layer standards.
- The *computer* or *transport layer* allows two computers to exchange messages despite the fact that they come from different vendors. There are few transport-layer protocols.
- The *transmission layer* is responsible for moving streams of bits between two stations on a network. We subdivide the transmission layer into a subnet layer, which governs transmissions within a single network (subnet), and an internet layer, which governs transmissions across multiple networks.

Standards are created by *standards agencies*. Each agency has a broad plan for standards setting. This plan is called its *architecture*. In general, standards at the same layer from different architectures are incompatible.

ISO and *ITU-T* created the *OSI* architecture. OSI is dominant for *subnet standards*.

At the internetting layer, *TCP/IP* standards are the most widely used, including its main standard, the Internet Protocol (IP). At the *transport* or *computer layer*, the *TCP/IP* architecture is again the most widely used. TCP/IP offers inexpensive, rapidly developed products.

In PC networking, *Novell's NetWare* networking software is the most widely used. This makes its *IPX/SPX* architecture widely used.

For mainframe communications, *IBM's SNA* is dominant.

Companies, faced with hundreds of standards from several architectures, create *implementation profiles*—limited lists of standards they will use in their networks.

Networks offer any-to-any connectivity among attached stations. Networks vary by geographical scope from *LANs* to *WANs*. As distance increases, speed generally increases, and cost generally rises. Most organizations create *internets*—networks of networks—to link their individual networks together.

The next chapter essentially continues where we have left off in this chapter. We have begun looking at the individual layers in our Basic Communication Model, starting with the subnet and internetting transmission layers. In the next chapter, we will look at the computer layer, the application layer, and the user layer. We will close that chapter with an integrative example that will show how these layers are implemented in practice.

CORE REVIEW QUESTIONS

1. Define *data communications*.
2. List some ways VeriFone uses data communications to build a competitive advantage over other companies in the market.
3. What are NuPools' two lines of business?
4. Define *mediated communication*.
5. List the four layers in the Basic Communication Model, and describe the purpose of each.
6. List the two layers into which we divide the transmission layer and explain the purpose of each.
7. List the major standards agencies and their architectures. List the market niche or niches in which each architecture is dominant.
8. List the names and abbreviations for transmission speeds in increasing factors of 1,000. Underline capital letters.
9. Describe how LANs, MANs, and WANs differ in geographical scope, speed, and cost. What is internetting?

DETAILED REVIEW QUESTIONS

1. What is a virtual corporation?
2. How does VeriFone use its geographical decentralization to its advantage? What business values does this require in its corporate culture?
3. What networking technology does NuPools use? What business benefits does this bring?
4. Why is networking becoming important in the 1990s?
5. What is mediation? In what ways is the communication in Figure 1.1 mediated?
6. What is the purpose of standards? What is the competing standards problem?
7. What is the defining characteristic of networks? What is an internet? What is the Internet? Does each computer have an address on both the subnet and internet layers?
8. Distinguish between *de jure* and *proprietary* standards.
9. What is an architecture? Why is it important?
10. What organizations manage OSI? What is the distinction between OSI and ISO? Describe the attractiveness of OSI standards. How do OSI layers match those in our Basic Communication Model?
11. What agency manages TCP/IP? Are TCP/IP standards only used on the Internet? Describe the attractiveness of TCP/IP standards. How do TCP/IP layers match those in our Basic Communication Model? What is an RFC?
12. What agency manages IPX/SPX? What agency manages SNA?
13. What is an implementation profile? Why is it important?
14. Using the information in the footnote on speeds, write the following in correct form: 9,600 bps, 56,000 bps, 523 Mbps, 1,200 Mbps. At 64 kbps, what is the duration of a single bit?

THOUGHT QUESTIONS

1. Do you think VeriFone's innovations are needed in other firms? Do you think it is possible to implement them in other firms?
2. Why do you think architecture developers divide the work of standards development into layers instead of just building one complete standard to handle all of the work done at all layers?

3. Which standards architecture do you think will dominate in the future?
4. Which of the layers shown in Figures 1.1 and 1.2 do you think is most important? Justify your reasoning.

PROJECTS

1. Study a firm's use of data communications. Focus on how it uses data communications to support its business strategies, its implementation profile, and the technologies it uses.

For online exercises, please visit this book's Website at: http://www.prenhall.com/panko

REFERENCES

BENNIS, W. G. and SLATER, P. E. (1968). *The Temporary Society.* New York: Harper & Row.

DAVIDOW, W. H. and MALONE, M. S. (1992). *The Virtual Corporation.* New York: Harper.

FREEDMAN, D. H. (1995, September 13). "Culture of Urgency," *Forbes ASAP*, 25–28.

GALAL, H. (1994). *VeriFone: The Transaction Automation Company*, Case Study 9-195-088, Harvard Business School, Harvard Business School Publishing, Boston, MA 02163.

MOLLOY, M. (1994, January 3). Technology Planning Survey 94: Pushing the Speed Limit. *Network World*, 5–8, 10.

PAPE, W. (1995, Summer). Beyond E-Mail, *Inc.*, 17(4), 27–28.

STEELCASE, INC. (1989). *The Office Environment Index: 1989 detailed findings.* Grand Rapids, Michigan.

CHAPTER 2

Beyond the Transmission Layer

Introduction

In Chapter 1, we looked at our Basic Communication Model (Figure 1.1), which divides mediated communication into four layers: the transmission, transport or computer, application, and user layers. In Chapter 1, we looked at these four layers in general and at the standards agencies that develop standards in these areas according to some standards architecture. We also looked specifically at the transmission layer, including both subnet transmission and internetting.

In this chapter, we will continue moving up the hierarchy of mediated communication layers. We will look at the transport or computer layer that allows computers to communicate despite different hardware architectures and operating systems. We will look specifically at the most important computer platforms: terminal–host systems, PC networks, and workstation networks.

We will then look at the application layer that allows two electronic mail or other application programs to communicate despite coming from different vendors. We will focus on client/server processing, which is a new and important paradigm for application program interactions. We will also look specifically at two important applications: electronic mail and the World Wide Web.

We will end this chapter with an integrative example: using a PC to link to the Internet in order to access the World Wide Web.

THE COMPUTER (TRANSPORT) LAYER

Although networking is about communication *between* computer users, networking specialists have to understand the types of computers that people are using. We call types of computer systems **computer platforms.** The most popular computer platforms are terminal–host systems, PC networks, and workstation networks.

FIGURE 2.1
Terminal–Host Platform

The terminal has little or no processing and storage. It is basically an input/output (I/O) device—a remote screen and keyboard. The host handles processing and storage functions. A host can serve multiple terminals simultaneously, rapidly switching its attention among the active terminals. The largest hosts are mainframes. Mid-size hosts are minicomputers.

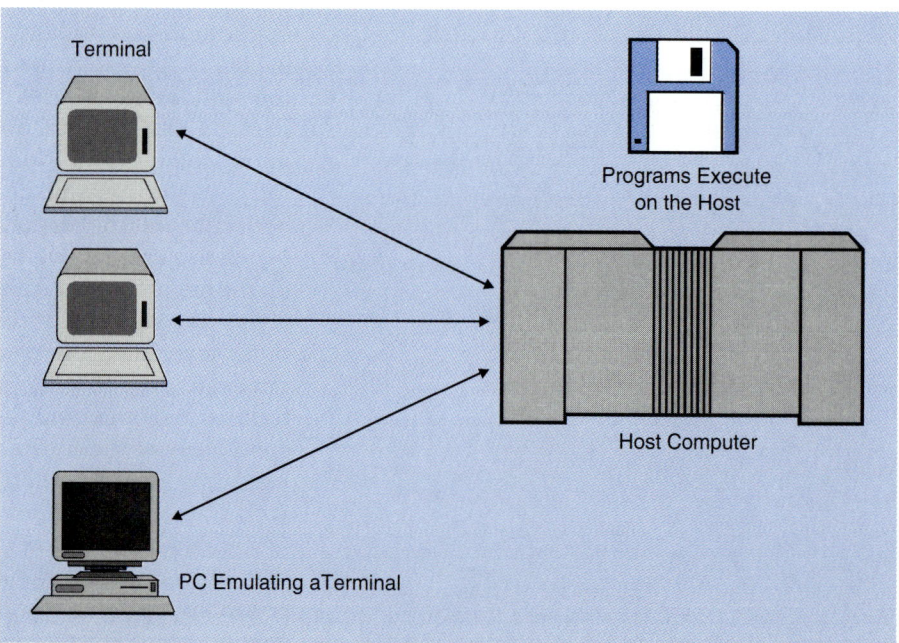

TABLE 2.1

Sales of Computer Platforms by Category, 1993

Computer Platform	Worldwide Sales, 1993 ($billions)	Trend
Personal computers	$70	Increasing
Mid-range systems	$49	Decreasing
Large systems	$26	Decreasing
Networking hardware and software	$18	Increasing
Workstations	$10	Increasing

Source: Data from International Data Corporation. Cited in Standard & Poor's [1993].

Terminal–Host Systems

Until recently, most users worked with terminal–host systems. Figure 2.1 illustrates a simple terminal–host platform. It shows that users work at **terminals.** Terminals allow you to type and to see things on the screen. However, they have little if any processing power or storage capacity. They look somewhat like personal computers, but they are far simpler. *For more on terminal–host transmission, see Module E.*

The *processing* and *storage*, in turn, are done on distant **host computers.** The host computer stores the programs and the data. It also executes the programs.

Mainframes and Downsizing ■

The first host computers were *mainframes*. These extremely large and powerful machines can serve hundreds of terminals simultaneously. Unfortunately, mainframes have extremely high purchase and upkeep costs.

Because mainframes are so expensive, most companies are **downsizing** many of their smaller applications, that is, moving them to smaller, less expensive computers.[1] For smaller applications, there is no need to use an expensive mainframe. While there will always be mainframes, Table 2.1 shows that large-system sales were already modest and were declining in 1993. Today they are much smaller. *For more recent data on computer platform sales, see* Standard and Poor's Industry Surveys *in your local library.*

Minicomputers ■

One alternative is to downsize applications to smaller host computers called *minicomputers*. "Minis" can serve a few dozen users simultaneously, which is

---

[1] The term *downsizing* is a bit awkward because it is also used to describe human layoffs in companies trying to reduce their costs. Many IS writers now prefer the term *rightsizing* to indicate that the company should choose the right size platform for the job. Sometimes the right size is a mainframe. In reality, however, the right size is usually smaller.

enough for many applications. The cost per user is far lower than it is on mainframes.

Once many people thought that the minicomputer would be the main downsizing platform. However, the phase of downsizing to minicomputers has largely run its course. Table 2.1 shows that minicomputer sales were large in 1993 but were declining. Today the market share of mid-size systems is even smaller. Downsizing is now moving increasingly to PC networks.

Networked PCs

Most computer users now have **personal computers (PCs).** Table 2.1 shows that the PC was already the largest platform in 1993.

Once these PCs were stand-alone devices, talking to nothing beyond their printers. But those days are gone forever. Today most PCs are already networked [Standard & Poor's, 1993]. By the end of the 1990s, the *networked PC* will be the dominant corporate computer platform. *For more on PC networking, see Chapter 8 and Module I.*

Client PCs ■

Figure 2.2 shows a small PC network. It shows that there are two types of PCs in PC networks. First, there are **client PCs.** These are the PCs that sit on the desks of ordinary users, such as managers, professionals, and secretaries. Some client PCs are even found on the factory floor and in delivery vehicles.

Almost all client PCs begin their lives as stand-alone PCs. Turning them into client PCs is quite simple. First, a technician installs a printed circuit board called a *network interface card* or *NIC*. Chapter 4 and Module I discuss NICs.

Second, it may be necessary to add software. As Chapter 8 discusses, newer operating systems, such as Windows 95, can turn a PC into a client PC without additional software. PCs with older operating systems need to install inexpensive or free *client shell* software.

Client PCs do not have to be extremely powerful. Some users do need top-of-the-line machines. Many client PCs, however, are one or two generations behind the state of the art. Just as stand-alone PCs span a broad spectrum of capabilities, there is no "standard" client PC hardware base.

Servers ■

The other machines shown in Figure 2.2 are **servers.** As the name suggests, servers provide services to the client PCs.

Servers sound somewhat like host computers. There are two main differences, however, between hosts and servers. First, host computers are designed to work with dumb or extremely low-power terminals. *Servers, in contrast, are designed to work with intelligent client PCs.* Indeed, we will see that the client PC does most of the processing in some services offered by servers. This dependence on intelligent desktop machines is ideal for attractive applications with

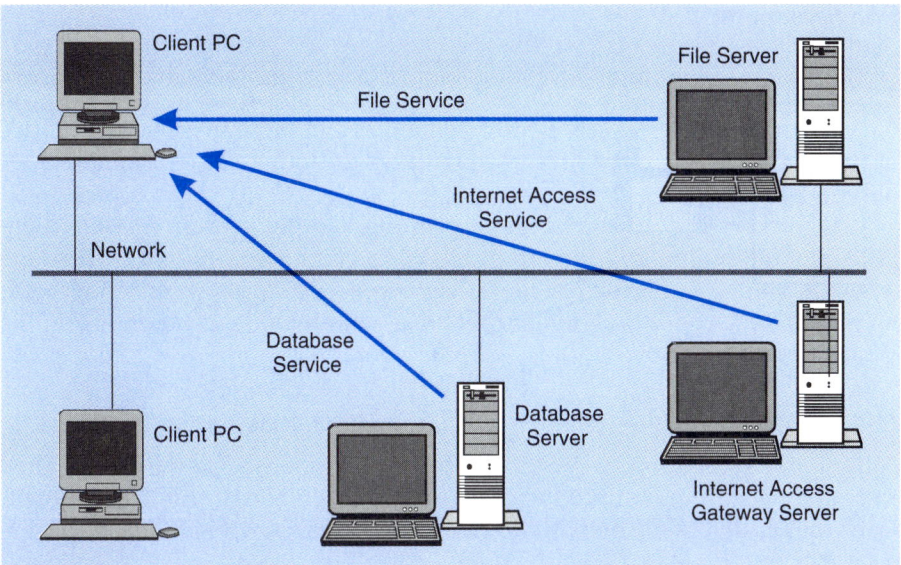

FIGURE 2.2

PC Network: Specialized Servers

In a PC network, client PCs sit on the desks of ordinary users, such as managers. Servers, in turn, provide services to the client PCs. Instead of one larger server to handle many applications, there are multiple servers with most supporting only one or a few applications. Most servers are personal computers. The specialization allows each server to be selected and optimized for its primary application. A single PC network may have hundreds of PCs and dozens of servers.

graphical user interfaces because graphics require extensive local processing power.

Second, host computers normally are very large machines that run dozens or even hundreds of applications simultaneously. This extensive *multitasking* makes them complex and costly. Have you ever attempted to do too many things at one time, jumping back and forth between them? Then you know that while multitasking is good, there are severe problems if you carry things too far. Mainframes and minicomputers waste a great deal of their processing power switching among their many tasks. In addition, their operating systems have to be very complex to handle this ultramultitasking.

In contrast, *servers are specialized machines*. Most servers only handle a single application service. At most, they offer a handful of applications. This **specialization** allows the network staff to choose and optimize each server for its particular applications. Database servers can be extremely powerful and can be optimized for disk drive access, which tends to be the bottleneck in database processing. Servers that link client PCs to mainframe computers, in turn, can be low-cost machines optimized for input/output (I/O) processing.

File Servers

Although we will see in Chapter 8 that there are many types of servers, every PC network has at least one file server. Figure 2.3 illustrates a file server. As the name suggests, a **file server** stores data files and program files. We will see in Chapter 8 that there are several advantages to storing at least some of your files on a file server.

File servers also support **remote printing.** Output goes from your client PC to the file server, and then from the file server to a printer located somewhere in your office area. The file server acts as a control center to link a PC network's dozens or hundreds of client PCs with the network printers on the PC network.

File Server Program Access

Although file servers store program files, they do not execute these files. They merely store a program until the client PC needs it. Then the file server downloads the file across the network to the client PC, as shown in Figure 2.3. It ex-

FIGURE 2.3

File Server and File Server Program Access

As its name suggests, a *file server* stores data files and program files. It also coordinates network printing, which allows client PCs to send output to printers scattered around the office area. Programs stored on file servers are executed through *file server program access*. In this process, the program is downloaded over the network to the client PC. It executes on the client PC. This is fine for smaller programs, but most client PCs are too limited to execute complex programs.

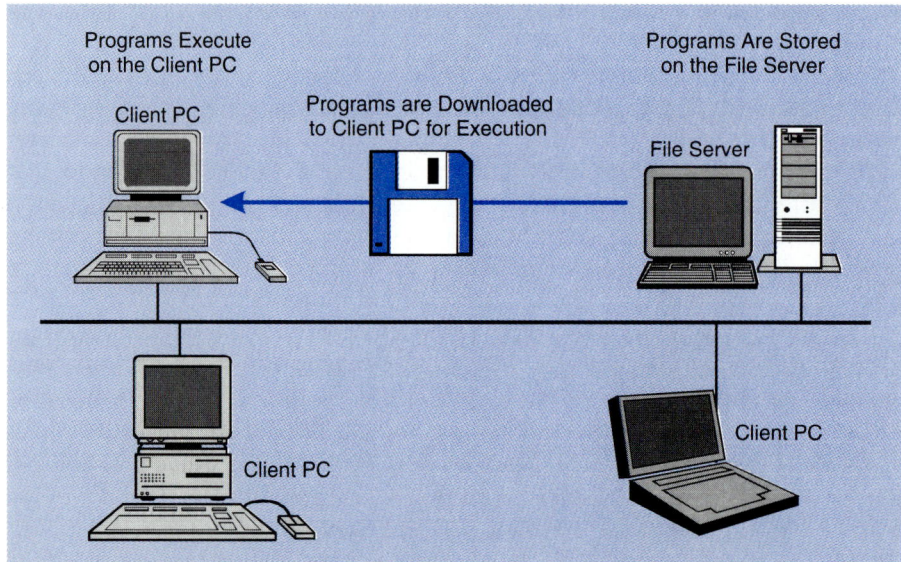

ecutes on the client PC. This storage of programs on the file server and their execution on the client PC is called **file server program access.**

For instance, many file servers offer electronic mail programs. These usually are file server program access programs. When you handle your electronic mail, the file server actually loads the mail program across the network to your client PC.

File server program access is sufficient for smaller programs, such as electronic mail and word processing programs. The client PC usually has ample processing power for such programs. For larger programs, such as sophisticated database programs, however, the client PC has insufficient RAM and processor speed. For complex programs, PC networks turn to the client/server processing discussed later.

PC Servers ■

When many people hear the term *server*, they think of specialized machines. In fact, like client PCs, most servers begin their lives as stand-alone PCs. As in the case of client PCs, the first step in turning a stand-alone PC into a server is to add a network interface card (NIC).

The second step is to add a server operating system (SOS) and application software. Popular server operating systems include Novell NetWare, UNIX, and Microsoft Windows NT Server.

Server operating systems normally come bundled with software for some applications, such as the file and print services discussed in Chapter 8. For advanced applications, such as database service, it is also necessary to buy specific application software.

Non-PC Servers ■

Not all servers, however, begin their lives as client PCs. Some are built from the ground up as special-purpose microcomputers. Others are more powerful than any PC. Mainframes and minicomputers can act as PC network servers. Even more common are the workstation servers we discuss next.

RISC Workstation Networks

PC networks are attractive because of their benefits and their moderate costs. However, many applications, such as computer-aided design (CAD), require more processing power than client PCs can provide. For such applications, many firms turn to more powerful machines called **RISC workstations.**

As in PC networks, there are both client workstations and workstation servers. Figure 2.4 shows a client workstation. As you can see, RISC workstations look like PCs. But there are big differences in processing power and, unfortunately, in price. Table 2.1 shows that workstation revenues were about

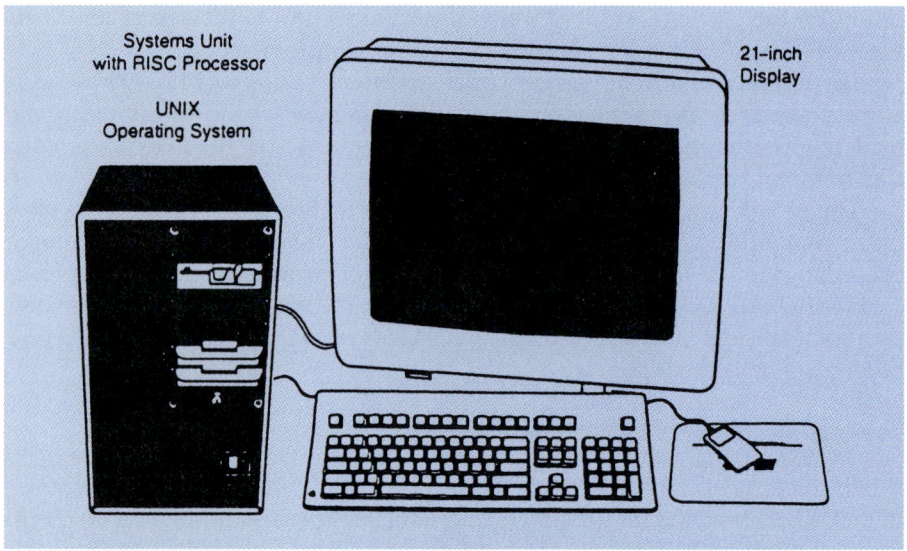

FIGURE 2.4

Client RISC Workstation

Client *RISC workstations* look very much like personal computers. However, they are far more powerful and expensive because they use high-speed RISC processors, ultrafast RAM, and ultrafast hard disk drives. They also tend to use expensive 21-inch displays. They usually run the UNIX operating system, which does not run standard PC software. They are used in such processing-intensive tasks as computer-aided design. RISC workstation networks have both client RISC workstations and RISC workstation servers.

one-seventh of PC revenues in 1993, making the workstation an important platform. However, RISC workstations have not closed the gap much since then, so they are still niche products. In addition, because workstations are so expensive, they are even smaller than PCs on a unit basis.

RISC workstation hardware is expensive partly because of the machine's microprocessor. Workstations use ultra-high-speed RISC processors instead of the high-volume Intel microprocessors used on PCs or the non-RISC Motorola microprocessors found on many Macintoshes. These RISC microprocessors are very expensive. In addition, to keep up with the processor's speed, these machines have to push the state of the art in RAM speed and hard disk speed. All of this costs money.

Another cost factor is the workstation's display. Workstations tend to have 21-inch displays, which cost $2,000 to $3,000.

Most RISC workstations run the UNIX operating system. Although UNIX is a good operating system, application programs written for MS–DOS and MS–Windows will not run on workstations. This makes upgrading to workstations even more expensive. Microsoft has made Windows NT available on some workstations, however, so the application software problem may be less severe in the future.

Multiple Computer Platforms and Networks

Figure 2.5 emphasizes that if you have multiple computer platforms at a site, this does not mean that you must have multiple LANs, one for each type of platform. A physical highway can mix residential, commercial, and government cars, trucks, and busses. LANs and larger networks, in turn, are like general data highways. As Figure 2.6 shows, there is no problem mixing terminal–host traffic and PC networking traffic on the same network. Hosts, terminals, client PCs, and servers are merely stations on the network. Each sends its messages to an address, and the network carries the messages there.

THE APPLICATION LAYER

The highest technical layer in Figure 1.1, our Basic Model of Communication, is the application layer. There is an almost infinite number of networked applications that take place on LANs, MANs, WANs, and internets. In this sec-

FIGURE 2.5
Networks Carry Traffic for Multiple Computer Platforms

A network can mix traffic from terminals, minicomputers, mainframes, client PCs, server PCs, RISC workstations, and other devices. Here a mainframe is sending a message to a terminal over a network at the same time the network is carrying traffic between a PC server and a client PC.

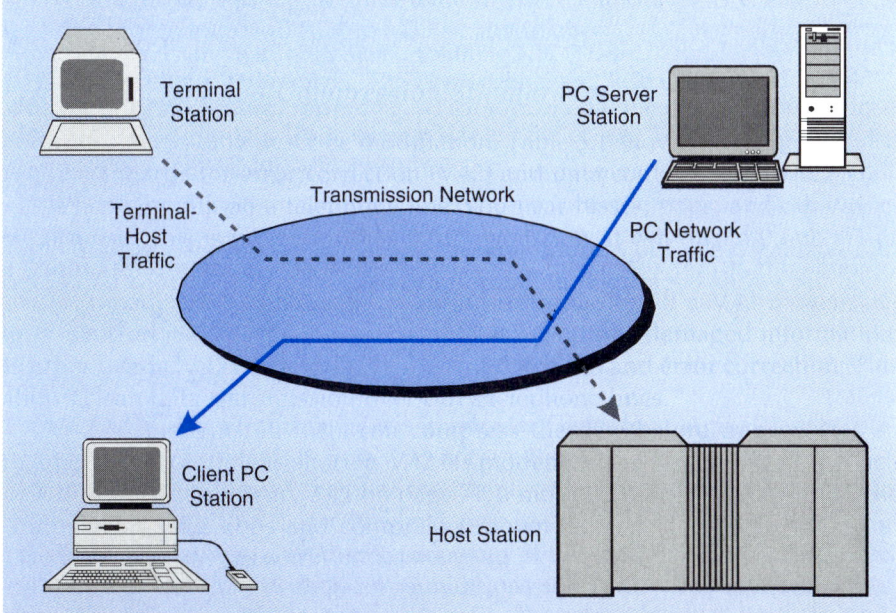

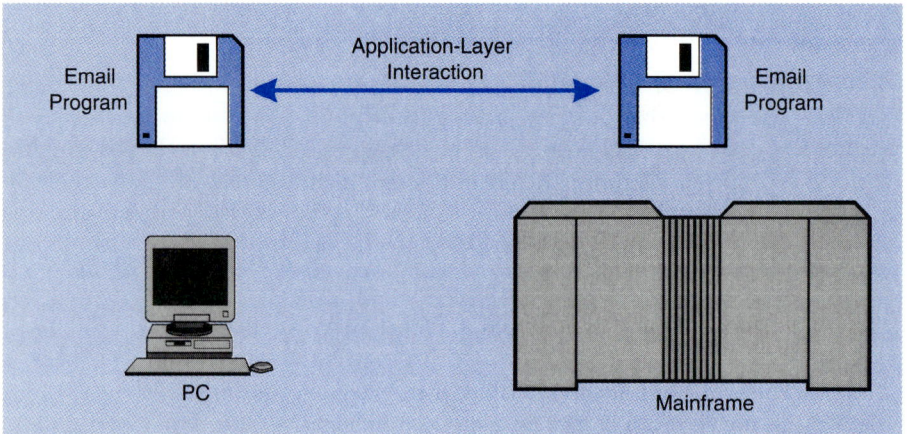

FIGURE 2.6

Application-Layer Standards

Application-layer standards allow application programs from different vendors to work together. Here a mail program from Microsoft is communicating with an electronic mail program on a mainframe from IBM. Of course, the two machines must be linked at the transport and transmission layers as well.

tion, we will only look at two: electronic mail and the World Wide Web. Chapters 10 and 11 discuss a number of additional applications.

Standards

Figure 2.6 shows that standards at the application layer allow application programs to work together even if they were created by different vendors. For instance, you can be using an electronic mail program from Microsoft while the intended receiver is using a mainframe-based electronic mail program from IBM. Of course, the two computers executing the application programs must be linked at the transport and transmission layers as well.

Note that application standards are platform independent. There is no requirement that both electronic mail programs run on the same computer platform. Here an email program on a PC server is communicating with a PC on a mainframe. As long as the two machines are linked at the transport and transmission layers, the two application programs will be able to communicate as equals.

Client/Server Processing

In Figure 2.6, the two application programs communicate as equals. Either can initiate interactions, and either can provide services to the other. There is another way for application programs to interact, however. This is client/server processing.

Interactions on Stand-Alone PCs ■

On stand-alone personal computers, application programs interact with the operating system, as shown in Figure 2.7. When the application program receives a command to save a file to disk, it does not do the job itself. Instead, it sends a request to the operating system, asking the operating system to carry out the work. This request is usually referred to as a **call.** The operating system carries out the command and sends a reply to the application program, confirming that the requested action was completed. If the command could not be completed, the operating system sends a reply indicating that the desired work could not be carried out.

In such interactions, the application program and the operating system do not act as equals. They have specialized roles. The operating system is a server (service provider). The application program, in turn, is a customer (client) of the operating system's services.

For this to be effective, the interactions must be standardized. In other words, there must be standardized request messages and standardized response messages for each request.

FIGURE 2.7

Program Interactions on a Stand-Alone PC

An application program sends a request for service to the operating system. This is a request to save a file to disk. The operating system tries to carry out the work and sends a response indicating its success or failure. The operating system is a server (service provider) to the application program. The application program is a client (service customer) for the operating system.

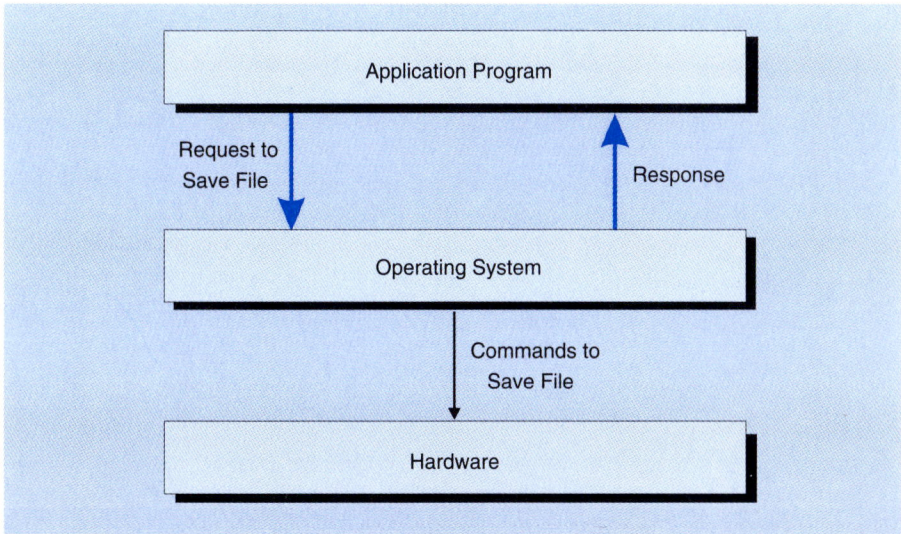

Client/Server Processing on a Network ■

Figure 2.8 shows that client/server processing extends this idea of clients and servers to network environments. Now the client and server programs are on different machines.

The principles, however, are the same. One machine has a **client program.** This might be a spreadsheet program on a client PC that needs a certain piece of information. The other machine has a **server program.** In this case, it is a database program running on a minicomputer. The database server program has the required information in its database.

Interaction between the client program and the server program begins with the client spreadsheet program sending a **request message** to the database server program. The server program then does the required work to produce the requested information. It sends back a **response message** containing the desired information. If it cannot find the required information, its response message explains why it did not succeed.

If request and response messages are standardized, then either program can come from any vendor. The database program might come from Oracle or Sybase. The client spreadsheet program might come from Microsoft or IBM.

Most Internet applications use client/server processing. We will see this later in the case of the World Wide Web.

FIGURE 2.8

Client/Server Processing

In client/server processing, the work is done by two programs. The client program operates on a client machine, usually on the user's desktop. The server program runs on another machine located somewhere else. When the client program needs services, it sends a request message to the server program. When the server program finishes the requested work, it sends a response message to the client machine.

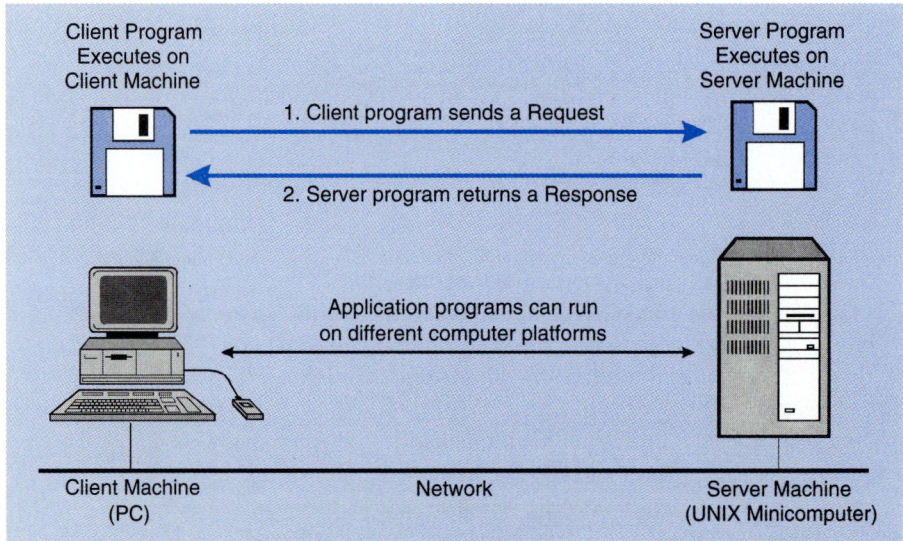

Electronic Mail

Probably the most widely used network application today is **electronic mail.** In fact, electronic mail should soon be the most widely used computer application of any type, if it has not achieved this position already. *For more on electronic mail, see Chapter 10.*

Header ■

As Figure 2.9 indicates, when you send an **email** message, you basically fill out a form that looks like a memo. At the top is the **header.** Here you type the address of the receiver, a subject, and perhaps other fields, such as who should get a copy of the message. The rigid format of headers allows you to retrieve messages later. You might, for instance, search for all messages from "Pat Lee" in the last month. The mail system would be able to give you a list of messages fitting that description.

FIGURE 2.9

Electronic Mail Message

An email message has two parts: the header and the body. The header consists of a number of mandatory and optional fields that specify such things as to whom the message is addressed, who else should get copies, and the subject of the message. Each user has an email address. "Sidhu" is one such address. In turn, "Staff" is a distribution list that contains the names of everyone in the staff group. Everyone on this distribution list will get a copy of the message. The body contains the text of the message. At the bottom of the body is a "signature." Clarence Whyte's email program adds this automatically to every outgoing message. This message also has an attachment. This is a spreadsheet file that is delivered along with the message.

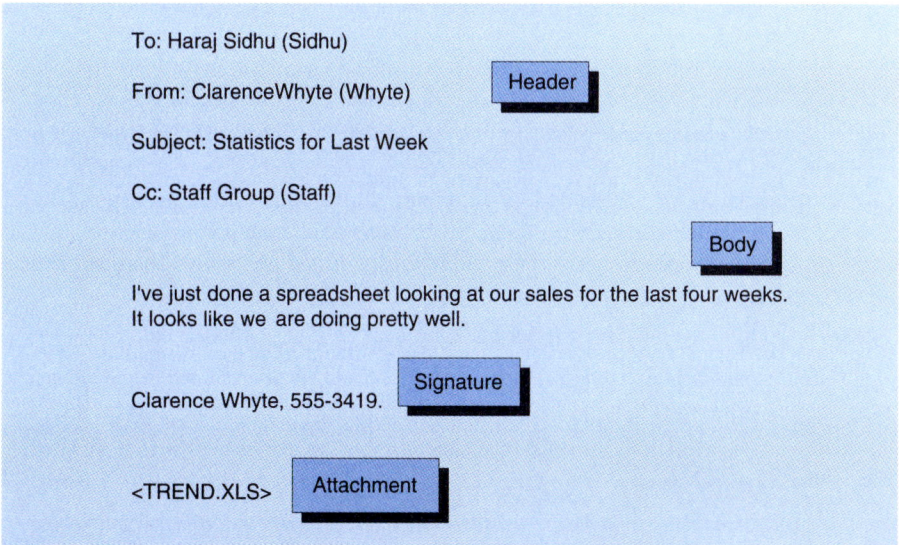

Body

Next comes the free-form text in the **body** of the message. Here the body is short, as it is in most messages, but it may be several pages long.

At the end of the body is Clarence Whyte's **signature.** The email program automatically places this signature on the bottom of every outgoing message. This is similar to fixed-format letterhead when you send letters.

Attachments

Finally, Figure 2.9 carries an **attachment.** You can attach spreadsheet files, database files, graphics files, or almost any other type of file to messages in most electronic mail systems. This means that email is not just for message communication. It has become a generic way to transfer files as well. When users read the message, they execute an extraction command, and the file is placed in a directory on their computer or file server.

Receiving Mail

Figure 2.10 illustrates that when a message arrives, the mail system places it in an electronic in-basket. The user can then scan through summaries of incoming messages, reading them in any order. After reading a message, the receiver can discard it, file it away, forward it to someone else for action, or dash off a brief reply. Email allows you to handle dozens of messages in a few minutes.

Here Message 1 is from Larry Ochoa, the NuPools controller. It is a progress report. It has an attachment.

The next message is from Haraj Sidhu. The Subject: field begins with "Re:". This means that the message is a reply to an earlier message, probably one that

FIGURE 2.10

Summary List for New Messages

When a message arrives, the receiver's electronic mail program places it in an electronic in-basket. The figure shows a summary screen for new messages. If the user selects a message, its full header and body appear. After reading a message, the receiver can discard it, file it away, forward it to someone else for action, or dash off a brief reply. The first message contains an attachment, as does the fourth. The second message is a reply to an earlier message, probably from Clarence.

New Messages for Clarence Whyte				
No.	From	Date Time	Subject	Attachment
1	Ochoa	5/12 9:17a	Progress report	+
2	Sidhu	5/12 8:50a	Re: New item in stock?	
3	Whyte	5/11 9:18p	Friday meeting	
4	Martinez	5/11 7:20p	Report on the Chen Account	+

Clarence sent to Haraj. A large number of all messages are replies, which are extremely easy to make.

Message 3 is from Clarence Whyte! He placed his address in the copy (Cc:) field to be sure he got a copy after sending it.

Finally, the last message is from Luz Martinez, the contracting manager. She was working late, and Clarence had gone home, so she sent him an email message. It also has an attachment.

Distribution Lists ■

Another critical service in electronic mail is **distribution lists.** If there are a number of people to whom you may have to send mail repeatedly, you can write down all of their email addresses in a distribution list. If you then type the name of this distribution list in the address field of a message, the message will go to everyone on the list. For instance, if a manager wants to call a staff meeting, he or she can send a message to the *Staff* distribution list, announcing the meeting and asking if anyone has conflicts. In the message shown in Figure 2.7, the user has typed a distribution list name (Staff) in the Cc: (copy) field.

The World Wide Web

One of the most exciting Internet services is the **World Wide Web (WWW).** *The Web* has tens of thousands of webservers that provide an almost bewildering variety of information services.

The World Wide Web is based on the idea of **hypertext.** As shown in Figure 2.11, a page in hypertext may contain **links** that point to other documents or to other spots in the same document. In Figure 2.11, links are underlined. *For more on the World Wide Web, see Module C.*

When users click on a link with the mouse, they jump to the other document or the other place in the same document. In this way, users can navigate the Web easily, pursuing their interests as those interests appear.

Browsers ■

Figure 2.12 shows that to read Web pages, you need a client program on your PC or other client machine. This program is called a **browser** because it allows you to browse through information in World Wide Web format as well as in a number of other formats discussed in Module C. You merely type a page's **URL (uniform resource locator)** in the URL box, and you jump to that page. In this case, the URL is "http://www.cba.hawaii.edu/panko/home.htm".[2]

---

[2] The "http" says that the information will be accessed using the WWW HyperText Transfer Protocol. The "www.cba.hawaii.edu" is the server. "Panko" is the directory. "Home.Htm" is a file in WWW's Hypertext Markup Language.

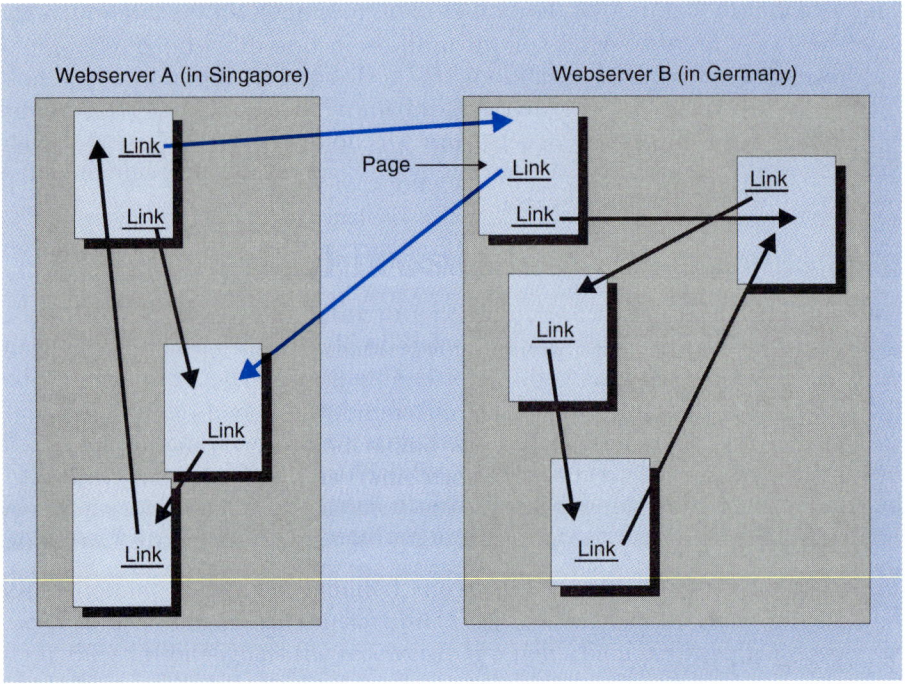

FIGURE 2.11

Hypertext in the World Wide Web

In hypertext, a page contains links that point to other documents or other spots in the same document. Links are shown here as underlined. When the user clicks on that link, he or she is taken to that place in the document.

Client/Server Processing ■

The World Wide Web uses client/server processing. Your browser is your client program. It allows you to ask for web pages and presents them to you. In turn, a **webserver** stores the information and responds to requests from your browser. The webserver is the client program.

Platform Independence ■

Remember that the computer platform that the client and server programs use is unimportant to how they interact. It does not matter whether your browser runs on an IBM PC compatible, a Macintosh, or a RISC workstation. It will still be able to talk to any webserver.

Nor does it matter what machine the webserver software uses. It might be a UNIX mainframe, a PC network server, or even a simple PC running the Windows 95 operating system. In fact, there is no way to know whether the webserver you have reached is a large or small machine.

NuPools and the World Wide Web ■

NuPools recently created a **home page** (main page) for the company. This home page, which is available on the Internet, lists the company's hours, department phone numbers, additional information on product offerings, and other information that would be much too detailed to put in printed telephone directories. If you are interested in a product, you can click on its link, and you will jump to another webpage with detailed information on that product. You can even order the product online, using a credit card.

VeriFone and the World Wide Web ■

VeriFone uses the Web extensively for its internal use. For instance, VeriFone has a number of research and development centers. Yet the company needs to keep its engineering drawings compatible. John Warner, manager of technical computing resources, developed a World Wide Web application to create a standard set of engineering part descriptions and symbols. Any engineer in the company can access them via any browser.

FIGURE 2.12
Browser Showing World Wide Web Page

A browser is a client program that allows you to see World Wide Web pages and pages in other formats. Each page has a universal resource locator (URL). You type this in the URL box, and the browser contacts the appropriate webserver. That webserver downloads the page to your browser so that you can view it. Note that links are underlined. Note also that the page can contain graphic elements as well as text.

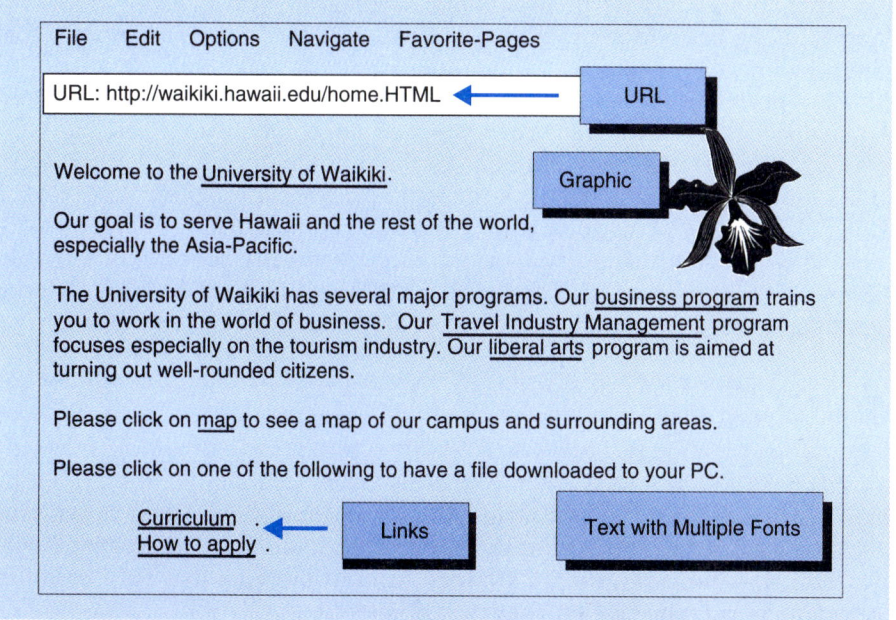

Actually, the application is fairly complex. The webserver does not store the information as webpages. Instead, it executes a script that gets the information from a database management system. To the user, however, all of this is transparent. *For more on scripts on webservers, see Module C.*

When companies use the World Wide Web and other Internet technologies internally, this is called using an **intranet.** In many ways, intranets are the cutting edge of Internet service use. Residential users of the Internet often have to deal with very slow transmission lines, as discussed at the end of this chapter. This means that content providers tend to make comparatively light use of graphics, animation, and other information that would take too long to download from the webserver to the user. In contrast, business users often access intranet services through their local area networks. This may allow them to access these services at megabit speeds. At such speeds, more intensive graphics, voice, video, and more sophisticated services are feasible.

Intranets are very attractive for several reasons. First, intranet applications use standard Internet browsers, such as Netscape Navigator. Many users already have these programs, reducing the need to install new software on client PCs when introducing new applications. In addition, browsers run equally well on personal computers, Macintoshes, and UNIX workstations, and they can all deal with information in the same format. There is no need to develop separate versions of new applications to run on these three desktop computer platforms.

As discussed in Module C, browsers can be used as front ends to traditional database applications and other applications. This means that a user loads a form to the browser, fills it in, and sends it back. The webserver deals with the mainframe or other database engine and then presents the results to the user as a new webpage.

There are even new programming languages, such as Java, whose programs can be embedded in webpages. When you load the webpage, you load the program. Although speed limitations currently make this useful only for small applications, called **applets,** many applications are fairly small. Indeed, some analysts feel that even when we used large applications, such as word processing programs, we will not store these programs on our local PC hard disks. Rather, instead of using a large word processing program at all, we will download a series of modular word processing applets over the network as we need their functionality. A number of vendors already produce low-cost **internet PCs** that have no hard disk drives at all because they download all programs they need from the network.

THE USER LAYER

In this book, we will focus on the three technical layers in Figure 1.1, but the real benefits come at the user layer. All organizations have access to the same technologies, but not all exploit the opportunities raised by new technologies. At their best, technologies are **strategy support systems** that help organizations implement effective business strategies.

A good example of success at the user layer is VeriFone's use of *round-robin* projects. It is best to illustrate this with an example. In Greece, a VeriFone salesperson was encountering problems with a customer. A competitor had told the customer that VeriFone had no experience with security for pen-based systems. On Friday, the customer expressed this concern. The VeriFone salesperson promised to reply on Monday.

That evening, the salesperson in Greece broadcast a message on the company's *ISales* distribution list. About 600 people got copies of the message. There were over 150 responses in the next 24 hours.

Now the real work began. In London, a sales professional put together a PowerPoint presentation to document VeriFone's expertise in this area. At the close of the day on Saturday, this analyst passed the presentation to another sales professional in Atlanta. Taking advantage of the time zone difference, this second salesperson reviewed the presentation and continued to work on it. Later, this second sales professional passed it on to the originator in Greece. On Monday morning, the salesperson gave the promised presentation on VeriFone's expertise in security for pen computing.

What is so extraordinary about this example is that, at VeriFone, it is not extraordinary at all. Will Pape, the company's CIO, is involved in one or two of these round-robin projects every week.

This does not mean that this business innovation is easy to implement. It requires a corporate culture that demands openness and support whenever it is needed. It also requires a great deal of logistics to be sure that only one person at a time has the master copy of the presentation so that everybody knows what they are to do and when. On a larger project, the project manager often begins with a telephone conference to clarify what is needed, what each person should do, and when each one should do it.

There is nothing special about the information technology that VeriFone uses in its round-robin projects. In fact, its email system and the way it transfers files are almost primitive in its simplicity. The *power is in the strategy*, not the technology.

Some years ago, the chairperson of the Stanley Works told a shocked board of directors, "Last year we sold ten million drills that nobody wanted." He went on to explain that drills are expensive and clumsy to use. People merely tolerate drills because they need drills to satisfy an underlying need, namely to produce holes. Data communications professionals must understand that what they offer their firms is like drills. While it may be technologically exciting, what organizations need are solutions for their business needs. New technologies may lead to new strategies, but there is nothing automatic or even technology driven about that process.

Internet Access via CSLIP and PPP

These first two chapters have discussed layering. We will finish with an integrative example showing how layering works in practice.

Devices

When NuPools employees access a World Wide Web server using a browser, the actual connection is relatively complex, as shown in Figure 2.13.

The Internet Service Provider (ISP) ■

To connect to the Internet, you must go through an **Internet Service Provider (ISP).** This might be a private company. It might also be a university. In a large firm, the firm can be its own ISP, handling connections to the Internet.

The User PC ■

The user must have a personal computer. If the user is going to reach the ISP via the telephone, he or she will need a device called a modem. The user will also need an account with the Internet Service Provider. *For more on modems, see Chapter 3 and Module D.*

FIGURE 2.13

Internet Access via CSLIP or PPP

The user must have a point-to-point connection to the *Internet Service Provider (ISP)*. This normally uses a telephone line and requires a modem (see Chapter 3). There must be *CSLIP* or *PPP* software to pass the data over this point-to-point link. The ISP's *access computer* is at the end of the CSLIP or PPP link. This access computer attaches to a device called a *router*, which connects the ISP to the Internet. Via a series of Internet routers, messages ultimately get to the webserver.

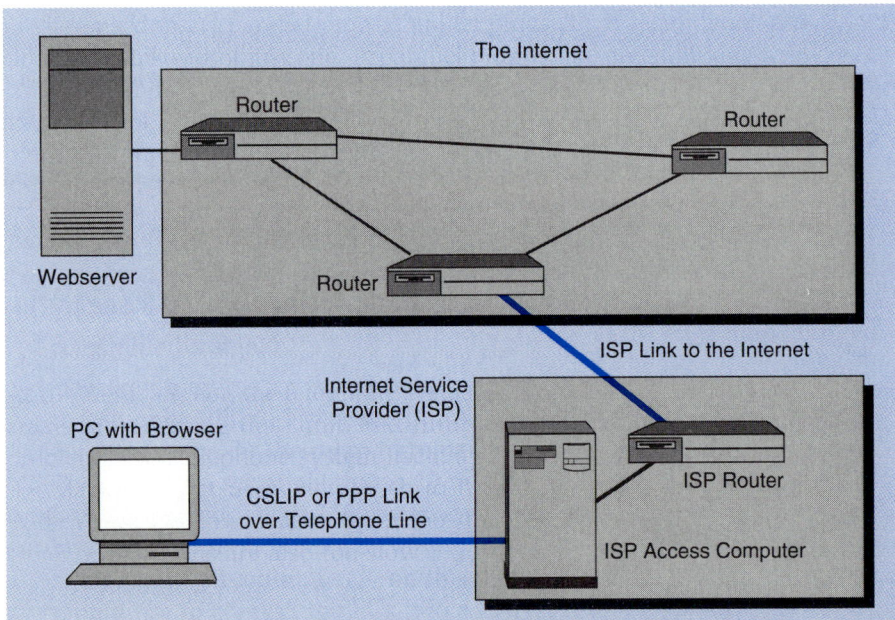

The CSLIP or PPP Link

The telephone system is prone to errors in transmission. The user will need software for transmitting messages across unreliable telephone lines. Some newer operating systems can do the work without additional software. Older operating systems require additional software. Some ISPs provide the additional software. Others require the user to buy the software from third parties.

Two standards are in common use for managing point-to-point data flows over telephone lines. One is **CSLIP,** the **Compressed Serial Line Internet Protocol.**[3] A newer and superior standard is **PPP,** the **Point-to-Point Protocol.** Both work quite well. Some ISPs support both. Others support only one. The user must select software that follows the proper standard(s).

The ISP Access Computer

The telephone line ends at the ISP **access computer.** This specialized machine manages the user interactions with the Internet Service Provider. Some access computers are PCs.

Router and Internet Access Line

As discussed in Chapter 6 and Module H, the Internet uses special switches called routers to route messages from one end of the Internet to the other. The ISP must have a router. The ISP must also pay for an access line to the Internet.

The Internet

The Internet consists of many subnets linked together by routers. Requests from the browser to the webserver may take many hops before reaching their destination. The same is true for responses from the webserver to the browser.

The Webserver

As noted earlier, the hardware of the webserver is unimportant because computer (transport) layer standards shield transmission from such details. The important thing is the webserver application software. This software manages the hyperbase at the site. It also reads browser requests and sends responses.

Layering

Figure 2.14 shows another way of looking at the situation. This figure shows who the user PC talks to directly or indirectly. This figure brings us back to the layering in Figures 1.1 and 1.2.

---

[3] There actually is an older protocol, SLIP, the Serial Line Internet Protocol. Almost all implementations today follow the new CSLIP protocol.

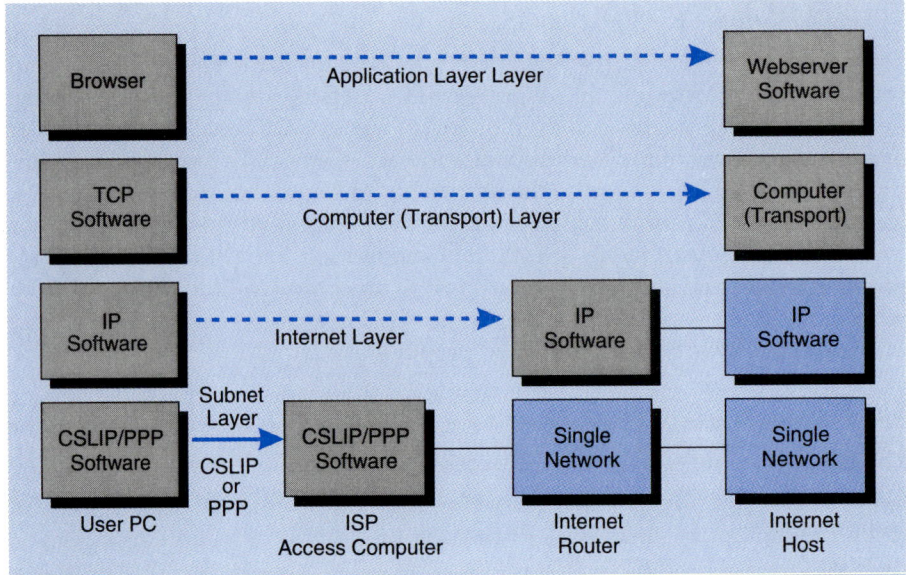

FIGURE 2.14
Layered Interactions in Internet Access

There is a subnet transmission layer link between the user PC and the Internet Service Provider access computer. There is an internetting transmission-layer connection between the user PC's IP program and the ISP's router. There is a computer-layer connection between the user PC's TCP program and the webserver's TCP program. Finally, there is an application-layer connection between the browser and the webserver application software.

The Subnet Transmission Layer ∎

In Figure 2.14, the subnet transmission layer consists of the telephone line, the modem, and the CSLIP or PPP connection between the user PC and the Internet Service Provider's access computer.

Internetting Transmission Layer ∎

For the internetting transmission layer, the user PC communicates via the IP protocol. This is the main internetting protocol in the TCP/IP architecture, which is used by routers on the Internet.

Note that when the user PC sends IP messages, it does not send them to the ISP access computer. Instead, the ISP access computer delivers these IP messages to the ISP router. Routers know how to handle IP messages.

The ISP router transmits the IP message to other routers on the Internet. They finally route the IP message to the router near the webserver. That router delivers the IP message to the webserver. The subnet-layer transmission by which it delivers the message to the webserver is irrelevant. It may be CSLIP/PPP, or it may use some other protocol.

Computer (Transport) Layer ■

Everything we have discussed to this point has merely provided a way of sending bits from the user PC to the webserver. Now these two machines need to be able to exchange requests and responses. They are likely to use different hardware architectures and operating systems. Fortunately, computer (transport) layer standards hide such details from the two computers.

Almost all Internet hosts use *TCP (Transmission Control Program)* as their computer-layer standard. This means that the user PC must do so as well. Newer operating systems, such as Windows versions beyond 3.1, can handle TCP communication without additional software. *For more on TCP, see Module H.*

Application-Layer Communication ■

Now that the machines can exchange messages, the two application programs can communicate. The browser sends requests to the webserver. The webserver sends its response messages back to the user PC.

Each application has a different standard. For the World Wide Web, the application standard is the **HyperText Transfer Protocol (HTTP).** This protocol defines requests and responses.

Message Transmissions

We have talked about various programs and computers sending "messages." Now we will look at this in more detail. We will focus on the client PC, as shown in Figure 2.15.

Protocols ■

Obviously, messages need to be standardized. Standards for *horizontal* message exchanges like those in Figure 2.15 are called **protocols.** The messages themselves are called **protocol data units (PDUs).** There must be a protocol at every layer to link processes on different machines. This means that there also has to be a PDU at each layer.

The Application Protocol Data Unit (APDU) ■

First, the application program creates its request message. This is the **application protocol data unit (APDU).** It is likely to be a complex message.

The Computer Protocol Data Unit (CPDU) ■

Figure 2.15 shows that the application program passes the APDU down to the TCP program. This program adds a header, which contains information for the TCP program on the other machine. The APDU plus the computer header (CH) is the Computer PDU.

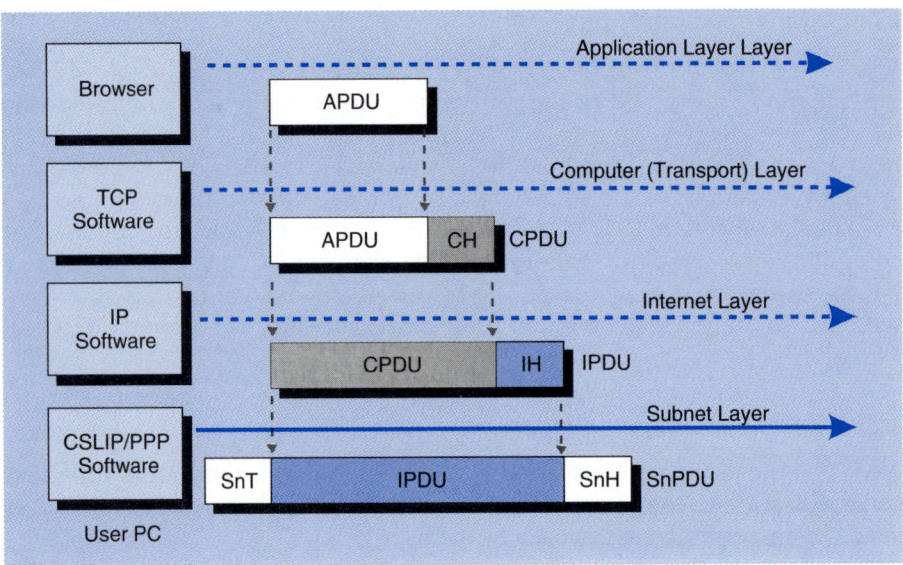

FIGURE 2.15

Protocol Data Unit (PDU) Transmissions from the User PC

Protocols are standards for horizontal message exchanges between peers on different machines. Messages are called protocol data units (PDUs). First, the browser creates an application PDU (APDU). It passes this to the TCP program. The TCP program adds a computer header (CH) to the APDU to create a computer PDU. The TCP program passes the CPDU down to the IP program. The IP program adds an internetting header (IH) to the CPDU to create an IPDU. Finally, the IP program passes the IPDU down to the CSLIP/PPP program. This creates a subnet transmission PDU (SnPDU) by adding a header and trailer. Headers and trailers provide delivery or supervisory instructions to the peer process on the other machine.

The Internetting Protocol Data Unit ■

Now the TCP program passes its message down the IP program, which handles internetting. This program adds an internetting header (IH) to the CPDU. The IH contains instructions for the routers on the Internet.

Another way to look at this is that the Internetting PDU (IPDU) consists of the APDU, the CH, and the IH.

The Subnet Transmission Protocol Data Unit ■

Finally, the CSLIP or PPP program gets the IPDU. This time it adds both a header and a trailer aimed at the access computer. We will call these the SnH and the SnT. This becomes the subnet PDU (SnPDU).

The SnPDU, then, contains the APDU, the TH and TT, the IH, and the CH. Headers and trailers give delivery and supervisory instructions to the peer process on another machine.

Reception of the PDUs

Figure 2.16 looks at what happens to the CSLIP or PPP PDU after the user PC transmits it.

The Access Computer ■

Although the figure does not show what happens at the other end, the access computer reads the CSLIP or PPP header and trailer, which are aimed at it. It passes the Internetting PDU that remains to the router.

The Router ■

Next the routers on the Internet use the information in the IH to route the IPDU from the Internet Service Provider to the webserver.

On the Webserver ■

The IP program on the webserver removes the CPDU from the IPDU and passes the CPDU to the TCP program. The TCP program reads the computer

FIGURE 2.16
Receiving the PDUs

The access computer receives the TPDU and interprets the TH and TT for instructions. It passes the IPDU on to the router, which receives the IPDU and interprets the IH. The routers on the Internet deliver the IPDU to the webserver. The IP program on the webserver passes the CPDU to the TCP program on the webserver, which interprets the CH. The TCP program on the webserver passes the APDU on to the webserver application software, which interprets the request.

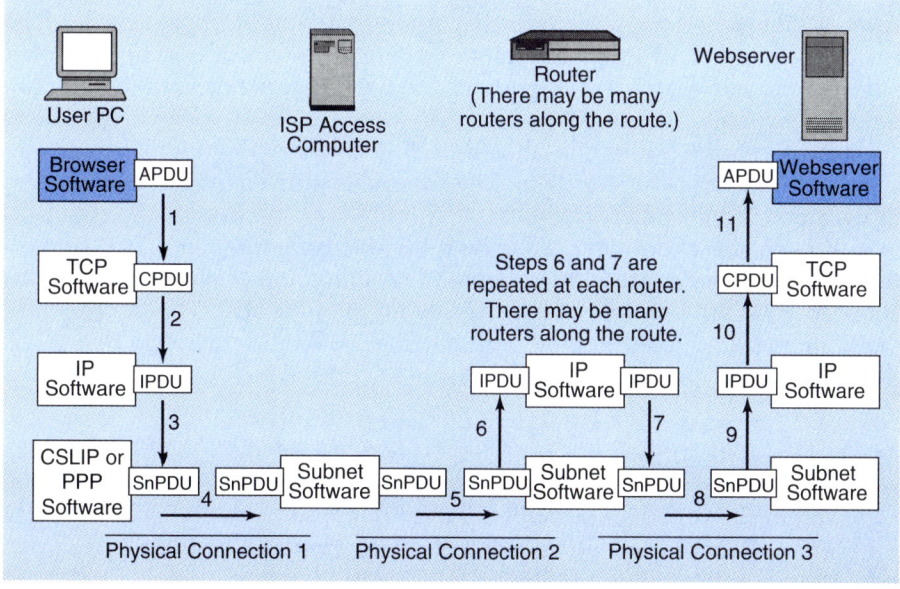

header, which is addressed to it. When it is finished, it removes the APDU from the CPDU and delivers the application PDU to the webserver application software.

Finally, the webserver application reads the APDU and realizes that this is a request. It does the required work and sends back its response message in the form of a new APDU. You should be able to trace what happens to it from this point.

Interfaces

We saw earlier that *horizontal* standards that control information between peer programs on *different* machines are called *protocols*. But there also need to be standards to govern the passing of PDUs *vertically*, between programs on the *same* machine. These are called **interface standards.** For instance, the standard governing the application program's interactions with the transport-layer process is called the *application program interface (API)*.

Flexibility

Although we have looked at the communication that takes place at all three layers, we should note that the actual situation during a CSLIP or PPP connection actually is more complex than we have made it appear.

Switching Webservers ■

After reading a webpage, the user may jump to another webpage on a different host computer. Perhaps the initial webpage contains a link to that other page on another computer. Or perhaps the user merely types the URL of another page on another server.

The telephone connection and the CSLIP or PPP connection remain unchanged at the transmission layer. The PC also continues to send IP messages to the same initial router. However, that router will now route the messages to a different host computer. Your own PC will have to open a TCP connection to that computer. It will also have to be able to open WWW connections between your browser and the webserver on the new host. This process normally happens so rapidly, however, that users often are unaware that it has happened.

Multiple Applications ■

Once you connect to a server, you are not limited to working with the World Wide Web. You can use any service that server provides, including electronic mail and the Internet services discussed in Module B.

Internet Access at VeriFone and NuPools

Internet Access at VeriFone ■

VeriFone began providing internal internet access through PPP servers in its individual offices. These usually used PCs as access computers. More recently, they have purchased PPP access computers from Shiva. The Shiva Boxes are complete systems that include hardware, software, and several modems.

Because VeriFone has engineering, manufacturing, and sales offices around the world, there is a very good chance that any traveling VeriFone employee will be within a local telephone call of a Shiva Box. If not, however, there is very likely to be a local access node for CompuServe. From their CompuServe accounts, VeriFone employees can get access to both electronic mail and browser services.

Internet Access at NuPools ■

Although NuPools has its own home page on the World Wide Web, and while two of its computers in the administrative office have PPP connections to the Internet Service Provider that maintains its webpages, it does not use these connections very much. Its suppliers do not use the Internet yet, although this is likely to change soon. In addition, their two traveling site managers are always a local call away and so can simply dial into the company's remote access server.

Internet at Home? ■

If you are interested in getting access to the Internet at home, see the box, "Connecting a PC at Home to an Internet Service Provider."

CONCLUSION

In this chapter, we have looked above the transmission layer to deal with the transport (computer), application, and user layers.

Stations on a network can be anything from mainframes to PCs. Most organizations are now *downsizing*—moving smaller applications down from mainframes to smaller machines. The *PC network* is now the major focus of downsizing.

In PC networks, there are multiple servers instead of just one mainframe or a few minicomputers. Each server runs only one application or a few applications. This allows each server to be optimized for its application. Among the most important servers are file servers. File servers store files, control network printing, and download programs to client PCs for execution in file server program access.

Connecting a PC at Home to an Internet Service Provider

What would you have to buy to get access to the Internet from home via a CSLIP or PPP connection? The answer is that you need to buy surprisingly few things.

The Telephone Connection

First, of course, you need a telephone connection. If you do not use it too much, you can use an existing line without inconveniencing anyone too much. If you are going to be using the Internet a great deal, however, you will probably need a second telephone line to keep harmony at home.

You will also need a device called a modem. As discussed in Chapter 3 and Module D, modems allow computers to communicate over telephone lines. A modem will cost about $50 to $250.

Core Communication Program

You will also need a core communication program. Generally, this will be a single program that handles the CSLIP or PPP connection, the IP communication with the router, and the TCP connection to the target Internet host. Because TCP/IP standards are so simple, this program usually fits on a single floppy disk. Some programs, such as Trumpet Winsock and Chameleon are sold as shareware or at very low cost.

Newer operating systems, such as Windows beyond Version 3.1, contain all the

---

In applications, we focused first on standards. We noted that standards allow two applications to communicate even if they are on computer platforms. We looked at client/server computing, in which one program sends out a request, and the other program sends a response. We also looked at two extremely important applications: electronic mail and the World Wide Web.

At the user layer, we focused on how one of our two case study organizations, *VeriFone*, sets standards at the user layer to achieve its complex business strategy. As we saw, power lies in the strategy, not the technology. Our

code you need for CSLIP, PPP, TCP, and IP connections. If you have one of these operating systems, you do not have to buy a core communication program.

Application Programs

You would like to have multiple application programs, such as browsers, electronic mail programs, and so forth. You might want to have one or two dozen application programs to take advantage of the Internet's many offerings. At a minimum, you need a browser and an electronic mail program.

Kits

It is usually possible to buy kits that contain both the core communication program and multiple application programs. A typical price is about $100 for the software in the kit.

The Internet Service Provider

Of course, you must also pay the Internet Service Provider. Maintaining a link to the Internet is relatively expensive, and it is expensive to provide services when users are having problems. Typically, access provider charges vary from about $10 per month to about $30 per month, depending on the level of services and how much time the user spends connected to the access computer. There is likely to be an initial charge as well.

This basic price is likely to include a reasonable amount of disk storage to hold electronic mail messages and other information. (For electronic mail, the access computer is likely to act as a mail server.)

---

other case study, *NuPools*, has also changed the way it works to take advantage of new possibilities, but not as radically.

We closed this chapter with an integrative example: using the CSLIP or PPP protocol to reach an Internet Service Provider in order to get access to World Wide Web services on the Internet. This allowed us to look at the complexities of layered communication, including the distinction between protocol (horizontal) standards and interface (vertical) standards. We also looked at how layer processes handle the messages of other adjacent layers.

In the next chapter, we will begin working our way up the layered hierarchy in Figure 1.1, our Basic Communication Model. Chapter 3 deals with point-to-point transmission. Later, Chapters 4 and 5 will take us into local area networks. Chapters 6 and 7 take us into local internets and enterprisewide internets.

CORE REVIEW QUESTIONS

1. Where is the processing done in terminal–host communication? In file server program access? In client/server processing?
2. What is the dominant computer platform today? Distinguish between PCs and workstations.
3. Do you need a separate transmission network for each computer platform? Explain.
4. Is client/server computing limited to PC networks? What are the two main types of messages in client/server computing? Distinguish their roles. Which program initiates the interaction?
5. List the parts of an email message. What is an attachment, and why is it useful? What is a distribution list, and why is it useful?
6. Define hypertext. What is the WWW? What is a URL? Does the World Wide Web use client/server processing? What is the client program? The server program?
7. Draw the TPDU in terms of the APDU, headers, and trailers.
8. What is a protocol? What is a PDU? What is an interface?
9. What is an intranet? Is every internal corporate internet an intranet?

DETAILED REVIEW QUESTIONS

1. In client/server processing, does the client machine have to be a PC? Does the server have to be more powerful than the client?
2. List the major pieces of hardware and software processes when a browser on a PC sends a request to a webserver on an Internet host. Assume a PPP connection to an Internet Service Provider.
3. At what layer do each of the following standards operate: hypertext transfer protocol (the World Wide Web standard for client/server interactions), CSLIP, IP, PPP, and TCP.
4. Trace the PDU creation process on the user PC in WWW access via PPP. Give the actual name of the protocol at each layer.

THOUGHT QUESTIONS

1. Why do you think the networked PC has become the dominant computer platform?
2. Describe how VeriFone's round-robin projects give it a competitive advantage. What user-level strategies, aspects of corporate culture, and technology are needed for these projects to succeed?
3. File server program access uses client PCs and servers. Is it client/server processing? Explain.

PROJECTS

1. Log into a **PC network.** Execute a program stored on the server's hard disk drive. Where does this program execute?
2. Send yourself an electronic mail message. Note the fields in the header in the message you receive. Add a signature and, if possible, an attachment. Send an electronic mail message to someone in the same network. Send an email message to someone via the Internet. Create a distribution list and send a message to it.
3. Access a webserver via a CSLIP or PPP connection.

For online exercises, please visit this book's website at: http://www.prenhall.com/panko

REFERENCES

STANDARD & POOR'S (1993). Computers: Basic analysis, *Standard & Poor's Industry Surveys 1994*, 162(7), Section 1, C75–C125.

CHAPTER 3

Terminal–Host Communication

INTRODUCTION

Chapter 1 noted that the defining characteristic of networking is any-to-any communication. Chapter 4 will begin our discussion of networking. This transition chapter looks at a simpler topic, point-to-point communication. It will introduce many concepts we will need to discuss full networking.

Terminals and Hosts ■

Specifically, this chapter focuses on a single situation, shown in Figure 3.1. This is terminal–host communication. As we saw in Chapter 2, programs execute on a large host computer. This is either a mainframe or a minicomputer. The user's desktop machine, in turn, is a terminal. A terminal looks like a PC, but it is much simpler. It has only enough intelligence to transmit keystrokes to the host computer and to display text (and sometimes graphics) on the terminal's screen.

Terminal Emulation ■

Actually, the desktop machine in Figure 3.1 is not really a terminal. Very few of us have

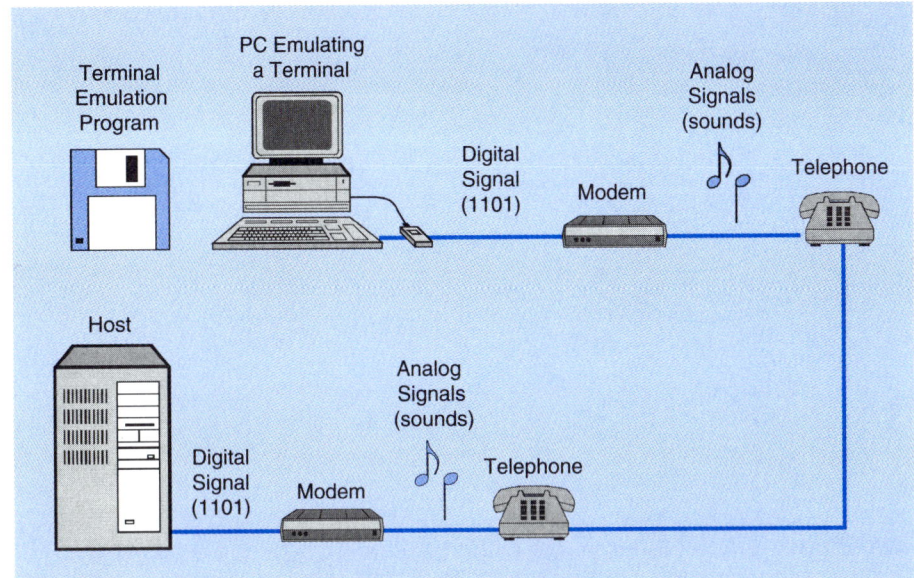

FIGURE 3.1

Terminal–Host Communication

Programs execute on the *host computer*. *Terminals* have almost no processing power. PCs can *emulate* (act like) simple terminals using software alone. In addition to terminal emulation software, the PC user will need a *modem* if communication with the host will take place over a telephone line.

terminals on our desks. Instead, we have personal computers. We would like our PC to be able to *emulate* (act like) a terminal. In this way, we will not have to add a terminal to our already crowded desks and offices. We will see that terminal emulation is very inexpensive for simple terminals.

Telephone Transmission ■

There are many ways to link a terminal to a host computer. One of the simplest is to use the *telephone system*. Telephone lines are not perfect, but for the low transmission speeds of simple terminals, they do a good job. You can turn your desktop telephone at home or work into a data transmission system.

Figure 3.1 shows that telephone transmission requires you to add a piece of hardware at each end of the telephone line. This hardware is called a *modem*. We will see that computers transmit *digital information* (1s and 0s). In contrast, the telephone system is designed for *analog information*, namely sounds. The modem turns the computer's 1s and 0s into sounds.

Asynchronous ASCII Transmission ■

When we write in English, we have to spell our words using the 26 letters of the alphabet. In digital transmission, we have to "spell out" everything in 1s and 0s. For instance, the ASCII coding system used by simple terminals spells each character as a string of seven bits (1s or 0s). Figure 3.2 shows the ASCII code for "Happy Birthday." Note that each letter is represented by a seven-bit

FIGURE 3.2
ASCII Codes for "Happy Birthday."

H	1001000
a	1100001
p	1110000
p	1110000
y	1111001
	0100000
B	1000010
i	1101001
r	1110010
t	1110100
h	1101000
d	1100100
a	1100001
y	1111001
.	0101110

Notes:
Every character to be printed needs a seven-bit ASCII code. Capital "H" and lower-case "h" have different codes so they will print differently. The space between "Happy" and "Birthday" needs a code so it will print. The period needs a code so it will print.

code. We also have to represent the period and even the space by ASCII codes. (Otherwise, neither would print or appear on our display screen.)

If we just sent the 105 bits of these 15 ASCII codes back to back without any space, how could the receiver tell where one character stopped and another began? Instead of just sending a stream of raw data bits, simple terminals send each character separately in a form of transmission called *asynchronous transmission*. We will see that asynchronous transmission places each character code in a ten-bit package called a *frame*.

TERMINAL EMULATION

In **terminal emulation,** a PC emulates (acts like) a terminal. In effect, it lies to the host computer. It says "I'm a terminal." It then carries out the lie very well. There is literally no way for the host to know that it is talking to a PC rather than a terminal. Nor does the host care that it is not talking to a real terminal.

VT100 Terminals

There are many types of terminals. Most host computers can only work with a few of these. So it is important to emulate a terminal acceptable to the specific host computer you wish to use.

Most host computers can work with a particularly simple type of terminal called the **VT100 terminal**. Figure 3.3 shows a VT100 terminal screen.

Note that it can only show text. It merely paints ASCII characters on the screen. There are no graphics whatsoever. Nor is there color.

In addition, a VT100 terminal cannot show boldface, italics, multiple fonts, or any of the advanced text formatting that we are used to on personal computers. The basic VT100 terminal design is a child of the 1960s—a much simpler age in computing.

VT100 terminals are also painfully slow. They have a top speed of 19.2 kbps, and many run at slower speeds. Even for simple text, this is rather leisurely. If most of today's data communications systems resemble high-speed jet travel, the VT100 is a bicycle.

Although these limitations make the VT100 terminal a poor way to serve users, the very simplicity of VT100 terminals makes them easy for hosts to support. As a result, the VT100 terminal has become a lowest common denominator for terminal–host transmission. Almost every host can support VT100 terminals today.

VT100 Terminal Emulation Software ■

Because VT100 transmission is simple, VT100 terminal emulation is very easy to do. All you need is terminal emulation software. Most newer PC operating systems come with built-in terminal emulation software. (In Windows 95, the program is *HyperTerminal*.)

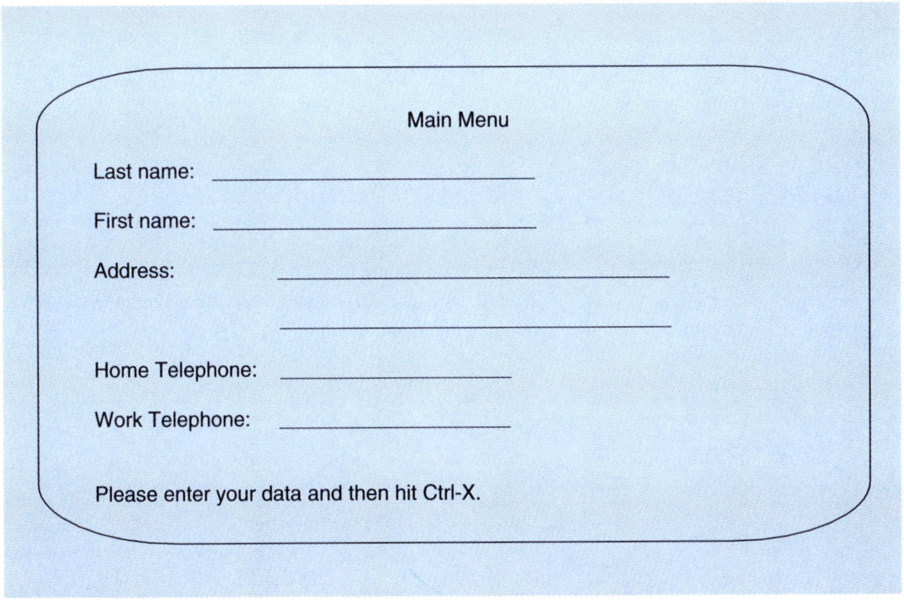

FIGURE 3.3

VT100 Terminal Screen

A *VT100 terminal* offers a *full-screen interface*. You can cursor up, down, left, and right. However, the VT100 does not support color, graphics, or multifont text.

You can also buy commercial terminal emulation software. This will cost anywhere from $75 to $250. For this extra money, you will get additional features. In fact, most of the terminal emulation programs that come bundled with operating systems are subsets of commercial programs.

Login Scripts ■

In addition to painting information on the screen, terminal emulation programs allow you to automate login and sometimes other operations. Figure 3.4 shows a **login script**. It anticipates what the host computer will say and gives responses automatically. This way, the user merely starts the terminal emulation program and starts the login script. There is no need to memorize the arcane series of commands needed to log in.

Actually, the login script in Figure 3.4 is an example of bad practice. Note that it even types the user's *password*. This means that anyone with access to the person's computer can log into his or her account. This security violation is remedied by ending the script when the host asks for the password. The user can then type the password. Some more sophisticated scripting languages allow the script to pause and prompt the user to type his or her password.

Scripts (sometimes called macros) can be used for other purposes. For instance, once you are logged in, you can execute another script to handle an application. One might take you into your electronic mail program. It might then upload outgoing messages from your PC to the host. Afterward, it might download new electronic mail from the host to your PC. This would allow you to

FIGURE 3.4

Login Script

You can create *scripts* (also called macros) that automate the process of logging into the host computer. The script anticipates what the host will say and issues appropriate responses at each step.

Remark: Dial the telephone number
Output: ATDT9,5553270

Remark: Interact with the network
Input: Enter terminal type:
Output: VT100
Input: Enter network ID:
Output: Y202430
Input: Host name:
Output: Voyager

Remark: Log into the host computer
Input: Login:
Output: PLee
Input: Password:
Output: PASXA

Input is what the host computer transmits to the PC emulating a terminal.

Output is what the script sends back in reply.

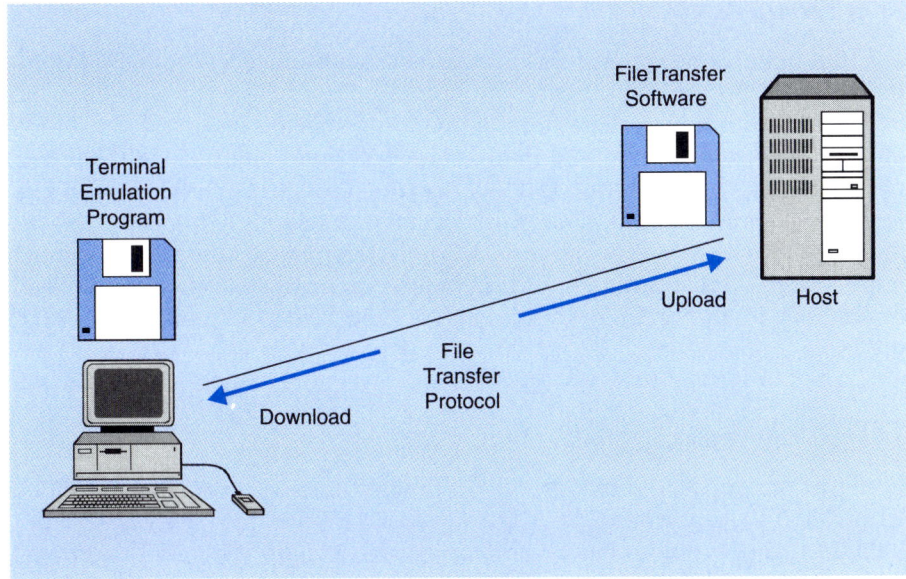

FIGURE 3.5

File Transfer

For *file transfer*, the host must have a *file transfer program* that follows a *file transfer protocol* that the terminal emulation program also supports. Files are *downloaded* from the host to the PC or *uploaded* from the PC to the host.

do your mail preparation and reading *offline* (while not connected to the host). For long-distance transmission, this saves money.

File Transfer ■

Figure 3.5 shows another feature of terminal emulation software. This is **file transfer**. Sometimes you want to transfer whole word processing files, spreadsheet files, or other files between your PC and the mainframe. File transfer supports this. It even corrects for errors during transmission, so that you have a clean file transfer.

The host is the dominant machine in the transaction. So when you transfer a file from the host to your PC, this is **downloading**. When you transfer a file from your "lowly" PC to the "lofty" host, in turn, this is **uploading**.

File transfer requires the host to have a **file transfer program**. In addition, your terminal emulation program and the file transfer program must communicate according to some **file transfer protocol (FTP)** standard. In addition to the Internet FTP standard, there are many other protocols. These include XMODEM, ZMODEM, Kermit, and several others. So it is important to pick a terminal emulation program that is compatible with the file transfer program on the target host. *For more on File Transfer Protocols, see Module E.*

Other Terminals

Although most hosts support VT100 terminals, there are many other types of terminals.

3270 Terminals

For IBM mainframe computers and mainframes from several other vendors, there are **3270 terminals**. These can communicate at over one million bits per second, although a more normal operating speed is about 64 kbps. Most models support graphics, and many models support color as well. For people who work with computers all day, such as reservation clerks, slow VT100 terminals would hamper their work. It pays to use higher-speed 3270 terminals to reduce overall work costs. IBM 3270 terminals, however, cost much more than VT100 terminals and are more expensive to emulate as well. Emulation, in fact, requires adding a printed circuit expansion board to the PC.

TTY Terminals

Going in the opposite direction, TTY terminals are even simpler than VT100 terminals. TTYs are essentially VT100s without a full-screen interface. With a VT100 terminal, you can cursor up, down, left, or right. With a TTY terminal, however, you can only type on a single line, and the only cursor movement is backspace/delete. If you have ever used MS-DOS, you have used a TTY terminal interface. Once TTYs were the lowest common denominator in terminal emulation. Almost all hosts that can support TTYs today, however, can also support VT100 terminals.

Newer VT Terminals

The VT100 terminal was created by Digital Equipment Corporation. The VT100 was actually a rather early member of the Digital "Virtual Terminal" line. The newest VT terminals are in the 300 series and are much more capable than the VT100 design. Some terminal emulation programs can emulate newer VT100 designs. However, newer VT designs are not widely supported by non-DEC computers. In addition, when ANSI (the American National Standards Institute) standardized the design for a low-end terminal, it chose the VT100 design. (This is why VT100 terminals are sometimes called **ANSI terminals**.)

TELEPHONE TRANSMISSION

As noted earlier, the telephone system is an ideal low-end transmission system. Almost every desktop has one, and the telephone is fine for transmitting low-speed data.

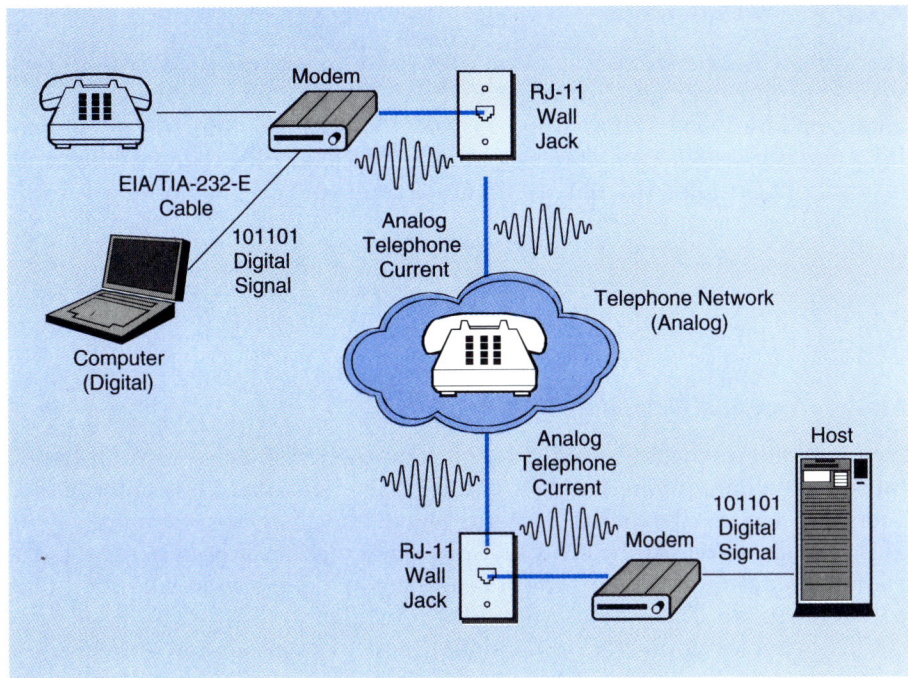

FIGURE 3.6

Modem Transmission

Modems translate between digital communication signals and the analog telephone system. The sending modem *modulates* the computer's digital signal into an analog signal that can travel over the telephone line. The receiving modem *demodulates* the analog signal back to digital. *Digital signals* have only a few states—often only two (1 and 0) and jump abruptly between those states. *Analog signals* vary smoothly within some range, taking on any value within the range.

Digital and Analog Transmission

Digital Transmission ■

Figure 3.6 shows that computers transmit 1s and 0s. This form of signaling that changes abruptly from one "state" (1 or 0, in this case) to another "state" is **digital** signaling.[1] All computers today are digital.

---

[1] More generally, digital transmission consists of abrupt transmissions between a discrete (countable) number of states. For instance, in a digital watch, each position (digit) can only be in one of ten possible states, from 0 through 9. Each position changes abruptly when it changes. Ideally, the change would be absolutely instantaneous, although this is seldom possible.

Analog Transmission ■

In contrast, the telephone system was designed to carry **analog** signals, which vary smoothly over time. Analog signals vary smoothly within some range, taking on any value in that range. Figure 3.6 also shows this situation. Analog transmission is ideal for the human voice, which changes smoothly instead of abruptly. (Except in science fiction, people do not speak in 1s and 0s.)

Modems

Modulation and Demodulation ■

Figure 3.6 shows that to transmit digital computer information over an analog transmission line, you need a device called a **modem**. At the transmitting end, a modem converts the digital computer signal into an analog telephone signal. This is **modulation**. At the other end, the other modem *demodulates* the signal back to digital form. In two-way transmission, modems must both modulate and demodulate. This is why we call them modems.

This simplest form of modulation is **frequency modulation**, in which a 1 is represented by one sound frequency (pitch) and a 0 is represented by another sound frequency. When the receiving modem hears a sound, it transmits either a 1 or a 0 to its attached computer. The box, "Modulation," discusses the much more complex forms of modulation found in today's high-speed modems. *For more on modulation, see the box, "Modulation."*

Modem Standards

Obviously, the modems at the two ends must communicate according to standards if they are to be able to work together. Table 3.1 shows a number of common modem standards.

Speed Standards ■

The most obvious thing to standardize is transmission speed. Table 3.1 shows that all newer modem standards have been created by the ITU-T. Today's fastest modems follow the **V.34** standard, which can transmit data at 28.8 kbps. *For more on ITU-T, see Chapter 1.*

Speed standards actually standardize modulation. The box "Modulation" shows that high-speed modems use extremely complex forms of modulation. In fact, V.34 specifies multiple complex forms of modulation for use under different conditions. The two modems talk back and forth to find the best possible form of modulation for the line conditions that exist at that time.

TABLE 3.1

Modem Standards

Speed Standards

Name	Speed	Origin
V.34	28.8 kbps	ITU-T
V.32 *bis*	14.4 kbps	ITU-T
V.32	9600 bps	ITU-T
V.22 *bis*	2400 bps	ITU-T
V.22	1200 bps	ITU-T
212A	1200 bps	AT&T

Error Correction and Data Compression Standards

Name	Type	Origin
V.42	Error correction	ITU-T
V.42 *bis*	Data compression	ITU-T
MNP (Microcom Network Protocol)	Error correction and data compression	Microcom

Facsimile Modem Standards

Name	Speed	Origin
V.14	14.4 kbps	ITU-T
V.29	9600 bps	ITU-T
V.17 *ter*	4800 bps	ITU-T

What if there is a newer 28.8 kbps V.34 modem at one end and a slower 14.4 V.32 *bis* modem at the other end? The answer is that newer modems can also communicate using older modulation/speed standards. The V.34 modem will slow down automatically to V.32 *bis* modulation.

Error Correction and Compression Standards ■

Speed standards only describe modulation. Table 3.1 shows that there are also ITU-T standards for error correction (V.42) and data compression (V.42 *bis*).

When you talk on a telephone line, you hear hisses, pops, and other electrical **noise**. This is very bad for data transmission. It can turn a 0 into a 1 or a 1 into a 0.

V.42 modems check for errors during transmission. If a V.42 modem detects an error, it asks the other modem to retransmit the damaged information. In other words, V.42 modems do both **error detection** and **error correction**. This allows clean data transmission over dirty telephone lines.

V.42 *bis* modems, in turn, can **compress** the data before transmission so that transmission time is shorter. V.42 *bis* modems can compress data as much as 4:1. As shown in Figure 3.11 on page 71, a modem can receive data from the computer at 115.2 kbps and compress it down to 28.8 kbps for transmission over the telephone network.

Modulation

Here we will look at the main forms of modulation in use today.

Frequency Modulation

Modulation essentially converts 1s and 0s into sounds. Sounds consist of waves. Figure 3.7 illustrates a sound wave.

Figure 3.7 illustrates that a pure wave oscillates (vibrates) in a regular way. The **frequency** of the wave is the number of times per second it travels through its entire cycle of rising, falling, and rising again. The frequency is measured in **hertz (Hz)**. One hertz is one cycle per second. For higher speeds, we use metric notations.

The wave's **wavelength** is the physical distance between comparable parts on adjacent waves. Ocean waves have wavelengths of many meters; a violin's vibrations have a very small wavelength.

Frequency and wavelength are related. The wave's wavelength times its frequency equals the speed of the wave in the transmission medium. So if you increase the wavelength, you decrease the frequency, and vice versa. Think about strings vibrating. A shorter string will produce a higher-pitch sound.

We can use frequency differences to represent 1s and 0s. For instance, we can use a high frequency to represent a 1. We can then use a lower frequency to represent a 0. So to send "1011", we would send a high frequency for the first time period, a low frequency for the second, and a high frequency for the third and fourth. Older modems use frequency modulation.

Amplitude Modulation

Frequency and wavelength are two of the four characteristics of radio waves. The third is **amplitude**—the level of intensity in the wave (see Figure 3.8).

In amplitude modulation, we represent 1s and 0s as different amplitudes. For instance, we can represent a 1 by a high-amplitude (loud) sound and a 0 by a low-amplitude (soft) sound. Then to send "1011", we would send a loud sound for the first time period, a soft sound for the second, and high-amplitude sounds for the third and fourth. Amplitude modulation is not used by itself in modems.

Phase Modulation

The last major characteristic of waves is **phase**. As shown in Figure 3.9, we call the starting point of the wave at zero amplitude and rising as 0 degrees phase.

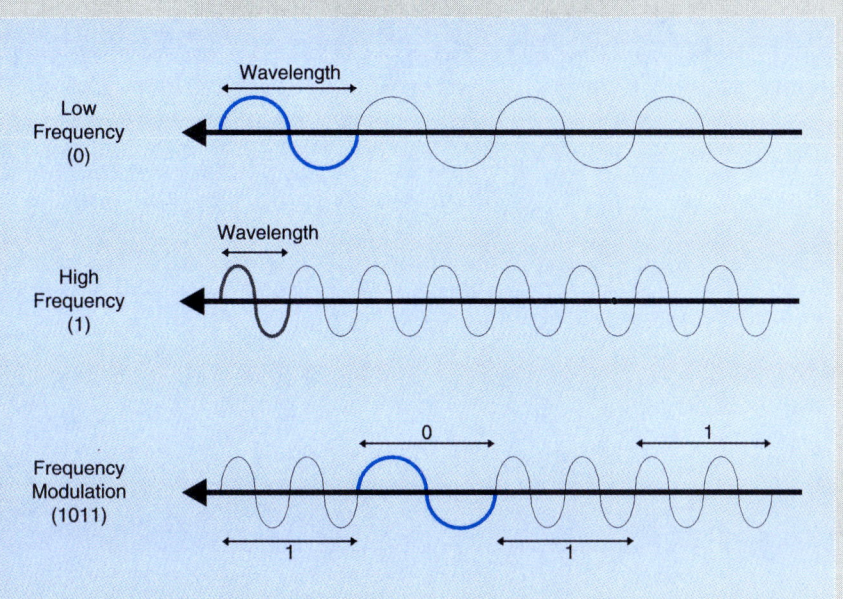

FIGURE 3.7
Frequency Modulation

A pure *wave* is a regular oscillation. The wave's *frequency* is the number of times it travels through a complete cycle per second. Frequency is measured in *hertz (Hz)*—cycles per second. The *wavelength* is the physical distance between successive waves. The wavelength times the frequency is equal to the speed of light in the transmission medium. So waves of higher frequencies have smaller wavelengths. In *frequency modulation*, we represent a 1 by one frequency, say a higher frequency, and we represent a 0 by another frequency, say a lower frequency. So the signal "1011" would be sent as a high frequency for the first time interval, a low frequency for the second, and a high frequency for the third and fourth. Older modems used frequency modulation.

The wave hits its maximum at 90 degrees, returns to zero on the decline at 180 degrees, and hits its minimum amplitude at 270 degrees. While the human ear can pick out frequency (pitch) and amplitude (loudness), it is not good at picking out phase differences. Electrical equipment, in contrast, is very sensitive to phase differences.

We will let one wave be our reference wave or carrier wave. Let us use the carrier wave to represent a 1. Then we can use a wave 180 degrees out of phase to represent a 0. The figure shows that to send "1011", we send the reference for the first time period, shift the phase 180 degrees for the second, and return

continued

68 CHAPTER 3 TERMINAL–HOST COMMUNICATION

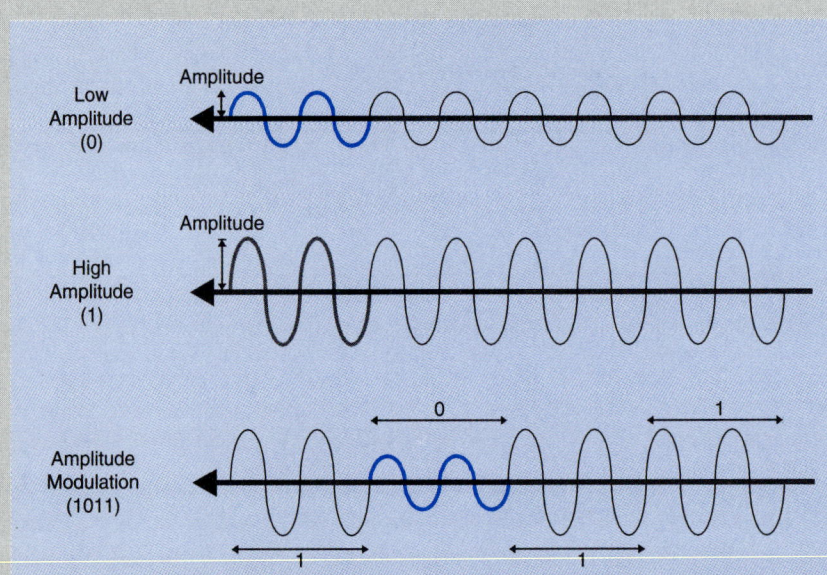

FIGURE 3.8
Amplitude Modulation

Waves vary in *amplitude* (intensity). In *amplitude modulation*, we represent a 1 by a high-amplitude sound and a 0 by a low-amplitude sound. To send "1011", we would send a loud sound for the first period, a softer sound for the second, and a high-amplitude sound for the third and fourth. Pure amplitude modulation is not used in modern modems.

to the reference wave for the third and fourth time periods. While this makes little sense in terms of hearing, it is easy for electronic equipment to deal with phase differences.

Complex Modulation

We have looked at rather idealized modulation schemes. But today's high-speed modems really combine multiple forms of modulation, giving **complex modulation**. Figure 3.10 shows that they combine amplitude and phase modulation. The sender varies both the amplitude and the phase of the transmitted signal with each transmission.

In the figure, there are two possible amplitude levels and four possible phase angles. (Real modulation standards use more combinations.) This gives eight

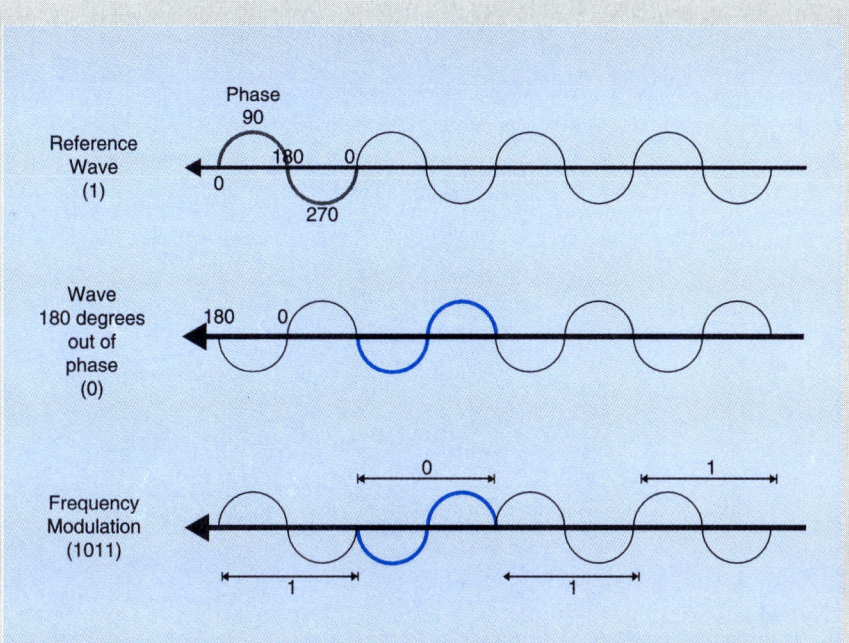

FIGURE 3.9

Phase Modulation

We represent the various *phases* of a wave as it passes over its complete cycle as lying between 0 degrees and 360 degrees. The figure shows two waves. The first is the reference or carrier wave. The second is a wave 180 degrees behind the carrier wave in phase. We can represent a 1 by the carrier wave's phase and a 0 by the second wave's phase. Sending "1011" would require us to send the carrier wave in the first time period, the second wave in the second, and the carrier wave in the third and fourth time periods. All newer modems use *phase modulation*.

possible signals to send in each transmission. With eight possibilities, each transmission can represent one of the eight possible sequences of three bits (000 through 111). The standard assigns a specific three-bit sequence to each combination.[5]

The transmission's **baud rate** is the number of times the line changes per second. Suppose our modem has a baud rate of 2,400. Then sending three bits per line change (baud) gives us a bit rate of 7,200 bits per second. In other words, baud rate

---

[5] One problem with complex modulation is that as you increase the number of possible combinations, you increase the likelihood of an error. To combat this problem, in real standards, not all combinations are valid signals. For instance, in our simplified case, only four of the eight combinations might be declared valid. Each of these four could carry two bits of data, then, giving the four combinations 00, 01, 10, and 11.

continued

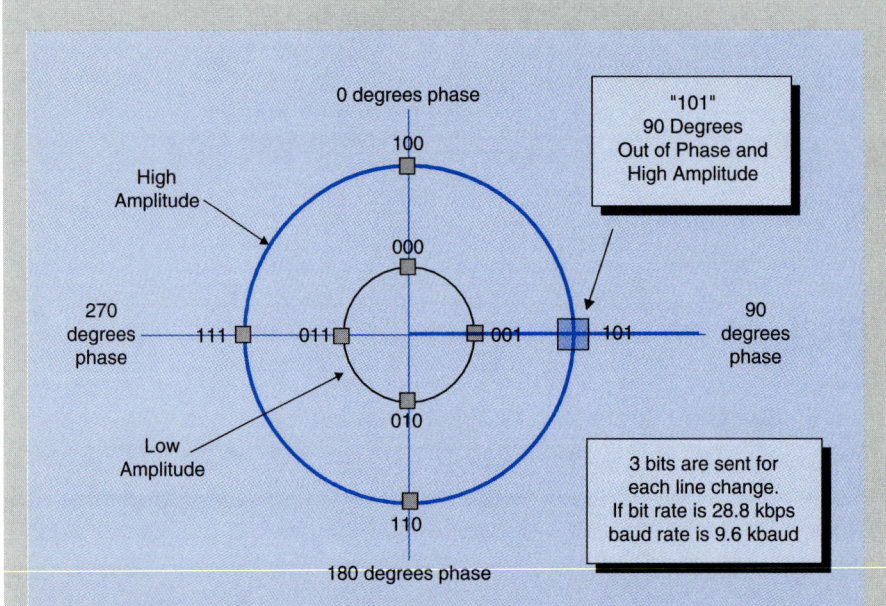

FIGURE 3.10

Complex Modulation

Complex modulation combines two modulation modes, usually amplitude and phase modulation. In addition, it uses multiple levels of both amplitude and phase. In this example, there are eight possible combinations. Each combination can represent three bits. If we assign a three-bit sequence (000 through 111) to each combination, for every line change (*baud*) we can send three bits of information. So *baud rate* is much lower than the bit rate in fast modems. A variation on this theme, trellis coding does not use all possible combinations. This makes it easier to detect errors.

is much lower than the bit rate in fast modems. This causes confusion because what modem vendors label as the baud rate is usually the bit rate. If they label the modem using one of the standards listed in Table 3.1, however, this will tell you the bit rate unambiguously.

Of course 4:1 compression is the *maximum*. In many cases, compression will be less effective. Fortunately, this does not cause problems, even if the computer transmits at 115.2 kbps to the modem. If the modem cannot keep up with the data stream, it can tell the computer to pause its transmission. This is flow control.

Data compression makes no sense unless you also do error correction. As a result, most modem boxes only say V.42 *bis*. These modems do, however, support V.42 as well.

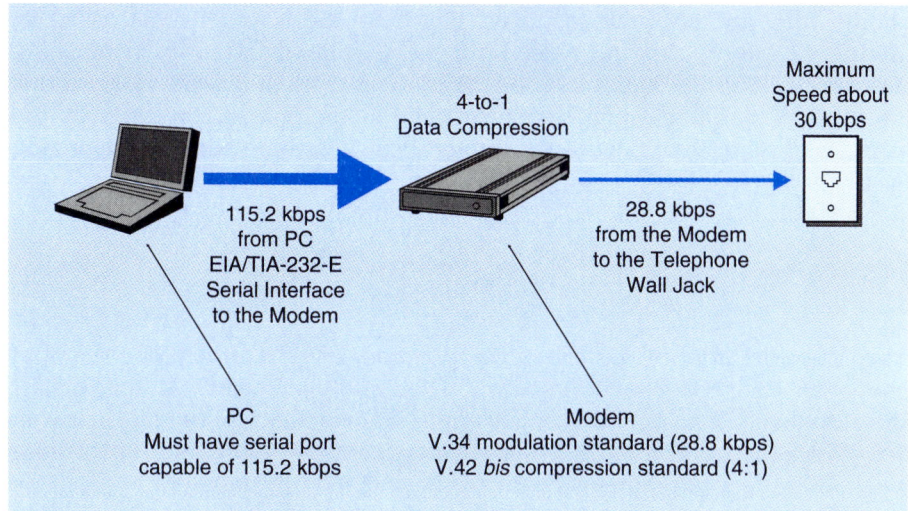

FIGURE 3.11
V.42 *bis* Data Compression

A V.42 *bis* modem can *compress* data as much as 4:1. This allows the computer to transmit as fast as 115.2 kbps. The modem will compress this to 28.8 kbps. If the modem cannot keep up, it tells the transmitting computer to pause.

Table 3.1 shows that the ITU-T standards are not unchallenged in error correction and compression. Microcom Network Corporation has produced the **MNP** family of proprietary standards, which also handles correction and compression. Many modems can obey both ITU-T and MNP standards.

Note that speed and correction/compression standards are independent. If someone tells you that a modem is a V.42 *bis* modem, you cannot assume anything about its speed. And if you hear that someone has a V.34 modem, this does not guarantee correction and compression. You must check the package carefully.

Other Standards ■

Table 3.1 shows other modem standards. For instance, there are different standards for facsimile modems. *For more on modems, see Module D.*

Limits of the Telephone System

Although the telephone system is convenient, it is very limited. The telephone system can only transmit sounds between about 300 Hz and 3,400 Hz. This means that it has a **bandwidth** (frequency range) of only about 3,100 Hz.[2] Tele-

---

[2] Some telephone systems have only 3 kHz of bandwidth, while others have as much as 3.4 kHz. V.34 modems cannot transmit at 28.8 kHz in many areas because of limited bandwidth.

phone lines are also rather noisy. For theoretical reasons discussed in Module D, bandwidth and noise limit the telephone system to a transmission speed of about 30 kbps. The V.34 modem is already very close to this limit. As a result, modem transmission will not get much faster in the future. For faster transmission, we must use something besides the analog telephone network.

Asynchronous ASCII Transmission

As discussed earlier in this chapter, you cannot merely send a long string of 1s and 0s. The receiver will not know where each character's 1s and 0s start and stop. Instead hosts and terminals must frame (package) the bits in a way that allows the receiver to understand where characters stop and start. Figure 3.12 shows the **asynchronous frame** used in VT100 transmission.

Start and Stop Bits ■

When nothing is being transmitted, the sender keeps the line in the 1 state. So each character begins with a **start bit**. This is a single 0. By changing the line state, the start bit signals the receiver that a new character is beginning.

In turn, each character ends with a single stop bit, which is always a 1. This returns the line to the no-data condition for at least a single bit time. This ensures that the next start bit will change the line state, signaling the start of a new bit.

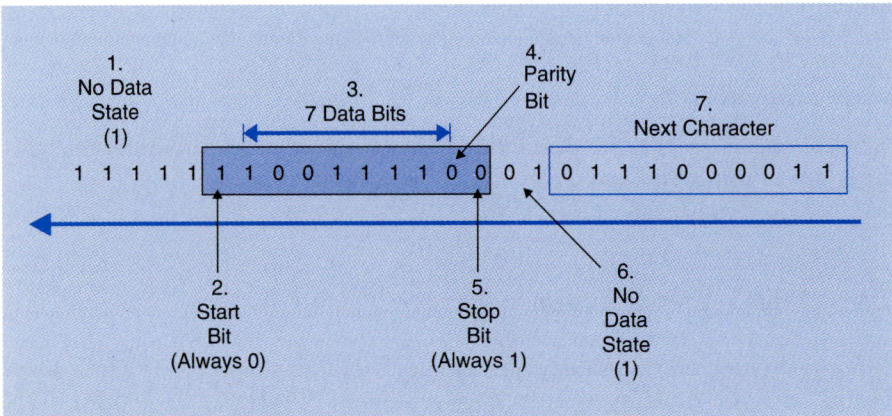

FIGURE 3.12
Asynchronous Character Frame

Each *ASCII* character is sent in its own 10-bit *asynchronous frame*. There is a single *start bit* (0) and a single *stop bit* (1). In between are either seven *data bits* plus a *parity bit* or eight data bits and no parity bit.

Data Bits

After the start bit comes the frame's data bits. In the *ASCII standard*, there are seven data bits. However, most PCs use **extended ASCII,** in which there are eight data bits per character. Some hosts use seven data bits, others eight. You must set your terminal emulation program to meet the host's requirements for the number of data bits and for all other parameters as well.

Parity Bit

If you are using seven data bits, there is still an eighth bit sent. This is the **parity bit**, which offers a crude form of error detection. Figure 3.13 shows that in even parity, the parity is set so that there is always an even number of 1s in the seven data bits and the parity bit. In odd parity, the parity bit is set to give an odd number of 1s.[3]

The receiver can use parity to detect errors. In even parity, a character with an odd number of 1s in the data and parity bits must be incorrect. Unfortunately, asynchronous transmission does not have a way of asking for the retransmission of the damaged character. The receiver either marks the character as incorrect, painting an error character on the terminal screen, or throws the character away. So parity offers *error detection* but not *error correction*.

FIGURE 3.13
Parity

The asynchronous *parity bit* offers error detection by making the total of the 1s in the data bits and the parity bit always *even* or *odd*. This is a weak form of *error detection* and does not offer *error correction* at all.

Even Parity

Character	Number of 1s	Odd or Even	Parity Bit
1110001	4	Even	0
1110000	3	Odd	1

Odd Parity

Character	Number of 1s	Odd or Even	Parity Bit
1110001	4	Even	1
1110000	3	Odd	0

---

[3] Note that the start and stop bits are **NOT** included. If there are an even number of 1s in the data plus parity bits, there will be an odd number of 1s in the entire frame because the sum of the start and stop bits is 1.

In fact, parity is not even good at error *detection*. If two bits are changed instead of one, the changed character will still have the correct parity. The receiver will think that the character is correct.

When the transmission uses eight data bits, there is no room in the ten-bit, asynchronous frame for a parity bit. So parity is only used with seven-bit ASCII.

Limits of Asynchronous Transmission ■

If you use VT100 terminal emulation, you are stuck with asynchronous transmission, which has poor error detection and no error correction. In addition, having to send each character in a separate frame limits your transmission speed.

Also there are no standards for speed, number of data bits, or parity. Some systems even use more than one stop bit. So users have to set up each host separately. They may even get information about hosts in cryptic ways, such as "9600E1" (9,600 bps maximum transmission speed, even parity, one stop bit). In general, for novice users, VT100 terminal emulation setup is a little daunting.

OTHER VT100 TRANSMISSION SETTINGS

Half-Duplex and Full-Duplex Transmission

Asynchronous transmission (and other forms of transmission) can be either half-duplex or full-duplex. **Half-duplex** is like a one-lane bridge. Traffic can travel in two directions, but the traffic in the two directions must take turns using the bridge. In walkie-talkies, too, the two users must take turns transmitting. In data communications, the partners in a half-duplex transmission system must take turns transmitting. Figure 3.14 illustrates this situation.

In **full-duplex** transmission, in contrast, both sides may transmit at once. This allows, for instance, one side to ask the other side to pause because it cannot keep up with the volume of data it is receiving. Asking the other side to pause creates **flow control**. Without flow control, data will be lost.

Full-duplex transmission is also needed in error correction. If the receiver finds an error, it must be able to interrupt the sender to ask for a retransmission.

Almost all modern hosts use full-duplex transmission. So do almost all other modern data communications systems.

The Serial Plug (Connector)

The **serial plug** at the back of your PC either has 9 pins or 25 pins (see Figure 3.15). The modem connects to the serial plug via a serial cable.

For our purposes, it is important to note that the serial plug only uses *one pin* for transmission in each direction. This means that in one clock cycle (time period), a serial plug can only transmit a single bit. Figure 3.16 illustrates this

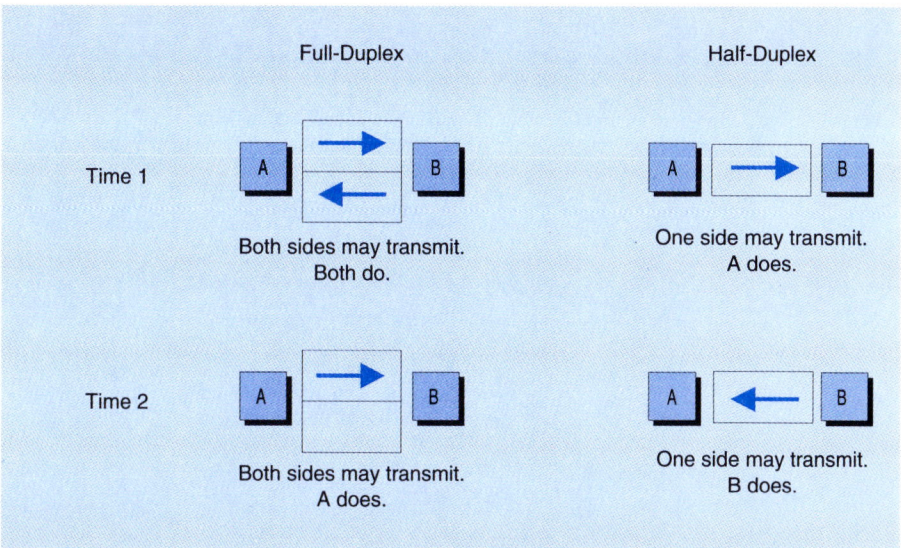

FIGURE 3.14
Half-Duplex and Full-Duplex Transmission

In *half-duplex* transmission, both sides may transmit, but they must take turns. In *full-duplex* transmission, both sides may transmit at the same time. Full-duplex transmission permits one side to interrupt the other, say for *flow control* or error correction. Almost all modern hosts today use full-duplex transmission.

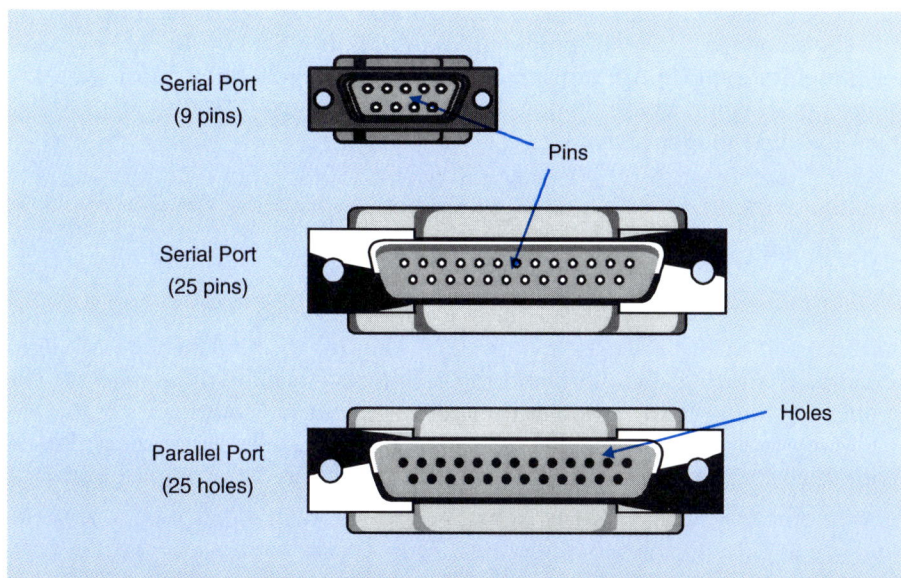

FIGURE 3.15
Serial and Parallel Plugs

Serial plugs have either 9 or 25 male pins. The *parallel plug* has 25 female connections.

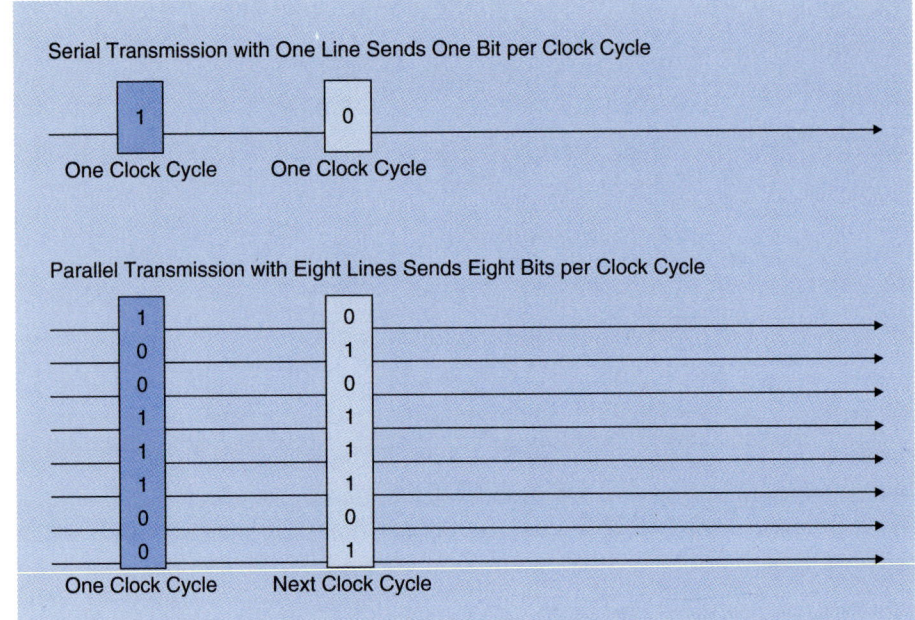

FIGURE 3.16

Serial and Parallel Transmission

Serial transmission sends only one bit per clock cycle. *Parallel transmission* sends multiple bits in each clock cycle. For a given clock speed, parallel transmission will be much faster.

form of **serial** (one-bit-at-a-time) transmission. Serial transmission is very slow. It is like a single-lane highway.

In contrast, PCs also have **parallel plugs**, which can transmit data along multiple wires simultaneously. Figure 3.16 illustrates the eight-pin **parallel** transmission used in PC parallel plugs. Parallel plugs can transmit eight bits in each clock cycle instead of just one. A parallel plug is like an eight-lane freeway.

The standard for serial plugs, EIA/TIA-232-E (formerly RS-232-C), actually specifies a maximum transmission speed of only 20 kbps. Most PC serial plugs can transmit faster than 20 kbps, but not all can transmit at the 115.2 kbps that V.34 modems can use.[4]

The ISDN

As discussed earlier, the telephone system is limited to about 30 kbps. Modems are already nearing this limit, and future improvements will be modest. The problem is the telephone system itself, as discussed in Module D.

Many Internet users now wish to have higher speeds. For them, World Wide Web data downloads are far too slow for pages with complex graphics. Other applications are also too slow.

---

[4] Serial plugs using the 16550 UART chip can transmit at 115.2 kbps.

The Basic Rate Interface (BRI)

One possibility is to replace the telephone line with an ISDN (Integrated Services Digital Network) transmission line. As its name suggests, this is an all-digital transmission network. As a result, there is no need for a modem to convert a computer's digital signal into analog transmitted signals. Figure 3.17 illustrates ISDN service.

More specifically, the figure illustrates the **Basic Rate Interface (BRI).** This is the basic service that ISDN brings to each desktop including three transmission channels. There are two **B channels** operating at 64 kbps. There is also a **D channel** operating at 16 kbps.

These three channels are **multiplexed** (mixed) over a single set of wires leading to the desktop. There is no need to have three separate sets of wires.

The B channel speed of 64 kbps is more than twice as fast as a 28.8 kbps V.34 modem. In fact, V.34 modems often have to drop back to lower speeds if line conditions are not ideal, so ISDN lines often are three to four times faster than modem transmission. Furthermore, ISDN lines are very clean, so error correction does not consume a significant amount of transmission time.

FIGURE 3.17

ISDN Service

The *ISDN's Basic Rate Interface (BRI)* brings three transmission channels to the user's desktop. There are two 64 kbps *B* channels and one 16 kbps *D* channel. These three channels are *multiplexed* onto a single wire bundle to the desktop. Computers need *terminal adapters* or *internal DSUs* to communicate over the ISDN. Analog telephones need *codecs* to translate their analog signals into digital ISDN signals.

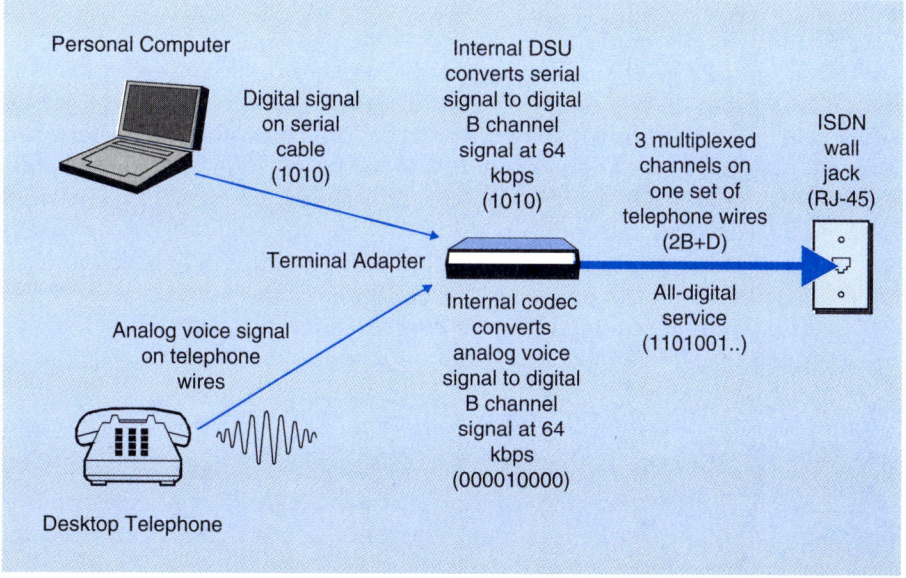

Using one ISDN B channel to link to an Internet service provider (ISP) still leaves one ISDN B channel free for telephone calls. (The ISDN is integrated with the regular telephone system.) With a regular modem connection, you cannot call out while using the line for data, and incoming calls either get a busy signal or disrupt the modem connection.

Alternatively, some Internet service providers (ISPs) allow you to connect to them with both B channels simultaneously, giving a speed of 128 kbps.

Cost

In general, ISDN is three to four times faster than modem speeds on the telephone network. For Internet users who find that they need a little more speed and are willing to pay somewhat more to get it, the ISDN is very attractive. However, users who wish ISDN service have to be aware of all costs they will incur.

Installation Cost

There usually is an **installation charge** by the transmission carrier to set up and install an ISDN line. This varies widely from carrier to carrier. It is often in the range of $50 to $200.

Monthly Service Cost

In addition, you must pay **monthly service charges**. There is normally a base rate about double the base rate of a telephone line. In addition, there may be per-minute charges based on distance. Again, different transmission carriers have very different pricing systems.

Connecting the Computer

Although the computer does not require a modem to transmit over the all-digital ISDN, the ISDN's digital formatting is not the same as that of the PC serial port. One alternative, shown in Figure 3.17, is to purchase a **terminal adapter**. One end of the adapter connects to the PC via a serial cord. The other end attaches to the ISDN via a telephone cord. The adapter translates between the computer and the ISDN line.

As an alternative, you can buy a **digital service unit (DSU)** board to place inside your PC. This DSU does the digital format conversion inside the PC. DSUs, then, translate between digital source (computer) signals and digital transmission (ISDN) signals.

Connecting an Analog Telephone

Recall that a modem translates between digital source signals and an analog transmission line. To send voice over an ISDN channel, you have the opposite problem. The human voice is analog. The ISDN line is digital. You need the

CODEC Sampling

Encoding often uses a process called **pulse code modulation (PCM).** Figure 3.18 illustrates an underlying process called **sampling** that is used in PCM.

Recall that an analog signal rises and falls fairly smoothly. In sampling, you look at (sample) the intensity of the source signal many times each second.

FIGURE 3.18
Sampling for Pulse Code Modulation (PCM)

Pulse code modulation (PCM) divides each second into many time slices. In each time slice, it *samples* the source signal, recording the signal's intensity. In telephony, voice codecs take 8,000 samples per second. For each sample, they record an eight-bit number representing one of 256 possible intensity levels. Multiplying these two numbers together, simple PCM generates 64 kbps of data each second. Many transmission facilities operate at multiples of 64 kbps. Sometimes sampling only produces 56 kbps of data; this allows 8 kbps of control information to be sent in a 64 kbps channel.

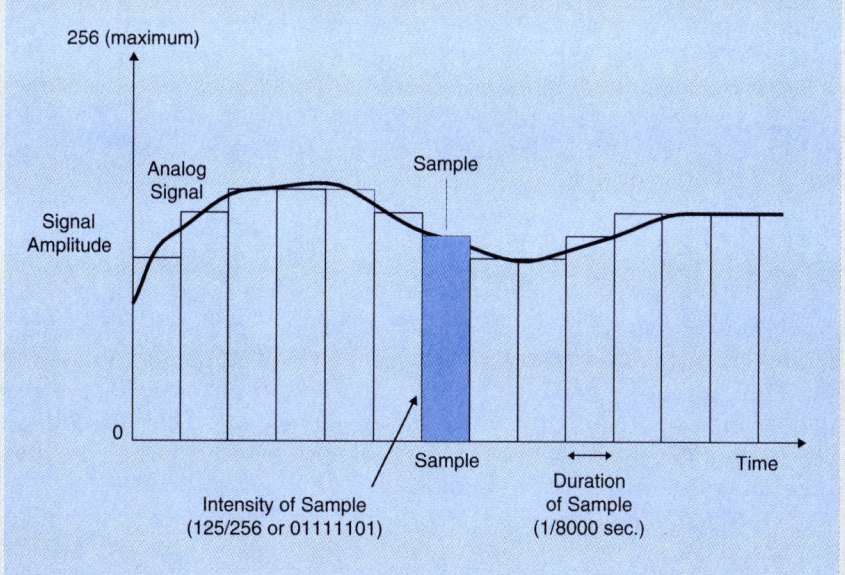

continued

> A general rule is that you must sample at twice the rate of the bandwidth. Recall that the bandwidth of a telephone line is 3.1 kHz. So PCM should sample the signal 6,200 times a second. Unfortunately, sampling at this rate still loses a little information [Hardwick, 1989], so the actual sampling rate is 8,000 samples per second.
>
> For each sample, you only need to record the intensity of the sound at that moment. If you use one byte per sample, this will allow you to represent 2^8 (256) possible sound levels, from -128 to $+128$. For the spoken human voice, this is adequate.
>
> If you multiply 8,000 samples per second times 8 bits per sample, you get 64 kbps. It is no accident that ISDN channels are 64 kbps. They were created precisely for digitized voice. In fact, we will see throughout this book that many transmission systems operate at 64 kbps or some multiple of 64 kbps.
>
> We will also see 56 kbps frequently because some 64 kbps systems "steal" 8 kbps from the data signal for control signaling.

opposite of a modem. You need a **codec**. A codec *codes* outgoing analog voice signals into digital ISDN format. At the other end, the receiving codec *decodes the digital signal to analog. For more on the encoding of analog signals to digital, see the box "Codec Sampling."*

Some telephones have internal codecs. In most cases, however, you connect your analog telephone to the "phone" line of the terminal adapter in Figure 3.17. The terminal adapter in this case has an internal codec.

CONCLUSION

This chapter introduced basic topics in point-to-point communication. It began with VT100 terminal emulation, which uses asynchronous ASCII transmission and typically uses telephone transmission using modems. This introduced such concepts as analog and digital signals, character framing, parity bits, duplex, and serial versus parallel transmission.

Although modem speeds are sufficient for VT100 terminal emulation, many applications need more than telephone speeds. One of these is the World Wide Web, which often uses graphic images, voice clips, and even video clips. For moderately higher speeds (and moderately higher costs), users can use ISDN service instead of the traditional analog telephone service.

CORE REVIEW QUESTIONS

1. What is terminal emulation? Why is it needed?
2. Why are login scripts useful? Why is file transfer useful? Distinguish between uploading and downloading.
3. Compare and contrast VT100 terminals and 3270 terminals in terms of transmission speed and the kinds of information that they can display.
4. Distinguish between digital and analog transmission.
5. Why do you need a modem? What is modulation? What is demodulation?
6. Describe what each of the following protocols specifies: V.34, V.32 *bis*, V.42, V.42 *bis*, MNP.
7. Will telephone modems get much faster in the future? Explain.
8. In asynchronous transmission, explain the purpose of the start bit, the stop bit, the data bits, and the parity bit. How many data bits are there in an asynchronous frame? What is the total number of bits in an asynchronous frame?
9. Most hosts today use full-duplex transmission. What benefits does this bring?
10. Why is parallel transmission faster than serial transmission?
11. You currently have a single telephone line for your home. You use a modem to access the Internet. Explain the benefits you will have if you move to an ISDN line.
12. Explain "2B + D" fully.
13. From various places in the chapter, explain why flow control is needed. Why does it require full-duplex transmission?
14. From various places in the chapter, distinguish between error detection and error correction. Which does asynchronous transmission do?
15. From the box, "Modulation," what do you vary in frequency, amplitude, and phase modulation? What do you vary in complex modulation?
16. From the box, "How Codecs Work," what is sampling? What do you measure in each sampling period?

DETAILED REVIEW QUESTIONS

1. What do you need if you want to reach a host computer via terminal emulation over a telephone line? What does the host need?
2. What will happen if your terminal emulation program can transfer files by

both the XMODEM and Kermit file transfer protocols but the host can only transfer files via Kermit and YMODEM?

3. Compare and contrast VT100 terminals and ANSI terminals. Compare and contrast VT100 terminals and TTT terminals.
4. What happens if the sending modem is a V.34 modem and the receiving modem is a V.32 *bis* modem?
5. What is the compression ratio of V.42 *bis*? Does it always achieve this? Is this an average, or is it a maximum?
6. Write out the bits in an asynchronous frame for the letter *H*. Assume even parity. The least significant bits of the ASCII code (the right-most bits) are transmitted first.
7. How many bits per second are multiplexed over the line to the user's desktop in ISDN's BRI?
8. List the costs you will face if you install an ISDN line to let your PC connect to the Internet.
9. From the box, "Modulation," explain the difference between the bit rate—the transmission speed in bits per second—and the baud rate.

THOUGHT QUESTIONS

1. Why do you think host operating systems generally cannot deal with PCs as full computers?
2. What happens if a host authority changes the way you log in and you try to use an old script?
3. Which of the following is digital or analog? A person's height. A day of the week. Your salary in dollars. The air temperature.
4. What happens if the sending modem is a V.42 *bis* modem and the receiving modem does not follow the V.42 *bis* protocol?
5. When a teacher lectures in class, is this half-duplex communication or full-duplex communication? Explain.
6. Why is parallel transmission used only for short distances?
7. Do you think that ISDN speeds are high enough to make it viable for at least another ten years?
8. Suppose you want a codec to sample music. Music has a bandwidth of 20 kHz. Instead of having 256 possible intensity levels for each sample, you want to have at least 64,000 possible intensity levels for precision. You also want stereo, so you want to sample two channels of music. How many bits per second of traffic will you generate?
9. With no spaces between bits, write out the sequence of bits a receiver would receive if you sent "Hi, Tad." Assume 7-bit ASCII and odd parity. Figure

3.2 shows most of the bit codes you will need. Others are in Module D. Remember that you transmit the least significant bit first.

10. Distinguish between modems and codecs, in terms of whether the stations are analog or digital and whether the transmission line is analog or digital.

PROJECTS

1. Set up a PC's terminal emulation program to communicate with a specific host. Get the communication parameters and an account from the host authority. If possible, set up a login script.
2. Determine the total cost of connecting a PC to the Internet using ISDN. First list the cost items. Then find a likely cost for each.

For online exercises, please visit this book's website at: http://www.prenhall.com/panko

CHAPTER 4

A Simple Ethernet LAN

Introduction

This chapter begins our discussion of local area networks (LANs). We will look at NuPools' initial LAN, which serves the company's back office. This LAN follows the Ethernet (802.3) 10Base-T standard. This is today's most popular LAN technology. Then we will then see how NuPools extended its LAN to serve the sales floor.

In the next chapter, we will look at the 802.5 Token-Ring Network standard that is today's chief competitor to Ethernet LAN standards. Then we will look at the new generation of 100 Mbps LANs, the long-term prospects for all-switched LANs and radio LANs. Module F covers some advanced topics in LANs.

IEEE LAN Standards

As discussed in Chapter 1, all standards architectures follow OSI standards at the single-network (*subnet*) transmission layer. This includes LAN standards.

The IEEE

So far, ISO and ITU-T have not created LAN standards themselves. Instead they have adopted LAN standards created by the Institute of Electrical and Electronics Engineers, the **IEEE.** The IEEE generates standards through its **IEEE LAN MAN Standards Committee,** which is better known by its number, the **802 Committee.**

To take things a step further, actual standards are set by 802 **working groups.** In this chapter, we will focus on what are commonly known as **Ethernet** standards. More precisely, these are 802.3 standards because they are set by the **802.3 Working Group.** Table 4.1 shows other important 802 working groups.

Layering

Figure 4.1 shows that OSI divides transmission standards for LANs into two layers.

- First, there is the **physical layer (OSI Layer 1).** As its name suggests, this layer standardizes things you can see or feel. Connector plug shapes, pin layouts, electrical levels, and other characteristics are standardized at this layer and so are transmission media, such as wiring. Also standardized at this layer are electrical signals—the voltage levels that define 1s and 0s.

- Second, at the **data link layer (OSI Layer 2),** there are standards for organizing bits for transmission. Chapter 3 explained a simple way of organizing bits into frames. This was asynchronous transmission, in which each character

TABLE 4.1

Working Groups in the IEEE LAN MAN Standards (802) Committee

Working Group	Sample Responsibility
802.1	Bridging, network management
802.2	Logical link control layer
802.3	CSMA/CD Bus (Ethernet) LANs
802.4	Token bus factory networks LANs
802.5	Token-ring LANs
802.6	MANs
802.7	Broadband communications
802.8	Optical fiber
802.9	Integrated voice and data
802.10	Security and privacy
802.11	Radio and infrared wireless LANs
802.12	Demand priority access method (100VG-AnyLAN)
802.13	No working group with this number
802.14	Cable television-based MANs

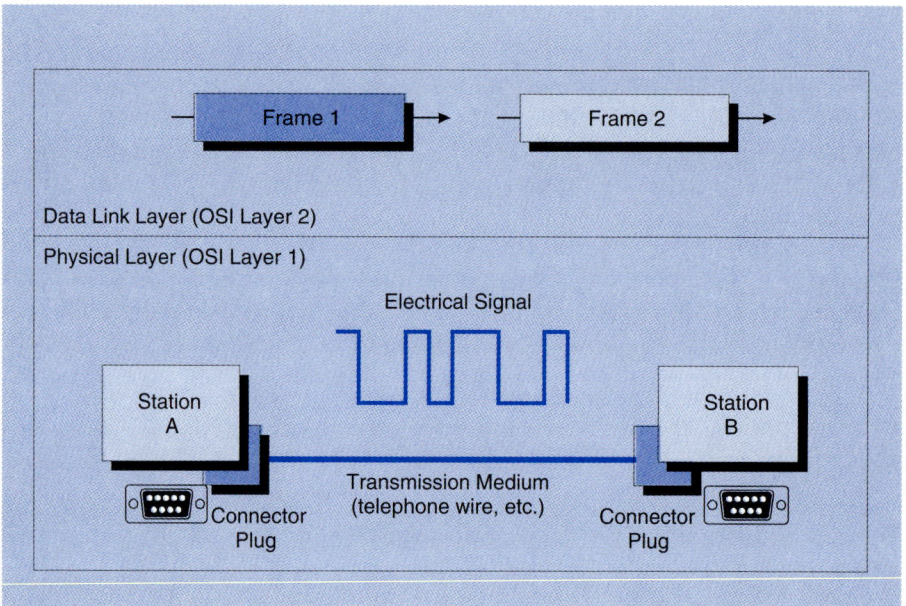

FIGURE 4.1

OSI Physical and Data Link Layers

The OSI *physical layer (Layer 1)* standardizes connector plugs, transmission media, and electrical signaling. The *OSI data link layer (Layer 2)* packages data into synchronous frames for delivery over a physical link.

was framed separately. In this chapter, we will see a more sophisticated form of framing called *synchronous framing*.

802 Physical-Layer Standards ■

There are many types of transmission media, connectors, and ways to represent 1s and 0s as voltage levels. Different 802 working groups have chosen different physical-layer standards. In addition, Figure 4.2 shows that most working groups have defined multiple alternative physical-layer standards. For instance, the 802.3 Working Group has defined several physical-layer standards besides the 10Base-T Physical-Layer standard that we will see in this chapter.

MAC Layer ■

Figure 4.2 shows that the 802 Committee further subdivided the OSI data link layer (Layer 2). First, there is a media access control (MAC) layer. This layer has several functions.

■ First, it implements **media access control.** If two or more stations were to transmit simultaneously, their signals would be scrambled together. Only one station may transmit at a time on the LANs we will see in this chapter. Transmitting is called accessing the transmission medium. Controlling when a station may transmit is media access control, which gives the layer its name.

■ Second, the MAC layer defines how to *frame* data to move it from one device on the network to another across a physical link.

LLC Layer ■

The MAC layer is for basic transmission. Figure 4.2 shows that the 802 Committee created a single **logical link control (LLC) layer** standard, 802.2, to control the reliability of transmission. This is used regardless of what MAC layer protocol a LAN uses. One 802.2 option specifies no error correction, leaving error correction to higher-layer processes. Another option specifies extensive error correction, in which lost or damaged MAC layer frames are retransmitted.

Having only a single LLC layer standard means that when the next-higher-layer process communicates with a LAN, it communicates via the interface with the 802.2 standard. It literally cannot see lower layers. As a result, all 802 LANs look the same to processes at the next higher layer. It is as if the IEEE only created a single LAN standard.

MAC Bridge Layer ■

As we will see in this chapter, LANs have maximum physical distance spans and can become congested if they have too many stations. Chapter 6 discusses how internetworking can reduce both problems. One internetting option is the

FIGURE 4.2
IEEE LAN Standards Framework

The 802 Committee divided the OSI data link layer into three (sub)layers. The *media access control (MAC)* sublayer controls when stations may transmit. The optional *MAC bridge layer* controls internetting if bridges are used. The *logical link control (LLC)* layer controls reliability and makes all 802 LANs look the same to the next higher layer.

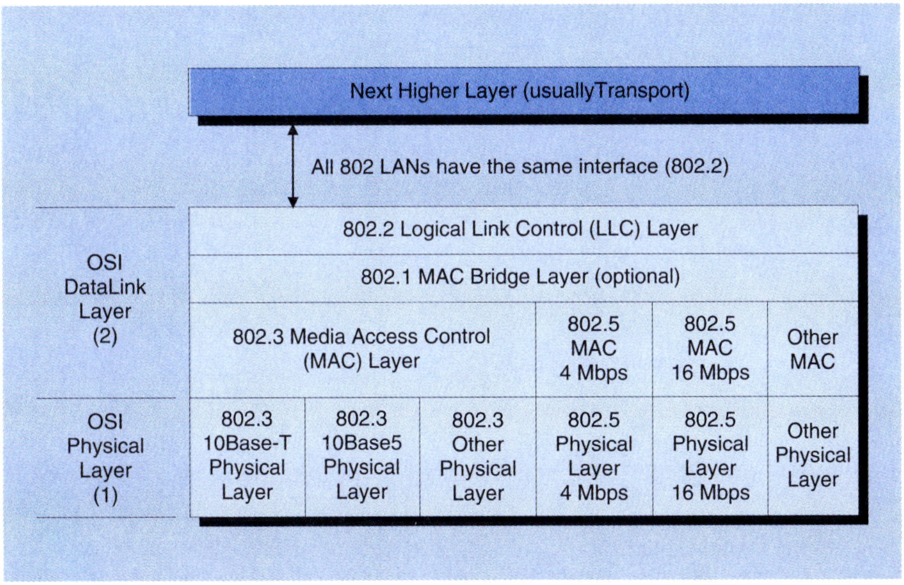

	Next Higher Layer (usually Transport)						
	All 802 LANs have the same interface (802.2)						
OSI DataLink Layer (2)	802.2 Logical Link Control (LLC) Layer						
	802.1 MAC Bridge Layer (optional)						
	802.3 Media Access Control (MAC) Layer			802.5 MAC 4 Mbps	802.5 MAC 16 Mbps	Other MAC	
OSI Physical Layer (1)	802.3 10Base-T Physical Layer	802.3 10Base5 Physical Layer	802.3 Other Physical Layer	802.5 Physical Layer 4 Mbps	802.5 Physical Layer 16 Mbps	Other Physical Layer	

bridge. The 802.2 Committee created an optional **MAC bridge layer** for use in LANs that use bridges.[1] Chapter 6 discusses bridges.

NuPools: Building a LAN with Ethernet 10Base-T: The Physical Layer

When NuPools first installed a LAN to network its PCs, it had six PC users in the back office. All had stand-alone PCs. Figure 4.3 illustrates the simple LAN technology that NuPools paid a contractor to install. This LAN follows the **Ethernet (802.3) 10Base-T** standard[2] [ISO/IEC, 1993]. This figure looks at physical-layer concerns because 10Base-T is a physical-layer standard.

This LAN has several components. We will look at each of them in some detail here.

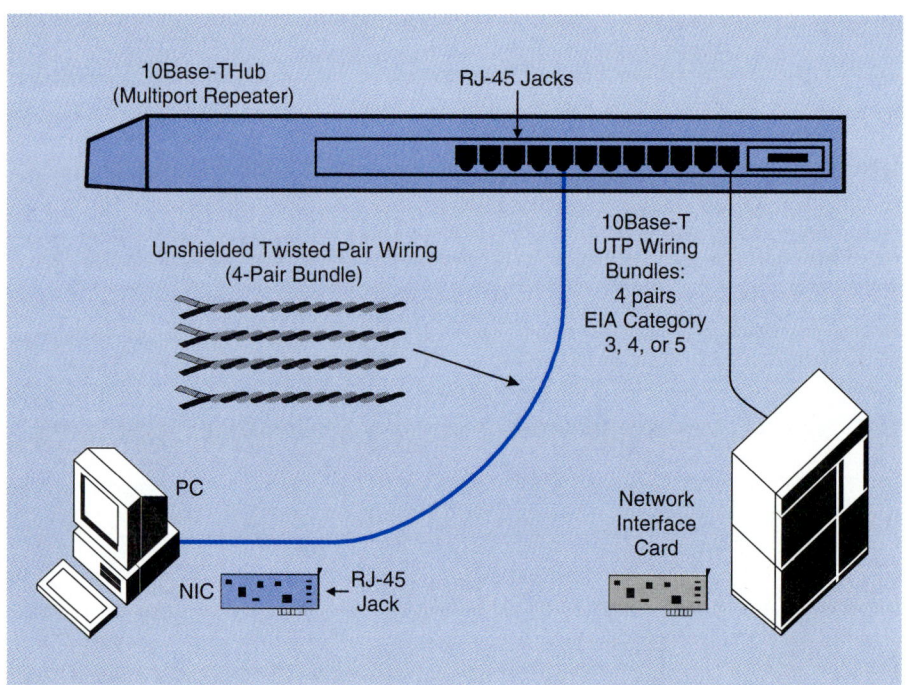

FIGURE 4.3

Simple PC Network Using Ethernet 10Base-T Wiring

Each personal computer must have a *network interface card (NIC)*. The *10Base-T hub* links the NICs together via *business telephone wiring*.

---

[1] Strictly speaking, the MAC bridge layer is a sublayer of the MAC layer, which we have seen is also a subnet of the data link layer.

[2] As Module F discusses, 802.3 standards are different from the Ethernet standards created in the 1970s by Digital Equipment Corporation, Intel, and Xerox. However, 802.3 standards are based on the original Ethernet standards, and almost everyone calls them Ethernet standards.

- First, each PC has an added piece of hardware—a *network interface card (NIC)*.
- Second, the LAN has a box called a *10Base-T hub*. This simple device connects the PCs.
- Third, simple *business telephone wiring* links each PC to the 10Base-T hub.

This is all the company needed for its back-office LAN. Of course, for PC networking—getting the PCs to work together at the computer and application layers—the company had to install more hardware and software. In Chapter 8, we will see that PC networking costs more than the underlying LAN. The purpose of this chapter, however, is to focus on networking.

Network Interface Cards (NICs)

As its name suggests, the **network interface card (NIC)** manages the interface (connection) between the PC and the network.

The NIC usually is an expansion board, which is installed inside the system's unit. Figure 4.3 shows an **expansion board NIC.** There can sometimes be serious problems when you install expansion board NICs. Fortunately, the contractor did not encounter too many problems when it installed the NICs at NuPools. *For more on PC technology and installation problems, see Module I.*

Unshielded Twisted Pair (UTP)

The wiring bundle that runs from the NIC to the 10Base-T hub consists of **unshielded twisted pair (UTP)** wiring. This is very similar to telephone wiring in your home. However, there are some differences.

Twisting and Interference ■

First, business UTP wiring is twisted several times per foot, as shown in Figure 4.3. For complex engineering reasons, this reduces interference from outside sources. A long wire can act like a radio antenna, and any electromagnetic signals in the area will add to the transmitted signal, "interfering" with the transmission and causing errors. Twisting the wires also reduces the amount of electromagnetic signal that the UTP wire can radiate outward, interfering with other systems.[3]

---

[3] In fact, signals on neighboring wire pairs in the bundle can interfere with one another. This is called *cross-talk*. Twisting each pair of wires also reduces cross-talk. In Category 5 wiring, it is important not to unwrap each pair more than a fraction of an inch at the end of a bundle, when you are connecting the wires to a plug. Unwrapping them too much will produce cross-talk errors.

Quality Standards ■

UTP quality standards are defined by the Electronics Industries Association (EIA). In terms of increasing quality, the EIA standards are **Category 3, Category 4,** and **Category 5** UTP. All three grades will work with Ethernet 10Base-T. Category 5 wiring is the most widely used in business and should be used in all new installations because it will be needed by the next generation of LANs.

In addition to quality standards, there are standards for fire resistance. If wiring has to run through airways, special **plenum wiring** is needed. Plenum wiring is more expensive than regular wiring, but it is less toxic when burned.

Wire Bundles ■

A second difference between UTP bundles and home telephone wiring is that your home telephone wiring has a single pair of wires. In contrast, 10Base-T wiring has four pairs in a bundle.

This allows the signal going from the NIC to the hub to use a different pair of wires than the signal going from the hub to the NIC. This prevents the signals from mixing together. In effect, home telephone wiring uses a single pair for full-duplex transmission, while 10Base-T uses four wires for full-duplex transmission. The other four wires go unused in 10Base-T.

RJ-45 Plugs and Jacks ■

A third difference between home telephone wiring and 10Base-T wiring is that home wiring uses RJ-11 plugs at the ends of the wiring, while 10Base-T wiring uses similar but slightly larger RJ-45 plugs. RJ-11 plugs are designed for up to three pairs of wires in the bundle. The RJ-45 plugs used in most business telephone wiring allow four pairs of wires. Naturally, the hub jacks and the jacks at the backs of the NICs follow the RJ-45 standard.

Distance Limitations ■

The 10Base-T standard specifies that UTP wires running from NICs to hubs must be no longer than 100 meters. There are several reasons for this.

- First, signals get weaker as they propagate (travel down a transmission line). This is **attenuation.** As signals attenuate, it is harder to be certain whether a bit is a 0 or a 1.
- Second, as signals propagate, they undergo **distortion** and look less and less like the square-cornered transmission shown in Figure 4.1. This also makes bits harder to identify.
- Third, as we noted in Chapter 3, **noise** (random energy on the transmission line) tends to turn 1s into 0s and 0s into 1s. As the signal attenuates toward the **noise floor** (the average noise level), errors increase.
- Finally, outside sources can produce **interference.** This consists of inter-

mittent signals that add to or subtract from the signal, changing 0s and 1s. In contrast, noise is roughly constant energy in the transmission medium.

All of these problems become worse with distance. While UTP is inexpensive and simple to install, signals degrade rapidly as they travel over UTP. By limiting transmission to 100 meters, however, the 10Base-T standard keeps these four **propagation effects** under control.

Operation of 10Base-T Hubs

At the other end of each wire bundle is the Ethernet **10Base-T hub.** Figure 4.3 shows that this hub is a box with multiple RJ-45 jacks on its front or rear panel. Most people call these jacks **ports.** A typical 10Base-T hub has a dozen ports. Some have 200 or more. The smallest have only two or four.

Another name for 10Base-T hub is **multiport repeater.** Clearly, it has multiple connections. In addition, as we will now see, it operates by repeating everything that one station says to all other stations. Although official IEEE Ethernet 10Base-T documents use the term *multiport repeater*, the term *10Base-T hub* is used more commonly in the marketplace.

Transmission of Signals ∎

Suppose Station A wants to transmit a message to Station B. In that case, Figure 4.4 shows that Station A transmits the message on wires 1 and 2. The message goes to the 10Base-T hub.

Inside the Hub ∎

Figure 4.4 shows that the 10Base-T hub takes the signal in from Station A and broadcasts it out to all the other PCs that connect to it. In effect, it reflects the incoming signal back out of all of its jacks. It repeats the message to all stations, giving it the name *multiport repeater*.

The hub transmits its signals on Pins 3 and 6. Note that transmission and reception only use four wires (1, 2, 3, and 6). So 10Base-T only uses four of the eight wires in the bundle.

Receipt of Signals ∎

Each NIC receives the broadcast signal on Pin 3 and Pin 6.

Bus Operation ∎

The way the Ethernet 10Base-T transmission process works is called **bus transmission.** Bus transmission means *broadcasting*. One station transmits. The 10Base-T hub then broadcasts the signal to all stations attached to it. All stations receive the signal almost simultaneously. Bus operation is very simple to implement, and it works reasonably well.

92 CHAPTER 4 A SIMPLE ETHERNET LAN

FIGURE 4.4

The Transmission of a Message in Ethernet 10Base-T

When a station transmits, it does so on a single pair of wires. The 10Base-T hub receives the message and then broadcasts it out to all attached NICs. This broadcasting is the essence of *bus transmission*.

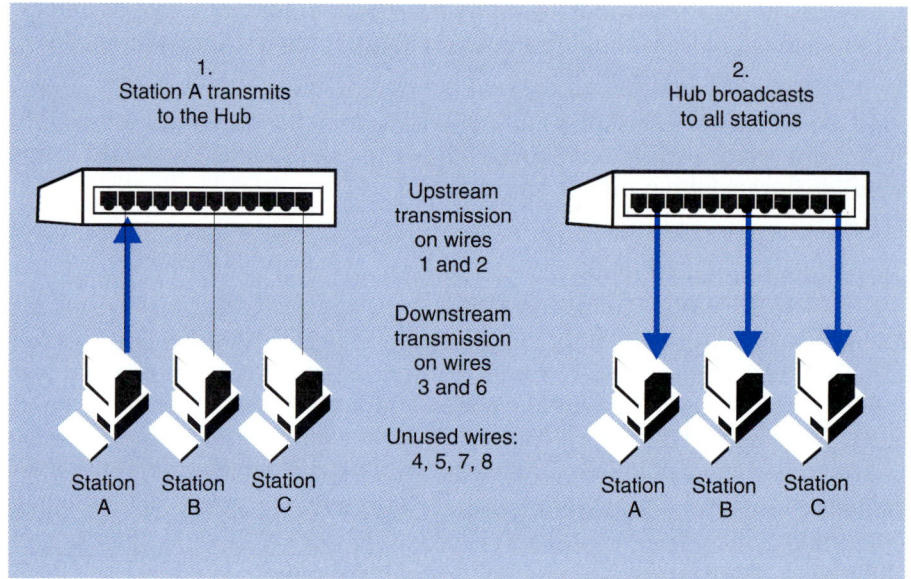

Electrical Signaling

We want to look in a bit more detail at the electrical signals that travel down the line.

Transmission Speed ■

The Ethernet 10Base-T standard transmits information at 10 million bits per second. (This is the reason for the "10" in 10Base-T.) This means that each bit is only one ten-millionth of a second in duration.

This is very fast. But this capacity is shared by all the stations on the network. If there are 100 stations on an Ethernet 10Base-T network, and if all are reasonably active, each might have an effective throughput of only about 1 Mbps, or even less.

Manchester Encoding ■

Figure 4.5 shows how the NIC *encodes* each bit as voltage levels. The figure shows that the Ethernet 10Base-T standard specifies *Manchester encoding*.

The figure shows that **Manchester encoding** represents a 1 as a low voltage (−2.05 volts) for half the time period followed by a high voltage (0 volts) for the second half of the time period. To send a 0, a station sends a high voltage (0 volts) for the first half of the period and a low voltage (−2.05 volts) for the second half. One way to remember this is that the ending tells the bit. A high ending is a 1, while a low ending is a 0.

This is a complex way to represent data. But at the high speeds at which 802.3 networks operate, this complexity is necessary. Suppose that a 1 were simply a high voltage. If the transmitter sends a long series of 1s, and if the receiver's clock is off by only a minute percentage, then the receiver will lose track of where each bit starts. For instance, it might expect bit 345 to appear at a time when bit 346 is appearing. By creating a change in the middle of every single bit, Manchester encoding effectively resynchronizes the receiver's clock with each bit. Unfortunately, this also reduces throughput.

Manchester encoding is safe, but it is not very efficient. The limiting factor in technology is how many times you can change the line per second. This is called the **baud rate.** The Ethernet 10Base-T standard specifies a baud rate of 20 Mbaud, which is very fast. However, it can only transmit 10 Mbps, which is rather slow for its baud rate.

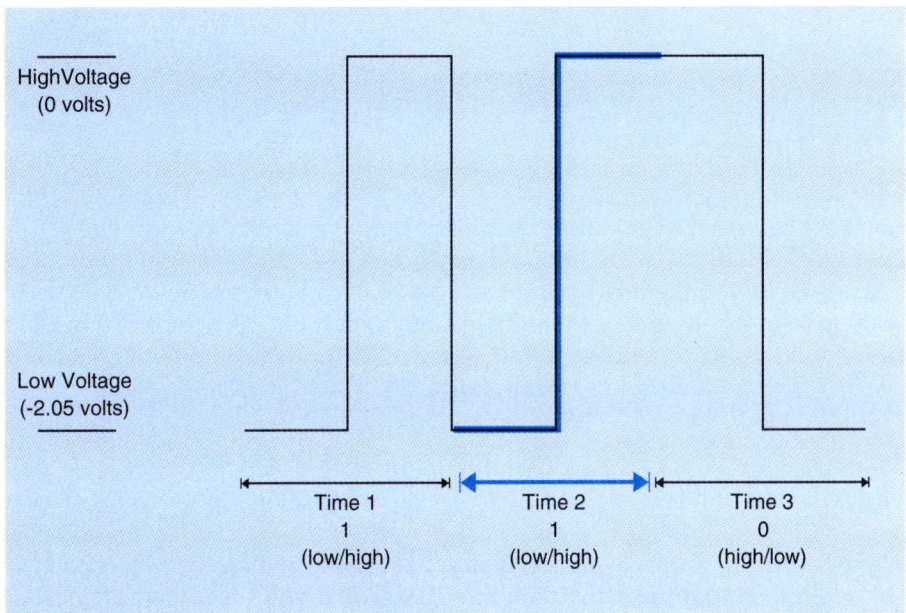

FIGURE 4.5

Manchester Signal Encoding

In Ethernet 10Base-T, stations transmit at 10 Mbps. This is total capacity, which is shared by all the stations. If there is even moderate traffic, each station's effective throughput will be much lower. For transmission, Ethernet 10Base-T uses *baseband signaling*, in which it injects its signals directly into the line instead of placing them in channels as in television and radio. The Ethernet 10Base-T standard specifies *Manchester encoding*, which divides each bit's time period in half. A 1 is a low voltage followed by a high voltage. A 0, in turn, is a high voltage followed by a low voltage. Requiring a line change in the middle of every bit cycle improves synchronization but reduces throughput. Manchester encoding in Ethernet 10Base-T has a baud rate of 20 million baud (20 million line changes per second) but only gets 10 Mbps of data transmission from this high baud rate.

Baseband Signaling ■

The Ethernet 10Base-T standard specifies that the NIC (and the hub) should inject the voltage changes directly into its wires. This simple technique is **baseband signaling.** Almost every LAN uses baseband signaling. The "Base" in 10Base-T stands for baseband.

Radio LANs (discussed later in this chapter) and a few other LANs, however, use **broadband signaling.** In broadband, there are multiple radio channels operating at different frequencies—just as commercial radio and television stations operate in different frequency channels. Signals in different frequency channels do not interfere with one another, allowing you to have many signals being transmitted at the same time. However, raising signals to a channel's frequency and lowering it back to its original baseband form at the other end are expensive. So broadband signaling usually occurs primarily in radio LANs.

THE MAC LAYER: CSMA/CD AND THE 802.3 MAC LAYER FRAME

So far we have looked at electrical signaling and wires—standards at the physical layer. This is the domain of 10Base-T. But as Figure 4.2 shows, you also need standards at the media access control (MAC) layer to govern when stations may transmit and how they organize data for transmission. Figure 4.2 shows that the 802.3 Working Group created a single MAC layer standard to be used with all physical-layer standards.

Media Access Control

While baseband bus transmission is simple, it does have one drawback. If two stations transmit at the same time, their 1s and 0s will mix together and become hopelessly scrambled. Receivers will not be able to pull them apart to make sense of them. In baseband bus transmission, you must never allow two stations to transmit at the same time.

To prevent this, we saw that there must be media access control. This means that there has to be a way to control access to the transmission medium, that is, limit when stations can place their signals on the transmission medium.

The specific way that Ethernet 10Base-T handles media access control is **CSMA/CD,** which stands for **carrier sense multiple access with collision detection.** This is a complicated name, but CSMA/CD is a fairly simple access control technique, as Figure 4.6 illustrates.

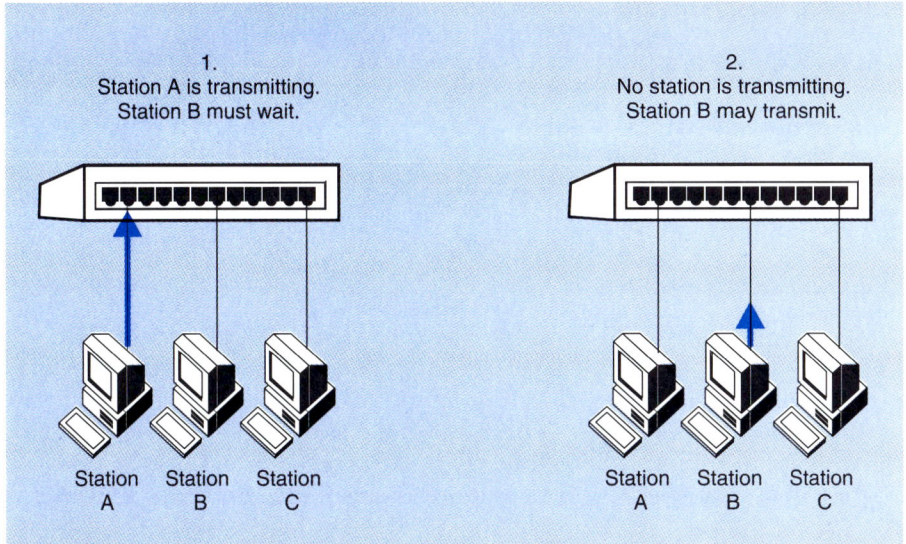

FIGURE 4.6

CSMA/CD Media Access Control in Ethernet 10Base-T

Each station injects its signal into the line to the 10Base-T hub. This is *baseband signaling*. In baseband signaling, if two stations transmit simultaneously, their signals will be scrambled. So there must be media access control, that is, control over when a station may transmit. The Ethernet 10Base-T standard uses the *CSMA/CD (carrier sense multiple access with collision detection) media access control* discipline. NICs always sense the carrier, that is, listen for traffic. For reception, they must do this to tell if messages are addressed to them. For transmission, if a NIC wants to transmit, it must wait until it senses no traffic. Then it may transmit.

Carrier Sensing ∎

The term **carrier sensing** simply means listening to (sensing) the signal (carrier) on the wire. Of course, a NIC must always be listening, so when a message arrives for it, the NIC can recognize the destination address as its own and accept the message.

Carrier Sense Multiple Access ∎

Multiple access means controlling when one of the multiple stations on the network can transmit (access the media). *Carrier sense multiple access* means that CSMA/CD provides access control via carrier sensing.

In CSMA, then, NICs follow a simple rule. If a station that wants to transmit hears traffic on the network, it must wait. But if it hears no traffic, it may transmit.

Collision Detection ■

Figure 4.7 shows that if two stations transmit at the same time, there will be a collision. Each station will recognize this collision because of carrier sensing. It will compare what it is sending with what it is receiving. If the two are different, there must be a collision. This is **collision detection.**

If a station senses a collision, it stops transmitting, waits a random amount of time, and then transmits if the line is free.

Limited Throughput ■

CSMA/CD is a simple media access control technique. As a result, NICs for Ethernet 10Base-T are inexpensive. Unfortunately, if the traffic load on the network becomes high, there will be throughput problems. Collisions will grow rapidly above about 30 percent utilization of the line's capacity [IBM, 1992]. So it is important to keep traffic moderate on Ethernet 10Base-T LANs. For networks with only a few PCs, such as the NuPools LAN, this is not a problem.

The MAC Layer Frame

In Chapter 3, we saw that asynchronous transmission sends each character separately in a ten-bit frame. However, this is only good for low speeds. All LANs use **synchronous transmission** in which many characters are packaged into

FIGURE 4.7

Collision Detection In ethernet 10base-t

When a station transmits, it listens to the traffic on the line and compares it to what it is transmitting. If the two are different, it detects a *collision*; that is, it knows that another station is also transmitting. If a station senses a collision, it stops, waits a random amount of time, and then resumes transmission.

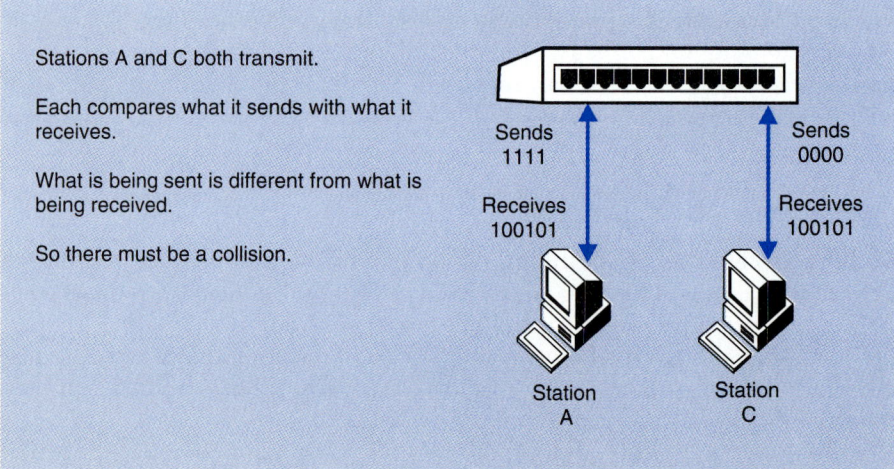

Preamble	7 Octets of 10101010 for synchronization
Start of Frame Delimiter	1 Octet of 10101011 to show start of frame
Destination Address	6 Octets for address of receiving station
Source Address	6 Octets for address of transmitting station
Length	2 Octets for length of Data Field in octets
Data	Data to be delivered
Pad	Padding if needed to make frame 64 octets long
Frame Check Sequence	4 octets of error checking information

FIGURE 4.8
802.3 MAC Layer Frame

Synchronous units at OSI Layer 2, the data link layer, are called *frames*. The *media access control (MAC) layer* is part of OSI Layer 2. So its message is also called a frame. The frame consists of multiple *fields*, each of which is several *octets* (bytes) long. Most fields control the transmission in order to deliver the data field, which holds the information the sender wishes to transmit. The *preamble* and *start of frame delimiter* synchronize the receiving NIC with the transmitting NIC. The *address fields* give the addresses of the sender and receiver. The *length* field gives the length of the data field in octets. The *PAD* field is for padding, which is inserted if the frame falls below its minimum size of 64 octets. (Its maximum size is 1,518 octets.) The *frame check sequence* field contains a number for error correction.

large and complex frames. Each synchronous frame's bits are organized into **fields** that contain different types of information. Figure 4.8 shows the fields in the **802.3 MAC layer frame.**

Octets, Fields, and Synchronous Transmission ■

The 802.3 MAC layer frame usually is hundreds of bytes long. Actually, data communications professionals typically do not use the term *byte* when talking about data transmission. Instead they use the term *octet*. An **octet** is a set of eight bits treated as a group. This is exactly what a byte is, so the terms *octet* and *byte* mean exactly the same thing. This dual terminology is unfortunate, but it is also a fact of life when dealing with data communications standards and publications.

The Preamble and Start of Frame Delimiter

The first field in the 802.3 MAC frame is the preamble. The **preamble** is like the start bit in asynchronous transmission. It alerts the receiver to the start of the frame.

While the start bit in asynchronous transmission is a single 0 bit, the 802.3 preamble is seven octets long. Each octet has the same pattern: "10101010." In other words, the preamble is a 56-bit series of alternating 1s and 0s.

Actually, the complete analogy for the asynchronous start bit is the preamble plus the following field, the **start of frame delimiter.** This field is a single octet whose pattern is "10101011." This is like a preamble octet, apart from its "11" ending. Together the preamble and start of frame delimiter form a 64-bit sequence of alternating 1s and 0s that ends with a 11 pair to mark the end of the sequence.

Replacing a single start bit with 64 bits may seem to be a source of high overhead. Yet in asynchronous transmission, the start and stop bits cost 20 percent of the bits transmitted. In contrast, in an 802.3 frame with 1,000 octets, the overhead represented by eight startup octets is less than 1 percent.

Why do we need such a long sequence? The answer is that this long sequence tells the receiving station the exact clock rate of the sender. At 10 Mbps, each bit is only one ten-millionth of a second long! Even slight differences in the clock rates of the sender and the receiver will cause chaos. The long pattern formed by the preamble and start of frame delimiter allows the receiver to synchronize its clock with the sender's clock. This is why we call this form of transmission synchronous transmission.

Address Fields

LANs combine the frames from many stations onto shared transmission lines. (In 10Base-T, this mixing is done in the hub.) In Chapter 3, we saw that when multiple signals are mixed onto a single line, this is called **multiplexing.** Multiplexing is almost universal in networking.

This is much like sending a letter. Your letter goes into a mailbox with many other letters. It travels to the final destination with even more letters via trucks and even airplanes. The address that you put on the envelope, however, allows the postal service to handle your particular letter appropriately.

Postal "multiplexing" requires that you put two addresses on your envelope: the receiver's address and your own return address. Network multiplexing requires the same thing. In 802.3 terminology, your own station's address is the **source address.** The address of the station to which you are sending the frame is the **destination address.**

The 802.3 standard specifies 48 bits (six octets) for each address field.[4] Two of these bits are reserved, but the remaining 46 bits allow an almost unlimited number of addresses.

---

[4] The standard also has an option for a 16-bit standard for MAC frames. However, the standard specifies that 48-bit addresses should be used for 10 Mbps implementations.

Each NIC vendor gets a block of addresses. This allows every NIC to have a unique access when you purchase it. Some organizations reprogram NIC addresses to create their own numbering systems.

Length Field

The 802.3 MAC standard does not specify a fixed length for frames. Their frames can be anywhere from 64 octets to 1,518 octets. The 802.3 MAC frame has a two-octet **length field,** which tells the number of octets in *the data field*.

Data Field

The **data field** holds the information being sent. Although we discuss the other fields in much greater depth, the data field normally is far larger than all of the other fields combined.

PAD Field

If the data field is very small, the frame would fall below the 64-octet minimum. If this happens, the sender adds a **PAD field** to bring the total length of the frame up to the minimum size. The receiver does not read the PAD field, so the sender can send anything.

Frame Check Sequence Field

To keep the error rate very low, the MAC frame specifies a four-octet error-handling field, the **frame check sequence** field. To compute this field, the transmitting station ignores the preamble, start of frame delimiter, and frame check sequence fields. A four-octet frame check sequence field will only miss about one error in 4 billion.

Although it is 32 bits long, the frame check sequence field introduces little overhead for error checking. In asynchronous transmission, the parity bit represents 10 percent of all transmitted bits. But in a frame size of 1,000 octets, the four-octet frame check sequence represents less than 0.5 percent of all bits sent.

The 802.2 LLC Layer Frame

At the beginning of this chapter we saw that to control the service, there is a second layer of framing in LAN standards. This is the logical link control (LLC) layer. As shown in Figure 4.9, the LLC frame is transmitted inside an 802.3 MAC layer frame. Recall that the LLC layer specifies whether there will be error control in LAN transmission.[5]

---

[5] Although the MAC layer process on the receiving system checks for errors, if it finds a bad frame, it simply discards it. It is up to the LLC process on the receiving station to recognize a missing frame and to ask the sending LLC process for a retransmission.

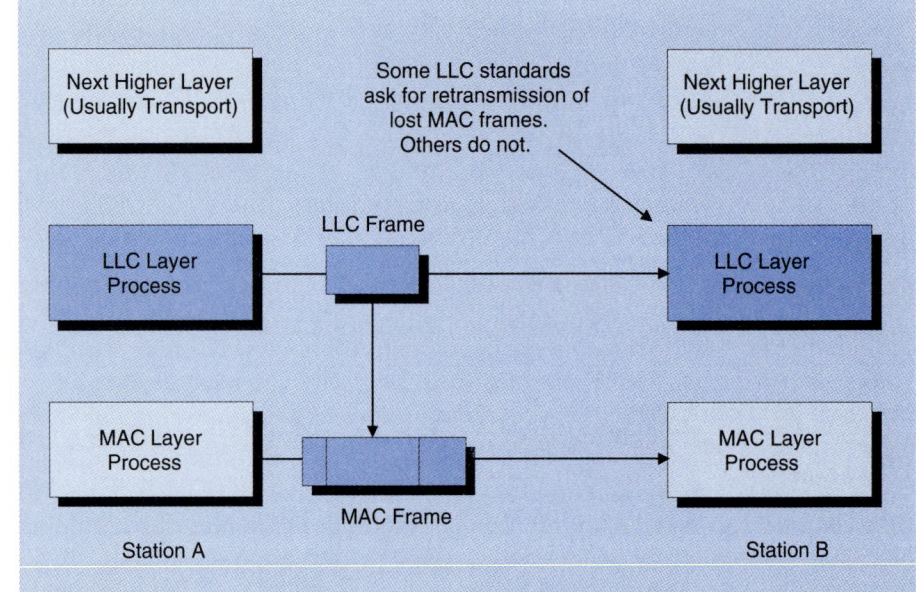

FIGURE 4.9
Logical Link Control (LLC) Layer Transmission

The *logical link control (LLC) layer* frame is carried as the data field of an 802.3 MAC layer frame.

LARGER NETWORKS

Building a Larger Network by Adding Hubs

The network in Figure 4.1 is very limited. The 10Base-T hub that NuPools selected can only serve a dozen PCs. In addition, there is a limit of only 100 meters for Ethernet 10Base-T wiring. Therefore, stations have to be within 100 meters of the single 10Base-T hub in the figure. As a result, the maximum distance between the two farthest PCs is only 200 meters. This was sufficient for NuPools' back office.

When NuPools became comfortable with their back-office LAN, they decided to extend the LAN to the retail sales floor. There NuPools would add three computerized point-of-sale (POS) terminals, plus a server to control the terminals. The company also wanted to add a PC for Haraj Sidhu, so that he could check electronic mail and inventory levels without going to the back-office area.

Adding a Second 10Base-T Hub ■

Figure 4.10 shows that NuPools addressed its needs by adding a second 10Base-T hub in the retail area. This second 10Base-T hub can serve up to 12 PCs. By adding a second 10Base-T hub, NuPools doubled the possible number of PCs on the LAN.

Adding a second hub also increased the maximum distance span of the network. A single **segment** of Ethernet 10Base-T UTP wiring has a maximum

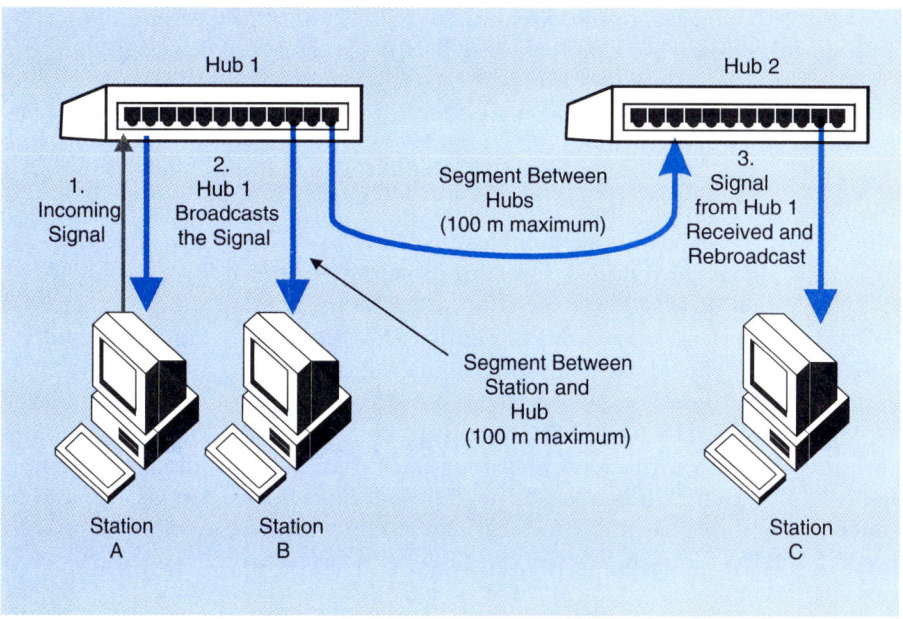

FIGURE 4.10
An Ethernet 10Base-T Network with Two 10Base-T Hubs

Adding a second 10Base-T hub with 12 ports doubled the possible number of stations. The maximum length of a segment of Ethernet 10Base-T wiring is 100 meters. Adding a second 10Base-T hub can also increase the maximum distance span of the LAN from 200 meters to 300 meters. Connecting the 10Base-T hubs is easy. You simply plug a length of Ethernet 10Base-T UTP wiring into one RJ-45 jack in each 10Base-T hub. Suppose Station A on Hub 1 transmits. Hub 1 will broadcast the message out all of its ports, including the one to Hub 2. Hub 2 will treat this incoming signal just like a signal from a station. It will broadcast it out to all stations, including Station C. So adding a second 10Base-T hub preserves broadcasting, which is the essence of bus operation.

length of 100 meters. This includes the segments from each PC to the 10Base-T hub. The segment linking the two 10Base-T hubs also has a maximum length of 100 meters. So adding a second 10Base-T hub, Haraj increased the maximum distance between the most distant stations to 300 meters. This was not important for NuPools, but for most firms the ability to increase distance by adding hubs is crucial.

Linking the two 10Base-T hubs is extremely simple. You take a length of Ethernet 10Base-T UTP wiring and plug its two ends into RJ-45 jacks in the two 10Base-T hubs. You use exactly the same wiring and plugs to connect 10Base-T hubs that you use to link a station to its 10Base-T hub.[6]

---

[6] The actual situation is slightly more complex. Recall that NICs transmit on Wires 1 and 2, while hubs transmit on Wires 3 and 6. Some hubs have an "out" port that transmits on the wires that NICs use (1 and 2). In addition, you can buy wiring that cross-connects Wires 1 and 3 and Wires 2 and 6 internally. You can use this with any port.

Operation is also simple. Recall that broadcasting is the essence of bus operation. In Figure 4.10, if only 10Base-T Hub 1 was present, and if Station A transmitted, the hub would broadcast its signal out on all of its ports, including the one to Station B. All stations attached to Hub 1 would receive the signal almost simultaneously.

Adding a second 10Base-T hub does not change anything. When Station A transmits, Hub 1 still broadcasts the signal out all of its ports. Now, however, one of these ports has the segment leading to Hub 2. This segment receives the broadcast signal from Hub 1, just as if it were a station.

The signal travels along the connecting segment to an input port in Hub 2. That hub has no way of knowing that the signal is coming from another 10Base-T hub. The signal looks exactly the same as it would if it were coming from a station directly attached to the second hub.

As you would expect, Hub 2 then broadcasts this signal out all of its ports, including the one to Station C. Although Station C receives the signal slightly later than Station B, this delay is negligible. In essence, Stations B, C, and all other stations receive the broadcast signal almost simultaneously. Having a second 10Base-T hub preserves the essence of broadcasting and therefore of bus operation.

Adding More 10Base-T Hubs ■

If adding a second 10Base-T hub is good, adding more 10Base-T hubs should be even better for firms with larger operations than NuPools. Figure 4.11 illustrates an Ethernet 10Base-T LAN with multiple hubs.

Linking hubs is easy. You merely run a 10Base-T wire segment between a jack on one repeater and a jack on another repeater.[7]

There are limits to your ability to add 10Base-T hubs, however. One rule is that the maximum number of hubs between the most distant stations is four. Each hub adds a little delay. It also increases signal distortion slightly. Beyond four 10Base-T hubs, these slight delays and signal degradations can cause problems in operation.

The second rule is that 10Base-T segments may not be longer than 100 meters. This limit applies to segments linking stations with hubs. It also applies to links between hubs.

Taken together, these two rules mean that the maximum distance span between the farthest stations is 500 meters. To see this, suppose that S represents a station and H represents a 10Base-T hub. Then the maximum arrangement is S-H-H-H-H-S. The number of dashes, corresponding to the number of wire segments, is five.

There is also a rule for connecting multiple 10Base-T hubs. This rule is that you must connect them in a **daisy chain,** which means that you cannot have

---

[7] Actually, you do something a little different. Otherwise, both hubs will transmit on one pair of wires and listen on another. Neither will hear the other. One solution is to purchase a wiring segment that reverses pins, so that each hub will transmit on the pair to which the other hub listens. Alternatively, some hubs have "out" ports that reverse the pins internally.

loops. If you had a loop, the rebroadcasting process would repeat endlessly around the loop. If you look at Figure 4.11, you will see that it indeed contains no loops.

Overall, while the ability to add 10Base-T hubs can extend the maximum distance span of an Ethernet 10Base-T network, this ability is limited. The maximum span only grows from 200 meters with a single 10Base-T hub to 500 meters with four of them. Even the largest Ethernet 10Base-T network is not very large.

Mixing Other 802.3 Physical-Layer Technologies

While a pure 10Base-T network is limited in size, there is no reason to have pure 10Base-T networks. As shown in Figure 4.2, the 802.3 Working Group has created other physical-layer standards.

All of these physical-layer standards use the same 802.3 MAC layer frame standard. In addition, segments following different 802.3 physical-layer stan-

FIGURE 4.11

An Ethernet 10Base-T LAN with Multiple Hubs

You can link multiple 10Base-T hubs. There are limits, however. First, there must be a maximum of four hubs between the farthest two stations. This is the same as saying that the maximum span of the LAN is 500 meters between the farthest stations. In addition, 10Base-T hubs must be connected in a daisy chain, meaning that there can be no loops.

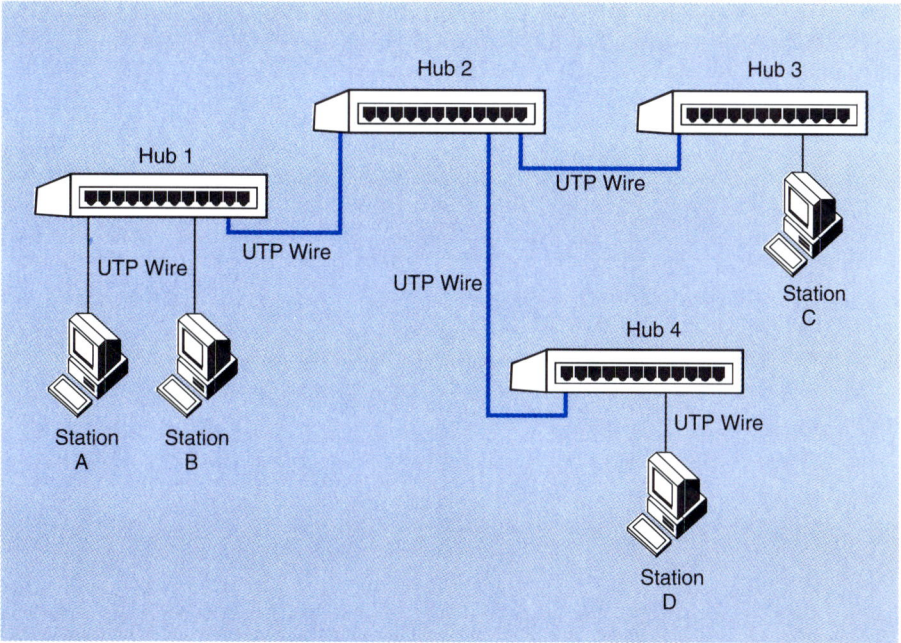

dards can be mixed together on the same LAN. They are connected by hubs (repeaters).

10Base5: Coaxial Cable ∎

Figure 4.12 shows that you can mix 10Base-T physical-layer technology with another 802.3 physical-layer technology, **10Base5.** Note that the name of this technology ends with a "5" instead of a "-T". This indicates that a single segment of 10Base5 trunk cabling can be 500 meters long—five times the distance of a 10Base-T segment.

All that this requires is the proper hub. As Figure 4.12 illustrates, some hubs come not only with the RJ-45 jacks used by 10base-T but also with 15-pin **DIX connectors** (AUI connectors) used by 10Base5. This allows you to link a 10Base5 segment with multiple 10Base-T segments. The stations connected to the LAN find this arrangement completely transparent.

Note that attaching the 10Base5 LAN to the hub is a bit complex. The 10Base5 segment does not attach directly to the 15-pin DIX connector on the hub. Instead a **transceiver (medium attachment unit)** taps into the segment. An **AUI drop cable** connects the transceiver to the hub's DIX connector. *For more details, see Module F.*

FIGURE 4.12

Mixing 10Base-T and 10Base5 on a Single LAN

A segment of *10Base5* cabling can produce a longer run between hubs than can a UTP wiring segment (500 meters versus 100 meters). The hub connects to the 10Base5 cable via an *attachment unit interface (AUI) drop cable*. At the hub, there must be a port with a *25-pin DIX connector* instead of an RJ-45 plug. At the 10Base5 trunk cable, there must be a *transceiver* (*medium attachment unit or MAU*).

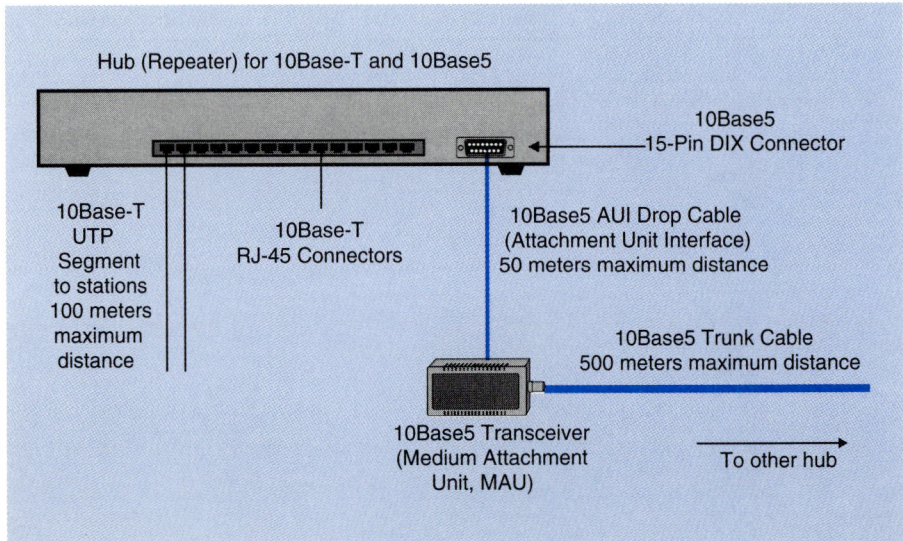

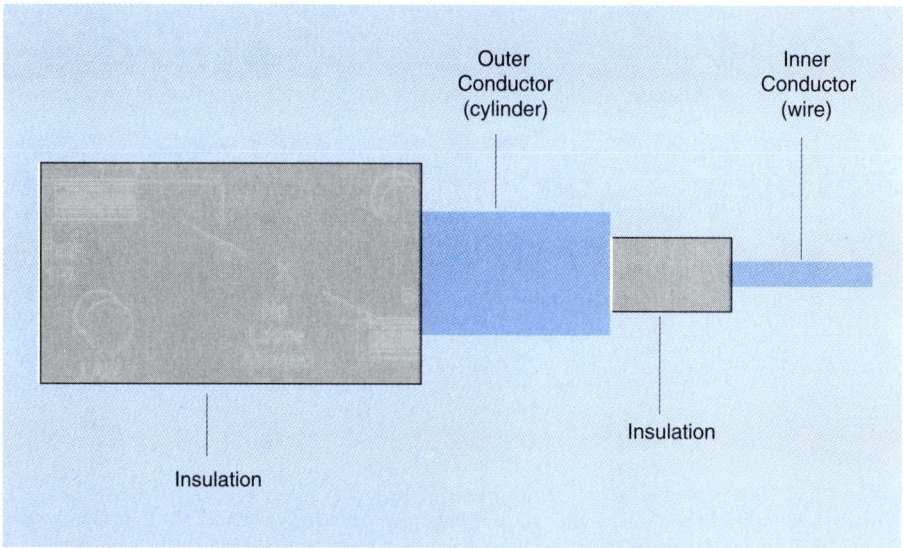

FIGURE 4.13

Coaxial Cable

Coaxial cable's two conductors consist of an inner wire and an outer conducting cylinder. The signal tends to be trapped within the cylinder, so signals can travel far without attenuation. While a 10Base-T UTP segment cannot be more than 100 meters, a co-ax 10Base5 segment can be up to 500 meters long.

Coaxial Cable ■

Instead of using UTP wires for transmission, 10Base5 uses **coaxial cable.** This is similar to the co-ax that connects your VCR to your television set, although 10Base5 uses a thicker form of co-ax. As Figure 4.13 shows, coaxial cable has a central wire plus an outer conducting cylinder. These form the "two wires" needed for a complete electrical circuit. The signal tends to become trapped within the cylinder. Signals can travel farther over co-ax than they can on UTP before propagation effects begin to produce too many errors.

10Base-F

Even co-ax is limited in distance. For longer transmission, there is optical fiber. As Figure 4.14 shows, in optical fiber, a light beam is injected into a thin glass or plastic rod. Light is injected into a thin glass core. When light rays try to leave the core, they strike the cladding, which is a cylinder of glass with a different index of refraction. There is almost total reflection back into the core.

As a result, light can travel this way over very long distances, even at very high speeds. The 802.3 Working Group created the **10Base-F** standard for connecting hubs (repeaters). 10Base-F can carry signals about 2,000 meters—much

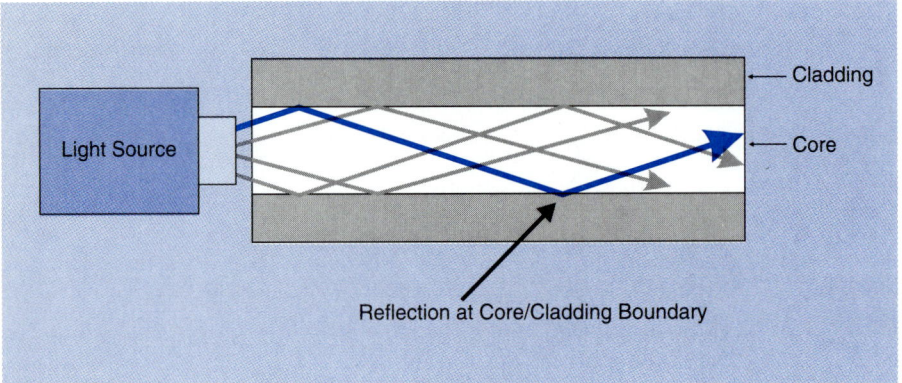

FIGURE 4.14
Optical Fiber

When light is injected into an *optical fiber*, the light rays become trapped within the inner core. When they strike the outer cladding, they are reflected back into the core. Optical fiber can carry high-speed signals over considerable distance.

farther than even 10Base5. This is far less than the distances that telephone fiber can carry long-distance calls, but the cost of 10Base-F fiber is much less expensive than long-distance fiber. *For more on optical fiber, see Module D.*

CONCLUSION

In this chapter, we have looked at the most popular LAN technology, Ethernet 10Base-T, which is standardized by the IEEE 802 Committee's 802.3 Working Group. These LANs use unshielded twisted pair (UTP) with RJ-45 jacks at the OSI physical layer. The most popular type of UTP is Category 5. If UTP is run through airways, special plenum versions of the wiring are needed. At the physical layer, they also use hubs (repeaters) to broadcast incoming signals out to all stations attached to the hub. This broadcasting is the essence of bus transmission. Finally, at the physical layer, they use Manchester encoding and baseband transmission.

At the data link layer's media access control (MAC) layer, CSMA/CD is used for access control, that is, controlling when a station may transmit. At the MAC layer, information is organized into synchronous units called MAC layer frames. These frames have several fields including the preamble, start of frame delimiter, destination and source address fields, the length field, the data field, the PAD field (if necessary), and a frame check sequence field to allow the receiving NIC to discard incorrect frames. All 802.3 networks use the same access control mechanism and the same MAC layer frame organization.

At the data link layer's logical link control (LLC) layer, 10Base-T follows the 802.2 standard. This allows the systems administrator to decide whether or not to do extensive error checking. All 802 LANs use the 802.2 LLC standard, so to the next higher layer, all 802 LANs look alike.

A 10Base-T UTP wiring segment is limited to 100 meters. For longer distances, it is possible to use multiple hubs. No station, however, may be more than four hubs from any other station, so the maximum distance between any two stations is 500 meters. This distance can be extended by adding 10Base5 coaxial cable segments between some pairs of hubs. It can be extended even farther by linking repeaters with 10Base-F optical fiber links.

CORE REVIEW QUESTIONS

1. What professional society has designed most LAN standards? What is the name of its committee that creates LAN standards? What is the name of the Working Group that created the Ethernet 10Base-T physical layer standard?
2. Distinguish between what is standardized at the physical layer and at the data link layer. Into what three (sub)layers is the data link layer divided in 802 standards? What is the purpose of each?
3. For what is *NIC* an abbreviation? What does a NIC do?
4. What is UTP? How many wires are in an Ethernet 10Base-T wiring bundle?
5. When an Ethernet 10Base-T hub receives a signal from one of the stations attached to it, what does it do? What is the defining characteristic of the *bus topology?*
6. How does Manchester encoding represent a 1? A 0?
7. How does CSMA use constant listening to determine when a station may transmit? Why is collision detection needed? What do stations do when there is a collision?
8. What does the preamble field do in an 802.3 (Ethernet) MAC layer frame? The start of frame delimiter field? The source address field? The destination address? The length field? The data field? The PAD field? The frame check sequence field? What is an octet, and how many octets long is each field?
9. How are the 802.2 LLC layer frame and the 802.3 MAC layer frame related?
10. What is the maximum distance between a station and a 10Base-T hub? What is the maximum distance between two 10Base-T hubs? What is the maximum possible distance between stations in a pure 10Base-T LAN?
11. How can distances in a 10Base-T LAN be extended by using segments following different 802.3 physical layer standards? Name the two standards mentioned in the text. Give the maximum distance of a segment for each.

DETAILED REVIEW QUESTIONS

1. What is *baseband signaling*? How does it differ from *broadband signaling*?
2. Why does Manchester encoding put a line change in the middle of every bit transmitted? What disadvantage does this have?
3. In CSMA/CD, why must devices always listen (sense the carrier) to *receive* information? What happens if the number of stations grows from 200 to 300?
4. Why is the preamble so long? How do the preamble and start of frame delimiter work together? How many *bits* long are the addresses in the 802.3 MAC layer frame? How does the receiving station know that there has been an error during frame transmission?
5. What is the minimum size of an 802.3 (Ethernet) MAC layer frame? The maximum size?
6. Are Ethernet 10Base-T hubs connected in a tree hierarchy with one master hub?
7. Describe the connection between a repeater hub and the 10Base5 trunk line.

THOUGHT QUESTIONS

1. Which do you think is likely to carry more traffic—LANs or WANs? Why?
2. Why do you think most LANs use baseband signaling?
3. Why do you think 10Base-T is becoming a popular LAN technology?
4. In Figure 4.11, how many 10Base-T hubs separate Stations A and B? What is the maximum distance between Stations A and B? How many 10Base-T hubs separate Stations A and C? What is the maximum distance between Stations A and C? How many 10Base-T hubs separate Stations C and D? What is the maximum distance between Stations C and D?
5. What would happen if you connected three 10Base-T hubs in a loop? Hint: Trace what would happen if a station attached to one of the 10Base-T hubs transmitted a signal.
6. You are designing an Ethernet 10Base-T LAN for a small airport. You are to use only 10Base-T technology. Assume that the firm will use 12-port 10Base-T hubs. In the main terminal, there is a single office with 12 PCs. The office is a rectangle that is 10 meters long and 15 meters wide. There is also a hanger about 150 meters away. In the hanger, which is a square building about 30 meters on a side, there are five PCs. Draw a diagram showing the situation. Draw on this diagram the devices and cabling you would use to connect the PCs into a LAN. Submit a spreadsheet showing

the cost of the transmission hardware. Hand in a paragraph giving your major design decisions. You may use the following costs, which are typical 1996 prices, or you may look up more recent prices: Ethernet 10Base-T UTP wiring is $0.04 per meter; Ethernet 10Base-T hubs are $20 per port. NICs for Ethernet 10Base-T cost about $50. Installation costs about $100 per PC (this varies widely in practice). Be realistic. Explain why your all-10Base-T solution is undesirable.

7. Redesign the LAN in the previous problem. This time you may use some 10Base5 cabling. This cabling costs four times as much per meter as 10Base-T wiring. AUI drop cabling costs five times as much as 10Base-T wiring. The maximum length of an AUI drop cable is 50 meters. MAUs (transceivers) cost about $100 apiece. Hubs consist of 11 RJ-45 ports and one DIX port.

PROJECTS

1. Cost out an actual network installation, including labor costs.
2. Actually install a network that uses one of the technologies that we discussed in this chapter.

For online exercises, please visit this book's website at: http://www.prenhall.com/panko

REFERENCES

IBM (1992). *Token-Ring Network: Introduction and Planning Guide.* (Document GA27-3677-05). Research Triangle, North Carolina.

ISO/IEC (1993). *International Standard 8803-3, Information Technology—Local and Metropolitan Area Networks—Part 3: Carrier Sense Multiple Access with Collision Detection (CSMA/CD) Access Method and Physical Layer Specification*, 4th ed., Geneva: International Organization for Standardization and the International Electrotechnical Commission.

PANKO, R. R. (1992). Patterns of Managerial Communication. *Journal of Organizational Computing*, 2, 95–122.

CHAPTER 5

Other LAN Technologies

Introduction

The previous chapter focused on a single LAN standard, Ethernet 802.3 10Base-T. In this chapter, we will look at other current and emerging standards.

We will look first at the 802.5 Token-Ring Network standard that is today's chief competitor to Ethernet LAN standards. Then we will look at the new generation of 100 Mbps LANs. We will then discuss switched LANs, which break many limits of earlier "shared media" LANs. We will close with a very brief discussion of radio LANs.

Module F covers some advanced topics in LANs.

The 802.5 Token-Ring Standard

Not all IEEE LAN standards come from the 802.3 Working Group. For instance, the second-most widely used LAN standard, the

802.5 Token-Ring Network standard, comes from the 802.5 Working Group. We will look at this standard in this section. Although 802.5 is an official 802 standard, it was developed primarily by IBM. The official 802.5 standards document is so skimpy, furthermore, that nearly everyone follows IBM designs and terminology [IBM, 1992].

Ring Topology

Figure 5.1 shows that Token-Ring Networks use a ring **topology** (physical layout) for their physical-layer standard. (Ethernet LANs all use a bus or broadcasting technology.) The figure shows that frames travel in a single direction around a loop, like a model locomotive on a simple track loop.

Note that each station only has two communication partners. It receives from one partner and transmits to another. There can be nothing like the broadcasting at the MAC layer in the 802.3 standard.

Note also that signals eventually travel all the way around the loop, back to the originating station. In this way the originating station can tell if its frame propagated through the network without error and was received by the destination station.

Note finally that a ring, like a bus, is a shared media topology. Every station hears every transmission. In addition, there is only one possible path between any two stations on the one-directional ring.

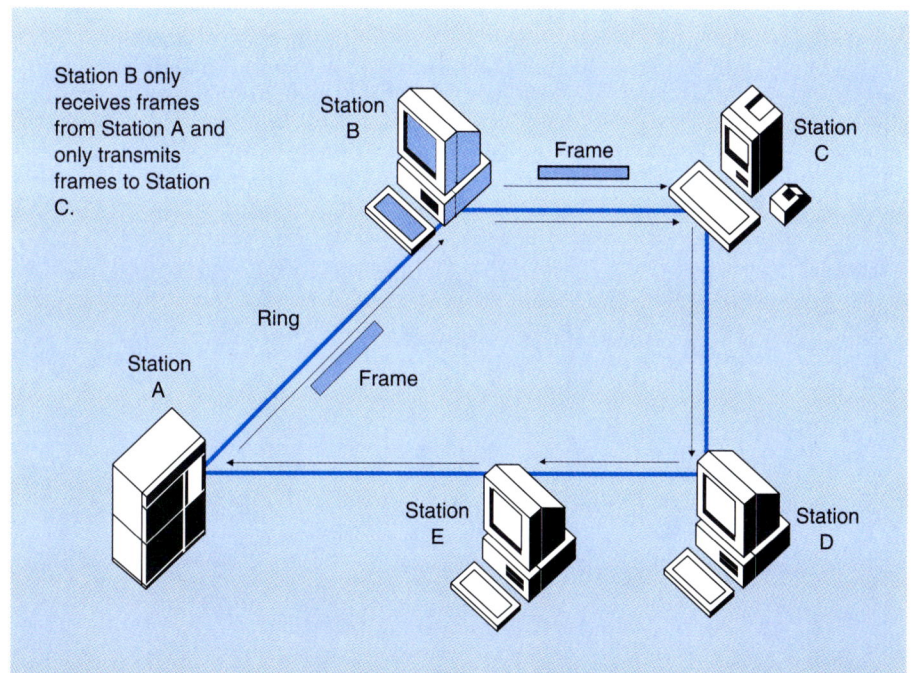

FIGURE 5.1

Ring Topology in 802.5 Token-Ring Networks

The *802.5 Token-Ring Network* uses a *ring topology* at the physical layer. Frames travel in one direction around the loop. Each station only receives from one station and transmits to one other.

Shielded Twisted Pair (STP) Wiring

As shown in Figure 5.2, Token-Ring Networks normally use **shielded twisted pair (STP)** wiring. STP is similar to UTP with one important difference. Each pair of STP wiring has a metal mesh sheath around it.

Most wire bundles, furthermore, come with at least two pairs of wires. STP places an additional metal shield around the entire bundle, giving a second layer of protection.

STP is attractive because the metal mesh shields around individual pairs and the bundle as a whole provides resistance to interference. Stray signals coming from other devices have a difficult time penetrating the shielding. As a result, STP can be used even in electrically hostile environments.

The bad news is that STP is much thicker than UTP. It is also more expensive and more difficult to lay.

Access Units

Because of the growing popularity of UTP wiring, the 802.5 standard now permits UTP on the runs between the stations and wiring concentrators. As shown in Figure 5.3, these wiring concentrators are called **access units**.

FIGURE 5.2
UTP and STP Wiring

As discussed in the last chapter, 10Base-T LANs use unshielded twisted pair (UTP) wiring bundles. In contrast, the normal transmission medium in 802.5 Token-Ring Networks is *shielded twisted pair (STP)* wiring. In STP, there is a metal mesh shield around each twisted pair and around the wiring bundle as a whole. This protects the transmission from interference. Unfortunately, STP is expensive to purchase. It is also thick and therefore expensive to lay.

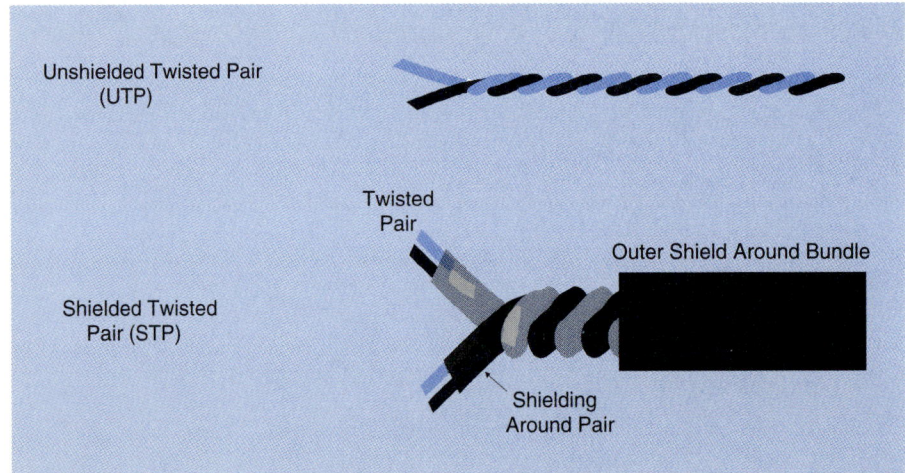

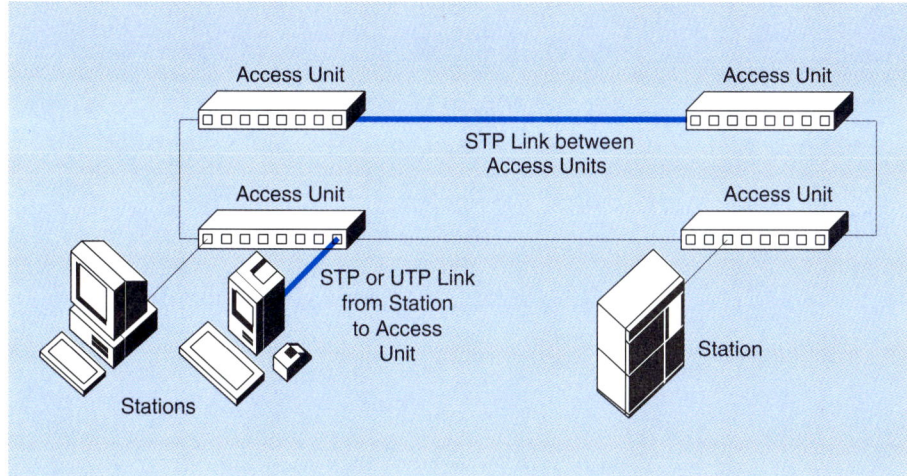

FIGURE 5.3

Stations Connect to Access Units in 802.5 Networks

Wiring concentrators in 802.5 Token-Ring Networks are called *access units*. The access units are organized in a ring.

Figure 5.3 shows that access units are themselves organized in a ring. As Module F discusses, even within an access unit, the ring topology (layout) is maintained.

The 802.5 standard does not permit UTP to be used to link access units. This still requires STP or, optionally, optical fiber.

Speed

The first 802.5 standard only specified a speed of 4 Mbps. However, IBM soon increased this to 16 Mbps, and the standard eventually followed. Most 802.5 Token-Ring Networks today run at 16 Mbps. Like 802.3 networks, they use *baseband signaling*.

Token Passing

Figure 5.4 shows that 802.5 handles media access control by sending a special frame called a **token** around the network. There is a simple access rule. If a station wants to transmit, it must wait until it receives the token. It then removes the token and sends its frames. After transmission, it retransmits the token, so that another station can use it.

In token passing, there are no collisions. This allows 802.5 networks to work much closer to their rated 16 Mbps capacities. This greater throughput at higher network loads (and the slightly higher speed of 16 Mbps) was a major factor in many companies' decisions to implement 802.5.

We will not look at the 802.5 frame in detail, but one interesting aspect of the frame is that it specifies three bits for the **priority** of the frame. This gives

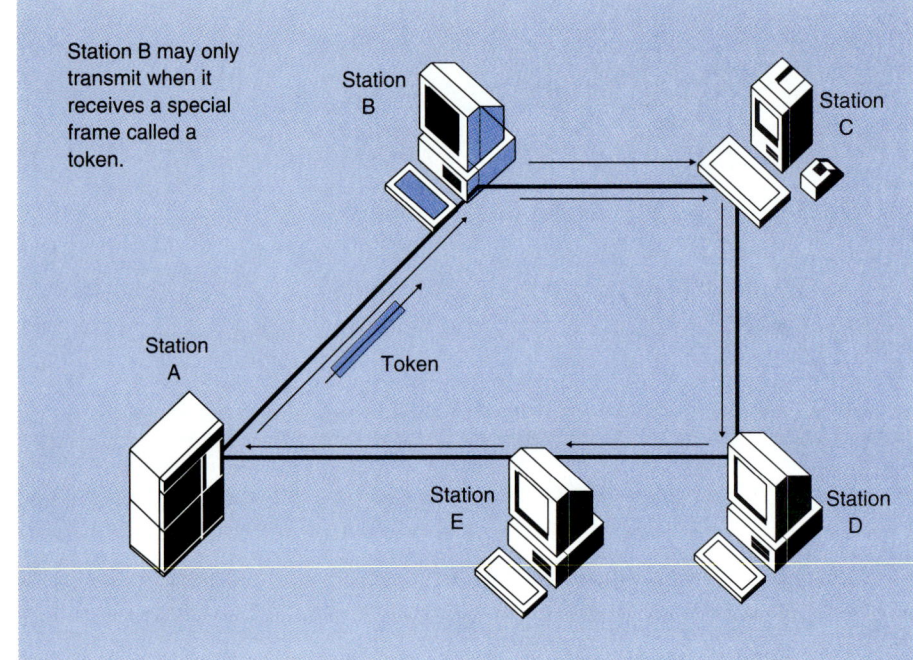

FIGURE 5.4

Token Passing in 802.5 Token-Ring Networks

To transmit, a station must wait until it receives a special frame called a *token*. It removes the token, transmits its data frames, and then retransmits the token.

eight levels of priority. High-priority frames, such as audio or video transmissions that need almost real-time delivery, always get first access to the network.

Market Status of 802.5

By almost any technical measure, 802.5 is a better standard than 802.3 10Base-T. Its STP wiring is more resistant to interference than Ethernet 10Base-T's UTP wiring. Its token passing access control mechanism performs much better than CSMA/CD when throughput is more than 30 percent to 50 percent of the LAN's rated capacity. Its raw transmission speed is 60 percent higher. Finally, it offers the priority levels that real-time multimedia applications need.

Nevertheless, while Token-Ring Networks are common, they have a much lower market share than Ethernet LANs. The basic problem is initial cost. Its wiring is more expensive, but this is a minor problem. More important, token-ring operation is more complex. So 802.5 NICs cost considerably more than Ethernet 10Base-T NICs. In a chicken-and-egg feedback loop, cost differences have limited 802.5's market size and have therefore limited the competition needed to get costs down rapidly. Ethernet 10Base-T works acceptably for most companies today, and its initial cost is modest.

At the same time, many companies that have installed 802.5 Token-Ring Networks have argued that life-cycle costs for 802.5 are lower because these networks require less maintenance.

100 Mbps LANS

Until recently, LAN standards from the 802 committee have been limited to about 10 Mbps (4 Mbps to 16 Mbps). But these "slow" LANs are beginning to give way to a new generation of LAN operating at 100 Mbps.

Problems with 10 Mbps LANs

Shared Media Operation ■

Figure 5.5 shows why 10 Mbps is not as fast as it first seems. As we noted earlier, this capacity is shared by all stations. For a burst of any length, a station cannot expect more than about 1 Mbps.

Congestion ■

When any station transmits on a shared media network, every other station on the network hears it. As the number of stations grows, so does traffic. At some

FIGURE 5.5

Congestion and Latency in 10 Mbps Shared Media LANs

A 10 Mbps LAN offers considerable throughput. But if multiple stations want to transmit, they must share the capacity. A speed of about 1 Mbps is about all a station can expect on a larger message. This produces *latency* (delays) when stations have to wait to transmit. Latency is a special problem in multimedia, especially video. As the number of stations grows, furthermore, the LAN becomes *congested*, much like an eight-lane freeway during rush hour.

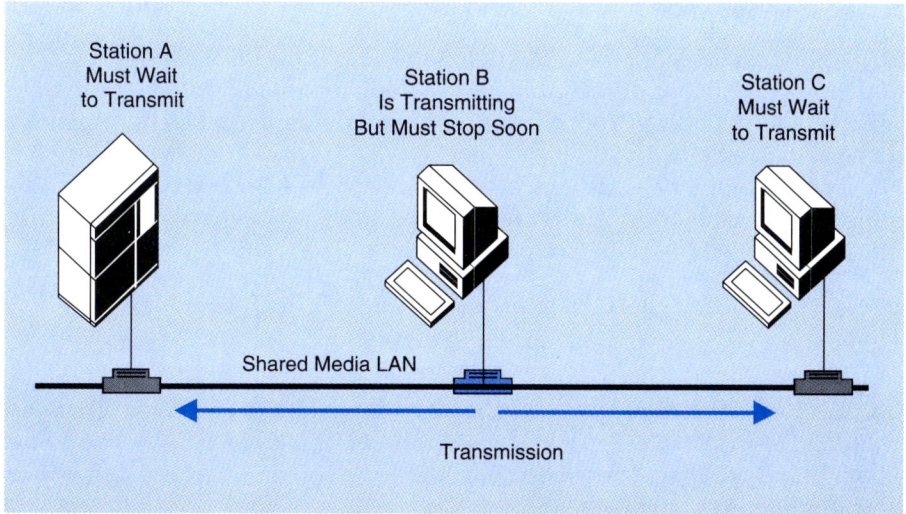

point, the demand for transmission exceeds 10 Mbps. There is **congestion** on the network, much like congestion on highways during rush hour. Even an eight-lane freeway can be brought to a halt if there are too many cars.

The maximum number of stations on a 10 Mbps network obviously depends on how much each station transmits. In general, however, an 802.3 Ethernet network begins to choke at about 200 to 300 stations. Token-Ring Networks running at 16 Mbps are also limited in the number of stations they can support. So even if companies can live with 802.3 and 802.5 distance limitations, congestion becomes a real problem as the number of stations grows.

The multimedia explosion is adding to the congestion problem. Now each station wants to transmit much more, on the average, than it did in prior years. So the number of stations you can have before congestion sets in is *decreasing*.

Latency and Priority Levels ■

As congestion on a freeway grows, it takes longer to get to your destination. In networking, delays are referred to as **latency**.

Although you can view latency as simply a natural consequence of congestion, latency needs to be addressed directly. For some real-time applications, such as videoconferencing, latency is unacceptable. Even for one-way video transmission, there is the problem of **variable latency**, in which delivery is sometimes in real time, sometimes delayed. Imagine watching a movie that stops every few seconds if traffic on the LAN is high!

LAN standards can address latency and variable latency by assigning priority levels to frames, as does 802.5. Real-time applications get high priority, allowing them to get through even during periods of high congestion. Applications with more tolerance for latency, such as downloading a file or sending an electronic mail message, would have to wait a bit longer during congested periods.

Increasing Speed ■

One solution to latency and congestion is to increase the LAN's speed, say from 10 Mbps to 100 Mbps. If each transmission only takes one tenth as long, there will be less congestion. You can support many more users before congestion sets in.

As technology has matured, standards agencies have become able to create 100 Mbps standards. These are expected to be very popular in the future, although they will have higher per-station costs.

FDDI

The first major 100 Mbps shared media LAN standard was **FDDI**, the **Fiber Distributed Data Interface**. As its name suggests, FDDI normally uses optical fiber instead of coaxial cable or twisted pair wire. We saw in Chapter 4 that optical fiber has very low attenuation.

Figure 5.6 shows that FDDI networks are token-ring networks. The FDDI standard borrowed from 802.5 technology, although it made many modifications in order to allow much higher speeds and longer distances. As a result, FDDI is incompatible with 802.5 Token-Ring Networks.

FDDI has many virtues. It supports priority levels, making it attractive for real-time applications. In addition, FDDI can support very large rings with circumferences of 200 km. Furthermore, FDDI is a *mature* standard, having emerged in 1987 and 1988.

The chief problem with FDDI is its high cost. Thanks to the complexity of token-ring operation, FDDI station NICs are three to five times as expensive as Ethernet 10Base-T and 802.5 NICs. In addition, optical fiber is more expensive to buy, splice, and install than UTP or STP wiring. A new standard for **FDDI over UTP,** popularly known as **CDDI** or copper distributed data interface, promises to lower media costs. It does not lower NIC costs dramatically.

802.3 100Base-X

The "10" in 10Base-T stands for a speed of 10 Mbps. Now the 802.3 Working Group is producing a family of **802.3 100Base-X** standards. These will run at 100 Mbps.

FIGURE 5.6
FDDI Network

FDDI networks are token-ring networks, although FDDI is incompatible with 802.5. FDDI can operate at 100 Mbps, although this is shared capacity. FDDI can operate at long distances—up to 200 km. This makes it attractive for large buildings and industrial parks.

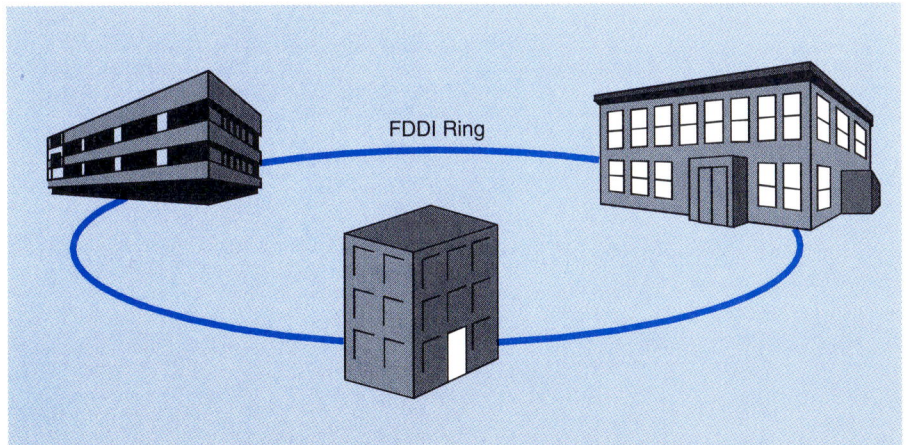

Wiring and Hubs ■

A major goal of the 802.3 Working Group was to create an easy upgrade for organizations with existing Ethernet 10Base-T networks. To this end, two versions of 100Base-X use the same wiring as Ethernet 10Base-T. This is crucial because it means that organizations do not have to rewire their buildings. They only have to replace hubs and NICs.

The most popular variant in 100Base-X to date is 100Base-TX, which uses two pairs of Category 5 UTP wiring. There is also a 100Base-T4 standard, which can use Category 3, 4, or 5 wiring. The 100Base-FX variant uses optical fiber to link stations to hubs.

Figure 5.7 shows that you can only have two hubs in 100Base-TX. In addition, these 100Base-TX hubs must be colocated or nearly so. Although wires from the hubs to the stations can be 100 meters, co-locating the hubs limits the distance span to only 200 meters on 100Base-TX LANs. Both 100Base-T4 and 100Base-X have similar distance limitations.

All of the 100Base-X standards use the same 802.3 MAC layer frame that 10Base-T and other 10 Mbps 802.3 networks use. This gives complete backward compatibility at the MAC layer. Unfortunately, it perpetuates CSMA/CD's inefficient use of the LAN's speed capacity. It also perpetuates Ethernet's lack of priority levels for real-time applications.

FIGURE 5.7

Hubs in 802.3 100Base-TX

The *100Base-TX* standard can use existing station-to-hub wiring as Ethernet 10Base-T, saving the need to rewire in upgrades. The existing Ethernet 10Base-T hubs are simply swapped for 100Base-TX hubs. The wiring remains unchanged. NICs need to be upgraded in the PCs. 100Base-TX does not support multiple levels of *priority* or other ways of reducing latency and variable latency. In addition, 100Base-TX LANs are limited to about 200 meters.

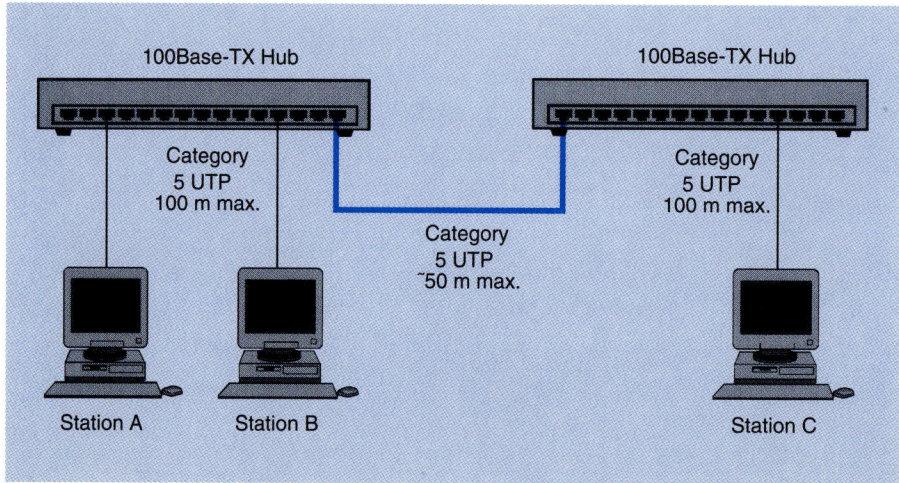

802.12: 100VG-AnyLAN

Another 100 Mbps shared media standard is emerging from a new 802 Working Group, 802.12. This 802.12 standard is popularly known as **100VG-AnyLAN**.

Like 100Base-X, 100VG-AnyLAN is designed for easy upgrading. If you have 10Base-T wiring, you merely replace NICs and hubs. The "VG" stands for "voice grade," indicating that you can use business telephone wiring.

If you have an 802.5 Token-Ring Network, you can also use 100VG-AnyLAN. You use existing 802.5 wiring but replace the NICs. You replace the 802.5 access units, in turn, with 802.12 hubs. This ability to upgrade both Ethernet 10Base-T and 802.5 LANs is the source of the "AnyLAN" in the name.

Hubs and Wiring ■

Recall that Ethernet 10Base-T hubs are daisy-chained together. But as shown in Figure 5.8, 100VG-AnyLAN organizes its hubs in a tree hierarchy. The top-

FIGURE 5.8

802.12 100VG-AnyLAN Hub Hierarchy

Hubs in *100VG-AnyLAN* are not daisy-chained together as in Ethernet 10Base-T. Instead they form a hierarchy, with the top-level master hub (Hub A) controlling the others. The hubs implement the *demand priority access method (DPAM)* instead of CSMA/CD. When a station wishes to transmit, it sends a request (demand) to the 802.12 hub. This request indicates high or low priority.

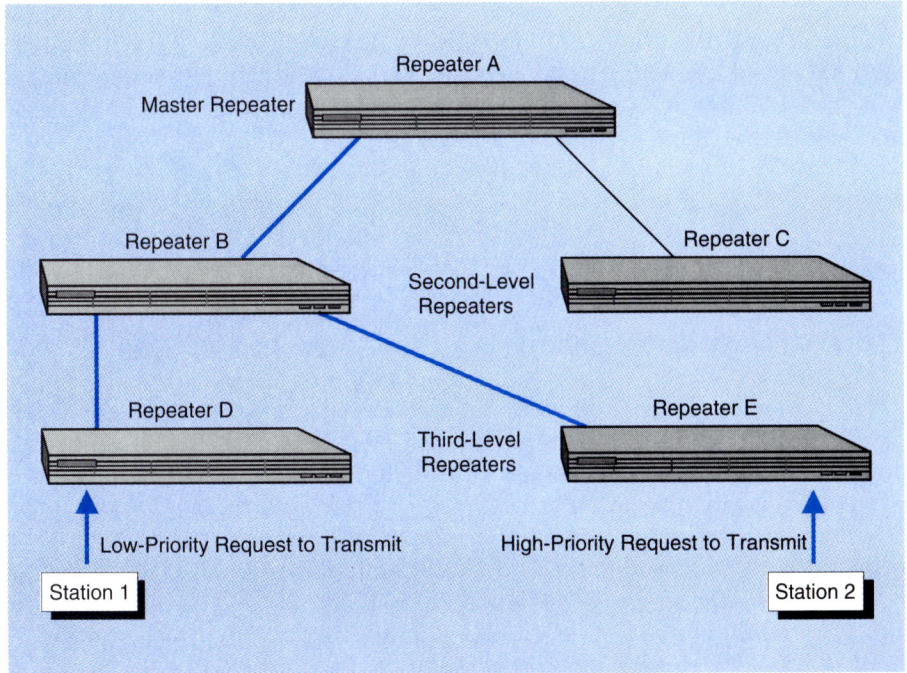

level hub controls the others. The draft standard recommends no more than three levels in the hierarchy.

Although the draft standard does not address the issue of distance, it appears that 802.12 will be able to support much larger distances than 100Base-X.

Demand Priority Access Method ■

For access control, 802.12 does not use either CSMA/CD or token passing. Instead it uses the **demand priority access method (DPAM).** When a station wants to transmit, it indicates whether or not it needs high priority. If it does, it gets to transmit before low-priority stations. This provides the priority levels needed for multimedia applications. It also makes better use of transmission capacity than CSMA/CD.

100 Mbps Shared Media LANs in Perspective

In general, the new 100 Mbps shared media LANs offer the ability to reduce latency and congestion. However, their costs are still somewhat high. In addition, 100Base-TX and 100Base-T4 have distance limitations and lack priority levels.

On the positive side, all operate at 100 Mbps, easing the congestion problems that are facing a growing number of companies. For many firms, the benefits will outweigh the costs even today. In the future, as NIC and hub prices fall rapidly in years ahead, 100 Mbps is likely to be the default speed for LANs.

At the same time, the developers of 100 Mbps LAN technologies are all looking beyond 100 Mbps, at 1 Gbps LANs. These gigabit LANs will use extensions of the 100 Mbps technology. Although relatively few organizations will need 1 Gbps LANs soon, each of the standards developers is attempting to dominate the "high ground" of technology.

ALL-SWITCHED LANS

The Death of Shared Media LANs

Recall that one reason for 100 Mbps shared media LANs is congestion. If too many stations transmit on a shared media LAN, there is bound to be congestion. Moving from 10 Mbps to 100 Mbps helps, but there will still be a point at which congestion occurs.

The basic problem is that every station must hear every transmission. For networks with many stations, this is always a disaster. For this reason (and for several others), the emerging generation of 100 Mbps shared media LANs will almost certainly be the last generation of shared media LANs. The future almost certainly belongs to a new generation of *switched* LANs.

Switching

Figure 5.9 shows four stations attached to a switch. Two stations wish to transmit. First, Station A wishes to transmit to Station C. Second, at the same time, Station B wishes to transmit to Station D. For reasons we will soon see, messages in switched networks are called *packets*.

If the switch were a shared media hub, then if Station A had begun to transmit its message, Station B would have to wait. In addition, all four stations would hear the message from A to C.

A switch is different. First, if Station A transmits its packet to Station C, the packet only comes out on the port for Station C. Stations B and D do not hear the packet from A to C at all.

Second, Station B does not have to wait until Station A is finished transmitting. Station B can transmit its packet to Station D at the same time Station A is transmitting its packet to Station D.

If you think of the telephone system, this is nothing startling. You and your next door neighbor can both place calls simultaneously, despite the fact that your calls both go through the same switch.

FIGURE 5.9

Switch

In a *switch*, multiple pairs of stations can communicate at the same time without interfering with one another. Here Station A is transmitting a packet to Station C at the same time that Station B is transmitting a packet to Station D. Only the destination station hears the packet. Because every station does not hear every packet, congestion does not increase as the number of stations rises. As long as switches have sufficient switching capacity, congestion is not a problem in switched networks.

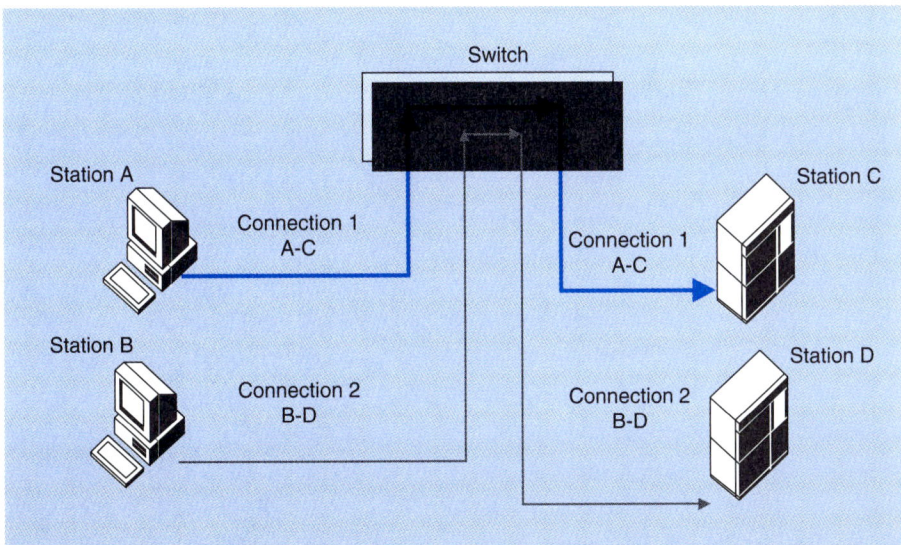

The most important implication of switching is that not all stations hear each packet. This means that there is no need for congestion as the number of stations increases. As long as switches have enough capacity, and as long as an organization installs enough switches, then there is no limitation on the number of stations in a network. Again, think of the telephone system, with its hundreds of millions of users around the world. Switching absolutely breaks the limit on the number of stations that can be on a LAN.

Switched Network

Figure 5.10 shows a **switched data network**. Here there is not a single switch but rather a group of interconnected switches. A switched network can have dozens, hundreds, or even thousands of switches. The lines that connect the switches are called **trunk lines**.

FIGURE 5.10

Switched Data Network with Multiple Switches

A *switched data network* has many switches connected by *trunk lines*. Stations are connected to the network via *access lines*. When a packet reaches a switch, the switch decides which out port to use. If there are many switches, there are many possible alternative routes between two stations. This permits failure recovery if a switch or trunk line fails. It allows the network to route traffic around congested switches or trunk lines. It also allows routes to be optimized for minimum cost, minimum delay, maximum security, or other quality of service measures.

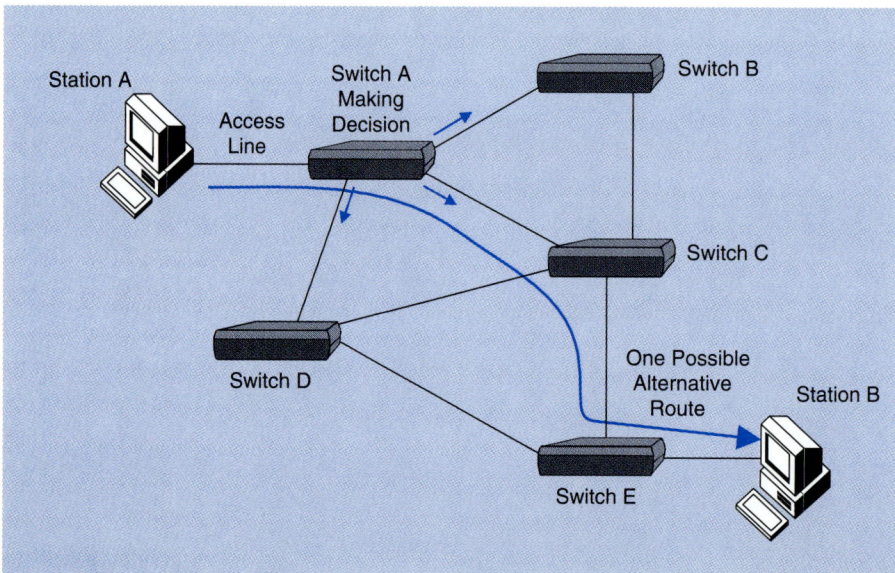

Switching Decisions

Suppose that Station A transmits a packet to Station B. First, the packet travels down an access line to the first switch, which is sometimes called an access node.

When the packet reaches Switch 1, the switch must decide what to do with the packet. There are two out ports on Switch 1. Each will carry the packet out in a different direction. In deciding what to do with the packet, Switch A may consider many things, such as security, cost minimization, or throughput. Switch 1, in other words, must make a **switching decision**.

Route

Eventually, the packet will reach its destination, which is Station B. In doing so, it will have traveled over a complex route that carried it across multiple switches and trunk lines. A **route,** then, is an end-to-end path between stations that a packet takes as it moves across the network.

Alternative Routing

Recall that in 10Base-T networks there can be no loops among the hubs. This means that there can only be one possible route from any station to any other station.

Figure 5.10 shows that switched networks are very different. In the figure, there are multiple loops. They actually are the key to the most powerful feature of switched networks, namely **alternative routing.**

Alternative routing means that there are many possible alternative routes between any two stations rather than a single possible route. We have already seen this. However, we have not yet discussed the importance of alternative routing.

Rerouting Around Failures

The most obvious advantage of alternative routing is the ability to route packets around failed switches and trunk lines. In Figure 5.10, if a trunk line fails, traffic is merely routed around the failure. While the longer path might introduce a delay (latency), this is likely to be a small price to pay for the ability to reach other stations despite failures.

In contrast, in a shared media network, there is only one route between any two stations. If a single trunk line fails along that one route, the two stations will not be able to communicate. In fact, the whole network will be cut in two.

Rerouting Around Congestion

A similar point is that switched networks can reroute around congestion as well as failure. If a switch or trunk line starts to be overloaded, the switches can adjust. They can achieve **load balancing,** evening the traffic load on various switches and trunk lines.

While failures are spectacular, they usually are infrequent. Congestion, in contrast, is a constant danger, and load balancing is a constant need as traffic loads change continually throughout the working day.

Route Optimization ■

As noted earlier, switches can base their switching decisions on a number of **optimization** variables. Cost minimization is one obvious candidate for optimization. Another is minimizing delays for time-critical traffic, such as voice. To give a third example, a packet may have to travel only over high-security lines. Different packets, furthermore, may have different optimization needs. Thus, their routes may have to be optimized differently.

Ethernet Switches

Although Ethernet was conceived as a shared media network, a number of enterprising vendors have created **Ethernet switches.** Ethernet switches look like Ethernet hubs, but they switch 802.3 MAC layer frames from one input port to one output port instead of repeating them out all output ports.

Many Ethernet switches are full-duplex. As we saw in Chapter 3, this means that a station can transmit frames to the switch and receive frames from the switch at the same time. This, of course, requires full-duplex Ethernet NICs. A number of vendors now sell such NICs.

Ethernet switches are more expensive than 10Base-T hubs, but they are less expensive than the ATM switches we will see next. Ethernet switches already have a large and growing installed base, and fast Ethernet switches are beginning to run at 100 Mbps to match 100Base-X NIC speeds. They are likely to grow more popular in the future.

The one weak point in Ethernet switching is standards. Despite the popularity of Ethernet switching, the 802.3 Working Group responsible for Ethernet LANs has not moved to standardize Ethernet switches. As a result, interworking among switches from different vendors is limited.

ATM Switches

Internationally the ITU-T is promoting another switching technology, called **asynchronous transfer mode (ATM).** In Chapter 3, we discussed ISDN service. Actually, the 2B + D service that we saw in that chapter is called **narrowband ISDN.** The ITU-T has developed a much faster service called **broadband ISDN.** ATM is the switching technology developed for broadband ISDN. While the full broadband ISDN service will take many years to develop, ATM switching is already mature.

One attractive feature of ATM is simply the fact that it is standardized. This allows equipment from different vendors to interoperate. In fact, vendors have created an ATM forum to ensure interoperability.

Another attractive feature of ATM is the fact that it is **scalable.** This means that it can handle a broad range of speeds. Some ATM switches operate as slowly as 1 Mbps. At the high end, ATM has been standardized at a speed of 2.4 Gbps.

This means that if an organization creates an ATM LAN, it can start with relatively slow and inexpensive switches. As demand grows, the organization can field upgrade its switches and the ATM NICs inside its computers. It can upgrade its network constantly without going through the expense of changing protocols.

Switching and the OSI Architecture

In Chapter 4, Figure 4.2 showed the IEEE 802 committee's LAN architecture. It involved the lowest two layers of the OSI architecture—the physical layer and the data link layer. These two layers are sufficient for moving data across a shared media LAN in which there is only a single possible path from one station to another. The sending station merely transmits its frames without being concerned with routing.

Switched networks, however, require a way of standardizing the routing of packets across the network. For this, the OSI architecture has the network layer (OSI Layer 3).

There must be a way to distinguish protocol data units (PDUs) at different layers of the OSI architecture. For instance, it is correct to speak about NPDUs (network-layer protocol data units). For simplicity, however, network-layer PDUs are called **packets.** Similarly, we saw in Chapter 4 that PDUs at the data link layer are called *frames*.

Radio LANS

Although future LANs are likely to be faster than today's LANs, there will be at least one major exception. A number of new LAN technologies use radio or some other wireless technology, such as infrared light. Figure 5.11 shows a **radio LAN.** It shows that some stations have their own transmitter/receivers (**transceivers**), while others share them. A hub links the radio LAN to the main corporate-wired LAN.

The obvious benefit of these LANs is that they can support mobile workers. Studies have shown that managers are away from their desks about half the time [Panko, 1992], so the need to support people away from their desks is very real. Just as we now carry our pens and paper notepads with us as we

FIGURE 5.11
Radio LAN

A *radio LAN* uses radio waves instead of wires to transmit signals. Some stations have their own transmitter/receivers (transceivers). Others share transceivers. A hub links the radio LAN to the main corporate-wired LAN.

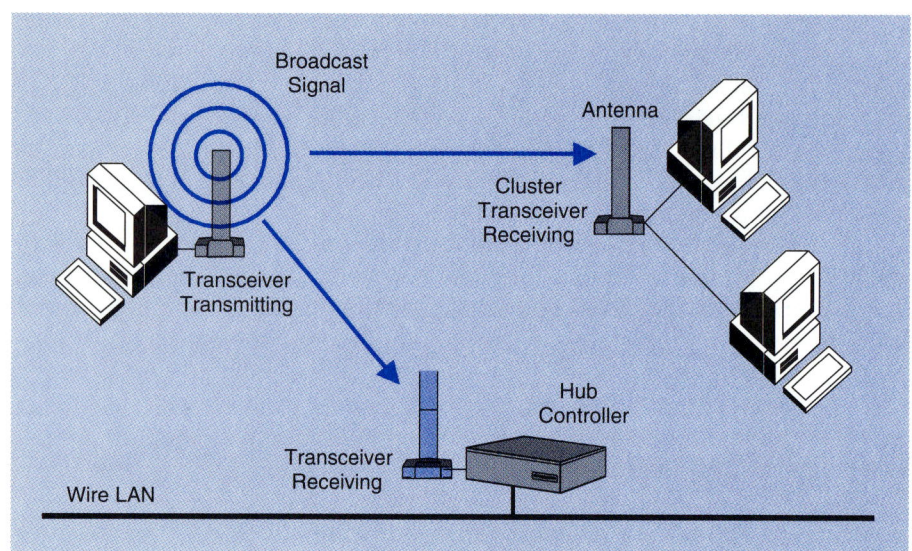

move around in our office areas, in the future we will be able to carry our pen computers and notepad computers and never be out of touch.

Because radio LANs will be interconnected with the main wired LAN, if you are traveling around a building with your wireless notepad computer, you will be able to communicate with wired desktops and servers. The next chapter discusses how to connect LANs together.

The main limitation of radio LANs is speed. Emerging standards are designed to work at around 1 Mbps. This is slow compared to today's 10 Mbps LANs. It is glacial compared to the emerging 100 Mbps LANs.

CONCLUSION

The previous chapter looked at Ethernet 10Base-T LANs. This chapter looked at other LAN standards, beginning with the 802.5 Token-Ring Network technology. Compared to 10Base-T, 802.5 is more reliable, faster, and has priority levels for time-critical traffic. Unfortunately, it is also more expensive, limiting its use.

Both 10Base-T and 802.5 networks are shared media technologies, meaning that every station hears every transmission. This inevitably causes congestion as the number of stations grows. One way to put off congestion is to move to 100 Mbps shared media LANs. Each transmission will take only one-tenth as long as it would take in a 10 Mbps shared media LAN, so congestion does not become a problem until there are many more stations. Faster LAN speeds are also good for multimedia applications, which tend to transfer very large files. We looked at three 100 Mbps shared LAN technologies: FDDI, 802.3 100Base-X (especially 100Base-TX), and 802.12 100VG-AnyLAN.

Eventually, even 100 Mbps shared media LANs become congested. In the future, we are likely to move to switched LANs. In switched LANs, only the source and destination stations hear messages, so congestion does not increase as the number of stations on the network grows. In addition, switched LANs offer alternative routing, which allows them to route packets around failed switches or trunk lines and congested switches or trunk lines. Alternative routing also allows the network to optimize message delivery for minimum cost, minimum delay, maximum security, or other quality of service measures.

We concluded with a brief discussion of radio LANs, which do not tie stations to wiring but are expensive and relatively slow.

CORE REVIEW QUESTIONS

1. What are the strengths and weaknesses of 802.5 Token-Ring Networks compared with 802.3 10Base-T networks?
2. In a ring topology, to how many other stations does each station *transmit* frames? From how many does it *receive* frames? Does every station receive every message?
3. What is the normal transmission medium for 802.5 Token-Ring Networks? When can UTP be used?
4. Explain how token passing is used to control access. Why is token passing superior to CSMA/CD in terms of performance?
5. In shared media LANs, why do you have congestion as the number of stations rises? What is latency, and how does congestion produce latency?
6. Among 100 Mbps shared media LAN standards, which have priority levels? Rank them in increasing order of maximum possible distance. Mark them "least," "middle," and "most."
7. What is the difference between a switch and a 10Base-T hub? Why can this eliminate the congestion problem?
8. Explain the advantages of alternative routing in switched data networks.
9. Compare the relative strengths and weaknesses of switched Ethernet and ATM.
10. What is scalability? Why is it desirable?
11. Distinguish between frames and packets in terms of the OSI architecture.
12. Why are radio LANs desirable? What is their chief limitation?

DETAILED REVIEW QUESTIONS

1. What company initiated the push for Token-Ring Network LANs?
2. Why is 802.5 technology more expensive than 802.3 technology?
3. In a STP wiring bundle with two wire pairs, how many mesh sheaths are there, and what does each surround?
4. Why is FDDI superior to the other 100 Mbps shared media LAN standards, and what are its limitations? What is FDDI over UTP? What are the limitations of 100Base-X? How does 100VG-AnyLAN compare to the other two standards?
5. Explain how 100VG-AnyLAN links hubs.
6. What 802 working group, if any, sets switched Ethernet standards? Why is this a problem?
7. What is the history of ATM?
8. Which 802 Working Group is creating radio LAN standards?

THOUGHT QUESTIONS

1. Comparing the costs and technical strengths of 802.5 and 802.3 networks, what does their relative market penetrations tell you about buyer behavior in organizations that use LANs?

PROJECTS

1. Prepare a report on 100 Mbps LANs.
2. Prepare a report on radio LANs.

For online exercises, please visit this book's website at: http://www.prenhall.com/panko

REFERENCES

IBM (1992). *Token-Ring Network: Introduction and Planning Guide.* (Document GA27-3677-05). Research Triangle, North Carolina.

PANKO, R. R. (1992). Patterns of Managerial Communication. *Journal of Organizational Computing,* 2, 95–122.

CHAPTER 6

Local Internets

INTRODUCTION

In the last two chapters, we looked at individual LANs. In this chapter, we will stay at the single site, but we will look at *local internets*. Recall from Chapter 1 that *internetting* involves linking multiple individual networks (*subnets*, in internetting terminology) so that any station on any subnet can send messages to any station on any other subnet. A **local internet,** then, is an internet at a single site.

The Need for Local Internets

Individual shared media LANs are fine for many smaller organizations, such as NuPools. But as we saw in Chapters 4 and 5, they have *distance limits* many organizations need to address. They also have *limitations on the number of stations* each LAN can support before congestion and delays become severe. We will see that organizations can address these problems through local internetting.

Increasing Distance Spans

Figure 6.1 shows a local internet at a single site that only contains two LANs. One LAN is in the headquarters building. The other is in a factory at the same site. Individually, the two LANs have limited distance spans. The total internet, in contrast, can span whatever distance the company needs at the site. Many sites have very complex local internets that connect dozens of individual LANs.

Although we did not discuss it in the preceding chapter, LANs have maximum distances because signals degrade as they propagate over long distances (see Chapter 4). Distance limits also have other causes. In CSMA/CD networks, for instance, if stations are too far apart, it will take too long for a signal to get from one end of the LAN to another. Thinking the line is free, a station at one end will start to transmit when another station at the other end has already started transmitting. There will be a large number of collisions.

The internetting device, in contrast, *regenerates* the signal, removing many propagation effects. It sends a clean new signal onto the next subnet. In addition, when distant stations on different LANs both transmit at the same time, internetting devices delay their insertion into each LAN on the internet until that LAN has no traffic.

Increasing the Number of Stations

For shared media LANs, internetting also serves a second purpose, *increasing the number of stations*. Recall that in shared media LANs, every station hears every transmission. Beyond about 200 to 300 stations on a 10 Mbps shared media LAN, there is too much traffic and congestion becomes a problem. In the last chapter, we saw that congestion produces *latency* (delays). Latency is especially a problem for multimedia applications and real-time applications, such as videoconferencing. In some cases, packets are lost completely.

As shown in Figure 6.2, internetting devices address this problem by being intelligent about what frames they pass to the other side. The figure shows

FIGURE 6.1

Local Internetting to Increase Distance Spans

In an internet, each LAN has a limited distance span. The distance span of the entire local internet, however, is ample for the organization.

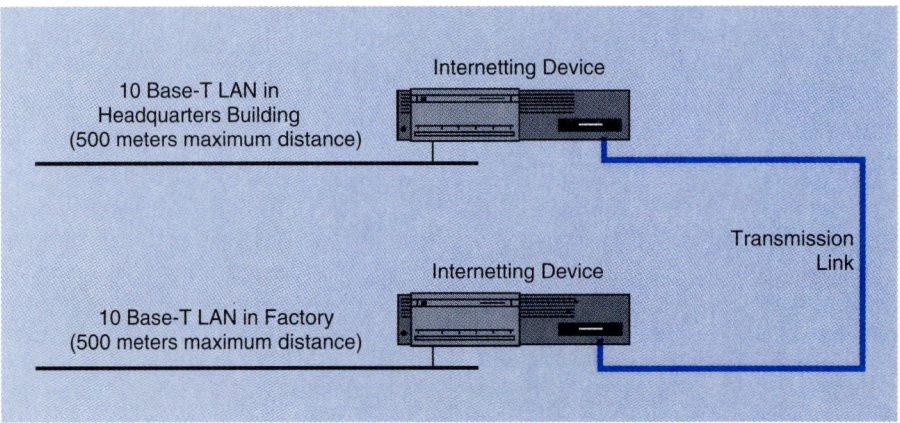

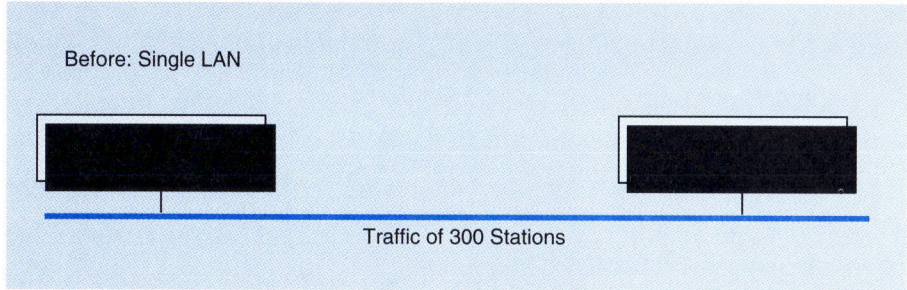

FIGURE 6.2
A Congested Shared Media LAN
Initially, there is only a single LAN serving both departments. This original LAN must serve the traffic from 300 stations. This is a congested situation.

a single LAN with two departments. Each department has 150 stations. So the total traffic on the network is the combined traffic for 300 stations, which is a situation of high congestion.

Figure 6.3 shows how to relieve this situation. The organization **resegments** its LAN by dividing its single initial LAN into two separate LANs. One is for Department 1. The other is for Department 2. The organization then adds an internetting device to link the two LANs.

Most traffic stays within departments. An internetting device does not copy intra-LAN traffic to the other LAN. It only passes frames if they are addressed to stations on the other LAN.

As a consequence, each department LAN subnet only carries the traffic of its own 150 stations, plus a small amount of cross-traffic from the other de-

FIGURE 6.3
Internetting Devices do Not Pass Local Traffic to the Other LAN
Resegmenting divides the original LAN into two departmental LANs and links the two LAN subnets via an internetting device. The internetting device does not pass frames to the other LAN subnet unless those frames are destined for a station on the other LAN. So each department LAN only has the traffic of 150 departmental stations, plus some cross-traffic from the other LAN. This is not a congested situation.

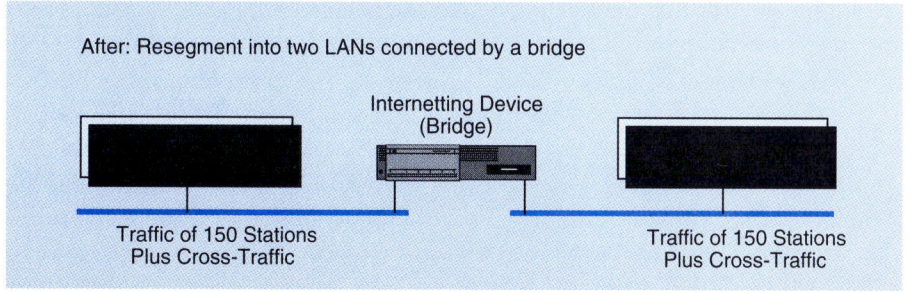

partment's stations. This is not likely to be a congested situation. Resegmenting is a popular strategy for network administrators facing increasing congestion.

Figure 6.3 shows the resegmentation of a LAN into only two LANs. In actual practice, most resegmenting efforts divide an original LAN into multiple LANs to further segment traffic.

Internetting Versus Switched LANs

With the switched LANs we saw in the last chapter, there is no need to internet to avoid congestion and to increase distance spans. Switched LANs do not become congested because all stations do not hear all messages. Only the source and destination stations hear any given message. Nor is distance a problem in switched LANs. Switches completely regenerate signals when they pass them on. Internetting to increase distance and to reduce congestion is a concern for shared media LANs.

However, even with switched LANs, there may be a need for local internetting. If you have multiple LAN technologies at a site—even multiple types of switched LANs, then you will need internetting to link the dissimilar LANs.

Internetting Technologies

In a limited sense, Ethernet 10Base-T hubs are internetting devices. They link the segments on which individual stations reside. They operate at OSI Layer 1, providing simple physical and electrical links. We do not consider them to be full internetting devices, however, because they are not at all selective about passing on traffic. OSI physical-layer connecting devices are called **repeaters**. At the electrical level, they merely repeat what they hear to all other segments. Every station on every segment hears every message. This does nothing for congestion. Nor, as we saw in Chapter 4, can you link an unlimited number of repeater hubs to increase the network's distance span.

Bridges ■

The simplest real internetting devices are **bridges**. As we will see later, bridges are simple devices that operate at OSI Layer 2. Their simplicity means that they are quite inexpensive. However, their simplicity becomes a burden for sophisticated local internets.

LAN Switches ■

Another internetting device is the **LAN switch**. In the last chapter, we saw switched LANs. However, all-switched LANs are still quite expensive. In this chapter we will look at the use of switches as internetting devices, perhaps as a first step toward all-switched LANs.

Bridges operate at OSI Layer 2. They do not add alternative routing to shared media LANs. Switches, in turn, operate at OSI Layer 3 (the network layer) for single networks. This adds the advantages of alternative routing: routing around failures, routing around congestion, and route optimization. However, because LAN switches are designed to operate at the subnet layer, instead of the internet layer, there are operational problems in using them as internetworking devices. These problems can be worked around, but they consume administration time.

Routers ■

At the next level of sophistication, **routers** offer extremely sophisticated interconnection. Routers are switches that were built to operate at the internetting layer. Routers have a rich set of functionality that allows them to be used in even the largest internets. In fact, the worldwide Internet uses routers to interconnect its millions of host computers across thousands and thousands of networks.

The Chapter

In the rest of this chapter, we will look at the most common bridging devices: bridges, LAN switches, and routers. We will see that local internet designers have a number of options that are appropriate under different circumstances.

We will end with a discussion of *backbone networks*. When you internet, you often create a *backbone LAN* to link the various subnets to one another. We will look at *distributed backbone* networks as well as new alternatives called *collapsed backbone* networks.

BRIDGES

The simplest internetworking device is the *bridge*. Bridges are simple, automatic, and inexpensive. For companies with modest local internets, bridges are an ideal way to extend LAN distances and the number of stations attached to the collective LANs.

Learning

One of the things that network administrators like best about bridges is that there is no need to spend time programming them to understand the organization of the internet. When a bridge is turned on, it listens to traffic and actually learns the organization of the internet very quickly. This learning process is very simple and completely automatic.

Bridge Learning

Figure 6.4 shows a bridge in operation. The bridge has multiple ports: A, B, and C. Each connects to a single LAN. The figure shows how most bridges go through a learning process to find which stations are attached to each port.

Initial Operation: Broadcasting

When you first turn on a bridge, it has no idea which LANs serve which stations. For instance, suppose Station 1 transmits to Station 2. Both are on LAN X, but the bridge does not know this yet.

FIGURE 6.4

Bridge Learning

When a bridge first begins to operate, it broadcasts each frame it receives out of every port. It always looks at the source address on frames, however, and it places this information in a table associating ports with stations. When it receives a frame from one station to a known station on the same LAN, it does nothing. If it receives a frame for a known station on a different LAN, it forwards that frame to that LAN's port only. To eliminate obsolete information, bridges erase their station–port tables every few minutes and start the learning process again.

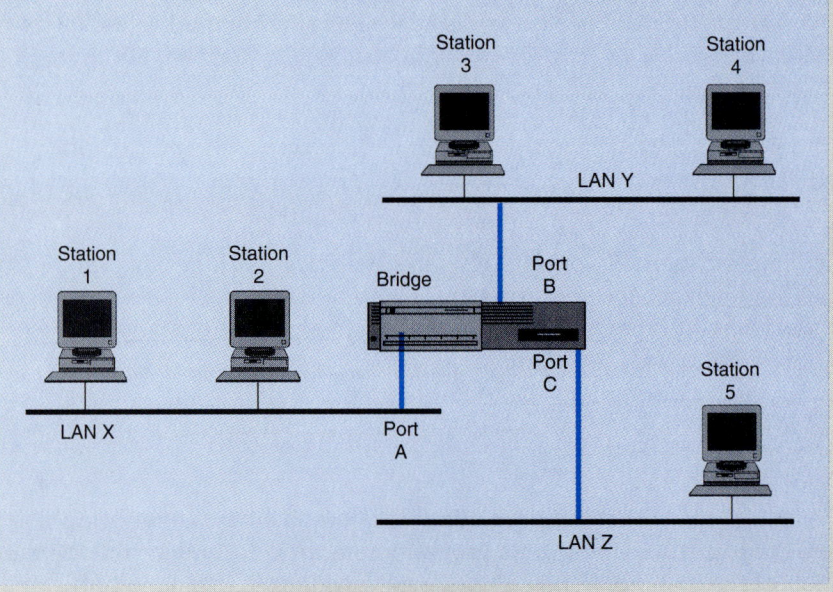

When the bridge receives the frame, it looks at the source field. From this it learns that Station 1 is out Port A. It places this information in a table.

It still does not know where Station 2 is, however. So when it finishes reading the frame, it broadcasts the frame out both Port B and Port C. (It does not send it out over Port A, of course, because it knows that any station on the LAN out Port A has already received the frame when Station 1 transmitted it over the LAN.)

At this stage, the bridge's operation is rather stupid. Because it broadcasts the frame out all other ports, both LANs Y and Z must bear the traffic. At this stage, the bridge does nothing to alleviate congestion. It is a mere repeater.

Station 2 Replies: Intelligent Silence

Now suppose that Station 2 replies to Station 1. This time, Station 2's address is in the source field. The bridge adds this information to its table.

When the bridge looks at the destination address, in turn, it sees Station 1. From its table, it knows that Station 1 is out Port A. Now the bridge acts more intelligently. It knows that the frame has come from Port A and that Station 1 is out Port A. So it knows that it does not have to do anything. It does not flood Ports B and C by broadcasting the frame. As a result, LANs Y and Z do not have to bear the traffic. Now the bridge actually reduces congestion because it does not transmit local traffic on LAN X to LANs Y and Z.

Station 3 Transmits: Selective Internetting

Now suppose that Station 3 transmits a message to Station 1. This frame arrives through Port B. First, the bridge notes the source destination. It learns that Station 3 is out Port B. It writes this information in its table.

Now it looks at the destination address. It sees that this is Station 1. From its table, it knows that Station 1 is out Port A. So it sends the frame out Port A. It does not send it out Port C.

As a result, LAN X has to bear the cross-traffic from LAN Y. LAN Z, however, does not have to bear this cross-traffic. So even where there is cross-traffic, the bridge minimizes its impact on other LANs.

continued

Forgetting

This learning process works well as long as the internet is stable. But what if a station drops off a LAN, or worse yet if an entire LAN is disconnected? In this case, there must be a way to flush out information that is no longer true.

The specific method that bridges use to eliminate obsolete information is simplicity itself. They simply forget everything they have learned every few minutes and then start learning again from scratch.

This complete "brain wipe" followed by relearning sounds like it would be very inefficient. But learning is extremely rapid. When two stations start to communicate, they typically send thousands of frames back and forth. Within a handful of frames, the bridge will learn where they are and will stop broadcasting.

Bridged Internets: No Loops Allowed!

Figure 6.5 shows a bridged internet with multiple bridges. Note that the bridged internet contains *no loops*. A prohibition on loops is universal in bridged internets. Given the way that bridges learn which port holds a station, frames would circulate endlessly around the loop. Fortunately, the 802.1d bridging standard discussed later prevents loops automatically.

Forbidding loops means that there can be only one possible path between any two stations. To see this, trace a path between any two stations in Figure 6.5. You will only find one possible path. Thus, in bridged internets, there can be no alternative routing.

Disadvantages of Bridges

With no alternative routing, there is no way to optimize the use of transmission lines connecting the bridges. This is a serious problem for long-distance communication, but for local internets it is rarely a problem.

In addition, without alternative routing, it is impossible to route around a failed or congested transmission line or bridge. Again, this is more of a problem for long-distance transmission. Long-distance lines are more prone to failure, and because of their costs, firms tend to use them close to their limits. This increases the likelihood of congestion. For local internets, this is not too much of a problem.

Often the only serious problem is accidentally creating loops. This becomes very likely if you have large numbers of bridges. However, most local internets only have a few bridges.

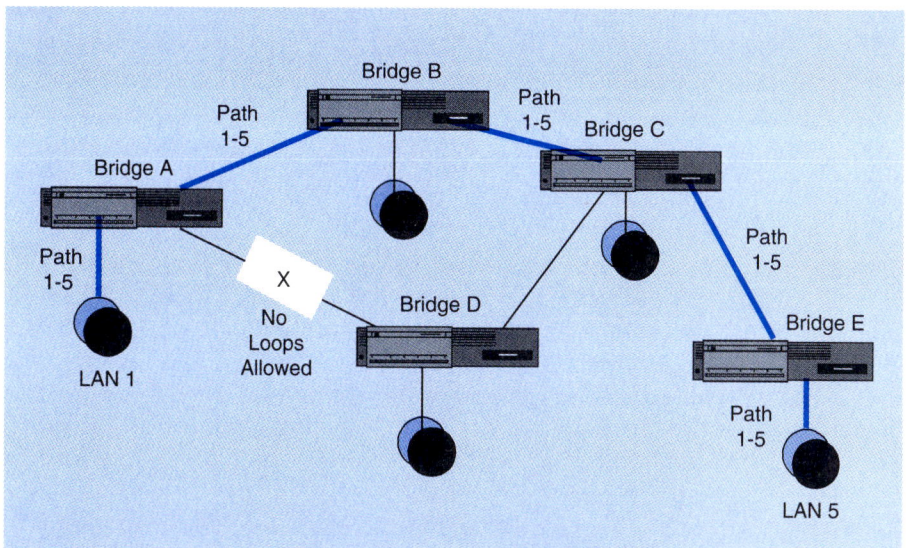

FIGURE 6.5
Bridged Internet with Multiple Bridges

In a bridged internet, there can be no loops connecting the bridges. This prevents the endless looping of frames around a ring of bridges. However, it also precludes alternative routing. There is only one possible path between any two stations on the internet. For instance, between Stations 1 and 5, the only possible path is 1-A-B-C-E-5. This prevents route optimization and automatic recovery if a bridge or LAN fails. However, it allows far simpler operation. Bridges have both low cost and high filtering rates compared to routers.

Advantages of Bridges

Low Cost ■

By eliminating alternative routing, bridges do not have to worry about route optimization. As a result, bridges are rather simple, and this makes them relatively inexpensive. A simple bridge may cost only $1,000 to $2,000. In contrast, a switch costs several thousand dollars, and a router can cost $10,000 to $20,000.

Speed ■

Another advantage of eliminating alternative routing is that bridges are *fast*. Because they do not have to do optimum routing calculations for each incoming frame, even low-cost bridges can handle a large volume of frames. We say that bridges have a higher **filtering rate** than routers, which must do more work deciding how to route each incoming frame.

Bridge Standards

When the IEEE 802 Committee first developed LAN standards, it did not envision bridging. As shown in Chapter 4, it eventually added a bridging layer between the MAC and LLC layers.

Interconnecting LANs with Different Standards ■

Because bridging sits on top of the MAC layer, it can link two LANs that have different MAC and physical-layer standards. To give an example, Figure 6.6 shows that bridges can link an 802.3 10Base-T LAN with an 802.5 Token-Ring Network. The bridge connects to the 10Base-T LAN using that LAN's physical and MAC layer protocols. The bridge also connects to the Token-Ring Network using that LAN's 802.5 physical and MAC layer protocols. The bridge then passes bridge-layer frames between them.

FIGURE 6.6

Bridging LANs with Different Physical and MAC Layers

Bridges can connect two LANs with different physical and MAC layers. An 802.3 10Base-T LAN can easily talk to an 802.5 Token-Ring Network via a bridge, as long as they use the same bridging standard. Of course, the bridge connects to each LAN using that LAN's physical and MAC layer protocol.

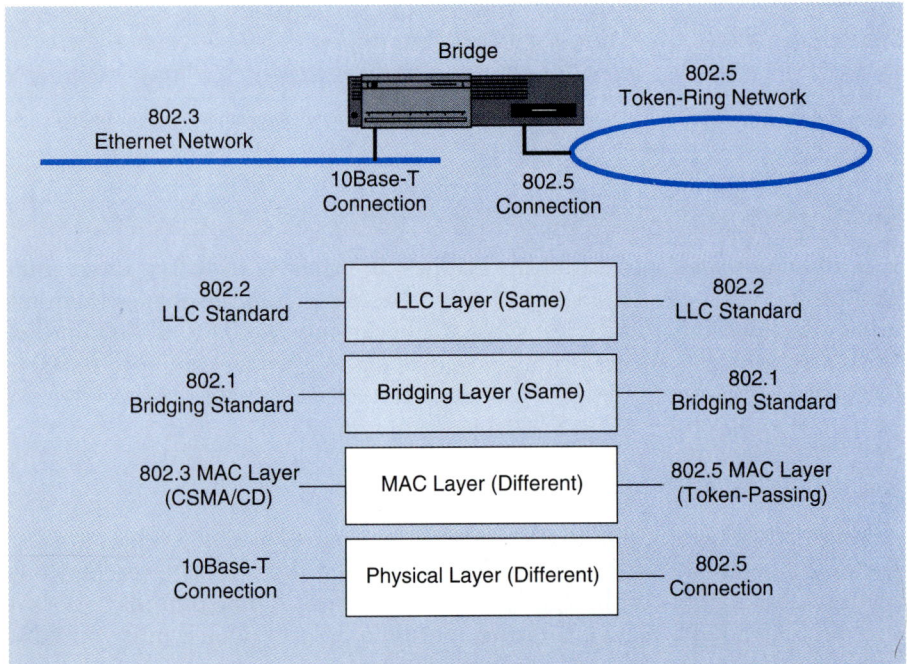

SWITCHED LOCAL INTERNETS

Although bridges are simple, their lack of alternative routing can be problematic. For this reason, many organizations are turning to LAN switches, including Ethernet switches and ATM switches, as internetting devices. In Chapter 5, we looked at all-switched LANs. Unfortunately, these all-switched LANs are extremely expensive. Adding one or two LAN switches to do internetting, however, brings considerable gains at small costs.

Switching

A Simple Internet: No Waiting ■

Figure 6.7 illustrates a simple switched internet. There are four LANs connected to a single *LAN switch*. In this figure, LAN A is transmitting to LAN C. LAN B, in turn, wants to transmit to LAN D. With switching, LAN B does not have to wait until LAN A is finished. It can transmit while LAN A is still sending.

FIGURE 6.7

Simple Switched Internet

On a switched internet, every LAN does not hear every transmission. The switch keeps the transmissions separate. Here LAN A transmits to LAN C. LAN B, which wants to transmit to LAN D, does not have to wait until LAN A is finished. It can transmit while LAN A is still sending. Because the switch keeps the transmissions separate, there is no danger of LAN B's signal interfering with that of LAN A.

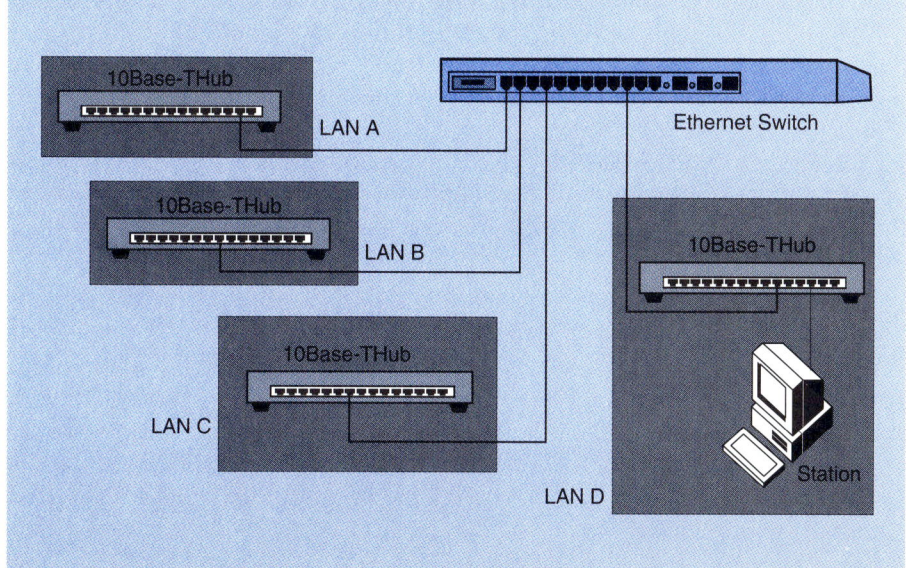

Because the switch keeps the transmissions separate, there is no danger that LAN B's signal will interfere with that of LAN A. Telephone switches already work this way. While you are talking to one person, hundreds of other conversations are also taking place over your local telephone switch.

Because there is no need to wait, LAN B can always transmit at the full speed of the access line to the switch for as long as the LAN needs to transmit. Recall that in a 10 Mbps shared media LAN, a station may only be able to transmit at about 1 Mbps because of waiting. On a half-duplex switch, LAN B will have to stop transmitting when LAN D replies, but in a full-duplex switch, even this delay vanishes. *Full-duplex and half-duplex transmission were introduced in Chapter 3.*

Multiple Switches

Figure 6.7 shows a switched internet containing a single switch, but many switched internets have multiple switches. This provides the benefits of alternative routing—routing around failures, routing around congestion, and optimizing routing.

Switched Ethernet Internets

The least expensive (and most popular) switched internet technology today is **switched Ethernet**. We saw all-switched Ethernet networks in the last chapter.

Simple Switched Ethernet Internet

Figure 6.7 shows how switching allows you to create simple internets. Here the organization is using Ethernet 10Base-T LAN technology. As we saw in Chapter 4, individual stations connect to Ethernet 10Base-T hubs through UTP wiring. Stations may be up to 100 meters away from the hub. The organization in this case is VeriFone. The figure shows the internetting used in VeriFone's Costa Mesa plant.

In Figure 6.7, the 10Base-T hubs connect to an Ethernet switch. The Ethernet switching hub segments the 10Base-T hubs and their stations into a number of smaller LANs. These LANs are small enough not to have congestion or distance limitations.

Because the Ethernet switch is a true switch, it can handle transmissions between several pairs of LANs simultaneously. Stations only have to wait on the occasions when a particular connection between LANs is busy.

Operation

One nice thing about switched Ethernet internetting is that there is no need to modify the 10Base-T hubs. To them, the Ethernet switch looks just like another 10Base-T station or hub. When a 10Base-T hub sends it a frame, the Ethernet switching hub accepts it as any 10Base-T hub would. The only difference is how it handles the frame after reception.

When the Ethernet switch sends a frame to a hub, in turn, it sends it as a 10Base-T station would send it. It waits to send the frame if there is already traffic on the LAN segment, just as CSMA/CD requires.

Virtual LANs

When you resegment a network, it is normal to base the resegmentation on workgroups, such as formal departments. People in the same workgroup tend to broadcast messages to one another, use the same servers, and have similar rights to access various resources on the network. Their traffic tends to be primarily internal. The simplest approach to resegmentation, then, is to place each workgroup on a separate physical LAN and then to attach the LAN to a switch. This way, the switch will only pass on some traffic to the larger internet.

Sometimes, however, members of a workgroup are in different locations. In this case, it is impossible to put them on a single physical LAN. Switch vendors approach this problem by allowing the network administration to define **virtual LANs** or **VLANs.** In a virtual LAN, the administrator defines a virtual LAN group and assigns members of a workgroup to that virtual LAN. If one station in a virtual LAN broadcasts a message, the message will go to all other members of the VLAN group—but not to all stations on the internet. This prevents congestion. Access rights, in turn, can be granted to the VLAN group. This is easier than assigning rights to each individual separately.

Unfortunately, although the IEEE 802.1 Working Group is currently defining a VLAN standard, it has not succeeded at the time of this writing. As a result, different vendors have different and incompatible approaches to defining and managing virtual LANs.

ROUTERS

Sophistication

Although switched internets are "smarter" than bridges, they are not designed to support extremely large and complex internets. They lack many of the features needed to implement and manage extremely large networks. For most local internets, they are sufficient. For complex local internets, however, and for the enterprise internets we will see in Chapter 7, they are insufficient.

For complex internets, organizations turn to routers. These machines are also switches. However, their standards are designed to allow fairly automatic operation, even in large and complex environments. While they are expensive to purchase, they save on network administration costs. As network size increases, these savings become crucial. The box, "TCP/IP Routing Standards,"

TCP/IP Routing Standards

Each standards architecture has its own family of routing protocols. However, the most widely used routing protocols are those of the TCP/IP architecture used on the Internet and in many private networks as well. We will use TCP/IP routing to discuss router sophistication.

Universal Addressing and Delivery

In the telephone network, you can reach anyone simply by dialing a telephone number. Telephone numbers are hierarchical. In the United States, for example, there is a country code, an area code, an exchange code, and a final four digits for a specific telephone within an exchange.

Routers also work on the basis of hierarchical addresses. For instance, Module H explains addressing in TCP/IP routing. Each host and router has a 32-bit **IP number.** It consists of a network ID to identify a specific network and a host ID to identify a specific host on that network. In addition, there is a **subnet mask** that allows a network to be subdivided into smaller networks called subnets. Module H shows how the IP number and the subnet mask work together in addressing.

As their name suggests, routers offer sophisticated routing across multiple routers, so that the sending host merely issues a packet with the address of the destination host. The intermediate hosts automatically handle the routing.

When you set up your computer for Internet access in Windows 95, you must specify your PC's IP number. (We will see later that there is an alternate way of doing this.) You must also set up the subnet mask, which will look something like "255.255.255.0".

Usually, a host is set up with a **default router**. This is the router to which outgoing messages normally will be sent first to begin their journey through the Internet. If another router on the sending host's network would be more appropriate for a specific transmission, however, the sending host can send the message via that other router. Windows 95 Internet setup allows you to specify a default router.

Error Reporting

In complex internets, there are bound to be problems. So router protocols include sophisticated protocols for sending messages about errors back to routers or to the source host, depending on the nature of the error. In TCP/IP, errors are handled by the ICMP protocol.

Peer Control Among Routers

One of the problems with switched internets (and all-switched networks) is that the network administrator must constantly update information in the switches, so that the switches will route messages correctly.

In contrast, nearby routers constantly communicate with one another, updating routing information in their peers. There is almost never manual intervention.

Autoconfiguration

Each host attached to a routed internet needs a unique IP address. Almost all networks have **autoconfiguration hosts**. When a host is booted up, it can call an autoconfiguration host to have a unique IP number assigned to it. This is unnecessary if the host has a permanent IP number. However, not all hosts have permanent numbers. When you log into an Internet service provider with a PC, your PC usually calls an autoconfiguration host to get a temporary IP number for use during that session.

For instance, Microsoft Windows 95 allows you either to specify a permanent IP number or the IP number of an autoconfiguration host. Unfortunately, that host must use the DHCP protocol. There are other autoconfiguration host protocols, and some autoconfiguration hosts do not support DHCP.

Domain Name Service

IP numbers (and addresses in other internetting protocols) are 32-bit binary numbers. These are very difficult to memorize.

Fortunately, TCP/IP host computers can also be given mnemonic (easy to remember) names, such as "voyager.cba.hawaii.edu." Special hosts called **domain name system (DNS)** hosts maintain a list of registered host names and the formal 32-bit IP number for each.

When you specify a destination host by name, your own host will contact the DNS host, find the 32-bit IP number for that host, and place this 32-bit number in the destination field of your outgoing packet.

In Windows 95, to set up your computer for Internet access, you must specify the address of a DNS host.

IP Packets

Figure 6.8 shows the packet organization for the most important TCP/IP routing protocol, the **internet protocol (IP).** An IP packet can hold data or supervisory messages.

Like the MAC layer frame in 802.3, which we saw in the last chapter, the IP

continued

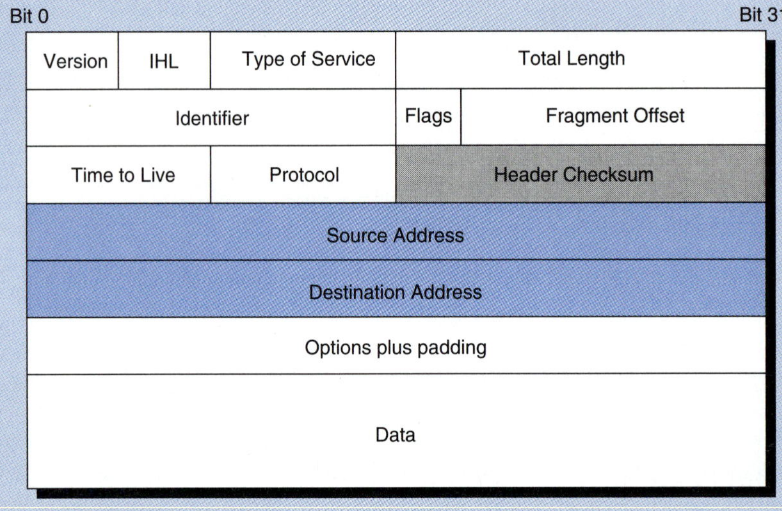

Bit 0					Bit 31
Version	IHL	Type of Service		Total Length	
Identifier			Flags	Fragment Offset	
Time to Live		Protocol	Header Checksum		
Source Address					
Destination Address					
Options plus padding					
Data					

FIGURE 6.8
The Internet Protocol (IP) Packet

The IP protocol data unit is a packet because it operates at OSI Layer 3. It has two address fields, but they are only 32 bits long rather than 48 bits long. This has proven to be too few, and the IETF is developing a new version of IP, Version 6. The packet only provides error checking for the header. A bad header can disrupt the internet's operation. Data errors, however, must be caught at upper layers.

has a header consisting of a number of fields. It then has a data field to carry the data that the IP packet is to deliver. However, unlike the 802.3 MAC layer frame, it has no trailer fields.

One interesting thing about IP packets is that they *do not do support error checking* on the data field. There is error checking for the header because a bad header can harm internet operation if a router tries to interpret its fields. If an IP process on a router or host detects a bad packet, furthermore, it simply discards it. It is up to the transport-layer process above the internetworking layer to deal with any discarded packets and data errors that occur during IP data transfers.

In addition, IP packets have 32-bit address fields rather than the 48-bit address fields of LAN standards. As Module H discusses, 32 bits is too few for the exploding number of networks and hosts on the Internet. The Internet Engineering Task Force (IETF) has standardized a new IP protocol (Version 6) which has much larger address fields, among other improvements. *For more on TCP/IP routing and transport protocols, see Module H. For more on IPv6, see Module H.*

discusses the sophistication of the TCP/IP routing protocols. It discusses both their complexity and their automatic operation. Routers, in contrast to switches, have the full functionality needed at the internet layer.

In addition router standards are scalable. This means that there is practically no limit on the sizes of routed internets. This is not surprising because routers were first created to support the Internet, which is the very model of a large, complex, and mixed-ownership internet. However, routers soon found their way into private networks because of the need to move beyond the limits of bridging. (For many years, bridges and routers were the only two choices. Switched internetting is actually a new development.)

Multiprotocol Routers

TCP/IP routing protocols are the most widely used routing protocols. But every standards architecture has its own routing standards. For instance, Novell's IPX/SPX architecture also has a sophisticated set of routing protocols. IBM's SNA was not set up for routing. However, there are ways of routing SNA messages across multiple routers.

Most routers today are **multiprotocol routers**. This means that they can mix TCP/IP, IPX/SPX, SNA, and other types of routed traffic. Therefore, you do not have to set up separate router networks for different types of traffic. Unfortunately, multiprotocol routers are more complex (and therefore expensive) than single-protocol routers.

INTERNETTING DEVICES IN PERSPECTIVE

We have looked at three internetting devices: routers, switches, and bridges. Each is best under different circumstances.

Routers

Routers produce very sophisticated interconnections among LANs. They can optimize routes and react quickly to failures and congestion. Routers have a rich set of supervisory messages to control the communication. (See Module H.) They can even handle multiple internetworking protocols simultaneously.

Router operation, furthermore, is rather automatic. For instance, routers talk to one another, so that individual routers can act intelligently when they forward IP packets to other routers. This minimizes the time that the network administrator must spend on routers.

Unfortunately, all of this is expensive in terms of both processing power and software development. Routers are powerful and automatic, but they are also very expensive.

Bridges

Bridges, which use far simpler algorithms than routers, cost only a fraction of the cost of a router. For the relatively simple interconnection among a few LANs at a local site, bridges are sufficient. In addition, not only do they cost less than routers but because they have to do less work on each frame, they are also faster than routers. Data communications professionals have long had a saying, "Bridge where you can; route where you must."

Mixing Bridges and Routers

Fortunately, bridges and routers can be mixed together to combine their strengths. Figure 6.9 shows a common example of how organizations mix different types of internetting devices.

Local Bridging ■

For local communication within each site, the company uses bridges to link the stations together. Most importantly, bridges are inexpensive.

In addition, bridges are sufficiently reliable for operation at a single site. If one bridge fails, usually only a small part of the local internet is affected. And there is local staff to bring the bridge back online rapidly.

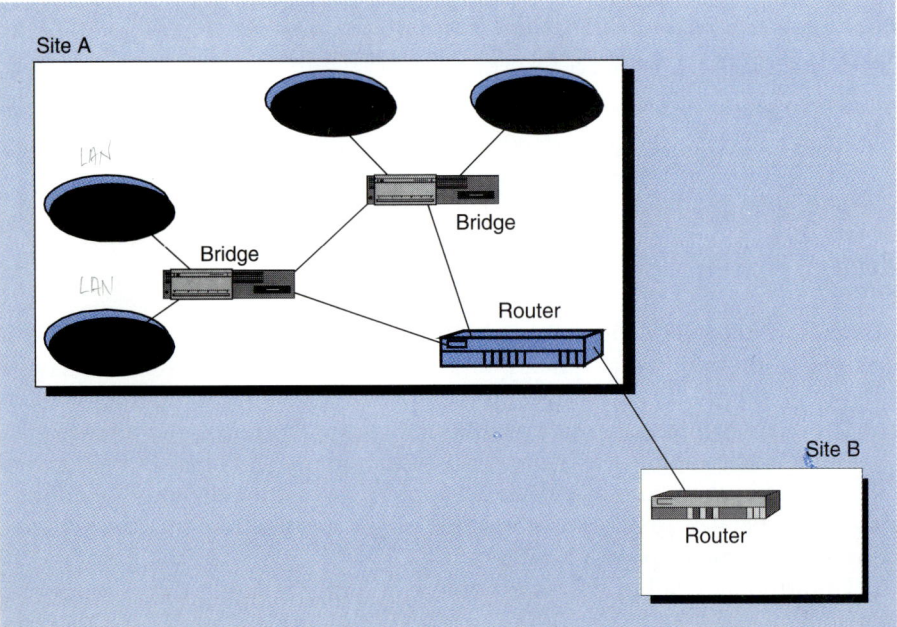

FIGURE 6.9

Mixing Bridges and Routers

For internetting within a single site, the organization uses bridges, which are inexpensive. It uses routers for transmissions between distant sites.

Within a single site, furthermore, route optimization usually is fairly unimportant. In local transmission, costs usually are not so high that optimization is critical.

Remote Routing ■

For intersite connections, the organization uses routers. These cost more, but their ability to optimize the use of long-distance transmission lines more than makes up for the cost difference.

In addition, long-distance transmission facilities may not be as reliable as local transmission facilities. Routers offer the ability to route around a failed or congested transmission link.

Switched Internets

In the last few years, switches have added a third tier of internetworking functionality. Switches bring the benefits of alternative routing (which bridges do not bring) at a smaller hardware cost than routers. However, switch operation is not as automatic as router operation. This can increase administration costs. Until switch sophistication increases, switches are likely to be most widely used in internets that are somewhat too complex for bridges but do not need the full sophistication of routers.

BACKBONES FOR LOCAL INTERNETS

In a local internet, the various routers, bridges, and switches need to transmit messages to one another. Typically, these interconnections are organized in one of two basic ways: distributed backbones and collapsed backbones.

Distributed Backbone Connections

Figure 6.10 shows **distributed backbone** interconnection. As you can see, a backbone LAN runs around the local site, passing the individual internetting devices. The internetting devices act like stations on the LAN.

Typically, the backbone runs at higher speed than the individual LANs it services. This allows it to handle a large volume of internetworked traffic.

Distributed backbone LANs are attractive because they are relatively insensitive to the failure of a single internetworking device. If a device fails, the

FIGURE 6.10

Distributed Backbone Network

In a distributed backbone network, the backbone comes to each of the individual internetworking devices. Typically, the backbone operates at higher speeds than the individual LANs it interconnects.

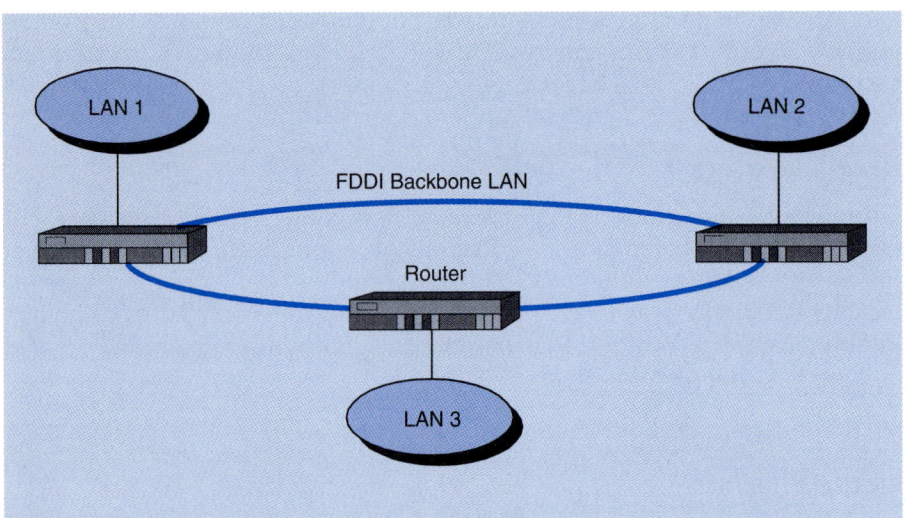

rest of the local internet remains intact. Even in the LAN with the failed router, traffic within the disconnected LAN will flow normally.

The most common technology for distributed backbone LANs today is FDDI. As we saw in the previous chapter, FDDI operates at 100 Mbps and has multiple priority levels, enabling it to handle real-time applications. More important, an FDDI ring can have a circumference of 200 km. This is more than ample for local internets.

Collapsed Backbone Internets

Many newer local internets use the **collapsed backbone** organization in Figure 6.11. Here all of the LANs connect to a central device, which may be either a router or a switch.

Single Point of Failure ■

Older network professionals wince when they see this figure because it has what they were always taught to avoid: a **single point of failure**. If the central device fails, all internet traffic fails. Although traffic will flow normally within the disconnected LANs, there will be no way to move traffic among the LANs.

Single Point of Service ■

Although a collapsed backbone creates a single point of failure, it also creates a *single point of service*. The central device typically sits in the network control center. This makes it easy to maintain and upgrade the device. It also makes certain types of problem diagnosis easier.

As technology has grown more reliable, having a central point of maintenance has become more important than having a single point of failure. Proponents of distributed backbones, however, argue that as distributed network management (see Chapter 9) becomes better, having a central point of service will become less important.

VeriFone at Costa Mesa ∎

Figure 6.7 showed VeriFone's facility at Costa Mesa, California. This is a collapsed backbone LAN, using an Ethernet switch. Physically, three of the four hubs are located within a few meters of the switch. From there, lines fan out throughout the facility to the individual client PCs. This allows a large amount of maintenance to be done without traveling more than a few meters from the switch.

The fourth 10Base-T hub is not colocated with the switch. It is in a relatively distant part of the facility. It links to the switch using an optical fiber transmission line.

FIGURE 6.11

Local Internet Using a Collapsed Backbone

In a collapsed backbone, all LANs connect to a central device (in this case, a router). This creates a single point of failure. If the central device fails, no traffic can flow among the LANs. However, this also provides a single point of service, making it easier to do maintenance and some types of problem diagnosis. The central device may be a single switch or router.

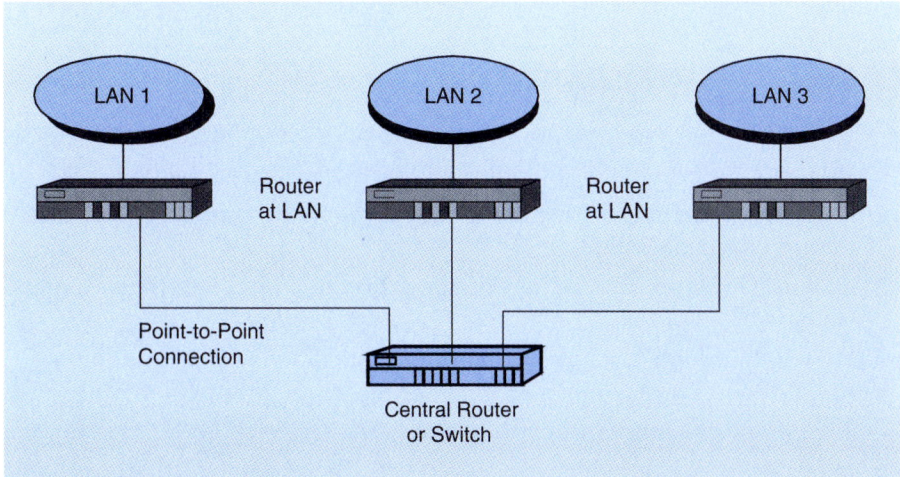

CONCLUSION

The last chapter looked at simple LANs. This chapter looks at the more complex local internets that most firms will use in the future.

Simple LANs use shared media technology, which becomes congested when they have about 200 to 300 stations. Congestion leads to latency (delays). Another problem with shared media LANs is distance. Shared media distance limitations are problematic for large buildings and for larger sites, such as industrial parks.

To address these problems, firms can build local internets that extend distances and resegment existing LANs to reduce congestion problems. Bridges allow fast and inexpensive internetting at the cost of routing inflexibility and scaling problems as the internet grows. Switches provide the benefits of alternative routing at a comparatively modest cost. Routers are more expensive forms of switches that provide very sophisticated and relatively low-maintenance internetting.

Finally, we looked at the advantages and disadvantages of distributed backbone and collapsed backbone internetting.

Most organizations use a mixture of the technologies we discussed in this chapter. For instance, many organizations mix bridges and routers. In addition, even where VeriFone uses all 10Base-T wiring within a site, it mixes bridges to link its distant sites.

The last three chapters have looked at networks and internets for individual sites. These topics are important because most of an organization's information remains within a site. However, it is critical to be able to connect sites together. In the next chapter, we will look at Enterprise internets that connect multiple sites into a single coherent internet.

CORE REVIEW QUESTIONS

1. What are the two main limitations of traditional shared media LANs?
2. Explain how an internetting device can allow you to support more stations through resegmentation.
3. At what OSI layer do repeaters operate? Bridges? LAN switches? Routers? Gateways?
4. How does bridge learning make bridge installation and operation easy?
5. Can you use a bridge to connect two LANs if one LAN is an 802.3 10Base-T LAN and the other is an 802.5 Token-Ring Network LAN? Explain.
6. How does a switch reduce congestion? Which is faster—a 10 Mbps shared media LAN or a 10 Mbps switched LAN? Explain.

7. Explain the difference between switched Ethernet internets and all-switched Ethernet LANs.
8. What is scalability? Why is it important to network administrators?
9. What is multiprotocol routing?
10. Compare bridges, switches, and routers in terms of speed (throughput), cost, and sophistication in routing. Use a table. You may use terms such as *least* and *greatest*.
11. What is the purpose of both distributed and collapsed backbones? Explain the difference between distributed and collapsed backbones. What are the advantages and disadvantages of distributed backbones? Of collapsed backbones?

DETAILED REVIEW QUESTIONS

1. In Figure 6.1, if the transmission link is 1,000 meters long, what is the maximum possible distance between two stations on the 10Base-T local internet?
2. How many loops can you have among the bridges in a bridged internet? What does this mean in terms of alternative routes between any two stations? What disadvantages does this elimination of alternative routing bring? Why do we often use bridges anyway?
3. In Figure 6.5, what is the single path between LANs 1 and 5? 1 and 4? 2 and 4? 4 and 5?
4. In Figure 6.6, what transmission medium is used to connect the subnet on the left to the bridge? The subnet on the right? Do both connections use the same MAC layer frame organization? The same bridging standard?
5. What is a virtual LAN? Why is it good? Why is it not good? What type of internetworking device uses virtual LANs?
6. Figure 6.7 shows the situation at VeriFone's Costa Mesa plant. Where are the hubs located—near the switch or near the stations they serve? Does this represent a traditional or collapsed backbone network? Explain. What advantage does the collocation of the switch and most hubs bring?
7. (From the box "Bridge Learning") In Figure 6.4, assume that the bridge has just forgotten all of its information. Explain the things that happen if Station 5 then transmits to Station 4. Explain the things that happen if Station 4 then replies.
8. (From the box "TCP/IP Routing Standards") List the major functions that are provided in sophisticated TCP/IP routing.
9. (From the box "TCP/IP Routing Standards") Compare IP with the 802.3 MAC layer frame in terms of addressing and error handling.

THOUGHT QUESTIONS

1. Explain the advice, "Bridge where you can. Switch where you should. Route where you must."
2. You are going to set up your Windows 95 computer to communicate with an internet service provider. What settings must you enter into the program?

PROJECTS

1. Collect data on current costs for bridges, routers, Ethernet switches, and ATM switches. Explain the main factors that produce price variations within each category.

For online exercises, please visit this book's website at: http://www.prenhall.com/panko

CHAPTER 7

Enterprise Internets

INTRODUCTION

Chapter 5 looked at traditional shared media LANs. Chapter 6 looked at more sophisticated local internets and all-switched LANs. This chapter goes beyond the local site. It looks at **enterprise internets,** which link a company's networks across multiple sites, sometimes internationally.

We have already looked at the routers that lie at the heart of enterprise internets. Although routers are more complex than bridges or switches, organizations need their level of sophistication to link multiple types of networks via expensive and sometimes unreliable long-distance links.

Now we must see how to interconnect routers not just within a site but among distant sites as well. At a local site, a company owns its wiring plant. For links to distant sites, however, companies must turn to transmission *carriers*, such as the traditional telephone companies. We will look at carriers, then at their two main transmission services: *leased lines* and *switched data networks*.

Transmission Carriers and Deregulation

As soon as the signal passes beyond the customer premises, the organization usually has to turn to a transmission carrier. For a price, the carrier transports the company's telephone calls and data to other sites and to other organizations. *For more on carriers, see Module G.*

Tiers of Carriers

As Figure 7.1 illustrates, transmission carriers provide different levels of geographical service.

Domestic and International Carriers ■

First, there is a distinction between domestic and international carriers. **Domestic carriers** provide service within a country. International carriers, called

FIGURE 7.1

Tiers of Carriers

Domestic carriers provide service within a country. *International common carriers (ICCs)* provide service between countries. In the United States, there are 161 geographical service regions called *local access and transport areas (LATAs)*. Some carriers provide service within LATAs, others between LATAs. Few other countries have this type of regional organization.

Geographical Scope	United States	Europe, Many Other Places
International	International Common Carriers (ICCs) Bilateral Negotiation	International Common Carriers (ICCs) Bilateral Negotiation
Between Regions	Interexchange Carriers (IXCs)	Public Telephone and Telegraph Authorities (PTTs) Ministries of Telecommunications
Within Regions	Regions called LATAS Local Exchange Carriers (LECs) Competitive Access Providers (CAPs)	
Customer Premises	Customer Premises Equipment (CPE)	Customer Premises Equipment (CPE)

international common carriers (ICCs), provide transmission service between countries.

Carriers in the United States ■

For historical reasons, the United States is divided into 161 geographical service regions called **local access and transport areas (LATAs).** A small state like Hawaii will only have a single LATA. A large state like California will have more than a dozen LATAs.

Within each LATA, there once was a single monopoly carrier called the **local exchange carrier (LEC).** This was the local telephone company. Now the LEC is being joined by numerous **competitive access providers (CAPs).** These include cable television companies, cellular telephone companies, and companies that once were restricted to providing long-distance service between LATAs.

Carriers that provide service between LATAs are called **interexchange carriers (IXCs).** The main IXCs are AT&T, MCI, and Sprint, but there are numerous smaller IXCs. Interexchange service in the United States has been highly competitive for many years.

Public Telephone and Telegraph Authorities (PTTs) ■

The alphabet soup we have just seen is peculiar to the United States. Traditionally, most other countries had a single national monopoly carrier for domestic service. This was the **public telephone and telegraph authority (PTT).** In Japan, for instance, the PTT is NTT. In England, it is British Telecoms.

Although these PTTs are still very important, almost all countries have introduced competition in domestic markets. The degree of competition varies considerably from country to country.

To manage the country's PTT, most countries established a government **ministry of telecommunications.** Note that the PTT *provides* service, while the ministry *regulates* service. Today the ministry also regulates competing carriers.

Deregulation ■

Previously, carriers had total monopolies over telephone and data carriage. You could not even own the telephone in your home or on your desk at work, much less modems and more sophisticated transmission systems. Today, however, governments have introduced a great deal of competition, although the degree of competition varies considerably from country to country.

The most deregulated tier in Figure 7.1 is the **customer premises.** This is your home or the land on which your company is built. Today in many countries the carrier is not even permitted to own equipment on your premises. Strong deregulation of the customer premises has allowed LANs and internal telephone systems to flourish.

The next most deregulated tier is the market for interregional transmission. Even in countries other than the United States, where there is no formal regional tier, competition has been introduced most strongly in long-distance service.

Next comes international transmission. Although there are international technical rules for interconnecting systems, all service issues, such as rates and the number of competitors, are decided through bilateral (two-party) negotiation between each pair of countries. In most cases, there are multiple carriers serving each pair of countries, but this is not always the case.

Finally, intraregional service usually is the least deregulated. Most countries fear that unrestricted local competition could leave many poor people without even basic telephone service. The United States, through the **Telecommunications Act of 1996,** has greatly expanded local competition. Other countries will be watching this experiment with great interest. *For more on carriers and deregulation, see Module G.*

Local Transmission Service

Figure 7.2 shows the organization's usual first link to the outside world in the United States.

The Local Loop and Switching ■

The **local loop** of the local telephone company provides a point-to-point connection between the customer premises and a **switching office** of the carrier. The local loop normally uses unshielded twisted pair (UTP) wiring, which we discussed in Chapter 4. *For more on the local loop, switching, and trunk lines, see Module G.*

From this switching office, the call may pass through a number of other switches and **trunk lines** connecting these switches. Eventually it reaches the switch serving the called party. That final switch connects the called party via the called party's local loop.

The local loop is owned by an intra-LATA carrier. This may be the local exchange carrier (LEC) or one of the newer competitive access providers (CAPs).

The Point of Presence (POP) ■

Obviously CAP customers need to talk to LEC customers, so there is a need to interconnect the LEC and the CAPs. As Figure 7.2 shows, this interconnection takes place at the **point of presence (POP).** The POP is located at one of the switching offices of the LEC. At the POP, each carrier places a switching unit that connects it to the other carriers.

Figure 7.2 shows that the POP is not limited to linking the LEC to CAPs. It also links the LEC (and CAPs) to interexchange carriers and to international common carriers. The POP is a general mechanism for linking carriers to one another.

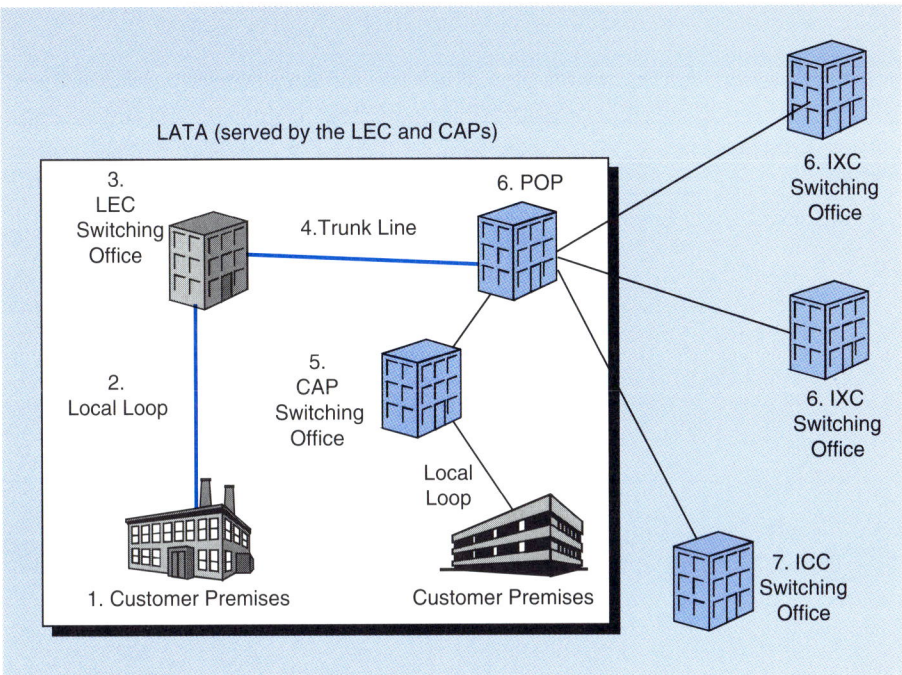

FIGURE 7.2

Transmission Service in the United States

The local exchange carrier (LEC) or a competitive access provider (CAP) provides a local loop running from the customer premises to one of the carrier's switching offices. Thereafter, the call passes through one or more switches and trunk lines connecting the switches until it reaches the called party. Local loop connections and end-to-end circuits come in a variety of speeds, ranging from 30 kbps analog voice-grade lines to gigabit digital lines. The LEC provides a point of presence (POP) at one of its switches. This links its customers with those of CAPs and long-distance carriers. Domestic long-distance carriers are interexchange carriers (IXCs). International calls are handled by international common carriers (ICCs). The POP also provides an access point for switched data networks. Again, this is the situation in the United States.

Circuits

End-to-end connections that carriers provide between two customers are called **circuits.** A single circuit may pass through a number of switches and transmission media connecting the switches to one another and to the customer premises. To the customers who are communicating, however, a circuit appears to be a simple link over which to communicate. Table 7.1 shows that carrier circuits come in a variety of speeds.

TABLE 7.1

Common Circuit Transmission Speeds

Line	Signaling	Speed
Analog Voice-Grade Lines		
Analog		up to ~30 kbps
North American Digital Hierarchy		
64 kbps	DS0	64 kbps
T1	DS1	1.544 Mbps
T3	DS3	44.7 Mbps
Fractional T1		
		128 kbps
		256 kbps
		384 kbps
		768 kbps
CEPT PCM Multiplexing Hierarchy		
E1		2.048 Mbps
E3		34.4 Mbps
Synchronous Optical Network (SONET)		
OC1	STS1	51.84 Mbps
OC3	STS3	156 Mbps
OC12	STS12	622 Mbps
OC24	STS24	1224 Mbps
OC48	STS48	2488 Mbps
Synchronous Digital Hierarchy (SDH)		
STM1		156 Mbps
STM4		622 Mbps
STM8		1224 Mbps
STM16		2488 Mbps

The slowest carrier transmission circuits are analog voice-grade lines designed to carry a single voice conversation. Faster circuits are digital, reducing error rates. Digital circuits run from as low as 64 kbps (sometimes 56 kbps) to as high as more than 1 gigabit per second. Higher-speed circuits can multiplex many telephone calls to the carrier's switching center. Or they can be high-speed video and data pipes.

Analog Voice-Grade Circuits ■

The slowest circuits are the analog voice-grade circuits that you use every day when you call people on the telephone. As we saw in Chapter 3, analog voice-grade circuits are slow and have high error rates. Faster circuits are always digital. This reduces errors dramatically.

Digital 64 kbps Circuits

Table 7.1 shows that the slowest digital circuits run at 64 kbps (often 56 kbps). Module D explains that if you digitize a voice signal, this generates a 64 kbps data stream. So the slowest digital circuits were designed to transmit voice in digital form.

T1 and E1 Circuits

Higher-speed digital circuits range from small multiples of 64 kbps to more than one gigahertz. For telephony, these higher-speed circuits allow the carrier to multiplex several 64 kbps voice conversations on a single circuit. Originally, the slowest multiplexed circuit ran at 1.544 Mbps in North America and some other countries (T1 circuit) or 2.048 Mbps in Europe and some other areas (E1 circuit). These circuits multiplex 24 and 30 voice calls, respectively.

Fractional T1 Circuits

The jump in speed between 64 kbps lines and T1 or E1 lines is extremely large. For many companies, a T1 line is still prohibitively expensive. For companies with lighter needs, many carriers offer **fractional T1** circuits. As the name suggests, these circuits have a fraction of a T1 circuit's capacity. Typical speeds are 128 kbps, 256 kbps, 384 kbps, and 768 kbps.

T3 and E3 Circuits

For companies with very heavy transmission needs, T3 and E3 circuits are attractive. (Carriers normally do not offer T2 and E2 service.) Until recently, only the largest companies needed such lines, and they only needed them on their most dense routes. These lines can multiplex large numbers of telephone calls.

SONET/SDH

For even higher speeds, there is a new series of high-speed digital lines. In the United States, these circuits are called **SONET (synchronous optical network)** circuits. In Europe, they are called **SDH (synchronous digital hierarchy)** circuits. However, although there are naming differences, SONET and SDH circuits are compatible.

High-Speed Circuits for Video and Data

We have talked so far about multiplexing voice calls, but customers also need high-speed circuits for video and data services. Video can operate at 64 kbps, but at such low speeds, the picture does not change smoothly, and voice quality is low. Most video services now run at fractional T1 speeds. This, of course, requires fractional T1 lines for the local loop.

In video, you do not break up the circuit's capacity into multiple telephone calls. In data transmission, too, you often use the whole circuit as a high-speed data pipe, carrying a single data stream to and from the carrier's switching office.

LEASED LINE TRANSMISSION LINKS

When you place an ordinary call from home, you are using the telephone carrier's switched service. You can dial any telephone number in the world, and you will be switched to that number.

Figure 7.3 shows that carriers also provide nonswitched, point-to-point circuits called **leased lines.** Leased lines restrict you to those two points, much as a

FIGURE 7.3

Network of Leased Lines

A *leased line* provides a nonswitched, point-to-point connection between two customer sites. Firms with multiple sites often have a mesh of leased lines connecting their sites. They match the speed of the leased line to the traffic on each link. Using site switches, they can connect any telephone or data station on any site to any telephone or data station on any other site. A call from Site A might go to Site B, where a switch would send the call on to Site C. Telephone companies now compete with networks of leased lines through *virtual private networks*. These networks restrict you to calling among a company's sites, much like a network of leased lines. However, the telephone company takes over the work of managing the network.

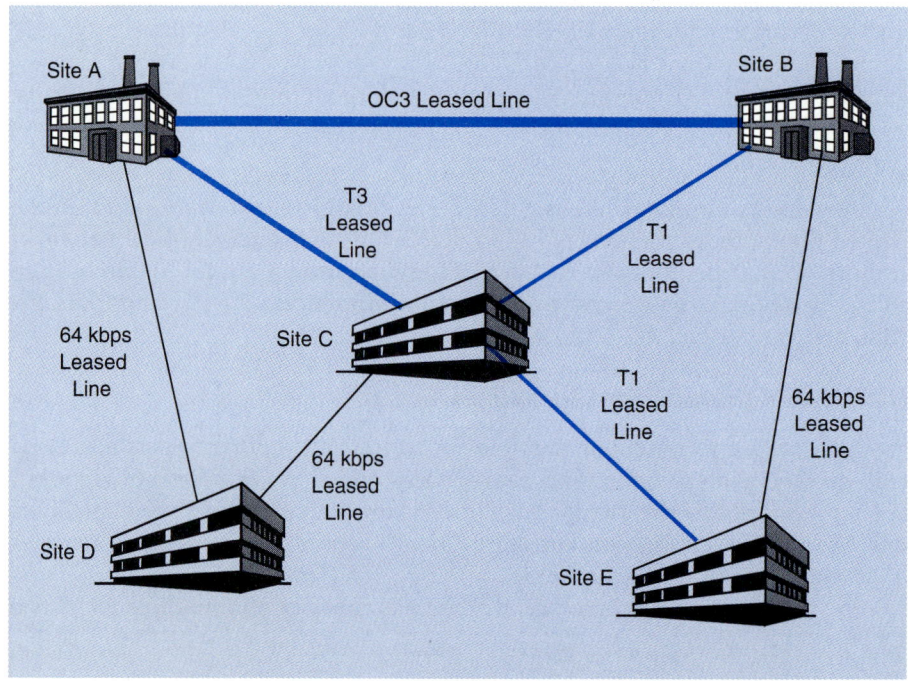

set of train tracks runs between two cities. To the customer, a leased line acts as a simple information pipe. Whatever the company puts into the leased line at one end comes out the other end. The actual circuit may pass through many switches and different types of transmission media, but this is irrelevant to the users.

Advantage and Disadvantage

Disadvantage ■

The obvious disadvantage of a leased line is its *inflexibility*. You can only talk between the line's two points. In contrast, with a switched service, like the telephone system, you can dial up any station on the network.

Advantage ■

Balancing this inflexibility is lower cost. If a company places many long-distance dial-up calls between a pair of sites, this will be very expensive. Sending these calls over a leased line will cost less. Just as railroad lines are less flexible than highways, railroad lines are also much more efficient for high-volume delivery.

Actually, leased lines and dial-up calls usually go over the same set of switches and trunk lines. Carriers do achieve certain administrative economies of scale, allowing them to charge lower rates. Generally, however, carriers charge less per minute for leased lines for competitive reasons. In reality, leased line service exists as a way for carriers to give discounts to high-volume customers. Leased line pricing focuses more on competition than on costs.

In carrier services in general, prices often have relatively little to do with real costs. As a result, cost comparisons between various carrier services are difficult because carriers reprice their services constantly in response to changing market conditions.

Networks of Leased Lines

Figure 7.3 shows that most companies have multiple sites. As a result, they often obtain leased lines between most or all pairs of sites. This allows them to handle most of their internal voice, video, and data traffic over leased lines.

Leased Lines Between Sites ■

Figure 7.3 also shows that companies look at the traffic between each pair of sites and then select a leased line of appropriate capacity. Note that leased lines are sold according to the line designations in Table 7.1.

Switching ■

Many companies also have switches at each site, which can route calls and data traffic through the mesh network. For instance, suppose a phone at Site A places

> ## Connecting Bridges and Routers to Digital Leased Lines
>
> Figure 7.4 shows that a router has two major functions: switching and input/output (I/O).
>
> The I/O function is handled by the router's **CSU/DSU**—its **channel service unit/data service unit.** We saw in Chapter 4 that sending a digital source signal over a digital leased line requires a DSU. The DSU formats the data in the way required by the transmission line. On a T1 line, for instance, the DSU sends outgoing signals in the DS1 signaling format. When it receives an incoming DS1 signal, in turn, it converts it into the native format of the router or bridge.
>
> Each type of leased line needs a different DSU. Before you buy a bridge or router, you need to know what lines you will use, so that you can buy a bridge or router with the correct CSU/DSUs.
>
> The CSU component, in turn, is a safety feature. It prevents a company from using voltage levels and other signal characteristics that could harm the carrier's transmission system.
>
> Fortunately, most bridges and routers are modular. Their CSU/DSUs are printed circuit expansion boards. These fit into slots in the bridge or router. A firm usually buys general-purpose bridges and routers. It then buys appropriate

a call to a phone at Site C. The switch at Site A would send the call to Site B. The switch at Site B would then pass the call on to Site C. Adding switches to a mesh of leased lines allows companies to build **private telephone networks.** *For more on site switches and private telephone networks, see Module G.*

For data traffic, the leased lines frequently are connected by bridges or routers. *For more information on bridges and routers in networks of leased lines,* see the box above, "Connecting Switches and Routers to Digital Leased Lines."

Management Services ■

The carrier's leased line service actually goes beyond bare transmission. It also includes the maintenance of the line. The customer merely plugs the line into a router. The carrier does the rest. Leased lines offer true *plug-and-play* service. The customer then has only the task of managing its switches.

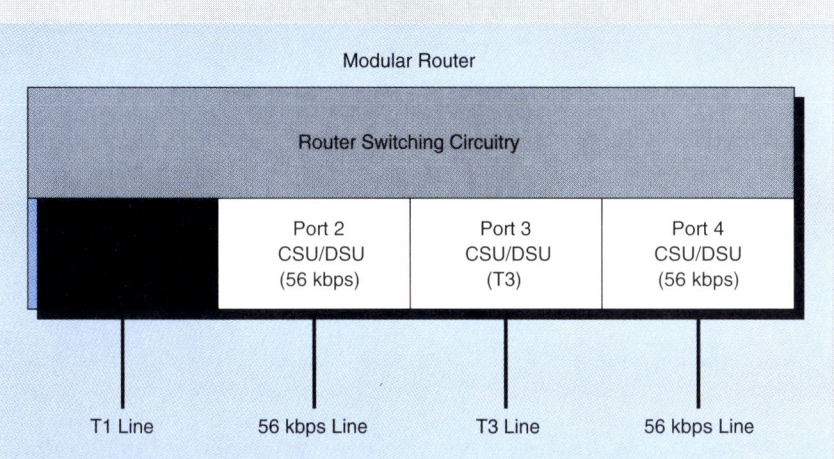

FIGURE 7.4
Linking Routers to Transmission Lines

A CSU/DSU links the router (or bridge) to each transmission line. If you change a transmission line, you must change the CSU/DSU. There is no need to change the router. Many routers are modular, meaning that the CSU/DSUs are expansion boards inserted into the chassis.

CSU/DSU boards for the transmission lines to be used. Router and bridge selection, as a result, depends at least in part on the vendor offering a wide range of CSU/DSUs

Virtual Private Networks

While networks of leased lines can save you money, they require you to purchase the switch for each site and to manage your network. As Chapter 9 discusses, managing a large network is a major chore. It is very different from simple dial-up service, in which the telephone company does everything for you.

Many telephone companies are now offering a service to compete with leased line networking called **virtual private network** service. Essentially, virtual private network service mimics the leased line network shown in Figure 7.3. It allows you to call any other telephone within the company, but you cannot place outside calls through it. The telephone company then handles all maintenance. While virtual private network service costs more than the leas-

Switched Data Network Technologies

All switched data networks route messages from one station to another. However, different switched data networks use different technologies for delivering these messages across a series of switches.

Circuit versus Packet Switching

Figure 7.5 shows a key distinction between circuit switching and packet switching.

Circuit Switching ■

In **circuit switching,** an end-to-end **circuit** (path) between the two stations is established at the start of a connection. In addition, a set amount of channel capacity is dedicated to the conversation.

Telephony uses circuit switching. When you dial a number, the telephone system establishes a connection to the other party. It then dedicates a voice circuit to the call. This circuit is yours for the duration of the call. If you do not say anything, the channel goes unused.

In voice conversations, there are rarely long silences. So telephone circuits are used about 50 percent of the time in voice calls (only one side usually talks at a time). Circuit switching is ideal for telephone calls.

Data transmissions, in contrast, tend to be *bursty*. There typically are long silences punctuated by bursts of traffic. For such bursty traffic, circuit switching's dedicated channel capacity is wasteful of channel capacity. That translates into high cost per bit sent.

Packet Switching ■

In **packet switching,** the data to be sent are first divided into small synchronous groups called **packets.** Some systems have maximum packet sizes of only 53 bytes. Others have maximum packet sizes of 1,000 to 10,000 bytes.

As shown in Figure 7.5, the packets from many different stations are multiplexed (mixed) onto the packet switched network's trunk lines between switches. This allows each transmission to use only the transmission capacity that it actually needs, greatly lowering costs.

Multiplexing's cost advantage is so large that almost all switched net-

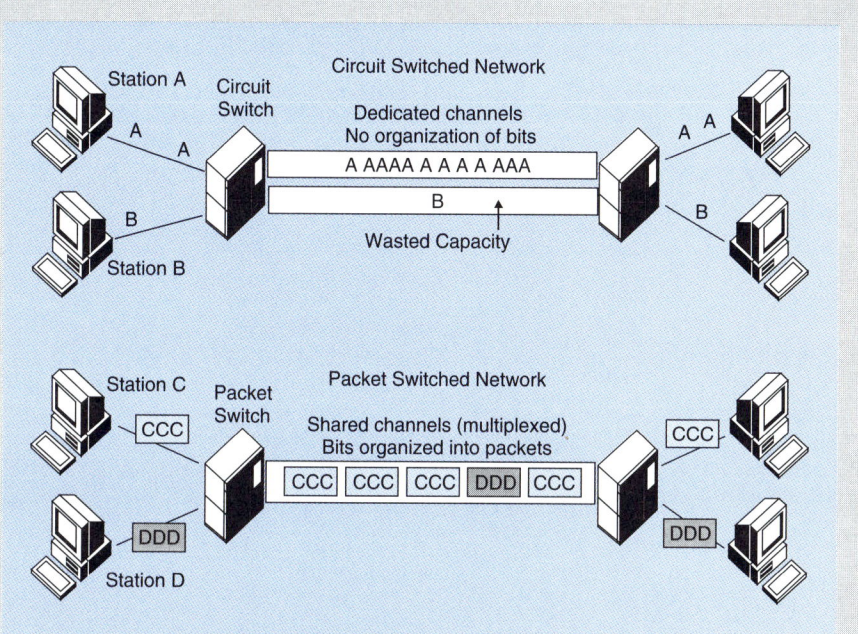

FIGURE 7.5

Circuit Switching and Packet Switching

In circuit switching, a *circuit* is established between the two stations at the beginning of the call. Channel capacity is dedicated to the circuit. If there are no messages, the channel capacity is wasted. In *packet switching*, messages are broken into small protocol data units (*packets*). The packets from several sources are multiplexed (mixed) over the same channels. Packet switching is good for data traffic, which tends to come in small bursts. Almost all switched data networks use packet switching.

works that you will encounter will be packet switched networks. This includes X.25, Frame Relay, SMDS, and ATM. ISDN's B channels, however, can be configured either as circuit switched or packet switched connections by the ISDN vendor. This is reasonable because ISDN can be used for voice or data.

Reliability

One design issue in packet switched networks is whether the network itself should do extensive error checking. As Figure 7.6 shows, networks are either reliable or unreliable.

continued

Reliable Packet Switched Networks ■

In **reliable** packet switched networks, each switch checks for errors after receiving the packet. If a switch does find an error, it asks the switch before it to retransmit the packet. So reliable networks lose very few packets and deliver very few damaged packets. The X.25 standard, which was created in the 1970s when transmission lines were unreliable, provides reliable delivery.

FIGURE 7.6
Reliable and Unreliable Packet Switched Networks

Reliable packet switched networks check for errors after each hop between switches. This produces nearly error-free delivery, but it places a heavy processing burden on switches. All newer packet switched networks are *unreliable*, leaving error checking to the two stations. In this way, error checking only has to be done once instead of many times. This greatly reduces the load on the switches.

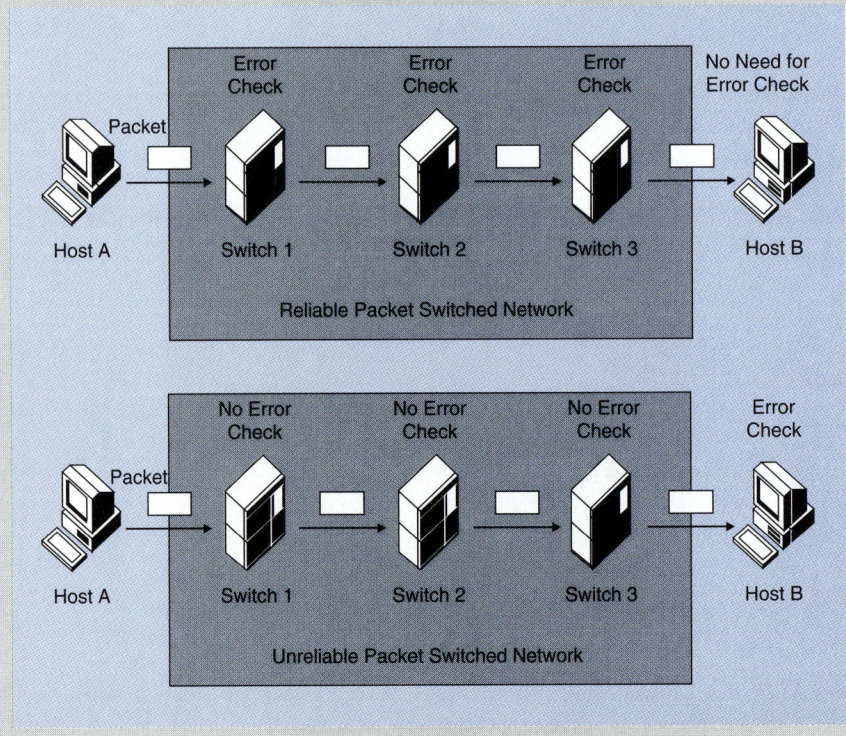

Unreliable Packet Switched Networks ■

Most newer packet switched networks are **unreliable,** including Frame Relay and ATM. This means that switches do not check packets for errors and ask for redelivery. Actually, they do check packet *headers* for errors because a bad header can lead to problems in the network. However, switches do not check the whole packet for errors. If they find a header error, furthermore, they merely *discard* the damaged packet.

There are two reasons why newer packet switched technologies are unreliable. The first is that transmission lines today are quite reliable. This means that there are few errors in the first place, so switch-by-switch checking is overkill.

The second reason why newer packet switched networks are unreliable is that not checking for errors greatly reduces the processing load on switches. Switches are computers, and it takes a great deal of code and execution time to check for errors and to handle retransmissions. The benefits are so great that slow X.25 switches can be upgraded to Frame Relay service using a software upgrade alone. Using Frame Relay, these switches can transmit information at much higher Frame Relay speeds.

Connectionless and Connection-Oriented Packet Switched Networks

Figure 7.7 shows another distinction among packet switched networks. Some are connectionless, while others are connection oriented.

Connectionless Service ■

In **connectionless service,** when the first packet is sent, the first switch (Switch A) must decide what to do with it, as shown in Figure 7.7. Based on network conditions, Switch A decides to send the packet to Switch B.

When the second packet arrives, Switch A must again decide how to route it. This time, it might send the second packet to Switch C.

In reality, however, the second packet is likely to arrive only microseconds later than the first. Since network conditions are unlikely to have changed, the second packet will almost certainly go to the same switch as the first. In fact, most streams of packets travel over exactly the same route.

Under these conditions, having each switch make a full routing decision for each packet is very wasteful of processing on the switches. Routing is a complex matter that consumes many CPU cycles on the switch.

continued

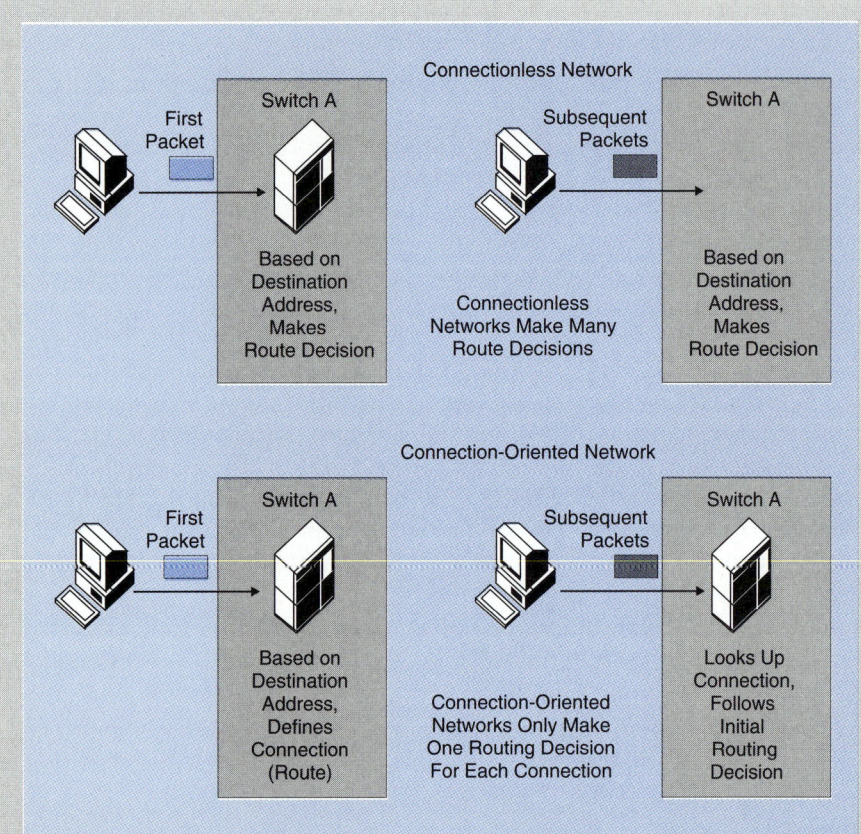

FIGURE 7.7
Connectionless and Connection-Oriented Packet Switched Networks

In *connectionless* packet switched networks, routing decisions are done for each packet separately. In *connection-oriented* packet switched networks, in contrast, a connection (route) is established before data are transmitted. All packets follow this route, which is called a *virtual circuit*. Connection-oriented service places a low processing load on switches.

Connection-Oriented Service ∎

Connection-oriented service takes a different approach. The first packet actually asks the switch to establish a connection or route between two stations. This places a considerable burden on the switch.

However, once a route is established, following packets are trivial to handle. As Figure 7.7 shows, the switch merely identifies the packet, looks up the decided route in a table, and passes the packet on to the next switch along the route. This table lookup requires only a fraction of the CPU cycles of a full routing decision for each packet. So connection-oriented service is much cheaper than connectionless service. Even the earliest packet switched network service, X.25, was connection oriented. So are Frame Relay and ATM.

Figure 7.8 shows an ATM packet. Note that there is no destination address in the packet. Instead there is a connection ID, which consists of two parts—the virtual path identifier and the virtual channel identifier. Therefore, an ATM packet merely tells switches its connection number for lookup in their tables.

Virtual Circuits ■

In a sense, a connection is like a circuit. All packets flow over the same set of switches to reach the other end. However, in real circuit switching, the channel capacity is dedicated. In connection-oriented service, there is no dedicated

FIGURE 7.8
ATM Packet (Cell)

An ATM packet does not have a destination address. Rather, its two fields, the virtual path identifier and virtual channel identifier, tell switches to which connection the packet belongs.

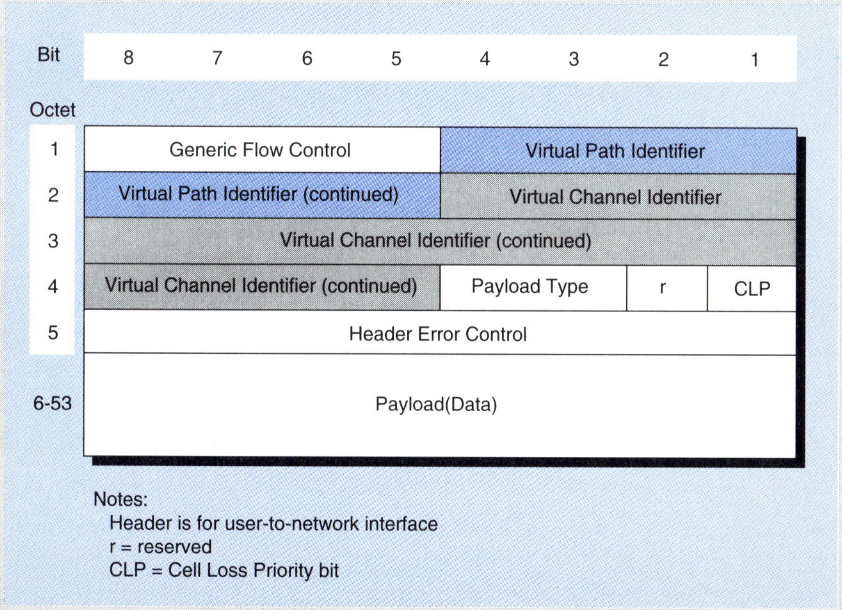

Bit / Octet	8	7	6	5	4	3	2	1
1	Generic Flow Control				Virtual Path Identifier			
2	Virtual Path Identifier (continued)				Virtual Channel Identifier			
3	Virtual Channel Identifier (continued)							
4	Virtual Channel Identifier (continued)				Payload Type			r	CLP
5	Header Error Control							
6-53	Payload(Data)							

Notes:
Header is for user-to-network interface
r = reserved
CLP = Cell Loss Priority bit

continued

channel capacity. Thus, we call connection-oriented routes **virtual circuits** instead of real circuits.

There are two types of virtual circuits. **Permanent virtual circuits (PVCs)** are established weeks or months in advance. For instance, a company with several sites will establish PVCs to each of its sites. As shown in Figure 7.9, however, this does not mean that a company needs multiple leased access lines to the switched data network. Traffic to and from several permanent virtual circuits can be multiplexed onto a single access line between each site and the network. Therefore, if a firm has four sites, it will need six PVCs linking the sites together. However, the firm will only need four access

FIGURE 7.9

Virtual Circuits

Connections between sites are called *virtual circuits*. As in real circuits, all packets travel over the same route between the sites. However, in virtual circuits, channel capacity is not dedicated to the connection. A single access line to the carrier can multiplex several virtual circuits. *Permanent virtual circuits (PVCs)* are established weeks or months in advance. *Switched virtual circuits (SVCs)* are established for each call.

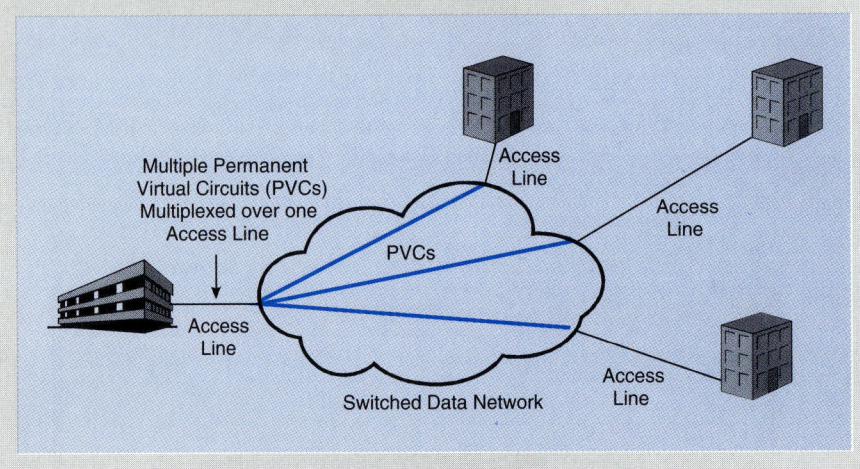

ing of lines, its cost difference is often small compared to the cost of managing the network internally.

Virtual private network service is primarily a pricing scheme. It uses the same telephone company switches and trunk lines. It merely uses software in the switches to restrict calling. (For this reason, these networks are often called software-defined networks.)

lines—one at each site. X.25, Frame Relay, and ATM all use permanent virtual circuits.

Permanent virtual circuits make life easy on the carrier because PVCs only have to be established once. However, users want to be able to establish new connections on demand, whenever they need them. For instance, if you want to send a message to a potential customer, you do not want to have to go to the carrier to establish a PVC. To meet this demand, a number of Frame Relay and ATM carriers now offer **switched virtual circuits (SVCs),** which are set up at the start of an individual call. This gives switched data network service customers the same flexibility that the telephone system has long given to voice callers.

Reducing the Load on the Switch

Optical fiber and other new technologies have brought vast improvements in transmission capacity at reasonable costs. However, switches are computers. They have input/output ports and a processor to do routing and error-handling computations. Although computer costs have been falling, they have not been falling as rapidly as the costs of transmission lines. As a result, the switch tends to be the bottleneck in improving performance and in reducing costs.

As noted earlier, both unreliable service and connection-oriented service greatly reduce the load on the switches. There does not have to be complex error checking and retransmission for each packet at each switch. Nor does there have to be complex routing for each packet in a series.

ATM introduces yet another way to reduce switching load. As Figure 7.8 shows, ATM packets are all 53 bytes long. This consists of 5 bytes of header and 48 bytes of payload (data). In contrast, the 802.3 MAC-layer frame in Chapter 4 was of variable length. X.25 and Frame Relay also have variable-length packets. By making all packets the same size, ATM makes it much easier for switches to handle traffic, for a variety of somewhat obscure technical reasons, including improved network predictability. Constant-length packets are called **cells.** Thus, ATM is called a **cell switched network.**

SWITCHED DATA NETWORKS

Leased lines restrict you to communication between two points. In contrast, the normal public switched telephone network allows any telephone to reach any other simply by dialing the receiver's number. As we saw in

Chapter 1, this any-to-any connectivity is the defining characteristic of networks.

The public switched telephone network is normally used for voice transmission. However, carriers also offer **switched data network** services, in which any computer can send a message to any other computer simply by giving the receiver's address. These networks offer multiplexing to reduce transmission costs.

This section focuses on the services that users receive from switched data networks, including circuit versus packet switching, reliable versus unreliable service, connection-oriented versus connectionless service, and cell switching. *For more information on the actual technologies of switched data networks, see the box "Switched Data Network Technologies."*

Switched Data Networks: Elements and Market Conditions

Figure 7.10 shows the main elements in switched data networks. These are the customer premises, the leased access line to the POP, the switched network's point of presence, the switches in the switched facility, and the trunk lines that link the switches.

FIGURE 7.10

Elements of a Switched Data Network

On the customer premises, the customer needs a router with an appropriate CSU/DSU. The access line to the POP usually is a leased line. The POP connects the user to the switched data network. Within the switched data network, there are many switches connected by trunk lines.

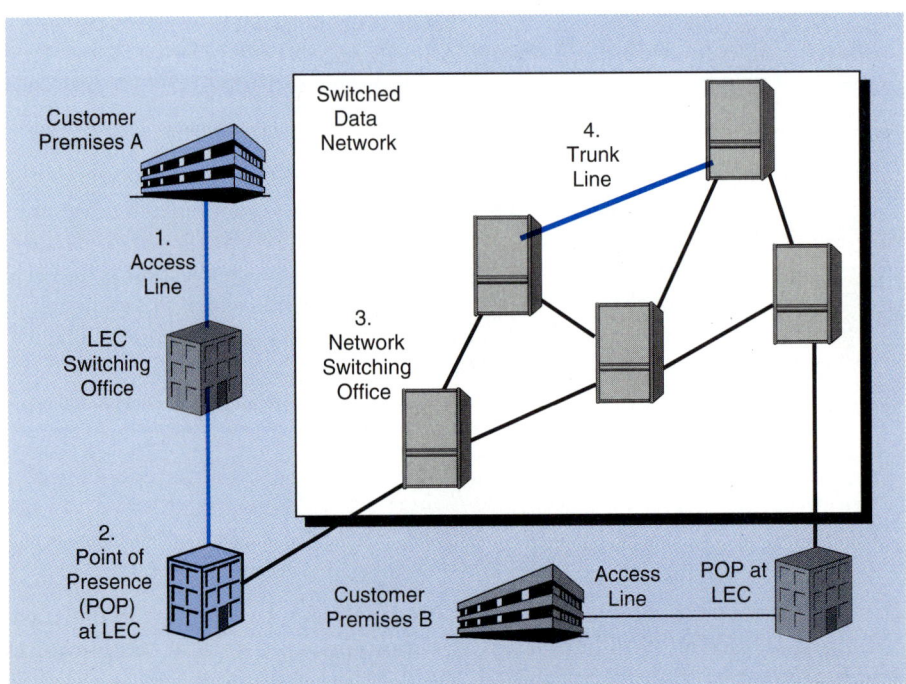

To use a switched network, the company usually has to deal with the LEC or a CAP. The company has to get a leased **access line** from one of these carriers. This is the line running from the customer premises to the POP. Its cost can be significant.

Of course, the leased access line to the POP must be fast enough to handle the data traffic. If you have a switched network operating at 1 Mbps, it would make no sense using a 64 kbps access line. You would only be able to transmit and receive at 64 kbps.

A T1 line would do nicely, however. Operating at 1.544 Mbps, a T1 leased line can easily handle 1 Mbps traffic. *The goal is to get an access line to the POP as fast as the switched data network's speed or a little faster.*

Switched network carriers understand this need to lease access lines to the POP. They typically gear their service speeds to the speeds of the leased lines that will be available to users. This is why many of the switching speeds that we will see in this section will be equal to or slightly slower than the leased line speeds in Table 7.1.

Cost–Flexibility Trade-Offs

We noted earlier that leased lines are cheaper than switched service for high-traffic routes between two sites. Leased line users give up the flexibility of any-to-any communication to get cost savings on these routes.

However, traffic usually is relatively light for many pairs of sites. For such connections, it will be cheaper to use a switched data network.

Even if costs are close, switched data networks are preferable to leased lines. If a company has many sites, it will have to maintain a sizable network management staff to manage its multiple leased lines and diagnose problems. In switched data network service, however, the company merely connects to the carrier. The carrier handles internal maintenance of the switching.

In addition, switched data networks are much easier to change than meshes of leased lines as the network grows and as you add and drop sites. As their names suggest, leased lines require you to sign leases of some duration. Ordering a new leased line, furthermore, can take some time.

Finally, while leased line meshes limit you to your own sites, switched data networks can connect you to other organizations.

In recent years, switched data network carriers have priced their services aggressively. This has made them quite attractive compared to leased lines. In fact, given current market trends, many companies will soon bypass all of the complexities of building meshes of leased lines as shown in Figure 7.3.

COMMON SWITCHED DATA NETWORK STANDARDS

Table 7.2 shows that there are several standards for switched data networks. The earliest networks understandably ran at comparatively low speeds. While

TABLE 7.2

Switched Data Network Services*

Service	Typical Speeds	Connection-Oriented?	Reliable?	Relative Maturity	Relative Cost
X.25	9,600 bps–64 kbps	Yes	Yes	High	Low
ISDN	2 64 kbps B channels; 1 16 kbps D Channel	Depends on Vendor	Depends on Vendor	Moderate	Low
Frame Relay	64 kbps–6 Mbps	Yes	No	Moderate	Moderate
SMDS**	1–4 Mbps	No	No	Moderate	Moderate
ATM	1 Mbps–2 Gbps***	Yes	No	Low	High

Notes:
*All are packet switched networks, except for ISDN, which can be either packet or circuit switched.
**SMDS is the Switched Megabit Digital Service.
***Officially, ATM speeds start at 156 Mbps, but many vendors offer lower speeds.

many companies have outgrown those speeds, some companies find them sufficient for their general needs. Even large companies use slower networks for their lighter needs and in countries where newer switched data networks are not available.

This table uses the terms *connection-oriented* and *reliable*. These concepts are discussed in the box, "Switched Data Network Technologies."

X.25 Data Networks

The first standard for switched data networks was X.25. More than 20 years old, X.25 is extremely mature. It is widely available throughout the world, and it usually works well wherever it is available. It is a proven technology. Its cost, furthermore, is very low compared to other switched data network services.

The bad news is that X.25 is slow. While X.25 can operate at 64 kbps, most X.25 networks limit users to 19.2 kbps or even 9,600 bps. These limits made sense initially because trunk lines were so expensive and because the demand for higher speeds was low. X.25 was fine for its initial target application: terminal–host communication. For LAN interconnection, X.25 is marginal. It may be sufficient to connect a small branch office LAN to a larger site, but to connect larger sites, it is far too slow.

ISDN Service

Today's telephone network provides analog service, which is slow and noisy. As noted in Chapter 3, the ITU-T has long been developing standards for an *all-digital* network that will handle both data and digitized voice equally well. This the *Integrated Services Digital Network* (ISDN).

Narrowband ISDN (NISDN) ■

There will be two forms of ISDN. Today there is narrowband ISDN. As shown in Figure 7.11, **narrowband ISDN (NISDN)** brings a single set of wires to each desk.

FIGURE 7.11

Narrowband ISDN

Narrowband ISDN (NISDN) offers a Basic Rate Interface (BRI) to each desktop. This interface multiplexes three signals. One is a 16 kbps D control channel. The other two are 64 kbps B channels, which can carry voice or data. You can link the two B channels to two stations, say, using one for voice and one for data. The connection from the customer premises to the ISDN carrier follows the Primary Rate Interface (PRI), which operates at 1.544 Mbps or 2.048 Mbps.

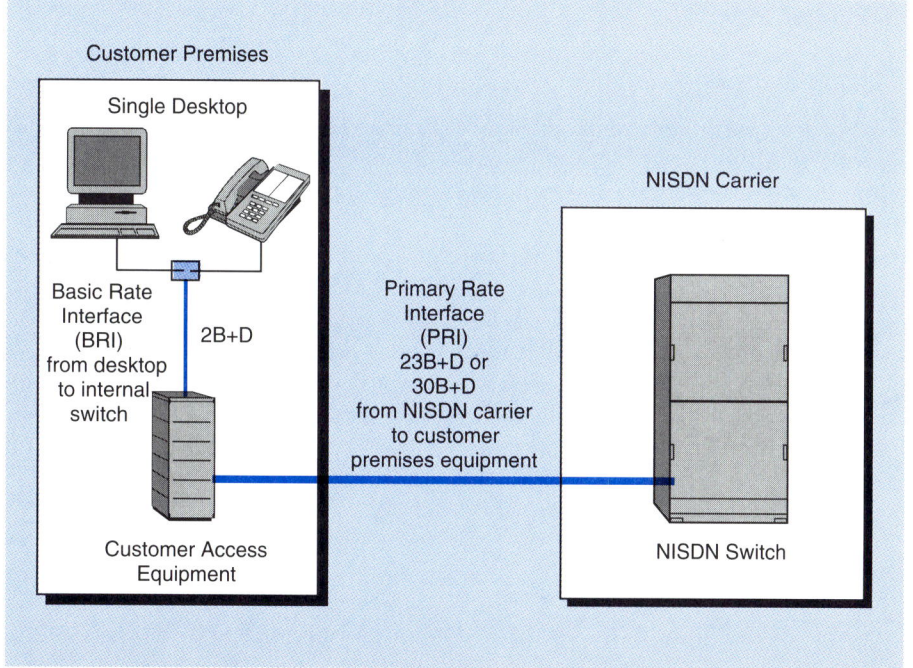

As discussed in Chapter 3, there will be three multiplexed channels on these wires. These three channels are known collectively as the **Basic Rate Interface (BRI).** This is narrowband ISDN's service to the desktop.

One of these three channels will be a 16 kbps **D channel,** which will be used primarily for signaling to control the transmission. The others will be 64 kbps **B channels,** which are for user transmissions.

For data, the 64 kbps speed of the B channel is about twice the speed of the fastest modem, and this is a welcome although modest speed boost for many users. In addition, you may be able to use both B channels to connect you at 128 kbps to a single station. This is a major gain over modem communication. In videoconferencing, for instance, sound will be better and pictures will be much less jerky than if you are merely using a modem over a voice-grade line.

ISDN is also interconnected with the traditional telephone network, so you can call ordinary telephone users as well. Of course, you cannot communicate at 64 kbps to these analog telephones, but you can carry on voice conversations.

The Primary Rate Interface (PRI) ■

The BRI is the service that ISDN brings *to the desktop*. Figure 7.11 also shows the **Primary Rate Interface (PRI).** This is the service that the ISDN brings *to the corporation*. The corporation provides the connection between the PRI and the desktop. The PRI runs at T1 speeds in the United States and at E1 speeds in Europe.

Market Factors ■

For the long term, narrowband ISDN may have limited appeal. At best, it only gives a fourfold speed increase over an analog telephone line. This is still too slow for adequate videoconferencing, good PC networking, and other growing needs.

ISDN originally flourished in Europe. U.S. carriers were slow to add ISDN service, sensing little market demand. It took U.S. carriers many years to upgrade their switches to handle ISDN control signaling. Now, however, ISDN is available even in the United States.

As we will see later, this is leading to a faster switched service, namely *broadband ISDN*. However, ISDN's superior speed compared to basic dial-up service and its relatively low cost should give it a healthy market niche in the short term.

Frame Relay

The ISDN uses an underlying technology called **Frame Relay.** Many vendors now bring Frame Relay service directly to corporations. Frame Relay speeds begin at the high end of X.25 (64 kbps) and extend to about 6 Mbps. Frame Relay is an attractive option for firms that have multiple sites that they now connect via 56/64 kbps and T1 leased lines.

Frame Relay provides switched moderate-speed service at costs that are usually much lower than multisite networks using T1 and E1 leased lines. This combination of reasonable cost and speed of about 1 Mbps puts it in the center of business needs today, and Frame Relay currently is the growing tip of switched data network service. Many firms are now ripping out their meshes of leased lines (see Figure 7.3) and replacing them with Frame Relay service.

Asynchronous Transfer Mode (ATM)

To most people, *ATM* stands for automated teller machine. But we saw in the last chapter that in high-speed data networking, **ATM** stands for **asynchronous transfer mode.** Just as Frame Relay grew out of early ISDN work, ATM grew out of later broadband ISDN technology.

Although Frame Relay is very attractive to corporations today, it probably has upper limits that will be reached soon. Even if it can be extended to speeds of about 45 Mbps, this will not be enough for many firms in the future. For true high-speed WAN service, many companies are looking forward to ATM. ATM has the potential to carry long-distance data traffic at more than 2 Gbps. Common speeds today are 156 Mbps and 622 Mbps, although most vendors also offer lower-speed services.

In addition, throughout the range from a few megabits per second to a few gigabits per second, ATM is perfectly scalable. There is no need to change protocols and to retrain the staff or to deal with multiple protocols.

For the time being, thanks to high cost, ATM is seeing limited use in long-distance communication, although it is beginning to see wider use in local internets (see Chapter 6). However, as demand for high-speed carrier services increases, and as ATM costs fall, many analysts expect ATM to grow rapidly and perhaps to become the premiere long-distance data service in the future.

Frame Relay and ATM Pricing

If you wish to buy Frame Relay or ATM service, Figure 7.12 shows that there are several costs involved. Most obviously, your CSU/DSU needs to be suitable for your leased line to the point of presence. If it is not, you will need to upgrade your CSU/DSU.

Hook-Up Charges ∎

Normally, switched data networks assess a one-time *hook-up charge* to connect you to their service. This handles the administrative work of adding you to their network.

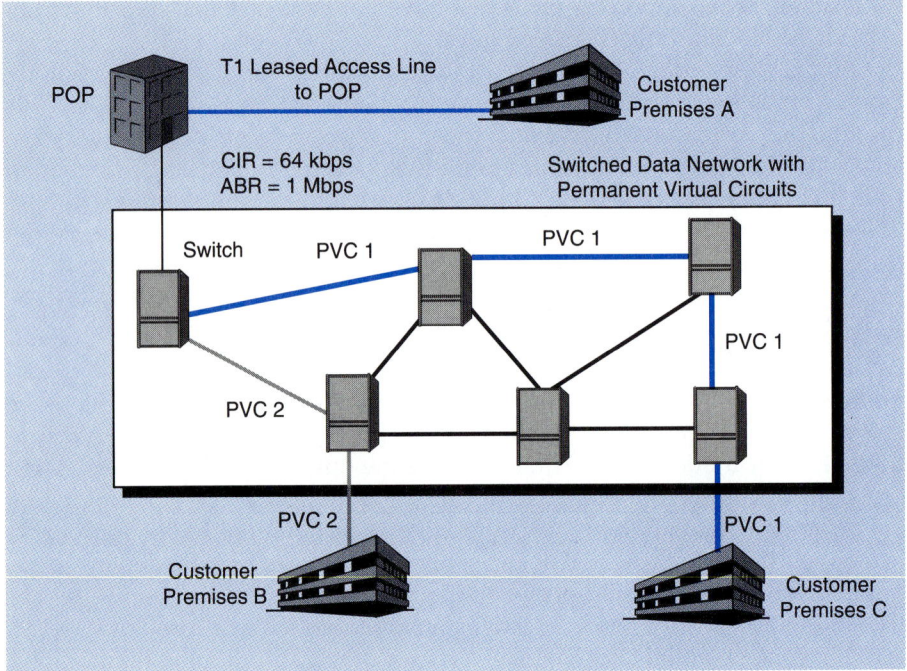

FIGURE 7.12

Switched Data Network Pricing

When you first sign up, there are initial hook-up charges from the switched data network carrier and also from the local exchange carrier that provides the leased line from the customer premises to the POP. Thereafter, there are ongoing charges for the switched data network. These depend on your committed information rate (CIR), which you are almost certain to get. Charges also depend on your available bit rate (ABR), which allows you to transmit fast bursts that will not get through during periods of high congestion. You also need the correct CSU/DSU on your router.

In addition, recall that you need a leased line from your premises to the point of presence (POP). Typically, you lease this line from a local exchange carrier rather than from the switched data network. There usually will be a separate hook-up charge for this leased line.

Leased Line to the POP

After the service begins, you have to pay the cost of the leased line to the POP each month. This line can be quite expensive, and it must be factored carefully into your cost analysis.

In addition, some switched data networks do not have POPs in smaller cities. If you have a site in a rural area, you need to find the distances to the POPs of various switched data network vendors. This will allow you to compare the costs of leased lines for reaching them.

Committed Information Rate (CIR) ■

The price of ongoing service is based on the speed of the leased line. In some cases, the only issue is the speed of the leased access line. In other cases, there are two speeds to consider.

When there are two speeds in the pricing, one is the **committed information rate (CIR).** As long as you stay within the CIR, your frames or packets are very likely to get through.

At the same time, Frame Relay networks with dual-speed pricing allow you to transmit at a rate higher than the CIR. If you do so, however, you must set a bit in the header to indicate that the frame or packet is disposable. If there is congestion on the network, the switches will throw these disposable frames away first.

In two-speed pricing systems, you can transmit faster than the CIR at the **available bit rate (ABR).** This is the speed of the access line to the first switch. For instance, if you use a T1 access line, you can have an available bit rate of 1.544 Mbps but a CIR of only 64 kbps.

Of course, there is nothing magical about available bit rates. The cost of service will depend both on the CIR and on the ABR. Also, if you transmit above the CIR, you lose the "commitment." If there is congestion on the network, your frames beyond the CIR will be discarded first.

Naturally, the lower the CIR, the lower will be the cost of service. Some vendors even offer CIRs of zero for companies that can live with brief outages during times of congestion.

Typically, switched data networks charge flat monthly fees, given a CIR and an ABR. Sometimes these flat monthly fees depend on distance, but even that is far from universal.

Permanent Virtual Circuits (PVCs) ■

The telephone system is a fully switched service. This means that you can dial any number you wish. There is no problem calling a telephone number you have never used before.

Requirements are different in X.25, Frame Relay, and ATM. As discussed in the box, "Switched Data Network Technology," these services require you to specify *permanent virtual circuits (PVCs)*. In a sense, these are like leased lines. They only connect two points, as shown in Figure 7.9. These PVCs are not truly permanent. You can add and drop them whenever you choose, although there normally are charges to do adds and drops.

As Figure 7.12 shows, however, having multiple PVCs does not mean having to use multiple access lines at each site. Traffic for multiple PVCs can be multiplexed onto a single access line for each site. If you are at a site and need to connect to ten other sites, you will need ten PVCs, but you will still need only one access line. Of course, that one access line has to have the capacity to handle the traffic of multiple PVCs.

There is a separate monthly charge for each PVC; however, the cost of an individual permanent virtual circuit tends to be small compared to the cost component based on the speed of the access line. Even if you have several PVCs, the total cost usually is small compared to the price of the access line. Essentially, the price of the access line is based on how many bits you will send into the network and receive from the network. This is what drives the network's cost. Individual PVCs merely add a little to switching complexity.

Some Frame Relay and ATM vendors also offer switched virtual circuits (SVCs). The box "Switched Data Network Technology" also discusses SVCs. Switched virtual circuits are established when you make the call. Although they do not completely make Frame Relay and ATM services into fully switched services, they come close.

CONCLUSION

In this chapter we have looked at the main elements in building a corporate enterprise internet: carriers, CSU/DSUs, leased lines, and switched data networks.

For transmission beyond the customer premises, you must deal with a transmission carrier. The carrier situation used to be very simple. Transmission service was controlled by monopoly or semimonopoly carriers. Now, however, deregulation has resulted in a flowering of new carriers, and this new competition has brought lower prices and a wider variety of services. Of course, this also means more complex service and pricing options.

We have spent most of our time on switched data networks. First, most organizations will use them more heavily than leased lines. Second, switched data network technology and market offerings are changing much faster than leased lines.

X.25 provides low-speed, low-cost service almost anywhere in the world. ISDN provides a modest upgrade to phone service at a modest increase in price. Frame Relay provides quite a bit of bandwidth at an attractive cost. ATM offers very high speeds and scalability.

Leased lines will always be around. Even with switched networks, you need them for the access lines that link the customer premises to the POP serving the switched network. More generally, on routes with very high density, leased line costs are still attractive. However, aggressive switched data network pricing will reduce the demand for leased lines between distant sites.

Pricing for switched data networks is rather complex. You must consider the cost of CSU/DSUs and leased access lines to the POP. In addition, there will be hook-up charges and monthly charges that depend on your committed information rate, the available bit rate, the number of permanent virtual circuits and, in some cases, distance.

This chapter finishes our discussion of the transmission layer in our Basic Communication Model in Figure 1.1. The next chapter looks at the computer and application layers involved in PC networking. As we noted in Chapter 2, the networked PC has become the dominant computer platform in corporations today. No introduction to data communications would be complete without a discussion of PC networking.

CORE REVIEW QUESTIONS

1. What is a carrier? What are the geographical tiers of carrier service? What is deregulation?
2. What is a circuit? What is the speed of a T1 circuit? List at least two common speeds of fractional T1 circuits.
3. What is a leased line? Why is a leased line limiting? Why do companies get leased lines despite this limitation? How do companies use them to build private networks?
4. (From the box "Switched Data Network Technology") Distinguish between circuit switching and packet switching.
5. What is X.25's greatest strength? What is its greatest weakness?
6. Describe the service that narrowband ISDN brings to each desk. Why would anyone need two B lines? What two entities does the BRI connect? What two entities does the PRI connect?
7. What is the speed range of Frame Relay service? Why does the answer to the first part of this question help explain Frame Relay's popularity?
8. What is the speed range of ATM services? What is scalability, and why is it good?
9. List the items that you must consider in Frame Relay and ATM pricing. The speed portion of the pricing is based on the speed of what line? Distinguish between CIR and ABR. Distinguish between fully switched service and PVCs. If your site needs seven PVCs to link you to other sites, how many leased lines will you need from your site to the switched data network?

DETAILED REVIEW QUESTIONS

1. For the United States, distinguish between LEC and LATA. Distinguish between LEC and CAP. Distinguish between LECs and IXCs. Distinguish between IXC and ICC.
2. In the United States, what is the function of a POP? Why is it crucial for competition?
3. Distinguish between PTTs and ministries of telecommunications. In what country mentioned in the text would you *not* find these entities?

4. Be able to give speeds for the following transmission lines: analog voice-grade, 64 kbps, T1, E1, T3, E3, OC3, STM1, OC12, OC48, and fractional T1. Are SONET and SDH technically compatible?
5. (From the box "Connecting Bridges and Routers to Digital Leased Lines") I have a router. What does a CSU/DSU do? How do I know which CSU/DSUs to use in a specific router?
6. (From the box "Switched Data Network Technology") Distinguish between connection-oriented and connectionless service. Why is the use of the former increasing?
7. (From the box "Switched Data Network Technology") Distinguish between reliable and unreliable service. Why is the use of the latter increasing?
8. (From the box "Switched Data Network Technology") Give a single reason why most newer switched network services are both connection oriented and unreliable.
9. (From the box "Switched Data Network Technology") Why were most early switched network services reliable? Why were they connection oriented? In terms of reliability and connection orientation, what are most switched networks today? How is this different from early switched networks? What has caused the change?
10. (From the box "Switched Data Network Technology") Why does the ATM header not specify the destination address?
11. (From the box "Switched Data Network Technology") What is cell switching? Why is it used at higher speeds?
12. What is the source of Frame Relay technology? What is the origin of ATM technology?

THOUGHT QUESTIONS

1. Why do you need to use a carrier at all? Why can't you just run your own transmission lines between sites?
2. Suppose you place a call from your home telephone to a telephone number in Japan. Describe how the call would travel in terms of the communications carriers that would handle the call.
3. Which do you think is changing more rapidly—leased line technology or switched technology? Why?
4. You are thinking of using a Frame Relay service to link four sites. You want every site to be able to communicate with every other site. You will have a CIR of 64 kbps. How many lines must you lease? What should be the speed of your leased line to the POP? Identify the main cost elements in such a network.

5. NuPools is thinking of opening a second retail center. The second center will have its own Ethernet 10Base-T LAN. NuPools wants to interconnect the two centers. Select a connection option and justify it.

PROJECT

1. Cost out the connections among a group of sites, comparing a mesh of leased lines and switches to Frame Relay or ATM service.

For online exercises, please visit this book's website at: http://www.prenhall.com/panko

CHAPTER 8

PC Networking

INTRODUCTION

Today most PCs in large organizations are already networked. By the end of this decade, it should be a challenge to find a true stand-alone PC. Radio links will even bring mobile computers into PC networks.

The networked PC has become the dominant computer platform in most organizations. In the future, it is likely to consolidate that position further, as networking becomes even more widespread. As discussed in Chapter 2, terminal–host sales are falling. Workstations will grow faster than PCs, but their unit base is too small to change the picture.

This chapter focuses on the basics of PC networking and the services that users receive. Module I looks at PC networking from the viewpoint of the network administrator.

ELEMENTS IN A PC NETWORK

Chapter 1 introduced the basic elements of a PC network, which are shown in Figure 8.1.

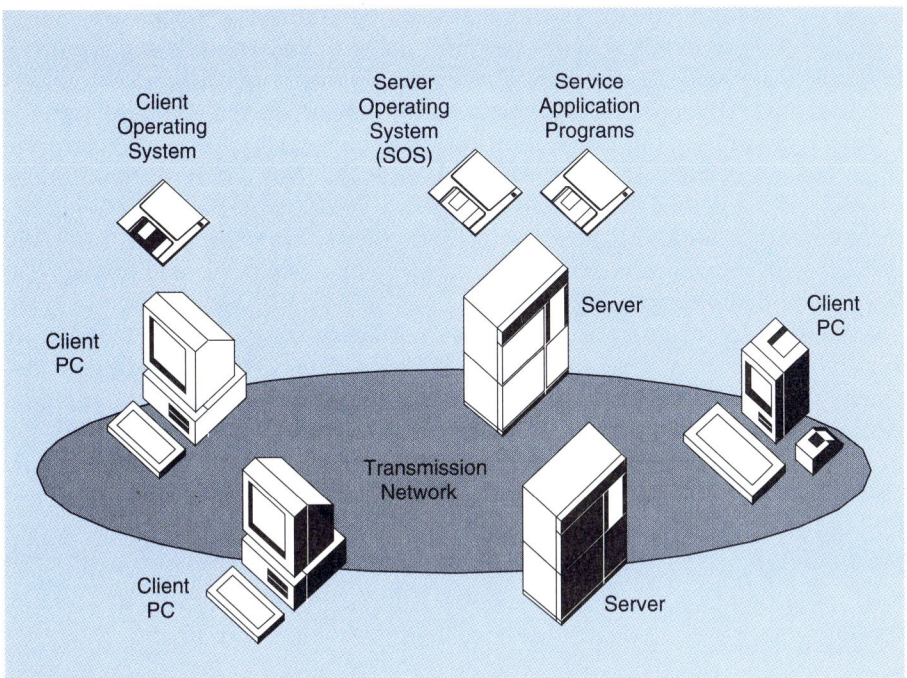

FIGURE 8.1

Elements of a PC Network

Client PCs sit on the desks of ordinary users. They are ordinary PCs with a *network interface card (NIC)* and sometimes with added software. *Servers*, which usually are PCs also, provide services to the client PCs. There are usually multiple servers, specialized by service. This *specialization* allows a firm to select the best machine for each service, optimize the machine for that service, and prevent a service program's crashing from taking down multiple services. While most servers are personal computers, they can also be workstation servers or even mainframes. Client PCs need *network-capable operating systems* or *client shell* software. Servers also need suitable *server operating systems (SOSs)*, plus *application software* for various services. For transmission, some small PC networks use a single LAN. Other PC networks use complex enterprise internets and have thousands of client PCs and servers at dozens of corporate sites.

We will review them briefly now. Later in this chapter and in Module I, we will look at some of them in detail.

Client PCs

The PCs that sit on the desks of ordinary managers, professionals, secretaries, and shipping clerks are called **client PCs**. They get this name because they are the clients (customers) of the network's various services. Most client PCs start their lives as stand-alone PCs. They then need a *network interface card (NIC)*, as discussed in Chapter 4, and perhaps some software, as discussed later.

In the old terminal–host world, the desktop terminals were dumb machines with just enough intelligence to communicate with the central host computers. Instead PC networking distributes processing power out to user desktops.

With the CPU connected directly to the screen via a high-speed cable, a client PC can present high-definition graphical images on the screen and change them instantaneously. In other words, a client PC user will have rich graphical user interface (GUI). Terminals, in contrast, can only portray what the mainframe or minicomputer transmits to them. Since this transmission normally uses low-speed communication lines, terminals have to be limited to text or at best text and simple graphics.

The one downside of using PCs as the chief desktop machine is the wide variety that exists among PCs in performance. Some PCs are high-end machines with ample processing power and RAM. Others are one to three generations behind the company's best PCs. Often services have to be built for the processing power of the lowest common denominator in machine power. At the very least, PC networks cannot limit themselves to serving only the fastest PCs in the firm.

Servers

In PC networks, servers take the place of host computers, as Figure 8.1 indicates.

Specialization: Selection and Optimization ■

In the old terminal–host world, there was a single large centralized host serving a wide variety of applications simultaneously. Having a single machine made data communications relatively simple. In many cases, all that was needed was point-to-point links between the host computer and its terminals.

Unfortunately, running many applications makes it *difficult to optimize* host computers. For database, you need a very fast processor. This requires the firm to buy a state-of-the-art mainframe or powerful minicomputer. Of course, machines that push the state of the art are expensive. Other applications might be able to get by with a much less expensive machine. Still other applications might need optimization along different dimensions, such as input and output (I/O).

In addition, hosts run dozens or sometimes hundreds of applications simultaneously, jumping back and forth among them. You know that as the number of tasks that you do increases, switching back and forth among them begins to take up most of your time. Hosts also expend many machine cycles when they *multitask* many simultaneous programs. In addition, their operating systems have to be very complex to do such ultramultitasking. This makes them expensive to purchase and requires a large systems programming staff.

PC networking, in contrast, uses server **specialization**. It does not use a single large host computer. Instead, as Figure 8.1 indicates, it uses multiple servers. Often servers are optimized for specific applications. A database server, for instance, will probably be a high-end machine optimized for processing power and hard disk transfer rate. A communication server that provides links to the outside world, in contrast, would probably be a low-end machine optimized for input and output. In addition, most servers will only have to multiplex a few tasks, which does not create a high processing load or the need for a complex operating system. Given today's technology economics, it usually makes sense to use multiple specialized servers instead of a single large host.

This does not mean that servers are incapable of handling multiple applications. A small business might use a single server to handle all services. We will see later that NuPools chose this alternative. Another small firm might put several services on a main server and put others on one or two specialized servers. Thus, a server may support a single application, a few applications, or many applications. How to distribute applications across servers is largely a matter of economics.

Server Technology ■

We have said earlier that client PCs are personal computers. Most, in fact, began their lives as stand-alone PCs. Servers, too, are normally PCs, although this is not always the case.

TYPICAL SERVERS. The typical server is nothing more than a medium- to high-power personal computer with added hardware and software. The hardware is the network interface card (NIC) discussed in Chapter 4. We will discuss software later.

HIGH-END SERVERS. Some PC servers are designed from the ground up to be high-end servers. They are likely to have several high-capacity hard disk drives, multiple power supplies, tape backup units, and integrated uninterruptible power supplies. Some even have multiple microprocessors for parallel processing. *For more on high-end PC server technology, see Module I.*

WORKSTATION SERVERS. As Chapter 2 discussed, workstations are also divided between clients and servers. Workstation servers are extremely powerful machines. Many have several times the speed of even the best high-end PC servers. As discussed later, PC networking is really an application-layer concern. As a result, there is no need for the hardware of the server to be the same as the hardware of the client PC. Workstation servers are often used where server performance is critical, most commonly in database service.

MAINFRAMES. The ability of a server to be something other than a PC extends to mainframes. Business mainframes, such as those from IBM, have always been optimized for database processing. As a result, they have state-of-the-art disk drive technology. Many types of PC network servers have the same requirements. Initially, IBM and other mainframe vendors spurned the use of mainframes as PC networking servers. Now they have moved to embrace this new role for mainframes.

Server Application Software

A customer does not buy products. Instead, the customer buys a stream of benefits stemming from the ultimate services that PC networks provide to the client PCs. As in stand-alone computing, PC networking services are embodied in application software. In the case of PC networking, this application software resides on the server, although it uses the intelligence of the client PCs. Later in this chapter, we will see a number of key PC networking services that are supported by server application software. These include *file service, print service, client/server processing, communication service, mail service,* and *network management service.*

Operating Systems

To use and provide these services, the client PC and the server both need operating systems capable of handling PC networking.

Client PC Operating System ■

The most popular operating system on client PCs is Microsoft Windows. Beginning with Windows 95, this operating system is **network capable**. This means that it is able to work with several types of servers without any additional software. At the time of this writing, a new network-capable Macintosh operating system is about to be released.

Older operating systems, such as MS-DOS, older versions of the Macintosh operating system, and older versions of Microsoft Windows, were designed in a prenetworking era. Fortunately, it is easy to add software to make them network capable. The main program in this software is usually called a **client shell** because it effectively fits on top of the client PC's basic operating system, adding networking capabilities. It is also called a *redirector*, for reasons discussed later.

Server Operating Systems (SOSs) ■

Every server needs an operating system. These **server operating systems (SOSs)**[1] have to be reliable to avoid server crashes that can leave dozens of people unable to work. They also need the sophistication to run complex applications.

Novell NetWare has dominated the server operating systems market for *file servers* that we will see later. NetWare also has a strong directory service, which is also discussed later. In the past, these file servers were synonymous with PC network servers. At the time of this writing, NetWare controls over

---

[1] In the past, server operating systems were often called Network Operating Systems (NOSs). This was appropriate because early NOSs, such as Novell NetWare, were custom-built for PC network servers. Most SOSs today, however, such as UNIX and to a large extent Windows NT Server, are general-purpose operating systems adapted to the server role.

half of the market for file servers and for servers in general. However, we will see that other types of servers are growing in importance, especially client/server database servers and World Wide Web servers. NetWare has fallen behind other competitors for use on such servers.

UNIX is an important competitor. UNIX is a robust server operating system capable of supporting complex application programs. It includes the **Network File Service (NFS),** which provides file service and other important server services. However, NFS has not proven popular on PC networks. Today UNIX is seen mostly on workstation servers acting as client/server servers. In the client/server market, UNIX is as dominant as NetWare has been in the file server marketplace.

A newer competitor is **Microsoft Windows Server NT**. Like UNIX, this SOS is sufficiently robust to be used on client/server servers. In addition, Microsoft offers *Back Office*, which is a broad and integrated set of application services. In addition, many companies find Windows Server NT attractive because it is similar to client versions of Microsoft Windows, making it easy for the technical staff to learn and support without extensive retraining.

There are other important SOSs. *Banyan Vines* is especially good for very large networks. The *Macintosh* operating system is used on most servers in Macintosh PC networks, of course. It is also fairly popular for World Wide Web servers.

The choice of which of these or several other SOSs to use is critical. Some are more mature than others and are more reliable. Others have a fuller set of features in critical areas such as file and print services. Some are more widely supported, even in small communities where the firm may have branch office sites. Some have more extensive support from third-party software developers.

Independent and Synchronized Server Operating Systems ∎

The first server operating systems were built in the days when most PC networks only had about a dozen PCs and a single server. SOS vendors in those days created **independent SOSs**. As shown in Figure 8.2, if there were multiple servers on a network, they would work independently. In fact, they would not even know that the others existed. If the user wished to use the resources on two servers simultaneously, he or she had to log into both.

In addition, to use any resource, the user first had to identify the server providing the resource. The user then had to get login permission from the PC network administrator. In general, independent SOSs become unwieldy for users as the number of servers grows.

In contrast, most newer SOSs are **synchronized server operating systems.** Figure 8.3 on page 191, shows a specialized server called the **directory server**. This directory server maintains a database of all resources on all servers on the PC network.

With a synchronized SOS, the user has a single password for the network. He or she can log into any server using this password. When the user calls for a specific resource, in turn, the directory server locates the resource and

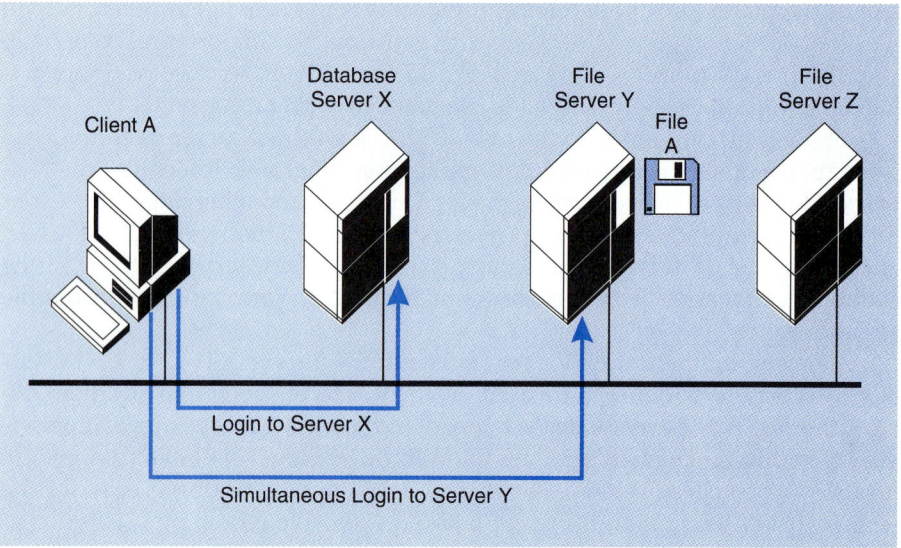

FIGURE 8.2

Independent Server Operating System

With an *independent server operating system*, the servers act independently. A user who needs services from multiple servers must attach to each server separately. To use an application program or access a file, furthermore, the user must know which server has the required resource. In this figure, the user must know that File A is on File Server Y and must log into that file server to get the file. In addition, such things as electronic mail addresses are tied to individual mail servers, so to reach someone via email, you must know the person's home mail server.

gives the user immediate access without logging into another server. This makes life much easier for users. The network administrator's life, however, becomes more complex because of the need to maintain the directory server's database.

Synchronized SOSs appear to be the wave of the future. Although Novell still maintains its older independent SOS, the 3.X series, it is putting most of its efforts into its synchronized SOS product, the 4.X series. UNIX is a synchronized SOS, thanks to its Network File Services (NFS) module. Microsoft Windows NT Server, finally, began its life as a synchronized SOS.

Although synchronized SOSs may be the wave of the future, different vendors have incompatible directory servers. To gain competitive advantage and to control the future of PC server SOSs, each vendor is trying to work with the directory servers of competitors, so that user organizations will adopt its directory server as their standard with other vendors' directory servers acting as subsidiary servers.

The Network

Figure 8.1 merely shows a network without specifying whether it is a single LAN or a complex enterprise internet. This vagueness is deliberate. The client PCs and file servers do not care at all how they are linked by transmission lines. In fact, they do not even know. PC networking is about application-level services that we will see later. How they are linked at the transmission layer (and even at the transport layer) is a mere implementation decision.

A typical early PC network consisted of only a handful of PCs connected by a LAN created for this purpose. Even today, this is the situation at smaller firms, such as NuPools. As a result, many people and even textbooks equate the terms *LAN* and *PC network*.

However, larger firms, such as VeriFone, have hundreds or thousands of PCs scattered over multiple sites throughout the world. Yet they may have only a single integrated PC network in which a client PC at one site can log into a server at any other site, as long as it has permission. "PC network" does not necessarily mean "small network" anymore.

FIGURE 8.3

Synchronized Server Operating System

In *synchronized SOSs*, there is a *directory server* that maintains a database of all resources on all servers. The user only has to log in once to any server using his or her network password. In addition, a user can access any resource from the server to which he or she is logged in. This makes the user's life easier. Maintaining the directory server database, however, causes headaches for administrators.

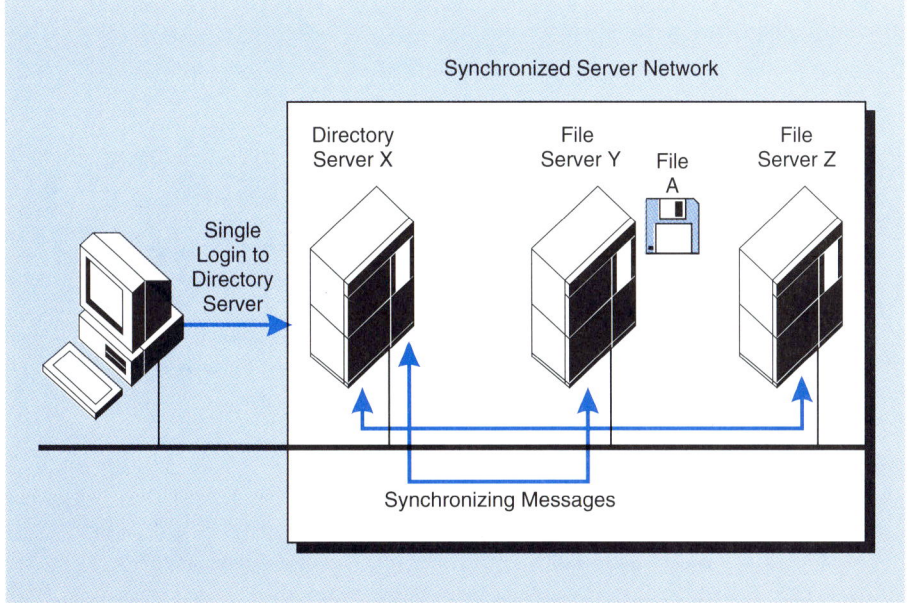

Communication Between Client PCs and Servers

Of course, client PCs and servers must be able to communicate with one another. There is nothing mysterious about how they do so. They merely implement standard protocols at each layer.

Communication Between a Client PC and a Single Server

Figure 8.4 shows the simplest case of communication between a client PC and a server. Here the client PC is communicating with a single server.

LAN Layer ■

At the bottom layer, there are LAN protocols. We have already seen these in Chapter 4.

FIGURE 8.4

Communication Between a Client PC and a Single Server

Communication between a client PC and a server follows the layering pattern introduced in Chapter 1. At the single-network (subnet) layer, the machines use standard LAN protocols. For the internetting layer, they both use IP. For the transport (computer) layer, they use TCP. Different application services require different application-layer standards. IP and TCP protocol data units can be exchanged between machines with different operating systems and hardware. This is why servers can be machines other than personal computers.

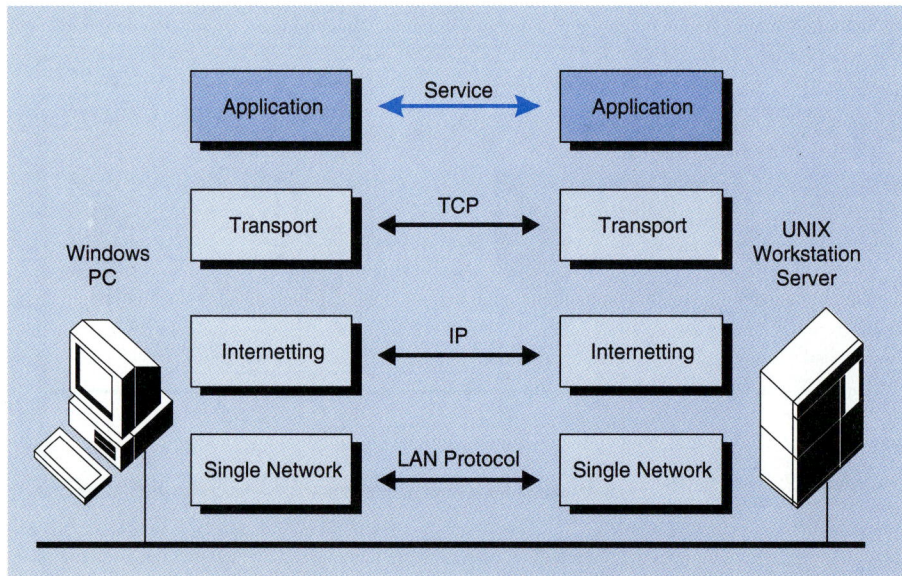

Internetting Layer

Recall from Chapter 1 that at the single-network (subnet) layer, OSI standards are dominant. At the internetting layer, however, this is no longer the case.

Generally speaking, the server operating system determines what internetting-layer standard must be used by the client PC. For instance, in Novell NetWare, the normal protocol is IPX. IPX is part of Novell's IPX/SPX architecture. However, NetWare can now support TCP/IP protocols.

UNIX servers, in turn, use the Internet Protocol (IP) from the TCP/IP architecture. Microsoft Windows NT servers typically use IP as well.

Transport Layer

At the transport layer in our Basic Communication Model (see Figure 1.1), the transport protocol follows the standard used at the internetting layer. For instance, if the server communicates via IP at the internetting layer, it normally uses TCP and UDP at the transport layer. Like IP, these standards are part of the TCP/IP architecture.

Novell NetWare has its own IPX/SPX transport-layer protocols. The most common is NCP, the NetWare Core Protocol. Another is SPX, the Sequenced Packet Exchange protocol, which gives the IPX/SPX architecture its name.

Application Layer

The application layer is the heart of PC networking. PC networking is really a set of application services, which we will see later. Application services are implemented at the application layer. Lower layers merely exist to link application-layer processes on client PCs and servers.

Lower layers are simple. At the subnet, internetting, and transport layers, there are only a few standards. The application layer, in contrast, is open-ended. Just as there are an infinite number of possible applications, there is an infinite number of possible application-layer protocols.

Simultaneous Communication with Multiple Servers

Figure 8.4 shows a client PC communicating with a single server. In reality, it is very common for a client PC to communicate with two or more servers simultaneously. This happens because the client PC user needs multiple services on multiple servers.

Figure 8.5 shows a client PC connected to three servers simultaneously. Note that it communicates with one via IPX/SPX and the other two via TCP/IP. The network-capable client operating system or the client shell software on the client PC orchestrates this simultaneous use of multiple protocols.[2]

---

[2] Windows 95 and other Microsoft products use the NDIS standard to support multiple protocols. NetWare uses the comparable ODI standard.

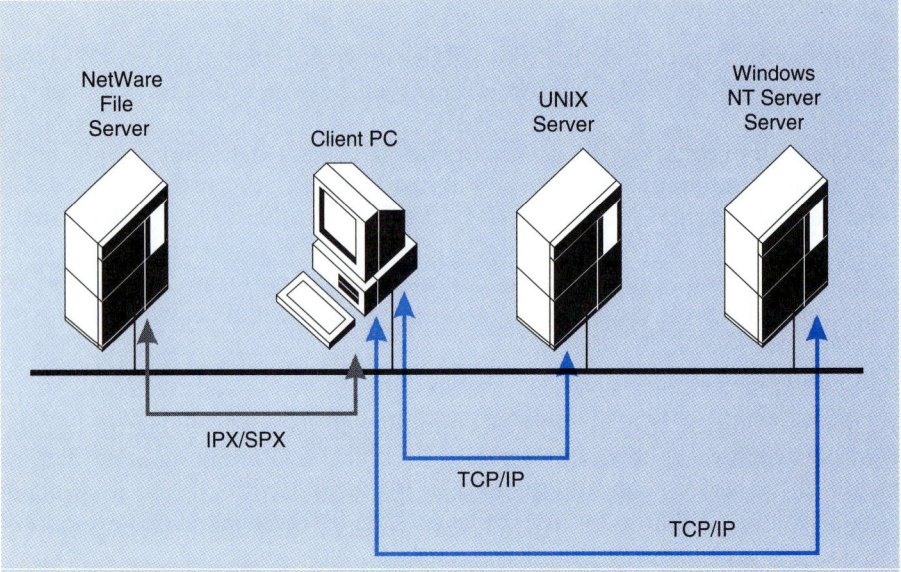

FIGURE 8.5

Communication Between a Client PC and Three Servers

A client PC can communicate with multiple servers simultaneously. Here the client PC communicates simultaneously with three servers. Two are TCP/IP servers. The other is an IPX/SPX server. The client PC must be able to communicate via multiple protocols.

FILE SERVICE

The remainder of this chapter looks at services. The most common type of service on a PC network is **file service**. We will see that file service essentially turns the server's hard disk drives into a shared storage space for network users. In fact, when some people think of PC networks, they sometimes think exclusively in terms of file service. On early PC networks, they would even have been correct. The earliest servers, in turn, were called **file servers** because this was their principal service. We still tend to call them file servers, even when they support other services as well, such as electronic mail.

File Service: The User's View

Figure 8.6 illustrates the idea of file service. Without connecting to a network, the client PC shown in the figure has three disk drives. Drive A: is a floppy disk drive. So is Drive B:. Drive C: is a hard disk drive.

The figure also shows that the file server makes parts of its disk drives available to client PCs. To avoid creating new concepts for the user to master, the client PC and the file server work together to make it appear that the user merely has some extra hard disk drives. In this case, the drives are F:, S:, Y:, and Z:. Each of these drives corresponds to a portion of a server's hard drive.

These are not real disk drives, of course. They merely look like drives to the user. To designate them as nonphysical drives, we will call them **virtual drives**. Drives A:, B:, and C: are real drives. Drives F:, S:, Y:, and Z: are virtual drives.

To the user the distinction between real and virtual drives is not very noticeable. Almost anything a user can do on Drive C:, he or she can do on Drive S: or on another virtual drive.

- *Storing Data Files*. The user can store data files on virtual drives, just as he or she can on a real drive.
- *Storing Programs*. Just as a user can store programs on his or her local hard disk drive, he or she can also store programs on virtual drives.
- *Executing Programs*. Of course, merely storing programs does little good if you cannot execute them. A user can execute a program on Drive Z: in exactly the same ways that he or she can execute programs on Drive C:.

FIGURE 8.6

File Service

A *file server* makes portions of its hard disk drives available to users for file storage and retrieval. To the user, each section looks like an additional disk drive. We call these *virtual drives*. Here Drives A:, B:, and C: are real local drives. Drives F:, S:, Y:, and Z: are virtual drives. Even most maintenance commands (copy, rename, delete, and so on) that the operating system can perform on real drives will work on virtual drives. You can even copy files from real drives to virtual drives (A: to S:) and from virtual drives to real drives (Z: to C:).

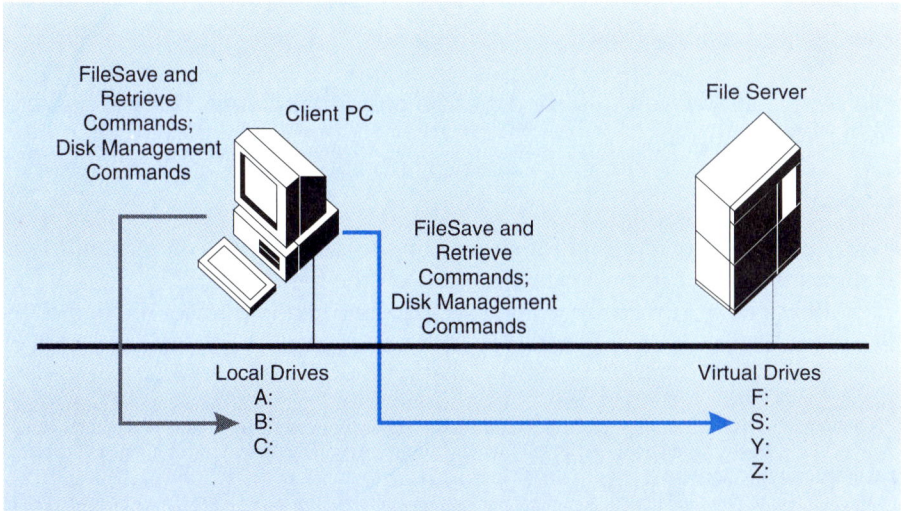

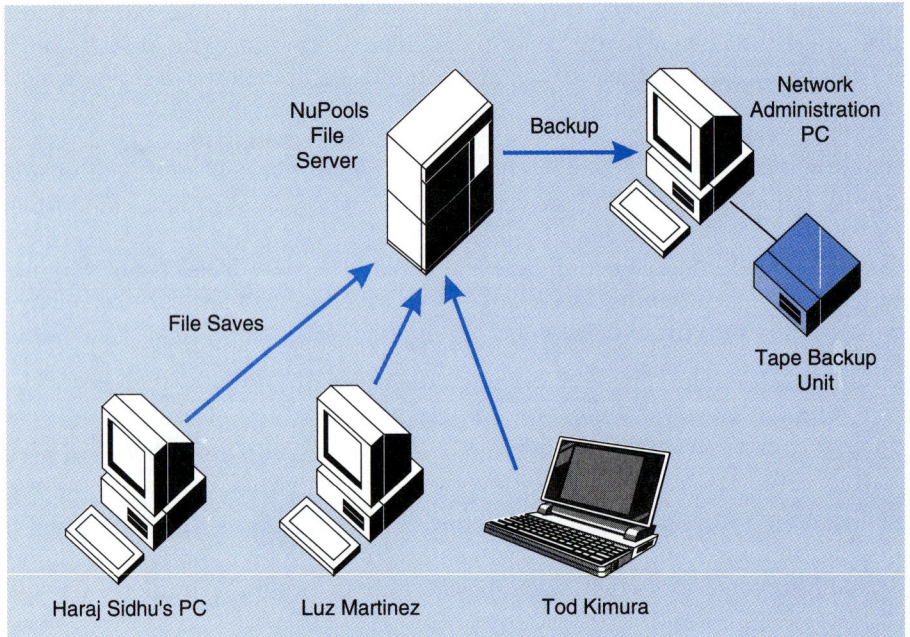

FIGURE 8.7

Data File Storage: Backup

Client PC disk drives are rarely backed up, but file servers are backed up regularly. This makes them safe places to store data files (if backup is done well).

- *Disk Maintenance.* On a normal drive, you must be able to create files, delete them, copy them, and do other disk and file management tasks. This is also possible on virtual drives. Almost all of the client operating system's file maintenance commands work on virtual drives. You can even copy files from real drives to virtual drives (A: to S:) and from virtual drives to real drives (Z: to C:).

File Service for Data Files

File service allows you to store data files on virtual drives. There are several reasons for doing so.

Backup ■

Figure 8.7 shows one reason for storing important data files on virtual drives. It shows three PCs storing their data files on the file server.

Although backup is important, users rarely back up the files on their client PCs. The file server, in contrast, is backed up regularly.[3] This makes it a much

---

[3] Note that the backup unit is not attached directly to the file server. Instead it is attached to a network administrator's client PC. In this way, a single network administrator client PC can back up multiple servers.

safer place to store critical information. Of course, if the backup system is not tested properly and fails in a critical recovery, the results will be worse than they would be if data files were stored on many client PCs. In general, however, storage on the file server normally is quite safe.

Access from Anywhere ■

Suppose you store all of your data files on your desktop PC's local hard disk drive. Then you can only **access** your data files if you are sitting at your desk. This is not always convenient.

Figure 8.8 shows the additional options you have if you store your data files on a file server. Of course, you can still access your data files from your desktop PC. But if you are in another office, you can log in from any available PC and get your files. You can also log in from your notebook computer on the road. And, of course, you can get to your data files from your home PC. (Later we will discuss dial-in access from the outside.)

Of course, access from anywhere has a disadvantage. Anyone who obtains your password can also get access to your files. Even if you do not lose control of your password, furthermore, the network administration staff typically has almost unlimited access to files.

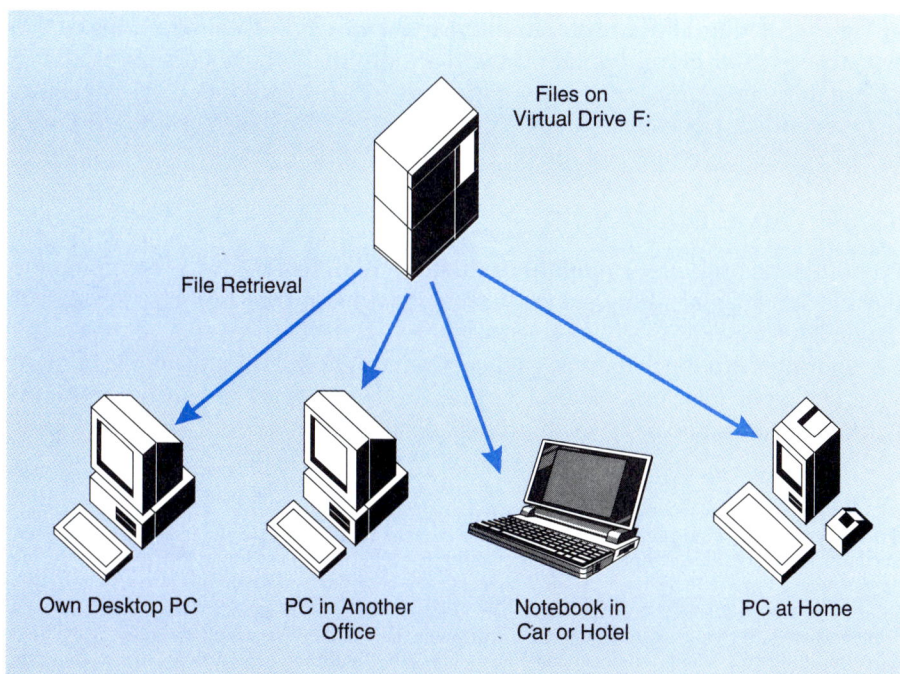

FIGURE 8.8

Virtual Drive Access from Anywhere

If your data files are stored on a file server, you can get access to your data files from your office desktop PC, a desktop PC in another office, a notebook computer on the road, or a PC at home.

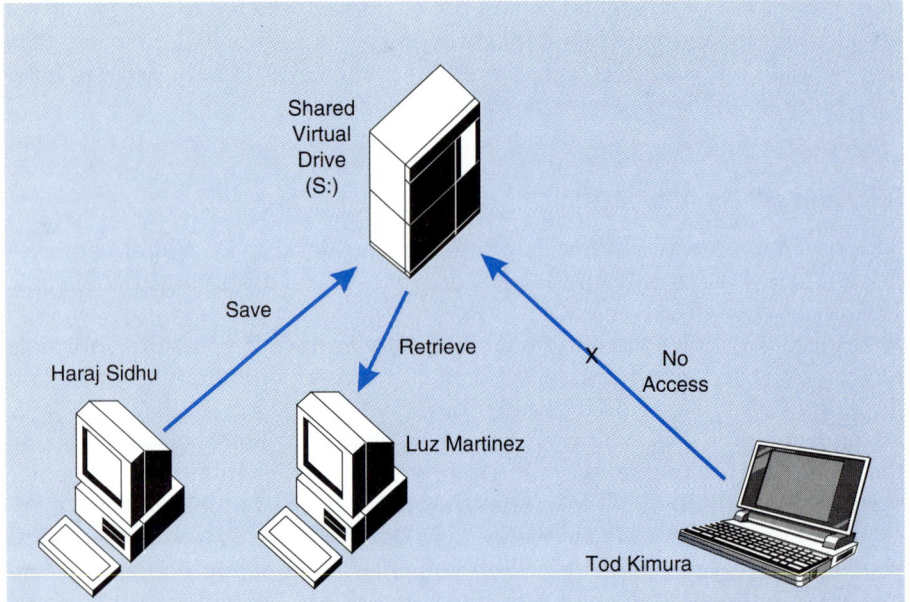

FIGURE 8.9

Shared Virtual Drive

Selected users have access to a *shared virtual drive*. This allows one to save a file there. Later others can read it and change it. This is ideal for group files.

Private Virtual Drives ■

In Figure 8.8, data files are stored on virtual Drive F:.[4] In this case, Drive F: is a **private virtual drive,** which means that only the user has access to it. This allows the user to store nonshared files there. As just noted, however, it is possible for others to get access to these files, so confidential files should not be stored there unless they are encrypted.

Shared Virtual Drives ■

What if Haraj and Luz wanted to share files? With real drives, it is impossible. But file servers also allow shared virtual drives. Figure 8.9 illustrates this situation.

Sharing virtual drives is a real benefit. In organizations, we work in many different types of teams. Shared virtual drives allow us to put our common work in a common part of the hard disk drive, so that we all have access to it.[5]

---

[4]By the way, there is no problem with having two or more different people use Drive F: as their virtual drive. Drive F: is simply a mapping between a letter and a section of the hard disk drive. At NuPools, both Haraj Sidhu and Luz Martinez may both have a Drive F:, but they would be mapped to different sections of the file server's real hard disk drive.

[5]Unfortunately, if two people try to edit a document file or other file simultaneously, one's updates could destroy the other's changes. For this reason, the first user to download the data file to a client PC is given total rights. Until the first person releases the file, the file is *locked*. Other users can only get read-only copies when they download the file.

Program File Service

File servers also provide services for program files as well as for data files. First, they can store programs. Second, they can execute them.

Program File Storage ■

In one sense, storing a program file is trivial. To an operating system, any file is just a stream of bits. Program files look no different from data files.

However, storing program files on virtual drives has some benefits that are different from those of data file storage. Note, in Figure 8.10, that there is only a single copy of the program on the virtual drive. This is important because PC network administrators constantly have to buy new software and updates for existing programs. It is far easier to install one copy of the program on a file server's Drive Z: than it is to install dozens or hundreds of copies of individual client PC hard disk drives.

FIGURE 8.10
Program Storage on a File Server

You can store program files on the file server, just as you can store data files there. In most cases, you can buy *network versions* of programs. These allow you to store a single copy for simultaneous use by multiple people. This is much easier than loading a copy onto each client PC. The *license fee* for the network version depends on the maximum number of simultaneous users it will allow. Sometimes you do have to install a *small stub* program on the client PC.

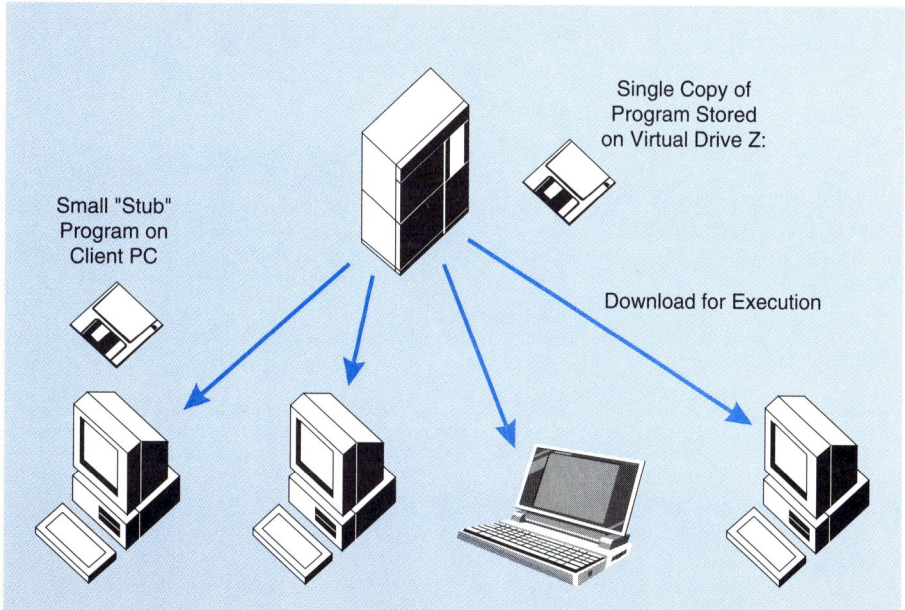

Of course, this does not mean that you can buy a single copy from a retail store and not pay for additional usage. The copies placed on the network are special **network versions** designed for multiuser access. The maximum number of people that they can serve simultaneously is determined by their **license fees**. Not surprisingly, versions that can support more users have higher license fees.

While the main part of the program is stored on the file server, a small portion of the program may have to be stored on the client PC. This is the program **stub**. Fortunately, adding stubs on individual client PCs is much simpler than adding large programs.

Program Execution on a Stand-Alone PC ■

Of course, storing programs is no good unless you can also execute them. In fact, file servers do allow program execution. However, they do so in a way that surprises many people. To understand why they execute programs the way they do, we will take a brief detour and look at how stand-alone PCs execute programs. Figure 8.11 shows this process.

Note that the program is *stored* on the local hard disk drive. It is not *executed* on the local hard disk drive, of course. Rather, for execution, the operating system copies it from the hard disk drive into the stand-alone PC's RAM. The stand-alone PC's microprocessor then *executes* the program.

Program Execution on a Client PC via a File Server ■

Figure 8.12 shows how the situation changes when there is a file server involved. In fact, it changes very little. The file server acts simply like a remote hard disk drive.

FIGURE 8.11

Execution of a Program on a Stand-Alone PC

The PC operating system copies the program file from the disk drive into RAM, where the PC executes the program. The disk drive merely *stores* the program file. It is *not* involved in execution.

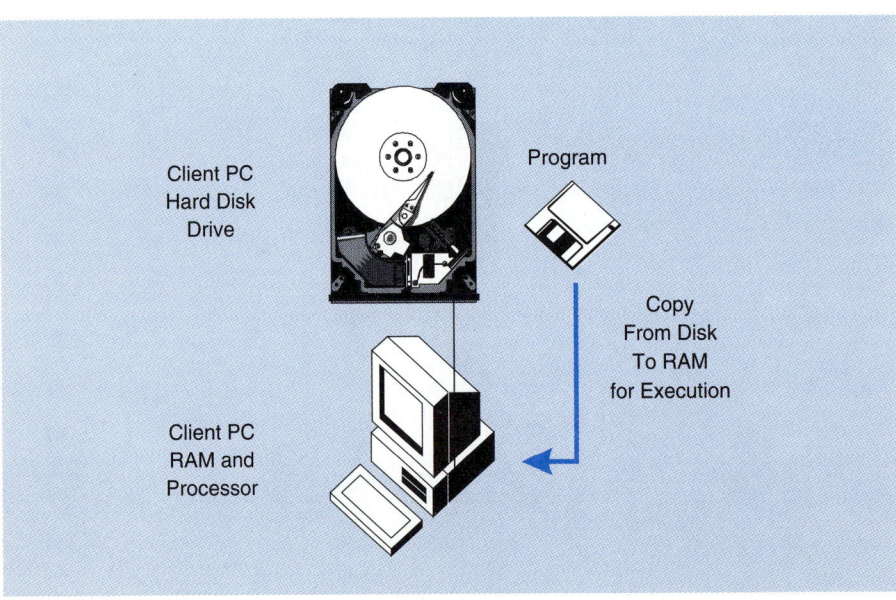

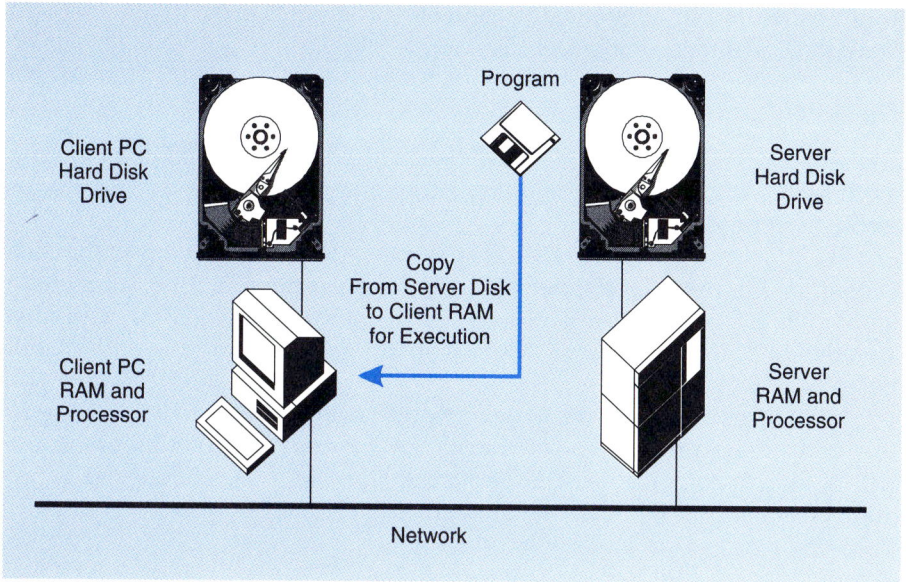

FIGURE 8.12
Execution of a Program Via File Server Program Access

The file server acts like a remote disk drive. It merely *stores* the program file. It does *not* play a part in execution. It merely copies the program file to the client PC's RAM. There the client PC's operating system executes the program. This is *file server program access*. An advantage is that programs that can run on stand-alone PCs can run via file server program access. The major disadvantage is that the size of the program is limited by the processing capabilities of the average client PC. This is acceptable for word processing and spreadsheet programs. For sophisticated database management programs, it is not acceptable.

Just as the program is not executed on the disk drive on a stand-alone PC, the program is not executed on the file server in a networked environment. The file server is merely a storage system. Instead the program file and related data files are copied over the network into the RAM on the client PC. There the program is executed by the microprocessor on the client PC.

Executing a program from a file server this way is called **file server program access**. The name emphasizes that file service merely provides *access* to program files. The server does not *execute* the program at all.

Advantage of File Server Program Access ■

File server program access has one major advantage: Programs created for stand-alone PCs do not have to be rewritten. You can put almost any program written for stand-alone PCs on a file server, and a client will be able to execute it via file server program access.

If you want more than one person to be able to execute a single copy of a program simultaneously, then the no-rewriting rule is broken, but only slightly.

Writing a multiuser networked version of a program is only slightly more difficult than writing a single-user version.

Disadvantage of File Server Program Access ∎

File server program access does have one major disadvantage: File server program access is not good for *very large* programs or programs that require *fast execution speed*.

The problem is low-powered client PCs. As noted earlier, many of the client PCs on a network are relatively old, slow, and lacking in large amounts of RAM. While some of the client PCs will be new, fast, and filled with large amounts of RAM, your client PCs will span a fairly broad range of capabilities.

In most cases, such as database, it would make no sense to buy a program that could only run on a few of your PCs. So you have to buy programs that will be able to run on at least most of your client PCs. This limits you to programs that can run on relatively medium-power client PCs.

Perspective on File Server Program Access ∎

The size limitation of file server program access is not bad for word processing, spreadsheet, electronic mail, or quite a few other popular types of software. In fact, quite a few mainstream programs run fine via file server program access. This is not accidental. Software vendors know that they cannot sell their products if their software only runs on very high-end PCs. So they size their software to run on somewhat older machines. In addition, because most PCs today are networked, vendors make sure that their programs can run via file server program access.

For some applications, however, file server program access is not acceptable. The most important example is database. Database programs have to be quite large to have high functionality and to have safety features that reduce the chances of data loss. For major corporate applications, file server program access is limited and unsafe. As we will see later, programs that cannot run well via file server program access have to run via another approach called *client/server processing*.

Of course, there are some database programs that run via file server program access. Two popular examples are *Access* and *Paradox*. However, these are really designed for personal applications or small-group applications. It would be unsafe to run large mission-critical corporate database applications using file server program access database programs.

Redirection

So far, we have looked at file service as it appears to users and administrators. Now we will look inside the client PC to see how the computer implements file service. We will see that it involves a process called *redirection*.

Redirection in Network-Capable Operating Systems

Modern operating systems, such as Windows 95, are capable of handling redirection by themselves without the addition of other software. Figure 8.13 illustrates file service redirection in such modern *network-capable* operating systems.

When an application program on the client PC wants to retrieve a file, it sends a **call** (a request for service) to the operating system. The call tells the operating system the drive, directory, and file name of the desired file.

The network-capable operating system looks at the drive name. If the drive is a real local drive, such as Drive C:, the operating system handles the retrieval itself. It goes to that drive, then to the directory named in the retrieval call. It locates the named file. Finally, it passes the file to the application program.

If the drive is a virtual drive, however, the operating system **redirects** the call over the network to the file server. The file server then handles the details

FIGURE 8.13

Redirection in a Network-Capable Operating System

Modern *network-capable* operating systems are capable of handling redirection without additional software. When an application program on a client PC wants to retrieve a file, it sends a *call* (a command) to the operating system. The operating system looks at the drive name. If the drive is a real local drive, such as Drive C:, the operating system handles the retrieval itself. If the drive is a virtual drive, such as F:, however, the operating system *redirects* the call over the network to the file server. The file server retrieves the desired file and returns it to the operating system. The operating system passes the retrieved file to the application program. Save commands work in a similar way.

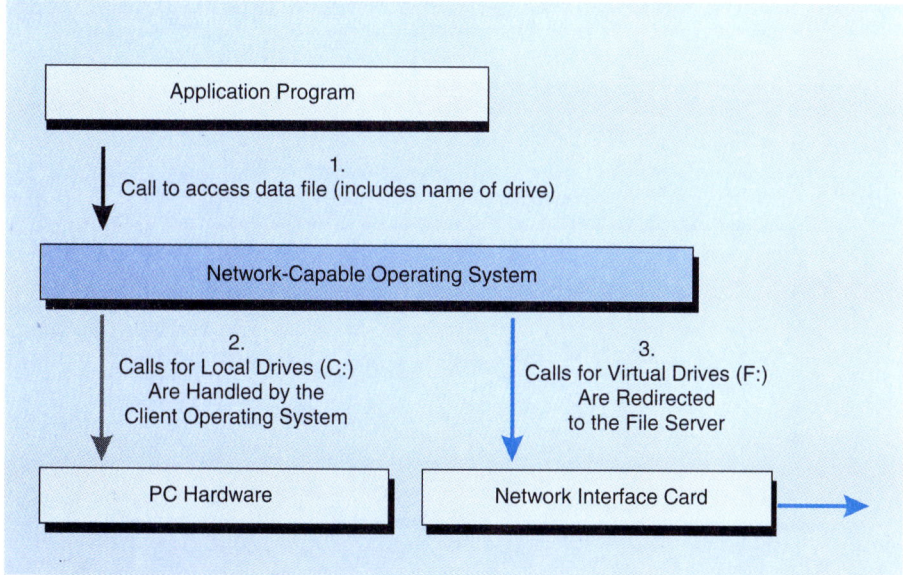

of the retrieval. It sends the file over the network to the client PC's operating system. The operating system passes the file to the application program.

File save commands work in the same way. However, this time the message from the client PC's operating system to the file service program on the file server contains the file to be saved. The response from the file server merely confirms the save.

Redirection in Client Shells with Older Operating Systems ■

Older operating systems, such as MS-DOS, older versions of Microsoft Windows, and older versions of the Macintosh operating system, do not have built-in redirection. In fact, they have no idea of what networking is at all. Fortunately, it is relatively easy to add client shell (*redirector*) software to older operating systems.

Figure 8.14 shows that this client shell handles the redirector function. The shell grabs application program calls to the operating system for file services. If the call is for a real local drive, the client shell passes the call onto the operating system, which can handle the call. For calls to virtual drives, in contrast, the client shell redirects the call to the file server.

OTHER SERVICES

Most of this chapter so far has focused on file service. File service is indeed the most widely used PC networking service, but it is not the only common ser-

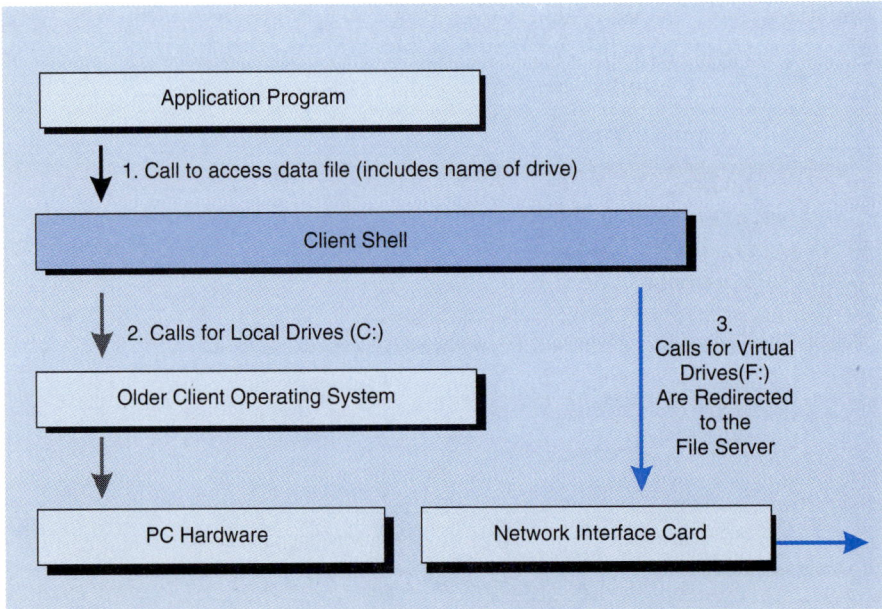

FIGURE 8.14

Redirection with a Client Shell

The *client shell* intercepts calls from the application program to the operating system. For calls to local drives, the shell merely passes the call to the operating system. For calls to virtual drives, the client shell redirects the call to a file server.

vice on PC networks. Any good-size PC network is likely to have several other types of servers.

Print Service

It is too expensive to give everybody a good laser printer in most firms. But if users can *share* high-end printers, everybody can have access to a very good printer. **Print service** implements this kind of sharing.

To the user, print service is very simple. The user simply gives the print command in the application program as usual. The only difference is that the user prints to a **virtual port,** much as they save files to virtual drives in file service. Just as you might print to LPT1: (the computer's first parallel port) for printing to a local printer, you might print to virtual ports LPT2: or LPT3: as virtual ports. The printout then comes out on a printer located somewhere in the general office area.

Although print service is simple from the user's point of view, it is fairly complex to implement. The box, "Print Service," discusses the details of network printing, which uses small and inexpensive print servers that attach remote printers to the network. *For more on print service, see the box, "Print Service," containing Figure 8.15.*

Client/Server Processing

We have looked at file server program access in which the client PC retrieves a program file from the file server and executes it, but there is another way to execute programs. This is **client/server processing**. We saw client/server processing in Chapter 2 as a general process. We will look at it in this chapter in the context of PC clients.

Limits of File Server Program Access ■

As stated earlier in this chapter, in *file server program access* the application program is not executed on the file server. Instead the client PC executes the program. As many client PCs have limited processing power and memory, complex applications such as a database cannot execute via file server program access.

Location of Processing ■

Figure 8.16 illustrates client/server processing. In client/server processing, there are *two programs*. One is the **client program**. As you would suspect, it executes on the client PC. The other program is the **server program**. It executes on the server.

Print Service

Figure 8.15 shows the steps that are involved in print service.

Virtual Ports

Recall that file service makes use of *virtual drives*, such as Drive F:. This allows the user to envision file service as nothing more than a set of additional drives. The client PC operating system or add-on client shell software works with the file server to map each virtual drive into a real section of the file server's hard disk drive.

Print service uses a similar concept called **virtual ports**. Suppose a client PC has a real parallel printer port. The first parallel printer port is called LPT1:. (The origin of the name is "line printer," but this is just an old name for a fast type of printer.)

FIGURE 8.15

Print Service

1) When an application program prints, the client operating system or client shell redirects the output over the network to the file server. 2) The file server places the output in a *print queue* until the printer is free. The file server then sends the output to a *print server*. 3) The print server sends the output to the parallel port of the printer, just as a personal computer would do if hooked up to the printer directly; this allows the use of ordinary printers.

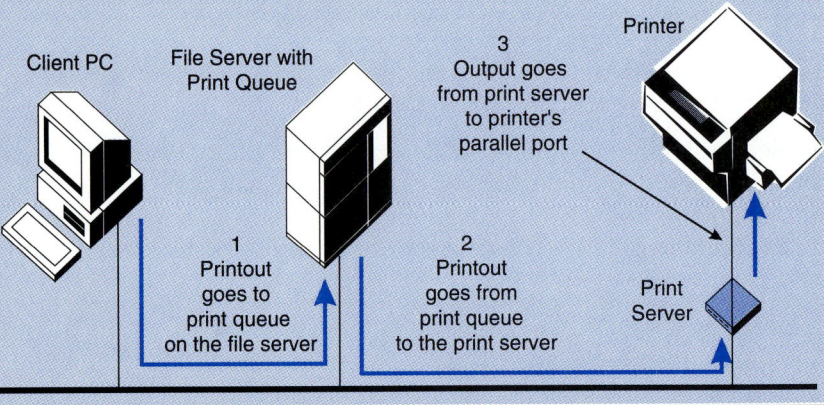

In addition, the user can have virtual ports LPT2: and LPT3:. When the user prints to a virtual port, the output will go over the network to the network printer.

File Servers in Print Service

When the client PC's operating system or client shell program redirects the print output to the network, Figure 8.15 shows that the output does not go to the printer directly. Instead it goes to the file server.

The file server places new printing jobs in temporary storage areas called **print queues**. These are simply holding areas to hold the output until a printer is free to print the output.

Specifically, a print queue is a directory. A file server may have multiple print queues for different network printers. This simply means that it has multiple print queue directories.

Print Servers

Figure 8.15 shows that when the file server finds that the printer linked to the print queues is ready to print the job, the file server sends the print job over the network. However, the job still does not go to the printer directly. Instead, it goes to a small device called a **print server**. The print server feeds the print job to the printer. Most print servers are small boxes about the size of paperback books.[6]

Network Printers

There is nothing complex about the printers that finally print the output. They can be ordinary printers. The print server handles all of the details of communication over the network. The printer gets the output just as it does from a directly attached PC.

Although any printer can be a network printer, several companies produce **network-ready printers**. These are printers with built-in print servers.

---

[6] If you are a small organization, you do not even need a separate print server box. File servers have software to allow them to act as print servers. You can simply attach printers to their parallel ports.

The Client Program ■

This division of labor allows both the client PC and the server PC to focus on what each does best. For instance, even relatively low-power client PCs can implement graphical user interfaces. This will make the application attractive and easy to use.

In Figure 8.16, the client program is Microsoft Excel, a spreadsheet program. Excel is running a spreadsheet model that requires the average salary of people in the accounting department. The spreadsheet model asks the server program for this information. The spreadsheet model then performs calculations based on the results.

The Server Program ■

The server, in turn, can have extensive processing power. If the server is large enough, it can handle database programs and other complex programs that once could only run on mainframes.

FIGURE 8.16
Client/Server Processing

In *client/server processing*, there are two application programs. One is the *client application program*, which runs on the client PC. In this case, it is Microsoft Excel. The other is the *server application program*, which runs on the client/server application server. This is Oracle. The client application program sends a *request message* asking for information to the application server program. Oracle does the processing and sends back a *response message* giving the desired information or notifying the client application program of an error. This is an application-layer process, so the client PC operating system and hardware can be different from the server's server operating system and hardware. It is not even necessary for the client/server application server to be a PC.

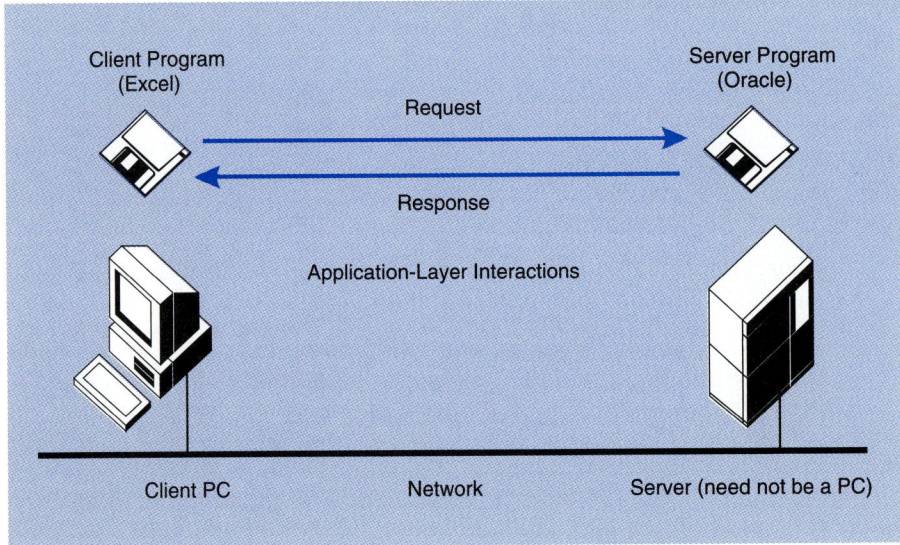

In Figure 8.16, the server program is Oracle. This is a powerful database program that offers a high level of data security. If the client machine, the server machine, or the network fail during a transaction, the database will be protected from corruption.

Request/Response Communication ■

Obviously, the two programs must communicate if they are to be able to work together. Figure 8.16 illustrates this communication.

Although the server is the more powerful machine, *it is normally the client program that begins a transaction*. In our example, the Excel model requests summarized data from the Oracle server program. In a sense, the server is the slave of the client program. It does what it is told to do. Although the client PC usually is the less powerful of the two machines, the client program is really in charge of the interaction.

Figure 8.16 shows that a transaction begins when the client application program, Excel, sends a **request** message to the server application program. This request message specifies what data it needs from the server application program's database.

The server application program, Oracle, reads the request and searches its database for the requested information. It then sends back a **response** message to the Excel program. Either this response message delivers the required information or it delivers an error message.

An Application-Layer Process ■

We noted earlier in this chapter that services really take place among application programs. Both the client program and the server program are application programs. As a result, the required protocols are application-layer protocols. The request message is an application protocol data unit (APDU). The response message is also an APDU.

Because interaction is an application-layer matter, there is no requirement that the client and server use the same hardware architecture or even use the same operating system. We noted earlier that the server could be a PC, a workstation server, or even a mainframe. In client/server database processing, the server is often a workstation server. If it is, it usually runs the UNIX operating system. Figure 8.16 shows this situation.

Recall from Chapter 2 that we have already seen client/server communication. In that case, the client application program was a browser and the server application program was a webserver. As we noted then, the World Wide Web uses client/server processing. So do most other Internet services.

Advantages of Client/Server Processing ■

The obvious advantage of client/server processing is that the "heavy" work—database retrievals, webserver retrievals, and so forth—can be done on powerful application servers. There is no limitation to the sophistication of the server application program. This allows very sophisticated database software to be used. Such software can offer both high functionality and safety.

Disadvantages of Client/Server Processing ■

The big disadvantage of client/server processing is that it requires a good deal of customized programming. Unlike file server program access, client/server processing cannot use existing application software. Many programmers, furthermore, are untrained in developing applications that involve two application programs on different machines that work by communicating.

Communication Servers

Sometimes the network must communicate with the outside world. This requires another type of server, the **communication server**.

Remote Access Server ■

Suppose you are working from home or from a hotel room. You want to log your client PC into the network remotely. Figure 8.17 shows that your client PC needs **remote access client software** and a modem. Your PC network, in turn, needs a **remote access server**. Once you log into the remote access server, you can do most things that a locally attached client PC can do. Of course, phone lines are very slow, so your service will not be as rapid as it is if you are locally attached.

Gateway Server ■

Remote access servers allow you to dial into the network *from the outside*. But suppose you are *inside the network* on a locally attached client PC and need to use a resource *outside your network*. Figure 8.18 shows that this requires a **gateway server**.

Gateway servers provide both a communication link and translation between your network's protocols and those of the target system. For instance, Figure 8.18 shows a **host access gateway** server that links a client PC to a mainframe. The client PC only communicates via IPX/SPX on its Novell NetWare network. The mainframe only communicates via SNA. The gateway server translates between them.

To communicate meaningfully with the mainframe, the client PC also has to emulate a terminal acceptable to the host computer. The host access gateway server in Figure 8.18 also supports terminal emulation on the client PC.

CHAPTER 8 PC NETWORKING 211

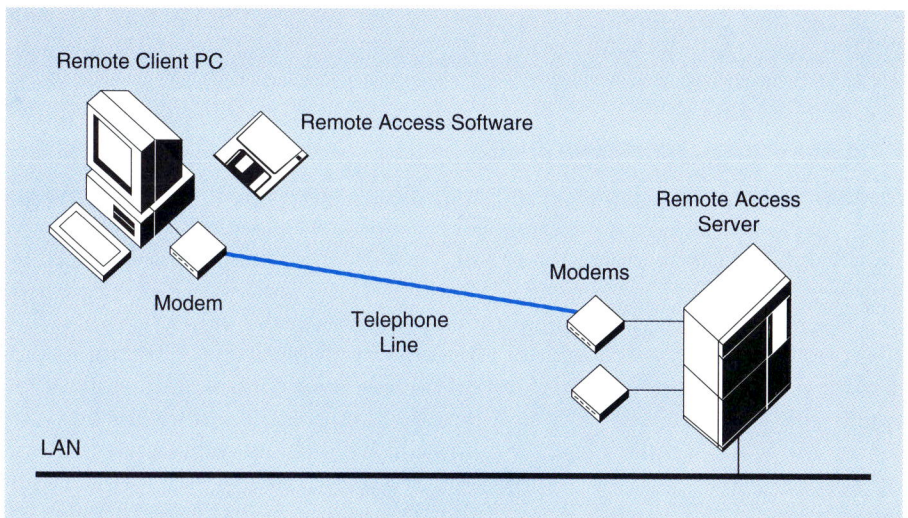

FIGURE 8.17
Remote Access Server

You use a telephone line and modem to connect your distant client PC into the *remote access server*. Once you log in, you can do most things that a locally attached client can do, although at a slower speed.

FIGURE 8.18
Gateway Server

A *gateway server* allows client PCs on a PC network to reach resources outside their network. Gateway servers translate between protocol differences. Here a client PC communicating via IPX/SPX protocols communicates via a gateway server with an IBM mainframe communicating via SNA protocols. The server also supports terminal emulation so that the client PC can communicate with the host as a terminal.

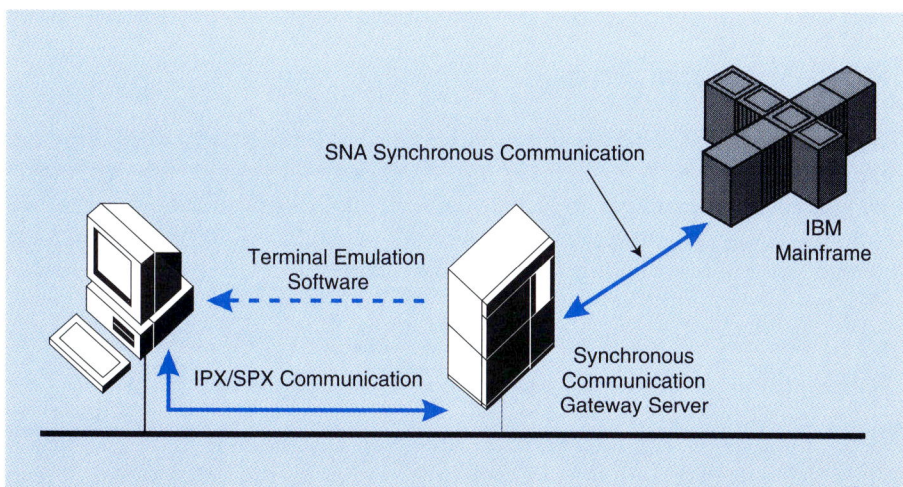

Of course, there are many types of gateway servers. For instance, Module B discusses *Internet gateway servers*, which connect a client PC to distant Internet host computers.

Communication Server Hardware ■

Most file servers and client/server applications servers are powerful (and expensive) machines. Communication servers, in contrast, usually are low-power PCs. Communication does not require heavy processing work in most instances.

Then why not just install communication software on other servers rather than having a separate machine? The answer is *reliability*. Communication servers are often at the mercy of events outside their control. As a result, they crash with some frequency. If you install communication software on a file server and it crashes the file server, this will inconvenience many people and may cause serious data loss. Although buying a separate machine for each communication service is expensive, many firms do it to reduce overall costs.

Electronic Mail Servers

After file and print service, electronic mail service is the most widely used application on PC networks. A **mail server** holds user mailboxes, so that client PCs do not have to be active when new mail arrives. The user logs into a mail server to read his or her mail and to send new mail.

Small PC networks almost always put the mail program on a file server. This is how NuPools handles its electronic mail. Larger firms have dedicated mail servers and often have multiple mail servers. Chapter 10 discusses mail service in more depth, including the need to synchronize multiple mail servers.

Network Management Servers

The last major type of server commonly found on PC networks is the **network management server**. As its name suggests, this server provides services to allow the network manager to manage the client PCs and servers on the network. It is impossible to separate PC network management from network management in general, which is discussed in Chapter 9.

THE NUPOOLS SERVERS

When NuPools first started its PC network, it only needed a file server. It purchased a fast PC and the Novell NetWare server operating system. The NetWare SOS comes bundled with software for file service, print service, electronic mail, and simple network management. It was very happy with that arrangement.

When NuPools purchased an inventory management program, however, the program required a client/server database application server. This required NuPools to make a choice.

One option was to purchase a second server for client/server database operation. This would have been expensive. However, it would have two advantages. First, client/server application servers often get very heavy use, and this could drag the file server to its knees. Second, having a separate application server allows a company to select the best operating system for the application server. Most larger organizations select this approach.

The other option was to run the server application program on the existing NetWare file server, making it a dual server acting both as a file server and also as an application server. NuPools chose this option because the cost to add a second server would have been prohibitive, given the modest loads that the inventory software would incur.

However, for NuPools to be able to select this second option, the vendor had to produce a version of the database server application program that could run on the NetWare server operating system. Server application programs, of course, have to be written for the server operating system of the server. Many vendors only offer a single version that runs on UNIX or some other operating system. Others produce versions that will run on two or three operating systems. Fortunately for NuPools, the inventory management vendor had a version that could run on NetWare. NetWare calls an application program that can run on NetWare a NetWare Loadable Module or NLM.

CONCLUSION

PC networking gives an ordinary desktop PC access to a wide array of application services. To turn a stand-alone PC into a client PC, you add an NIC. If the PC's operating system is not network capable, you must also add a client shell.

Instead of all executing on a large host computer (mainframe or minicomputer), the application programs that provide these services run on multiple servers. Most servers are specialized, running only one or a few application programs. This is more efficient than running the programs on a large host computer that is multitasking many simultaneous programs. Most servers are PCs, although some are workstations servers or even mainframes. Client PCs can log into multiple servers simultaneously.

Servers require server operating systems (SOSs), as well as application software. The most popular SOSs are Novell NetWare, UNIX, Microsoft Windows NT Server, and the Macintosh System. Different SOSs have different market niches. Older independent SOSs required users to log into multiple servers.

Newer synchronized SOSs allow a user to log in only once to reach all authorized resources. Synchronized SOSs require a directory service to centralize information about the location of resources on various servers.

The oldest and still the most widely used PC networking service is file service. File servers store programs and data files for users. This facilitates backup, access to files from anywhere, file sharing, and installing only a single copy of each program on the file server instead of installing many copies on client PCs. Executing a program on a file server requires file server program access in which the program resides on the file server but executes on the client PC. The file server also manages network printing services.

There are other types of services. In client/server processing, only some of the work is done on the client PC, allowing larger and more functional programs. Client/server processing is widely used in database processing and in Internet services. Communication services, in turn, allow access to the network from an external PC or provide access to resources outside the network. Electronic mail services provide person-to-person communication. Finally, network management servers allow the PC network administrator to manage the network.

CORE REVIEW QUESTIONS

1. Name the elements of a PC network. Are most client PCs personal computers? Are most servers PCs? Do they have to be? What is a network-capable operating system? What is a client shell, and when do you need one? What is specialization? What are specialization's benefits?
2. What are the relative strengths and weaknesses of independent file server networks and synchronized file server networks? Why are synchronized server networks easier on users but harder on network administrators? Why are directory servers critical in synchronized server networks?
3. What is a virtual drive? Why does the virtual drive concept make it easy for new network users to learn a network?
4. Name the benefits of file service for data files.
5. Explain the steps that occur when you run an application program from a local disk drive on a stand-alone PC. Repeat if the computer is a client PC instead of a stand-alone PC. Repeat for a program stored on a virtual drive.
6. Name the major elements in print service and how they work together.
7. Distinguish between file server program access and client/server program access. Where is the processing done in each? What are the advantages to each? What types of application programs are suitable for each?
8. List the types of communication servers discussed in this chapter. Distinguish between them. Why do most communication services have dedicated servers?

DETAILED REVIEW QUESTIONS

1. Some servers are high-end PCs. What are other alternatives?
2. Explain why PC networking is largely a set of application-layer issues.
3. Consider Figure 2-14 in Chapter 2. How would things change if you were using an 802.5 Token-Ring Network and Novell NetWare, which requires the IPX/SPX architecture?
4. Why is it inaccurate to equate *LAN* with *PC network*?
5. What are the main server operating systems? List the strengths and weaknesses of each.
6. Explain licensing for network versions of application programs.
7. Where is the processing done in traditional file server program access? What limitations does this create? How does client/server processing overcome these limitations? Do you need to rewrite stand-alone programs to run them via file server program access? Via client/server processing? Explain the purpose of the two types of messages in client/server program execution. Which program initiates each type of message? In client/server processing, does the client program or the server program initiate most interactions? Explain.
8. On my normal client machine, I have files saved on my real Drive C: and on my virtual Drive T:. I want to log in from another machine and work with some of my files. Should I store the files on Drive C: or Drive T:, or does it matter?
9. Describe what is needed for remote access. What are gateway servers? Name some common types of gateway servers.

THOUGHT QUESTIONS

1. Why do you log into a file server?
2. Which computer's disk drives store the server programs in client/server program execution? On which computers' disk drives can the client program be stored in client/server program execution?
3. Can you have a Novell NetWare file server and a UNIX client/server applications server on the same PC network, with a client PC communicating with both simultaneously? Explain.
4. A small firm has ten PCs. It wishes to network them together. Provide a list of what they will have to purchase. Assume that they need file service for data files and programs, print service, electronic mail service, word processing, and remote access service.

PROJECTS

1. Log into a PC network. List the real and virtual drives available to you. List two programs available from the local disk drive. List another two available from your file server. List any client/server programs. Copy a file between two real drives, between two virtual drives, and between a real drive and a virtual drive.

2. Prepare a report comparing competing server operating systems.

For online exercises, please visit this book's website at: http://www.prenhall.com/panko

CHAPTER 9

Network Management

INTRODUCTION

The first half of the 1990s was the "Great Age of Network Building." Most companies installed their first LANs and then many more. Most large companies at least started to build internets. Now, however, we have come to realize that building extensive networks is easier than managing them. In a real sense, many companies have built larger networks than they can manage effectively.

Part of the problem is *size*. When there are thousands of stations, routers, switches, bridges, twisted pair links, optical fiber runs, and carrier transmission facilities, a single failure can bring down large parts of the network. *Geographical distribution* adds to this problem. Even faults in distant buildings or remote sites can affect service. Identifying and fixing such faults can be very difficult.

Fortunately, we are beginning to see *network management software* to help network administrators deal with the complexity of modern networks. In effect, we are using complex software to help us deal with complex networks.

In this chapter, we will look first at *network management systems*—collections of programs and hardware devices that help the network administrator manage complex networks.

We will then describe several competing standards for network management systems. The first of these is the *Simple Network Management Protocol (SNMP)* of the TCP/IP architecture. The second is the *CMIP/CMIS* standards set of OSI. The third is OSI's *NetView*. We will also look at secondary standards, such as standards for managing client PCs.

In the rest of this chapter, we will look at the types of functions that network management programs serve. Figure 9.1 shows the five *functional management areas* defined in the OSI management framework. We will organize our discussion of functions around this framework.

Network management is becoming a major expense in networking. In 1993, for instance, network management software accounted for 16 percent of all capital purchases [Bruno, 1993]. This percentage has probably grown over time. In addition, network management consumes a great deal of the network staff's time. In these times of shrinking staffs and expanding network complexity, network management systems will have to bring large gains in staff productivity if service quality is to be maintained.

NETWORK MANAGEMENT SYSTEMS

Network managers need real-time control over their networks and internets. This requires a group of hardware and software components working effec-

FIGURE 9.1

Functional Management Areas in Network Management

The OSI network management standards define five *functional management areas*. These are management areas for which the network management program and other components of the network management system provide support.

Area	Description
Fault Management	Recognizing faults. Diagnosing their causes.
Configuration Management	Querying managed devices about their current configuration and status. Changing the configuration or status of a managed device. Repairing faults remotely. Electronic software distribution.
Performance Management	Measuring the performance of the network. Conducting what-if analysis on the network to determine the impact of changes. Planning for new networks.
Security Management	Limiting access. Providing confidentiality through encryption. Authentication. Virus protection.
Accounting Management	Providing accounting statistics that can be used for chargeback and other matters.

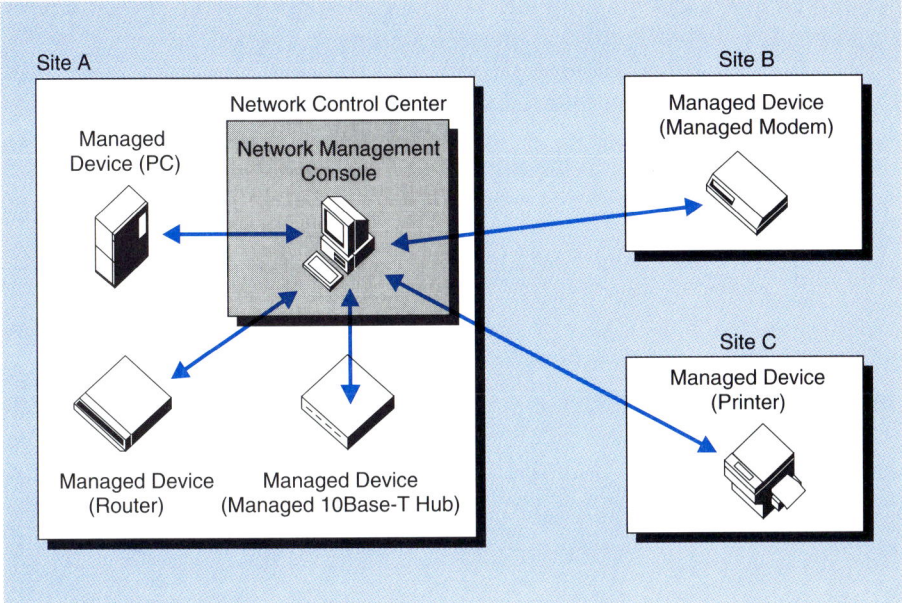

FIGURE 9.2

Overall Network Management System

The heart of the overall *network management system* is the *network control center*. Here the network administrator has one or more *network management consoles*. These consoles have screens and keyboards for displaying information about the network. They also allow the network administrator to make changes in the network, even at remote sites. The individual components of the network that the network console can work with are *managed devices*. These might be routers, carrier transmission lines, client PCs, or even managed modems and managed 10Base-T hubs.

tively with one another to help the network administrator manage the network. These components are called, collectively, a network management system.

The Overall Network Management System

Figure 9.2 shows that most firms have **network control centers.** These are rooms in which the network administrator has one or more **network management consoles.** These consoles may be PCs, RISC workstations, or even terminals attached to mainframes. They allow the network administrator to *collect data* on what is happening on the network or internet. They then *display the data* through printed reports and on-screen presentations.

In addition to allowing the network administrator to see what is happening on the network, a network management console also allows the network administrator to send *commands*. For instance, if a router at a remote site appears to be malfunctioning, the network administrator may be able to send it a command to do a self-check or even to shut itself down.

The individual network components that the network control center can manage are called **managed devices.** These might be routers, carrier transmission lines, client PCs, and even managed modems and managed 10Base-T hubs. Managed devices may be in the same office or thousands of miles away.

The Network Management Program

Figure 9.3 shows the software on the network management console. This is the **network management program,** which has three basic functions.

- The first function is to *collect data* from the hundreds or thousands of managed devices in the network and to organize this information.
- The second function is to *summarize information* for the network administrator. This may be done through on-screen displays, such as maps of the network with potential trouble spots highlighted in red. This may also be done through printed reports.

FIGURE 9.3

The Network Management Console

The program that runs on the network management console, called the *network management program*, has three basic functions: to collect data from managed devices, to present summarized data to the network administrator, and to send commands to the managed devices. It stores data from the managed devices in a *management information base (MIB)*.

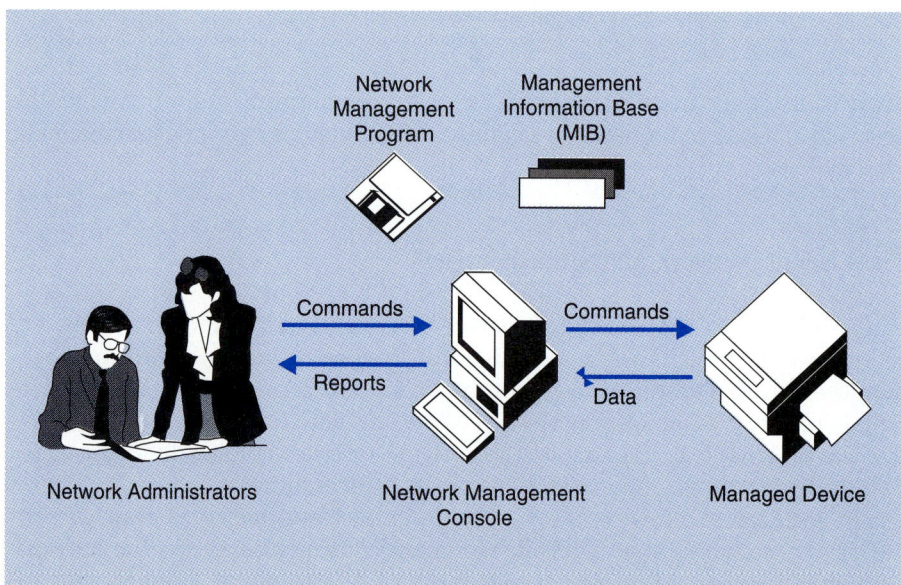

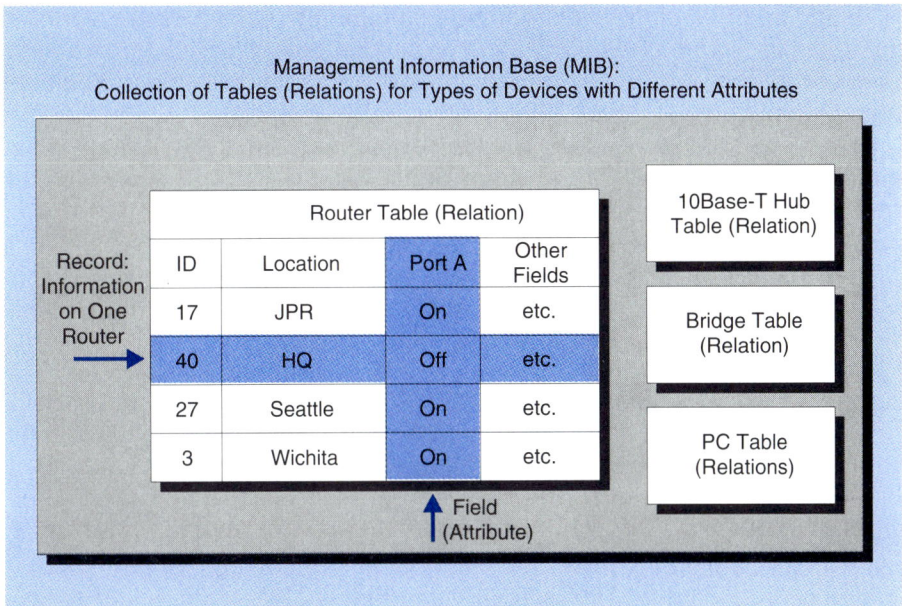

FIGURE 9.4
The Management Information Base (MIB)

The *MIB* is a database consisting of multiple files. There will be a *file* for each type of managed device (in this case, a router). Each time a router sends information on its status, the network management program will add a record to the MIB. This will consist of the router number, the time of data collection, and the value of each *attribute*. An attribute may be whether Port A is turned on.

- The third function is to *send commands* to the managed devices. These commands might be "give me data on your current status" to "turn off the network carrier on port 3 because we do not need its capacity at present."

Each managed device may send information about its status every few minutes or even seconds. This will result in a flood of information at the network management console. The network management program organizes this information in a database called a **management information base (MIB).**

The MIB usually is a relational database. It has a collection of tables (relations). For instance, there will be a table on routers. Every time a router sends information on its status, the network management program will add a record to the MIB, as shown in Figure 9.4.

For instance, suppose that the managed device is a router. Then there will be a table for routers. One record would hold information for one router at one moment in time. The attributes in the record might be the router number, the time of data collection, the overall status of the router, and the status of each port. For example, one attribute may be whether Port A is turned on.

Different types of managed devices will have different attributes. As a result, there has to be a table for each type of managed device on the network. Managed modems, managed hubs, and carrier transmission lines will all require separate files.

To summarize information for the network administrator, the network management program will have to search through the MIB and make sense of the pattern of data it finds in individual records in different files. Obviously, this is a major challenge.

Managed Devices

Suppose that the managed device is a 10Base-T hub. To turn a basic 10Base-T hub into a managed hub, you need to add hardware and software. In a hub, this may consist of a printed circuit board with software in ROM. As shown in Figure 9.5, this additional hardware and software is called a network management agent.

FIGURE 9.5

Network Management Agents

Managed devices must have *network management agents*. These agents act on behalf of the managed device in dealings with the network management program. It is the agent that communicates with the network management program in the network control center. The concept of network management agents allows us to add management capabilities to devices not basically designed as managed devices. On an unintelligent device, such as a 10Base-T hub, the agent may be a printed circuit board with a microprocessor and software in ROM. On an intelligent device, such as a client PC, the agent may be software alone.

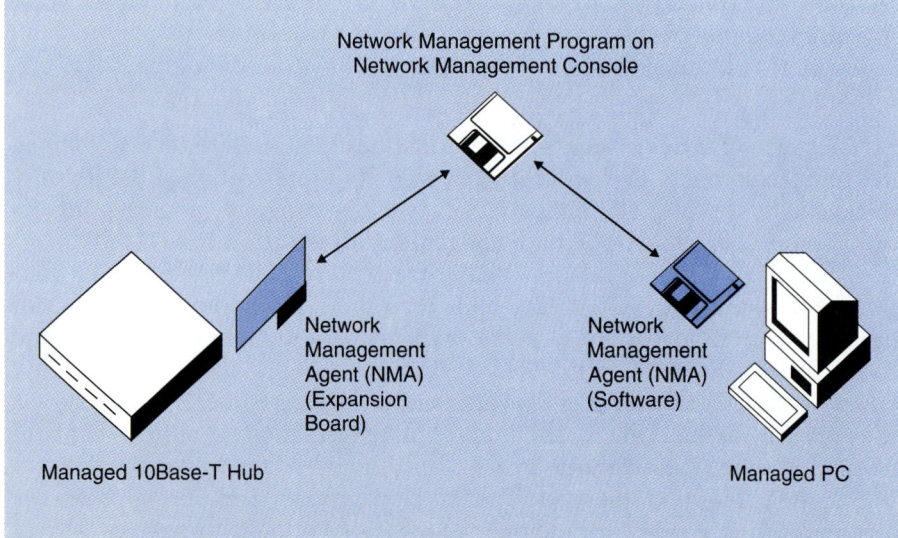

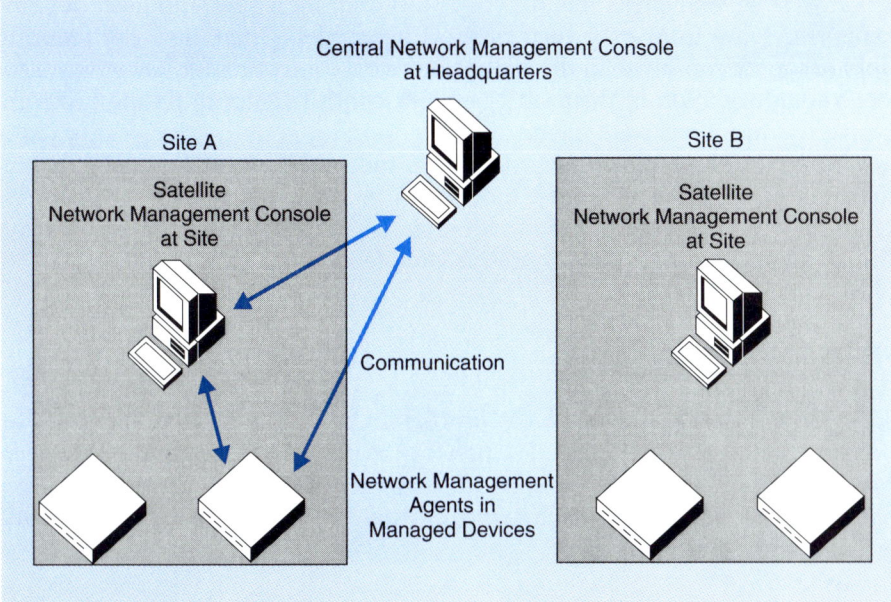

FIGURE 9.6

Distributed Network Management System

While some firms only have a single network control centers, many firms use *distributed network management*. There are *satellite network control centers* for individual departments, sites, or regions. There may also be a *central network control center*.

A human agent is someone who acts on behalf of someone else. In this case, a **network management agent** acts on behalf of the managed device in dealings with the network management program. It receives commands from the network management program. In turn, it collects information about the status of the managed device and passes this information to the network management program.

Figure 9.5 shows a managed 10Base-T hub. The basic design of the hub has little or no intelligence or processing power. This requires the agent to be a printed circuit board with a microprocessor and software in ROM.

Figure 9.5 also shows a managed client PC. In this case, the agent uses the PC's processing power. The agent may be a program stored on the client PC's hard disk.

A firm may have only a single network management console or only a few. However, it will have hundreds or even thousands of managed devices and network management agents.

Distributed Network Management

Figure 9.2 shows a network management system with a single network control center. While this may be the case, the situation in many firms is more complex. Figure 9.6 shows a **distributed network management system.**

Note that there are several network control centers. Most are **satellite network control centers.** These may serve a single department, a single site, or a region with several sites. This fits the reality that local sites and even departments also want to oversee their network resources. Often, they can identify problems more rapidly than the **central network control center.** They may also need the information in their local network control center to fix the problem and verify that it has been fixed.

The network management programs at the various sites should be able to exchange information with one another. In fact, the central network control center may not even speak to individual network management agents. It may simply collect summary data from the satellite network control centers.

Network Management Messages

Network management works on a *client/server* basis, although in the opposite direction than you might think. The network management program is the *client* in network management. The network management agents act as *servers* for their managed devices. So we have the unusual situation of one or a few clients and many servers.

Requests ■

Figure 9.7 shows the types of communication that must exist for network management to take place. First, there are **requests.** These are sent by the network management program to the network management agents. Some requests ask

FIGURE 9.7

Communication in Network Management Systems

The network management program is a *client* program. The multiple network management agents are *servers*. The network management program can send *request* messages to agents, asking for information or sending commands. The network management agents can reply with *response* messages, providing the requested information, acknowledging that the command has been executed, or explaining why it cannot comply with the request. The network management agent can also initiate messages, most typically *alarm* messages to indicate a fault or suspected fault.

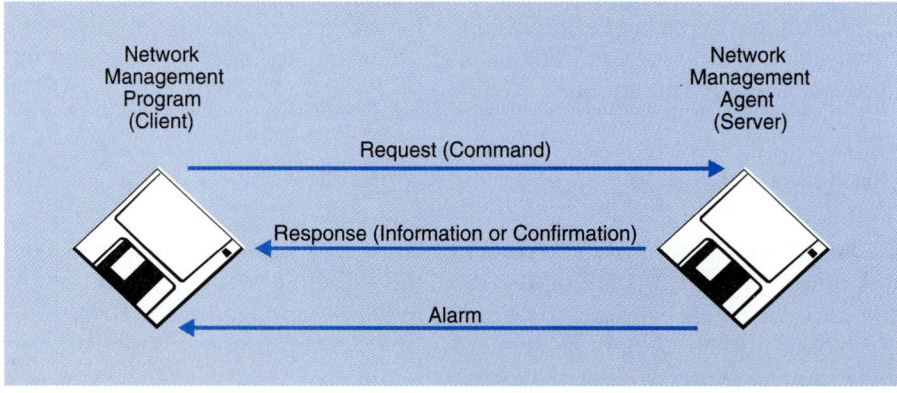

for *information*. Others send *commands* to the network management agents to change the configuration (status) of the devices they manage.

Response Messages ■

The network management agents can reply with **response** messages. A response message might provide requested information, or it might acknowledge that the command has been executed. If this is not possible, it might explain why it cannot comply with the request.

Alarms ■

The network management agent can also initiate messages, most typically **alarm** messages to indicate a fault or suspected fault in the managed device or of something attached to the managed device, such as a carrier transmission line.

NETWORK MANAGEMENT STANDARDS

For the network management systems to work, there must be standards. Otherwise, each network agent would collect data in a different format. In addition, network management agents would not even be able to recognize commands from the network management program.

Needed Standards

As shown in Figure 9.8, standards are needed in three main areas: the *schema* of the MIB, the definition of *services* to be provided, and the *protocol* by which the network management program and the network management agent communicate.

MIB Schema ■

Recall that the **MIB** is the **management information base**. It is a database consisting of multiple files, with one file for each type of managed device. The general design of any database is called its **schema**. This schema has several elements.

- ■ First, the schema defines which *tables* will exist. This is equivalent to defining which devices will be managed. There must be a file for each type of managed device—bridge, router, carrier transmission line, and so forth.
- ■ Second, for each file, the schema defines the *attributes* that will be stored for each *record*. For instance, if the managed device is a router, the schema might specify an attribute holding the result of the last self-test performed by the router.

FIGURE 9.8
Standards Requirements for Network Management

Requirement	Description
MIB Schema	The MIB schema defines how information is stored. It specifies files in the MIB. It also specifies the attributes of records in each file. Finally, it specifies the formats for individual attributes (text, numerical, and so on).
Services	The network management program and network management agent ask one another for services. These services must be standardized so that the other will act as expected.
Protocols	The network management program and the network management agent must exchange application-layer protocol data units to request services, provide responses, and send alarms. These protocols must be standardized. The standard must also specify at least the interface to the next lower-layer protocol.

- Third, for each attribute, the schema must specify the *format* of each attribute. Will the data consist of text information or a number? If the information is numerical, how many decimal places will it have? This format information must be standardized if the MIB is to be able to store the attribute information sent by the network management agent.
- Fourth, for each attribute, the schema may contain *validation rules* which give possible parameters for the data. If the data fall outside these parameters, the data will be rejected.

Services

The communication between the network management program and the agent can be defined in terms of **services,** that is, what the network management program will ask the agent to do.

One service might have the network management program asking the agent to provide a self-test on its managed device. Another service might ask the agent to transmit an entire record for the MIB. Yet another service might ask the agent to transmit the value for a single attribute, or to turn off the managed device.

Obviously, services need to be defined and standardized. Otherwise, if the network management sent a message requesting a service, the network agent would not even understand the message. Or, if the receiving agent did act, it might do something unexpected.

Protocols

Services define *what* should be done. Protocols define *how* to implement the required communication by defining request and response message protocol data units needed to carry out each service.

Figure 9.9 shows that the network management program and the network management agent are *application-layer* processes. Although agents collect data from all layers, they communicate with the network management program via *application protocol data units (APDUs). For more on application protocol data units (APDUs), see Chapter 2.*

The standard must also specify the interface to the next lower-layer process. In TCP/IP, this is usually the transport-layer process. In OSI, it may be the presentation-layer process.

Simple Network Management Protocol (SNMP)

Figure 9.10 shows that there are several competing network management standards. The core competitors come from the three major standards architectures: TCP/IP, OSI, and SNA. There are also important secondary network management standards.

The most widely used network management standard today is the **Simple Network Management Protocol (SNMP).** This is the network management standard of the TCP/IP architecture.

Simplicity ■

As its name suggests, SNMP is *simple*. This simplicity made it easy to implement and gave it a strong lead in actual usage.

FIGURE 9.9
Application-Layer Exchanges
Although the network management agent may collect data on transmission-layer or transport-layer processes, the network management program and network management agent software are themselves application-layer processes. Standards must define the application layer protocol data units that the two programs exchange.

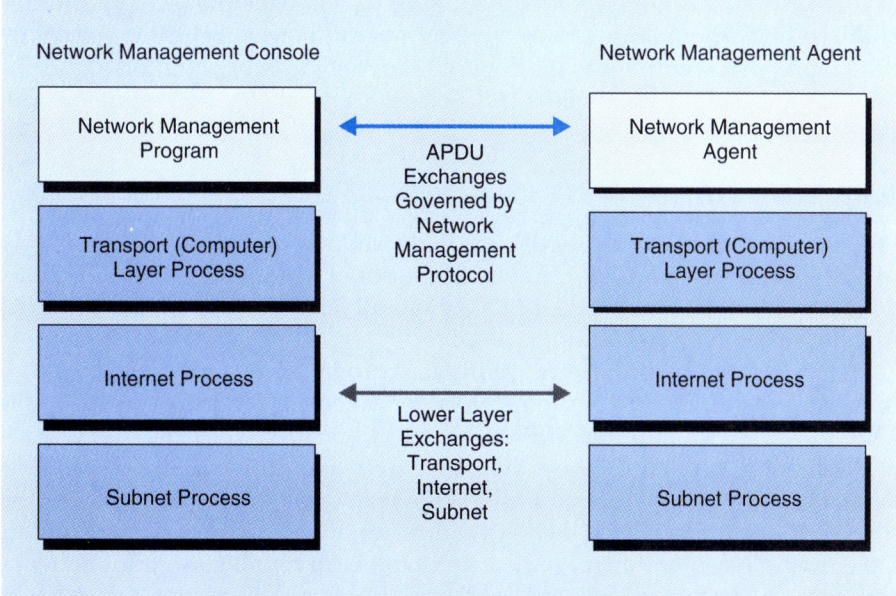

Standard	Description
SNMP	Simple Network Management Protocol. TCP/IP. Simple. Version 1 focused on fault management. Version 2 does more. Nonstandard MIBs. Limited security. RMON standard for remotely monitoring LANs.
CMIP/CMIS	Common Management Information Protocol/Common Management Information Services. OSI. Sophisticated, expensive.
NetView	From IBM. For SNA Networks.
Client PC Standards	Collect data on client PCs. Can remotely configure client PCs.
Server Systems Management Standards	These allow the network administrators to manage multiple servers, performing such tasks as backup and performance management. Server management standards are especially important in synchronized server networks.
Carrier Standards	Carriers have large transmission networks with their own network management systems. Some carriers are making their systems available to end user organizations.

FIGURE 9.10
Network Management Standards

At the same time, SNMP is somewhat limited. Its original version, SNMPv1 (Version 1) dealt primarily with *fault management*, that is, the detection of errors. Security features were almost nonexistent, and other features were weak. Fault management, however, was a pressing need, so many companies adopted SNMP. However, they were still uncomfortable with its limitations.

SNMPv2 (Version 2) is now under development. It will add many features to SNMP, including distributed network management. However, at the time of this writing, it is not clear whether SNMPv2 will have the security features required by corporations. This is a concern because even in 1993, a survey by BRG found that security was the biggest concern of network administrators for network management [Battelle, 1993]. Security certainly has not declined in importance since that time.

Proprietary MIBs ■

One specific problem with SNMPv1 is that it did not provide a systematic way of adding service standards for specific classes of products. As a result, when vendors introduce a new product, they often create **proprietary MIBs** to define their product's attributes.

This creation of proprietary MIBs has produced two problems. First, it makes it very difficult to create network management programs because they have to work with multiple MIBs.

Second, it is often the case that the only network management program that can deal with a proprietary MIB is the management program supplied by the vendor of the managed device. As a result, many network control centers have multiple network management consoles, each running a vendor network management program for a single class of devices. This means that when a problem occurs, the network administrator may have to jump from one console to another, getting only part of the picture from each.

RMON (Remote Monitoring) ∎

One of the strongest features of SNMP is its **RMON (remote monitoring)** MIB for LANs. This MIB offers a limited but important form of distributed management. As shown in Figure 9.11, each LAN can collect data on an RMON server. The central network work management console can then query these servers to get summarized data on the status of the LAN. This is more efficient than having to query each managed device on distant LANs. **RMON-2** (Version 2) will provide even more sophisticated tools for managing remote LANs. It will include information from the internetworking and transport layers, not just the data link layer and the physical layer.

CMIP/CMIS

OSI's network management standards, as usual, are much more complex than SMNP. And, as usual, they have been slower to evolve. Products, in turn, have been more complex and so, as usual, have been considerably slower to market and more expensive than SNMP products.

At the same time, OSI network management standards are more extensible, making them easier to change. They also cover more of the functional management areas shown in Figure 9.1.

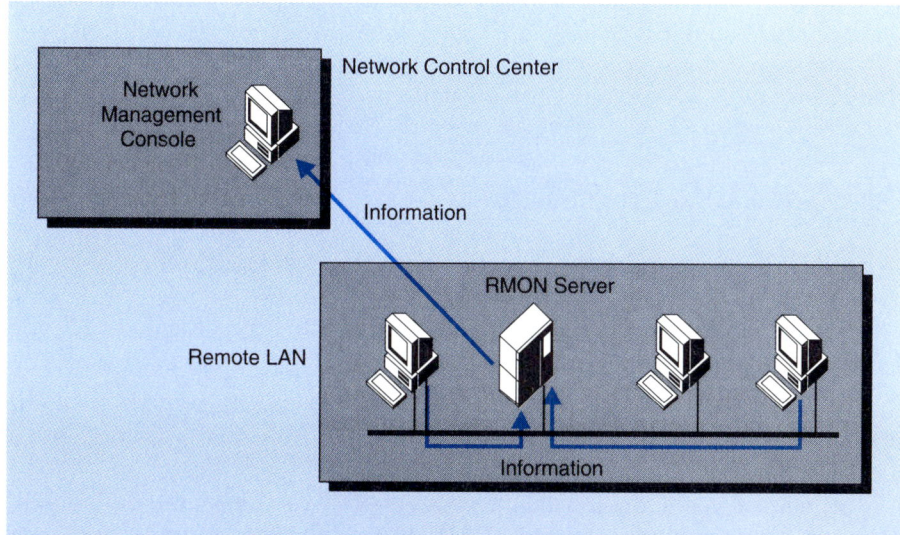

FIGURE 9.11

TCP/IP's RMON (Remote Monitoring) Standard

Each LAN has an *RMON server* that collects data on the local operation of the LAN. The central network management program can query this server instead of querying each managed device on the LAN.

CMIS

The OSI network management standards come in two sets. First, there are the **Common Management Information Services (CMIS)** standards. As the name suggests, these standards define the *services* that will be provided.

By separating services from the protocols that implement them, CMIS makes it much easier to add new services and to modify existing services. The services group can specialize in service-level concerns and can ignore the creation of protocol messages. In addition, the process for proposing and standardizing services is relatively well developed. These things have made it relatively easy to standardize management information bases—a major weak point in SNMP. In addition, CMIP/CMIS is object oriented, so changes in the definition of specific services do not cause widespread ripple effects in other parts of CMIS/CMIP.

CMIP

In turn, the **Common Management Information Protocol (CMIP)** standards define the application-layer protocols through which network management programs communicate with network management agents. The OSI network management protocols are often referred to simply as CMIP. We use the term *CMIP/CMIS* in this book.

NetView

IBM was the first company to face the need to manage large networks. Its mainframes often dealt with hundreds or even thousands of terminals over long distances via SNA networks.

The network management program in SNA is **NetView,** so SNA network management standards usually are called NetView standards. NetView actually comes in several versions. The most widely used versions run on mainframe computers. Others run on workstations.

Other Network Management Standards

In addition to the "Big Three" network management standards, there are many specific network management standards.

Client PC Standards

Most computers in a network are client PCs. In the past, there was no way to manage these machines remotely. Now, however, active efforts are underway, although the direction is not yet clear.

For several years, the *Desktop Management Task Force* has been defining the **Desktop Management Interface (DMI)** to standardize communication between management agents running on desktop computers and network management programs.

With Windows 95 and Plug-and-Play computing, Microsoft has adopted part of DMI but not all of it. This has led to confusion in the industry.

One reason for client PC standards is inventory management. Not only must firms be able to count their PCs. For planning and repair purposes, they must also know how each PC is configured. Client PC agents provide this information to central network management programs, as shown in Figure 9.12.

Obviously, the network management agent in the client PC provides *hardware* information to the network management program. If there is a problem, this information helps the network management program isolate the difficulty and helps a repair technician know how to attack problems on the PC or how to upgrade it.

More controversially, the agent will also tell the network management program what *software* is on the client PC. Objectively, this will help the critical **inventory management** process. For example, you would like to identify users who are working with old versions of software. Companies would like to be able to do **software metering** to determine actual use. This would help them know how many copies of the software to buy or license. It would also allow them to audit usage so that licenses are not exceeded.

FIGURE 9.12

Client PC Management

The client PC agent sends configuration information to the network management program. This configuration includes hardware information and what software is on the client PC. This information is important in inventory management. The network management program can send commands to the client PC agent.

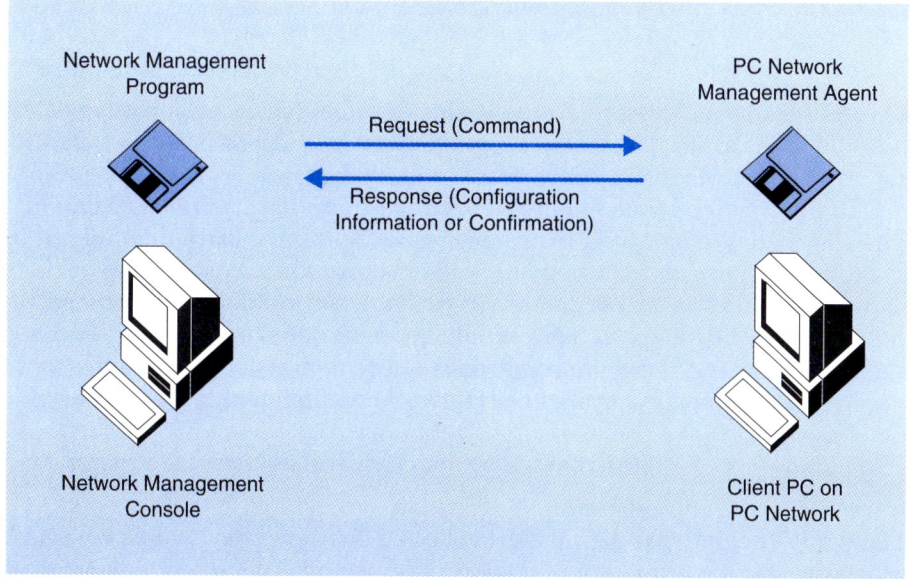

Software monitoring also means that the company can "snoop" at users by seeing what software each user has installed on his or her client PC. This can help the company identify games and other unauthorized software.

The network management program can also send commands to the network management agent on the client PC. It might use this capability to fix problems remotely, perhaps even before the user knows that these problems exist.

Server Systems Management Standards ■

Servers too need management. Organizations that have downsized to PC and workstation servers from mainframes are almost always disappointed by the quality of **systems management** tools that are available for servers compared to those available for mainframes and even minicomputers.

For instance, all servers must be *backed up* regularly. We saw in Chapter 8 that this makes them good places to store user data files.

In addition, servers must be monitored to ensure adequate performance. Most servers have **performance monitoring** software that can help a network administrator see, for instance, if the disk drives are getting too full. More subtly, they can tell the administrator if there is insufficient RAM for caching (see Module I). This will lead to slow disk access times. In practice, network administrators must know many things about their servers.

New problems exist when servers are synchronized (see Chapter 8). If there is a directory server or name server, there must be standards for communication between the directory server and devices.

Server management standards typically are controlled by the SOS vendors, such as Novell, or by consortia of providers, as in the case of UNIX.

Each of these vendors, seeing the growth of multivendor server environments, is attempting to extend its server network management program to include data from other vendors' servers.

Carrier Standards ■

Carriers such as AT&T have always managed large transmission networks. As a result, each carrier developed its own network management system, including managed devices and network control centers.

Until recently, these systems were closed to customers. The transmission line was merely a pipe, and management was left to the carrier. However, in many cases, a network management program needs to know such simple things as whether or not a carrier line is operational. It can infer this from the behavior of attached devices, but it would be much simpler and more certain if the network mangement program could query the carrier's MIB. Now some carriers are cautiously opening their network management systems to users.

Limits of Standards

Standards govern what data will be collected and how they will be collected. However, standards do *not* govern how the data will be summarized and what reports and screen displays will be presented to managers. Figure 9.13 shows this situation.

Of course, such reports and screens are crucial to managers. By leaving these things to vendors, however, standards agencies give vendors the freedom to be creative in what summarized information they will present and how they will present the information. There is no need to standardize such matters.

Cutting Across Standards

It would be nice if a company could decide on a standards framework and simply require that all agents comply with that framework. Unfortunately, this is rarely possible. We saw earlier that even when products follow SMTP generally, they often embody proprietary extensions. In addition, carrier, client PC, and server standards lie outside traditional frameworks.

Figure 9.14 shows that some products, such as Hewlett-Packard's OpenView and IBM's NetView, can work with the agents and MIBs of multiple standards agencies. These **multifamily network management programs** give a reasonably comprehensive view of the network to the network administrator. Still, much is always lost in the translation.

FAULT MANAGEMENT

Having looked at network management systems in general, we will now begin to look at the functional management areas in OSI. Figure 9.1 summarizes these functions. In this section, we will look at **fault management**. A **fault** is a

FIGURE 9.13

Limits of Standards

Standards govern what data will be collected and how they will be collected. However, standards do *not* govern how the data will be summarized and what reports and screen displays will be presented to managers. This allows network management program vendors to use the data creatively.

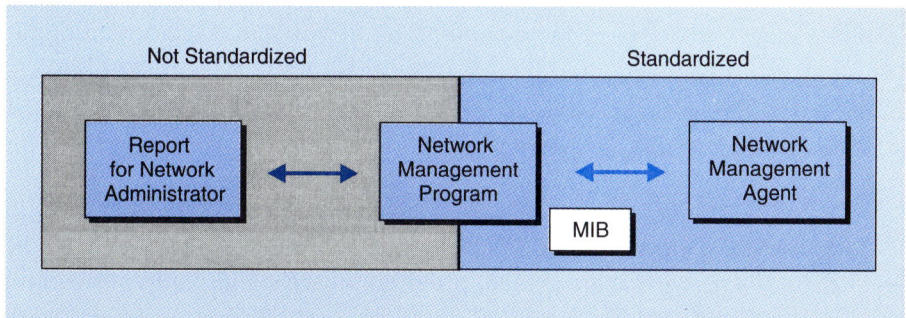

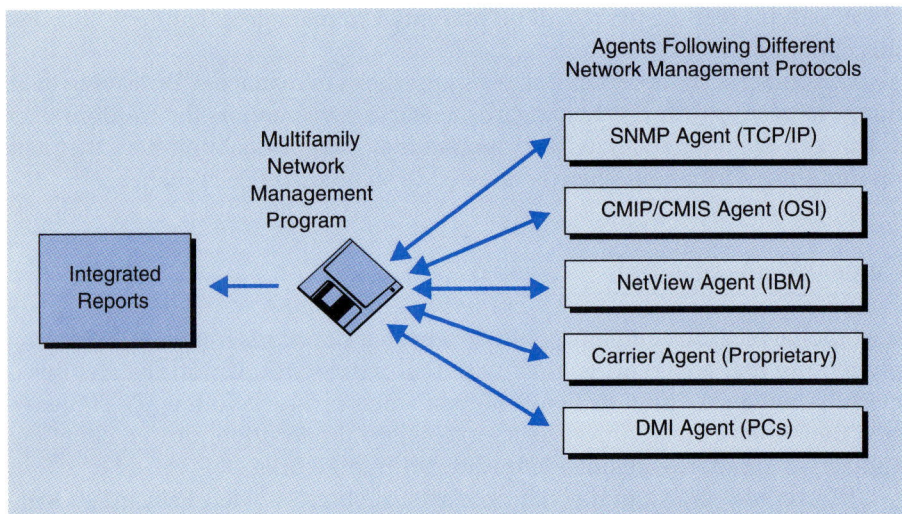

FIGURE 9.14

Multifamily Network Management Programs

Multifamily network management programs can work with agents from multiple architectures. They can combine the information into a reasonably comprehensive view of the network.

condition that has led to a failure or may lead to a failure. We will look at it first because it is the most obvious network management area and is in some ways the most basic. If you cannot identify and diagnose faults, you cannot begin to do more sophisticated aspects of network management.

Recognizing the Existence of Faults

There are several ways for the network management program to recognize the existence of faults.

Alarms ■

First, when a managed device detects an unusual condition, it can send an **alarm** message to the network management program, as noted previously. For instance, if a router detects the apparent failure of a carrier transmission line, it will send an alarm to the network management program.

MIB Analysis ■

In other cases, diagnosis is more subtle. The network management program will look at the parameters in the MIB record for a device and detect an unusual condition.

Diagnosing Faults

Once there is a realization that a fault exists, the network administrator with the help of the network management program must diagnose the nature of the fault or determine that the suspected fault is a false alarm.

Alarm Storms ■

One problem in diagnosing faults is that a single fault is likely to affect many devices. So a single fault, such as a transmission line failure, may generate many alarms from many devices. The network management program will literally be overloaded when such **alarm storms** occur. The network management program must sift through these many pieces of evidence to detect patterns in the alarms that it is receiving. Only this way can it identify the real problem.

Diagnostic Testing ■

Many managed devices have **self-test** procedures. One way to test hypotheses about problems is to send request messages to specific agents, asking them to perform their self-test procedures and report the results.

CONFIGURATION MANAGEMENT

Recall that the MIB keeps certain information about each managed device. The record for that device will contain certain fields or attributes. The device's configuration is defined by the values of these attributes at any moment.

Changing States

At any given moment, the configuration of a device is called its **state.** For a router, the state might include whether the router is turned on or off, whether each port is working, and the pass/fail result of the router's last self-diagnostic test.

The commands that the network management program can send to the network management agent include commands to tell the agent to change the state of the router. In the extreme, the command might tell the agent to change the router's overall state from "on" to "off." In a less extreme case, the network management program might tell the router to turn on a port or turn it off.

Fault Repair ■

One obvious reason to change a managed device's state is **fault repair.** If a network interface card is *jabbering* (transmitting constantly), this will clog the network with traffic. The device might have to be turned off. Or, if a carrier line appears to have a high error rate, its port on the router can be turned off remotely.

Efficiency ■

Another reason to change the state of a managed device is efficiency. If a router's leased line connection to another router is being overloaded, there may be a backup dial-up line. The port attached to this line can be turned on remotely to bring the (expensive) backup capacity into play. Or at night some lines may be shut down to lower costs.

Electronic Software Distribution and Control

One advantage of PC networking is **electronic software distribution (ESD)** in which the network manager installs software on only one server and then copies the software to other servers and even to client PCs. This greatly reduces the work needed to install new software and upgrades.

Especially for client PCs, this requires considerable knowledge of the PC's current configuration, including its currently installed software and the directories in which the programs reside. As noted earlier, the DMI standards can report such information remotely, as can Microsoft's variant of the DMI standard.

Remote **software inventory** tools can bring other benefits. For instance, a firm can **audit** software in use to ensure that license restrictions are not exceeded and to ensure that PCs do not contain games or other unauthorized software.

More positively, **software metering** tools can go beyond auditing to help firms understand how software is being used, so that the firm can buy only what it needs. Software often is a firm's largest investment in information systems and needs to be managed prudently. As software licenses become more complex, companies will need ever greater information on software usage.

One problem with software inventory (and hardware inventory) tools is that they tend to be platform specific. If a firm has a mixture of IBM compatible PCs, Macintoshes, and UNIX workstations, automating software inventories may require several different and nonconnected tools.

Current Network Map

One goal of configuration software is to give the network administrator a current network map. This will be an up-to-date graphical representation of all pieces of equipment and software. When there are additions, changes, and moves, it will be very easy to update this map; in fact, this should be done automatically as much as possible.

Remote Management

At smaller sites, it is difficult to have a network staff. Even in medium-size sites, there is seldom a night staff. With sufficiently good software, it will be possible to do a great deal of configuration management remotely. This is sometimes called *lights out* operation.

Directory Servers

As discussed in Chapter 8, if you have many resources on your network, you would like to have a directory server to keep track of these resources. These directory servers fall under configuration management because they document how directories and files are configured on the various servers, client PCs, and other computers on the network.

OSI has produced the **X.500** standard for directory servers. Most directory server products are based on X.500. Unfortunately, most vendors deviate from X.500 in their directory server products. This makes interoperability poor among directory servers from different vendors.

PERFORMANCE MANAGEMENT AND PLANNING

The network administrator has to understand how a LAN, WAN, or enterprise internet is performing in order to make plans for the future and to optimize the network or internet.

Vital Statistics

A key issue in performance monitoring is what to measure. Classically, performance monitoring has involved several statistics on *reliability* and *transmission speed*. Each vital statistic sheds different light on the parameter it measures. *For more on this topic, see the box "Reliability and Transmission Speed Statistics."*

What-If Analysis in Performance Testing

If performance is flagging, a firm may wish to take steps to improve it. In addition, network administrators may wish to forecast how adding a new server or how adding a transmission line will change performance.

Network design programs, fed with data on actual network performance, may allow the network administrator to do **what-if** analysis to forecast how various changes will help or hurt the network's performance.

Figure 9.15 shows that some network design programs, for example, show you a picture of the network with each station and device shown as an icon. It may be possible to simply drag a server icon from one network section or another, in order to determine the likely performance impact.

Using network design programs is useful because most networks are complex. Nemzow [1994] gives numerous examples showing how difficult it can be to anticipate bottlenecks during changes. In many cases, the results of a change can even be the direct opposite of what you would expect. A survey

Reliability and Transmission Speed Statistics

Availability

One issue in network reliability is **availability,** which measures the percentage of time a user can send data through the network. A figure of 98 percent or some other value may be set as a standard, depending on the cost of downtime.

Response Time

Another characteristic of reliability is **response time,** which is usually measured in seconds. This is how long the system at the other end of the network takes to respond to a message. A goal of having 0.25 seconds as a maximum response time may be selected.

Availability and response time are interrelated. If a network is congested, one way to improve response time is not to allow new users onto the network. This, of course, harms availability.

Outage Frequencies and Durations

Other measures of reliability focus on how frequently failures occur and how long it takes to fix them. **Mean time between failures (MTBF)** is usually measured in days or even hours. This tells how long the network will operate, on the average, between failures.

Another outage measure is **mean time to repair (MTTR).** This is the average of how long it takes to get the network back up if there is a failure. Hopefully, this is measured in minutes or at worst hours.

Error Rates

Network failures are extreme events. However, networks may constantly create errors in packets—although hopefully at a low rate. Error rates are measured as **incorrect packets per thousand packets, percent of packets with errors,** and so forth.

Measuring error rates normally requires a tool called a **protocol analyzer.** This device samples traffic on the network (or at least on its segment of the network). It analyzes each packet to determine its protocol type and whether it contains an error. This allows a firm to discuss error rates by type of protocol instead of having just a single number for error frequency.

Throughput

Up to now, we have focused on errors; but transmission speed is also important. Typically, speed is measured in **throughput,** which is the **number of bits per second (or packets per second)** really traveling through the network. Normally, throughput is much lower than the rated speed of the network. In a 10 Mbps Ethernet network, throughput over about 3 Mbps can create increasing latency in delivery.

Throughput is also rated as a **percentage of capacity.** For instance, an Ethernet supervisor will become wary if traffic reaches about 30 percent of the network's published speed.

Throughput is normally limited by a bottleneck. This might be a slow server, inadequate capacity on one transition line, and so forth. Often finding and fixing a bottleneck can give a substantial increase in throughput at comparatively low cost.

Peak Period Analysis

In measuring reliability and speed, firms often focus on **peak periods,** such as the morning rush to read electronic mail. Reliability and throughput problems tend to be worst during the peak period. Often figures are quoted both as overall averages across the business or clock day and as averages during the **busy hour,** which is the hour with heaviest traffic.

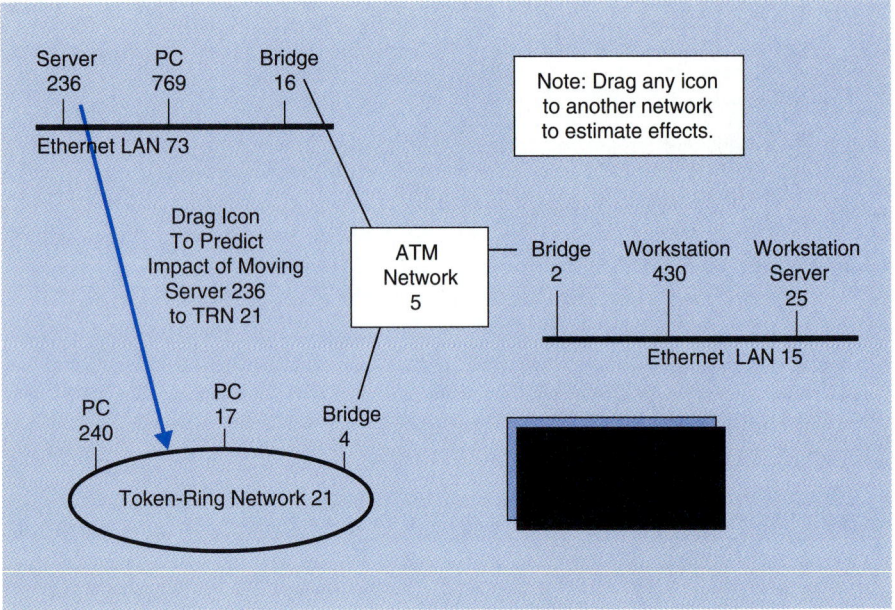

FIGURE 9.15

Network Design Program

A network design program usually gives the network administrator a schematic view of the network with individual devices appearing as icons. Some allow the user to drag an icon from one part of the network to another, in order to see the effect of the change. This is *what-if* analysis. Other forms of what-if analysis include adding a new device or transmission line or removing something from the network.

by IDG in 1993 looked at reasons for performance bottlenecks: 31 percent of the problems were due to design errors.

Network design programs provide a much more detailed picture of how the network performance is likely to change following a specific change to the network.

Planning New Networks

Of course, network design programs are also useful in designing *new networks*. The difference is that for new networks, there will be much more *uncertainty* because there will not be a solid base of actual use statistics.

Everyone agrees that the most crucial step in designing new networks is to understand the user needs to be served by the network. Unfortunately, this is always difficult, especially for new networks, where you do not have good base data on past usage patterns.

In addition, there is the **turnpike theorem,** which states that if you build a turnpike (fast road), traffic will always increase considerably compared to its level before the turnpike. So even if you have solid data on past behavior, if you install more network capacity, you can expect substantial increases in actual usage. It is important not to base upgrades on current usage patterns alone.

In new network design, *what-if analysis* is also needed. For instance, competing carriers typically offer different pricing and service arrangements. Comparing prices requires the network designer to see which carrier service would be better under the specific traffic statistics expected on the network.

Client/Server Performance Management

The blackest black hole in network management today is managing client/server processing in which there are dozens or hundreds of servers and an order of magnitude more client PCs. Even knowing how many of the servers are operational is difficult. Tuning their performance from a central site is even a more difficult problem. Tuning clients and even being sure that they are running the proper version of the client software are also impossible or extremely difficult.

SECURITY MANAGEMENT

Surveys of network managers almost always put security at the top of their lists of concerns in network management. The 1995 Ernst and Young Information Security Survey collected data from 1,293 network managers [Panettieri, 1995]. Seventy percent said that security threats had worsened in the last five years. Nearly half said that they had suffered a security-related financial loss in the previous two years. In 20 cases, the loss had been over a million dollars. Seventy percent had had a serious virus attack in the previous year. Five out of six were running key business systems on LANs, but 60 percent were dissatisfied with LAN security. One in five said that there had been a break-in or an attempted break-in, but these figures may have understated the actual threat. In one experiment, a government group was able to break into 80 percent of a sample of government installations, but only 10 percent of the sites realized that it had happened [*InformationWeek*, 1995].

Encryption and Authentication

When you send information to another person, you normally do not want others to be able to read the message. This is called **privacy** or **confidentiality.** One way to ensure that others do not read your message is to encrypt it. **Encryption** turns messages into unreadable bit streams, so that others cannot make sense of your messages if they succeed in intercepting them. Encryption is especially important in monetary transactions.

There are several forms of encryption. Although most current systems use older forms of encryption, the future probably belongs to **public key encryption.** In this form of communication, you have a private key that only you know and a public key that you publish widely. (See Figure 9.16.) Anyone can encrypt a message with your public key, but only you can decrypt it with your private key. No one else can decrypt the message because they do not have your private key.

The sender *encrypts* the orginal message (*plaintext*) using the receiver's *public key*. This public key is not secret. It is widely published, so that many people can send the receiver encrypted messages. However, the sender, after encrypting the message, cannot decrypt it using the public key. Neither can anyone else knowing the public key. The public key is only good for encrypting messages. The receiver decrypts the arriving *cyphertext* (encrypted message) using his or her *private key*. The receiver must keep this private key secret so that nobody else can decrypt messages that have been encrypted with the receiver's public key.

Another concern is **authentication,** that is, ensuring that the message really comes from the person who claims to be sending it. You especially wish

FIGURE 9.16

Public Key Encryption

The sender *encrypts* the original message (*plaintext*) using the receiver's *public key*. This public key is not secret. It is widely published, so that many people can send the receiver encrypted messages. However, the sender, after encrypting the message, cannot decrypt it using the public key. Neither can anyone else knowing the public key. The public key is only good for encrypting messages. The receiver *decrypts* the arriving *cyphertext* (encrypted message) using his or her *private key*. The receiver must keep this private key secret so that nobody else can decrypt messages that have been encrypted with the receiver's public key.

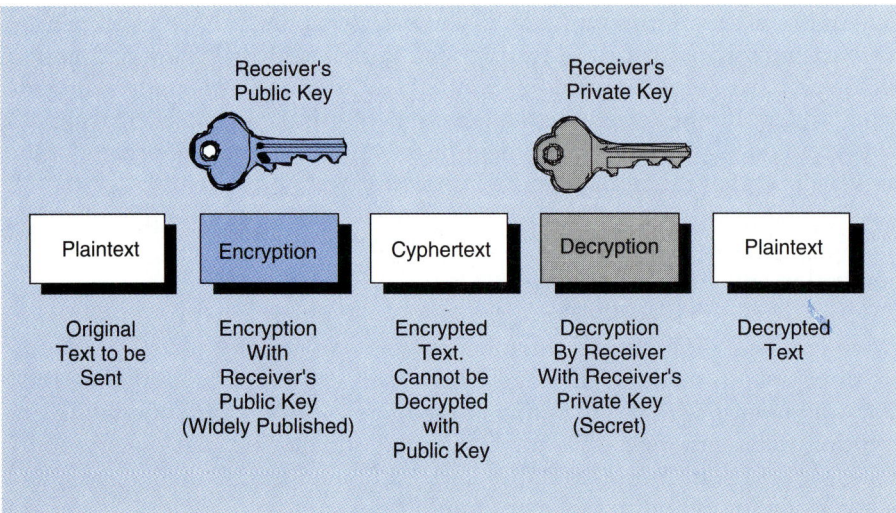

to guard against someone who sends you a message requesting a monetary transfer. There are now reasonably good schemes to allow you to authenticate incoming messages. *For more on encryption, see Online Module J.*

Access Control

Although encryption and authentication are interesting technically, most thieves use simpler techniques to break into a system or to read messages than breaking encryption methods and finding the correct keys. Companies must guard against these approaches too. In **access controls,** you take active measures to prevent others from reaching your system. People who attempt to access your system or some part of it without authorization are called **hackers.**[1]

Physical Control ■

The strongest access controls are **physical controls.** In PC networking, if you place all PCs and their server in a single room, getting access will be extremely difficult. If this is impossible, you can restrict the user to one or two client PCs, so that someone cannot log in from any free machine. You can also restrict usage times, say not allowing people access after 6 p.m.[2]

While various forms of physical security offer relatively strong protection, they can be a nuisance (for instance, if a person needs to work late some time or call in from a hotel room). The primary objective of security is to match the degree of control to the degree of risk. If you have low risk and high controls, this will be extremely inconvenient for users, and the added cost to users will not justify the security.

Logical Control ■

In logical control, you use nonphysical ways to establish the identity of the person wishing to access your system. The most common way to do this is to use **passwords** during login. You can strengthen this by requiring passwords to be a certain length, requiring that they not be dictionary words, and requiring that they be changed frequently. You can also add a second layer of passwords for sensitive parts of the system.

In higher-risk situations, you might use **challenge and response** identity verification. This is similar to challenge and response authentication. You give a person a test number. He or she performs some calculation on it using a secret method. In the simplest form, the person would add 1 to the number. So if you sent the number 23 and the person responded with 24, you would es-

---

[1] Many information scientists use the term *hacker* to refer to anyone with extensive computer skills who does something intricate and tricky. They call people who break into systems *breakers*.
[2] If the person is at home, you can use *dial-back modems*. After the person dials in, their dial-back modem hangs up. The server or host computer then calls the person back at his or her standard telephone number. This ensures that someone is not logging in from an unauthorized location.

tablish the person's identity. Of course, this algorithm is too simple. Many challenge and response systems use **smart cards** that look like credit cards but that contain microprocessors. They can execute complex algorithms.

Firewalls ■

As discussed in Module B, attaching a firm to the Internet can create threats from millions of people on the Internet. Figure 9.17 shows that many firms install a **firewall** between the Internet and their internal network. This limits access to certain IP addresses and may impose other limitations. There are several categories of firewalls. Higher categories are more expensive but can block more sophisticated threats. *For more on firewalls, see Module B.*

Audit Trails ■

The last element of access control that we will consider is the creation of audit trails. The idea is that every time that a user takes a potentially important action, such as logging in or accessing a sensitive file, the information is recorded. This creates an **audit trail,** meaning that an auditor can go back and trace a user's actions.

FIGURE 9.17
Internet Firewall

A *firewall* protects your company from access to your network from the millions of people on the Internet. The firewall is a computer that controls who may access your internal systems. There are several categories of firewalls, offering varying levels of protection at various costs.

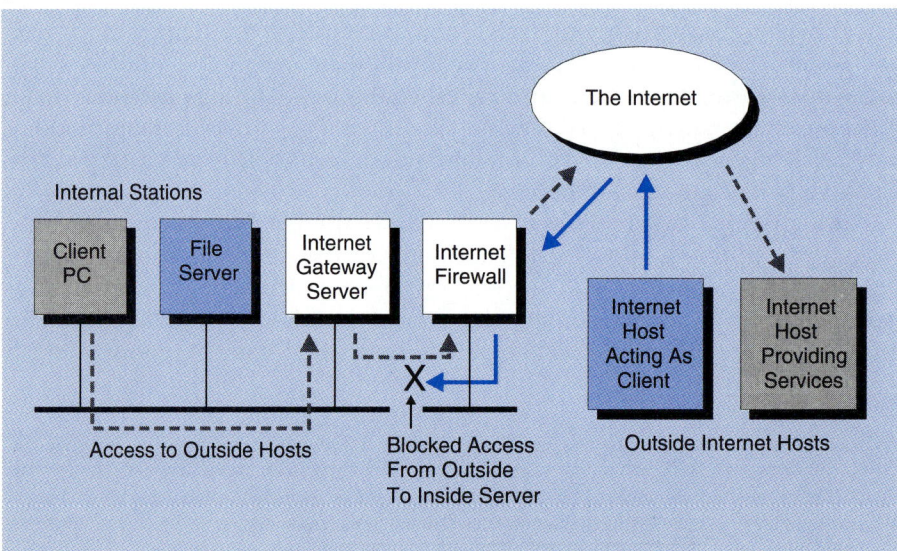

The best audit trail software scans its database for suspicious patterns. This can allow a systems administrator to identify users who may be posing threats, sometimes while they are still working on the system.

Virus Control

Virus control is one of the most vexing aspects of personal computing. Nearly every firm has been hit repeatedly by PC viruses. Viruses are becoming increasingly stealthy and difficult to catch. Many firms now require the execution of *PC virus control software* whenever someone turns on their computer or inserts a floppy disk.

Networks multiply the potential for virus infections. In file server program access (see Chapters 2 and 8), software is stored on a file server but is executed on the client PC. If an infected program is stored on a file server, it may infect dozens or even hundreds of client PCs.

Several companies now offer **LAN virus control software** to detect infected programs. Some execute on the server itself. Some are downloaded to the client PC and execute there.

Most server operating systems are more difficult to infect than client PC operating systems. As a result, viruses on file servers themselves are uncommon. However, this is likely to change in the future, as breakers become more sophisticated.

ACCOUNTING MANAGEMENT

Our final topic is **accounting management.** This will be our briefest topic because it is the least developed and used aspect of network management.

In the past, networking and computing in general typically was treated as overhead and was not charged to specific departments, projects, and users. But information technology (IT) is becoming rather expensive, and some firms now apply **chargeback** pricing to individual departments and projects. Departments and projects are charged a cost based on their degree of use of various resources. The basic goal is to encourage rational economic behavior.

Chargeback is particularly important in industries where customers pay by the project, as in the legal industry and in contract engineering. The client is billed for all costs incurred, including IT costs.

OUTSOURCING

Many firms are engaging in outsourcing for many of their business functions. In **outsourcing,** a firm contracts with another company to handle some aspect of the contractee's business functions. A basic rule in outsourcing is to retain *core functions* in which you have a competitive advantage and through which your firm adds value. You can outsource other functions, leaving management freer to focus on its core competence. A second basic rule is that you should think seriously before outsourcing a function whose poor performance could destroy your firm.

In recent years, many firms have outsourced much of their information systems functions. In firms that did not oursource all functions of IS, it was common to see networking and network management retained in house. It was felt that networking was a function that should not be outsourced. Now, however, lower-level networking has stabilized enough to permit outsourcing of at least some network functions. Many firms feel that network management, however, would be a poor area to outsource because of its potential for both good and harm.

CONCLUSION

In this chapter, we began with the basic elements of a network management system, including the network management console, the network management program, the MIB, managed devices, and network management agents. As an information systems professional, you should be intimately familiar with these terms and their interrelationships.

The Simple Network Management Protocol of TCP/IP is the most widely used network management standard. However, OSI's CMIS/CMIP and SNA's NetView standards are also important. There are also specific standards for managing client PCs, managing servers, and managing carrier networks. There is a growing number of multifamily network management systems that can work with network management agents from multiple network management protocol families.

We also looked at OSI's five functional management areas. Fault management is probably the most pressing concern in networks, but configuration management and performance management are also pressing concerns. Perhaps the greatest problem of all is security; unfortunately, this is one of the least mature aspects of network management today. Finally, accounting management deals with chargeback and other accounting matters.

CORE REVIEW QUESTIONS

1. What are the elements in a network management system? What are the roles of the network management console, the network management pro-

gram, the MIB, the managed device, and the network management agent?
2. How is network management communication an example of client/server communication? What is the client? The server? What is the function of alarms?
3. Distinguish between services and protocols.
4. At what layer does the client/server communication between the network management program and a network management agent take place?
5. Which is the most widely used network management standard? What is OSI's standard? How do the two systems mentioned in your first two answers in this question compare? What is the third network management standard from a standards architecture? Why is there a trend toward multifamily network management programs?
6. Name the five OSI functional management areas. Briefly characterize each.
7. Why is fault diagnosis difficult? How does configuration management work with fault management in failure recovery?
8. Why would you want to change the state of the managed device in configuration management?
9. Why are software inventory management and electronic software distribution important?
10. What do network design programs do? What is what-if analysis in these programs? What is a bottleneck?
11. What are encryption and authentication, and why do we need each?
12. What are the four elements in access control? (The first is physical control.) How does each help?
13. Why is chargeback desirable?

DETAILED REVIEW QUESTIONS

1. What are three basic functions of the network management program? Which of these are standardized?
2. In the MIB, there is a separate table for each type of _____. What does a record represent?
3. Does the network management agent consist of hardware, software, or both?
4. Why are distributed network management systems attractive?
5. What are the elements of the MIB database schema?
6. What is RMON, and why is it important?

7. What is the major network management standard for client PCs? Why do servers need network management? Why would it be desirable to incorporate carrier standards into network management data collection?
8. Distinguish between software auditing and software metering.
9. From the following box, "Reliability and Transmission Speed Statistics," what are the major vital statistics for reliability and transmission speed? Describe each briefly. Why are peak periods important in performance monitoring and planning?
10. How are planning changes in an existing network and planning a new network similar? How are they different?
11. What are the two basic forms of logical access control?
12. What is outsourcing? What are the two basic rules in deciding which business functions to outsource?

THOUGHT QUESTIONS

1. Why do you think concerns about network management are growing?
2. What isn't the reporting of summary information to managers standardized?
3. Why aren't a network's rated transmission speed and its throughput the same?
4. Do you think a great deal of networking will be outsourced in the future?

For online exercises, please visit this book's website at: http://www.prenhall.com/panko

REFERENCES

BATTELLE, J. (1993, June). "Special Report: Network Management," *Corporate Computing*, p. 122.

BRUNO, C. (1993, May 17). "Optimism Is Guiding Force in Prepping '94 Budgets," *Network World*, p. 1.

InformationWeek (1995, July 31). "Three Questions," p. 10.

NEMZOW, M. (1994). *Enterprise Network Performance Optimization.* New York: McGraw-Hill.

PANETTIERI, J. C. (1995, November 27). "Security," *InformationWeek*, pp. 32–40.

CHAPTER 10

Networked Communication Applications

INTRODUCTION

In a real sense, everything we have studied so far has had one purpose—to show how to link application programs on different machines. The real key to networking—and to computing in general—is application software. Application software is what users need to do their daily work.

The Two Application Chapters

In this chapter, we will focus on communication applications, including electronic mail and a complex set of applications called groupware, including videoconferencing, workflow, and electronic document management, among other applications.

In the next chapter, we will focus on database applications. We will also look at *reengineering* in Chapter 11. Reengineering consists of redesigning how you work instead of automating what you are already doing. Reengineering is the real key to enhancing productivity and performance. It is crucial for communication applications as well as database applications.

Standards

Obviously, we need to be able to buy two application programs (such as electronic mail programs) from different vendors and expect them to be able to work together. This requires standards for horizontal exchanges (protocols) between application programs on different machines. *For more on horizontal protocol standards and vertical interface standards, see Chapter 2.*

Some applications, however, such as electronic mail and workflow, often play a supporting role. For instance, you may compose a message in your word processing program. You might then select "mail" from the word processing program's main menu. As Figure 10.1 shows, your word processing program would then pass your message to the electronic mail program. The electronic mail program would deliver it.

In a sense, electronic mail, workflow, electronic document management, and some other applications can act as glue that holds other applications to-

FIGURE 10.1

Calls to Middleware Applications

Some applications, notably electronic mail, workflow systems, and electronic document management systems, can serve as delivery vehicles for other applications. They are called *middleware* applications. The user's application program sends a call to the middleware program using the middleware program's *application program interface (API)*.

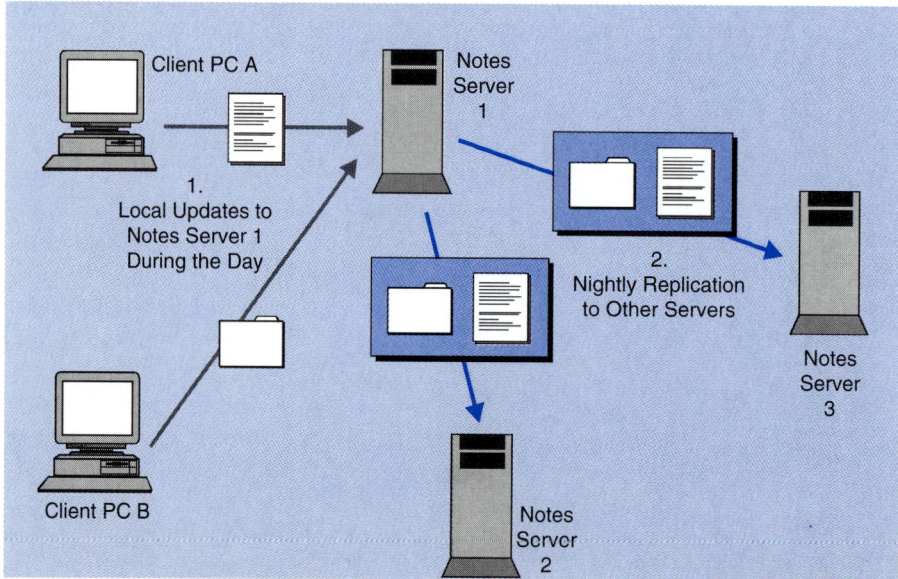

gether. Rather than building email into many applications, for instance, it is much cheaper for vendors to place calls to electronic mail programs when they wish to transmit their messages. We often call such applications **middleware** because, as shown in Figure 10.1, they stand between the top-level application program and the concerns of the transport and transmission layers of the network. When you "mail" your electronic mail message, the email middleware program handles all of the rest of the transaction.

When an application program sends a request to a middleware program, this is a **call**. On stand-alone computers, application programs call the operating system when they wish to print or save a file to disk. In middleware, calling is extended to programs other than the operating system.

From Chapter 2, standards for vertical communication are *interfaces*. The interface between the application program and a lower-layer program is called an **application program interface (API)**.

It is very desirable to create API standards for calls to middleware. In electronic mail, for instance, most electronic mail programs can recognize calls that conform to the **Mail Application Program Interface (MAPI)**. Other emerging APIs will serve electronic document management, workflow, and telephony.

Unfortunately, many types of middleware do not have API standards. In database applications, we will see, most middleware products are proprietary and only accept calls in their proprietary API formats.

Electronic Mail

Electronic mail is or very soon will be the most widely used computer application of any type. It is not hard to imagine why. First, it is extremely *inexpensive*. It is much less expensive than facsimile and may even be cheaper than postal or interoffice mail delivery. It also helps users prepare their outgoing messages efficiently and to dispose of incoming messages (read, print, file, retrieve, delete, and so on), thus reducing labor time.

In addition, email fills an important communication niche. The telephone provides instant communication, but only if the other person is available. Only about a quarter of all telephone calls reach the intended party.[1] Managers often play telephone tag for days, leaving messages for someone and then not being in when that person returns the call.

Facsimile provides instant delivery, but facsimile images are difficult to use in other applications, except as raw images. In contrast, email text can be copied into a document and edited after delivery if appropriate. Physical mail, in turn, is slow. On the Internet, it is called "snail mail."

---

[1]Managers are only at their desks half the day [Panko, 1992]. Even then, they may be in deskside meetings and other activities when they cannot be interrupted immediately to take a call.

Basic Features

We looked at the basic features of electronic mail in Chapter 2. We noted that an email message has a *header* with fixed-format fields. These fixed fields can be processed automatically, like fields in business forms. For instance, a user can search for a message from "Bob Lee" received in the last two weeks. The header is followed by a free-prose *body*. Many users end their messages with *signatures*, which give their full name, the name of their organization, their mailing address, their telephone numbers, their World Wide Web URL, and other information. This is like the information contained in corporate letterheads.

Research has shown that most electronic mail commands are issued in the *disposition* phase after a message is received. People can scan through their mail quickly, read specific messages, dash off replies, file messages for later retrieval, or simply delete unneeded messages. People often get ten to a hundred messages a day, so automating the disposition phase is extremely important for productivity.

One reason why people often get many messages is *distribution lists*. As discussed in Chapter 2, you can create a distribution list, say of members of your department. You give the distribution list a name, say *Department*. If you put *Department* in the To: field, your mail program will send the message to everyone on the distribution list. In a typical mail system, one-to-one or one-to-few messages will dominate the number of messages *transmitted*, while distribution list messages will dominate the number of messages *received*.

Advanced Features

Most mail systems today go beyond the "plain vanilla" capabilities of older mail systems. Mail vendors now attempt to differentiate their products by offering many advanced features.

Attachments ■

Chapter 2 discussed what is becoming one of the most important advanced features of mail systems—the ability to send *attachments*. An attachment is a word processing file, a spreadsheet file, an engineering drawing, or any other binary file. Attachments turn an electronic mail system into a general-purpose file transfer system. There is no need to log into another computer as in FTP (see Module B). There is no need to know the other person's password. You simply attach the file to an outgoing message. The message receiver removes the attachment and stores the file on his or her disk drive. *For more on electronic mail attachments on the Internet, see the box "Internet Electronic Mail Standards."*

Rules, Filters, and Agents ■

Although distribution lists bring many benefits, they tend to deluge users with more mail traffic than they can read. Fortunately, many systems now allow the user to automate how mail should be handled upon arrival. This capability is known by several different names, including **rules, filters,** and **agents**.

Whenever a new message arrives, the mail system looks up the **filtering rules** that the user has specified. For instance, one rule might be that if the From: field has the boss's name, then the message should be flagged with high priority.

Another rule might be that all mail from a certain distribution list should be automatically removed from the in-box and placed in a topical folder for reading when (and if dear) the receiver has time. Such a filtering rule might be used for distribution lists of low interest, such as those dealing with company social outings.

Most filtering rules follow the classic If-Then format. The "If" portion of the rule describes the fields to be considered and the content to seek. The "Then" portion describes what should be done with the message if it is flagged by the If portion.

Computer Conferencing ■

Electronic mail was created for private, one-to-one, and one-to-few messages. But *groups* also need to communicate. For example, suppose you are a member of a project team. In many cases, you have to send a message to everyone on the project team.

One way to do this is for each team member to create his or her own *distribution list* containing the addresses of all team members. Then, whenever someone sends a message, everybody gets it. This approach works well if group membership is static.

If group membership changes constantly, however, as it does in many groups, the mailing list approach breaks down. Even if everybody starts with the correct list, many lists will soon contain the addresses of people who are no longer in the group and no longer wish to get group mail. Worse yet, it will take some time before new members get onto most distribution lists. In the meantime, new members would only get some group messages.

Figure 10.2 shows how **computer conferencing** solves this problem. In conferencing, messages are sent to a conferencing system, which posts it to a common area. This area is often called a bulletin board, giving conferencing systems the synonym **bulletin board** systems. They are also called **forums** because members can broadcast their ideas to everybody, as in a physical political forum.

What happens next varies. Traditionally, when users wanted to see new postings, they had to log into the conferencing system and specify a conference name. Then they could look at the messages leisurely. On the Internet, **USENET newsgroups** work like this. So do traditional computer conferencing programs. *For more on USENET newsgroups, see Module B.*

More recently, many conferencing systems have been based on the **LISTSERV** program or programs that act like LISTSERV. LISTSERV essentially keeps a centralized distribution list that is maintained by a conference moderator. When a message arrives, the LISTSERV program broadcasts it out to

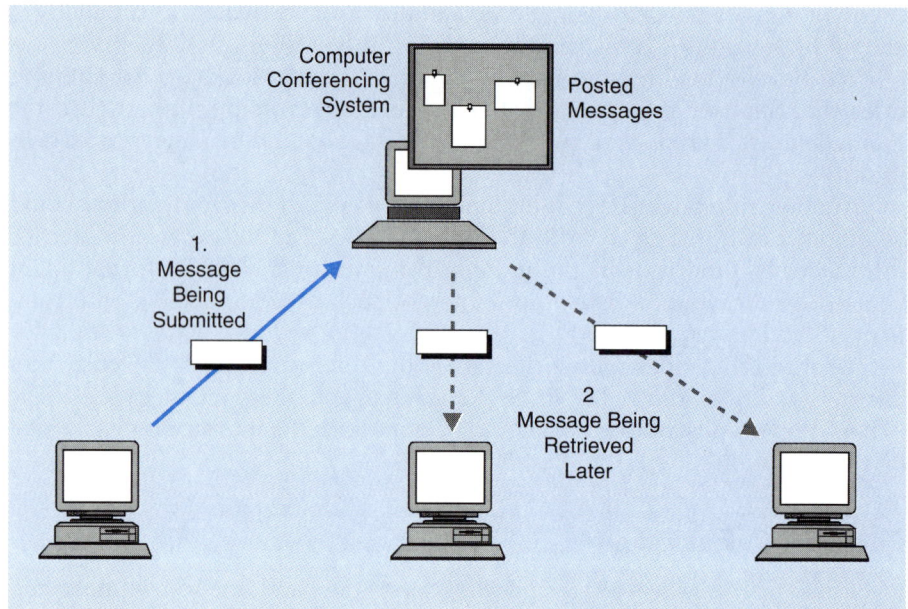

FIGURE 10.2

Computer Conferencing System

In *computer conferencing* systems, users post messages to the conferencing system. Some conferencing systems email the posting to participants. Others require participants to log into the conferencing system to see new postings.

everybody on the distribution list. Members of the group receive the message in their regular electronic mail in-box, along with other electronic mail. *For more on LISTSERV, see Module B.*

Whatever the conferencing system, there are almost always special **moderator tools** to help the moderator run the conference. Most basically, there are functions for adding and dropping members who can access the conferencing messages.

In many conferencing systems, furthermore, the moderator can decide whether messages will be posted automatically. The simplest approach is automatic posting. When messages arrive, they are posted automatically and immediately. To avoid inflammatory messages and other undesirable postings, however, the moderator may elect to screen incoming messages. In this case, messages are only posted after approval by the moderator. Such conferences are called **moderated conferences**.

Mature conferencing systems have many other features. For instance, there usually is an area for storing group documents, such as working drafts, finished reports, and common data files. They are storage systems for many aspects of *group memory*. Hiltz and Turoff [1993] present an extensive discussion of computer conferencing system features.

Electronic Mail Standards

Electronic mail began to grow rapidly in the late 1980s and early 1990s. Unfortunately, most companies soon found themselves with multiple mail systems, none of which would talk to any other. Some systems ran on PC networks. Others ran on workstation networks. Still others ran on mainframes or minicomputers.

Fortunately, there are now relatively good standards for electronic mail delivery. Although they are not everything we want, they typically provide basic connectivity among mail systems.

Internet Mail Standards ■

The most widely used electronic mail standards are those produced by the *Internet Engineering Task Force* discussed in Chapter 1. Even companies that are not on the Internet often use these standards. As shown in Figure 10.3, these TCP/IP standards assume the existence of **electronic mail hosts**. These hosts accept incoming mail and hold the mail until the subscriber is ready to read it. This allows the receiver to work on a client PC, which will often be turned off. Email hosts also transmit outgoing messages to other mail hosts.

Most electronic mail hosts exchange messages through the **Simple Mail Transfer Protocol (SMTP)**, which is in the TCP/IP Internet architecture. As discussed in the box "Internet Electronic Mail Standards," several other standards work with SMTP to standardize both delivery and message structure. *For more on Internet electronic mail standards, see the box "Internet Electronic Mail Standards" containing Figure 10.4: Internet Electronic Mail Standards.*

X.400 ■

In the OSI architecture, mail delivery and structure are defined by the **X.400** family of standards. X.400 and Internet standards have similar functionality. As usual, X.400 standards are more sophisticated and slower to market.

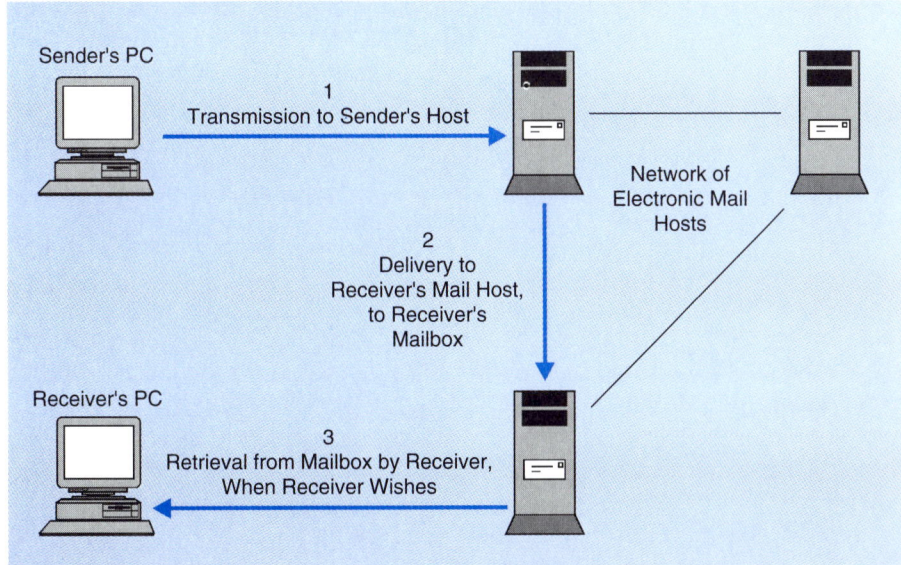

FIGURE 10.3

Electronic Mail Host

Electronic mail hosts receive incoming mail for mail subscribers. They store the mail until the subscriber logs in to read it. They can also transmit outgoing mail to other mail hosts. Hosts typically exchange messages using the *Simple Mail Transfer Protocol (SMTP)*.

Internet Electronic Mail Standards

Internet mail standards are the most widely used in industry. In the body of the chapter, we saw that the core TCP/IP mail standard is the Simple Mail Transfer Protocol (SMTP), which governs communication between mail hosts. There are, however, several other important Internet mail standards.

Post Office Protocol (POP)

There are two basic ways for the mail host to interact with subscribers. As Figure 10.4 shows, the traditional way was to place the mail application program

FIGURE 10.4
Internet Electronic Mail Standards

SMTP is the standard linking electronic mail hosts. The Post Office Protocol (POP) allows users to log into the electronic mail host, download messages to their PC for processing, and upload messages to the host for delivery over the network. Otherwise, the user logs into the electronic mail host during the entire reading, composition, and sending process. There are also standards for message structure, including a traditional all-text standard and a newer MIME standard for multimedia messages.

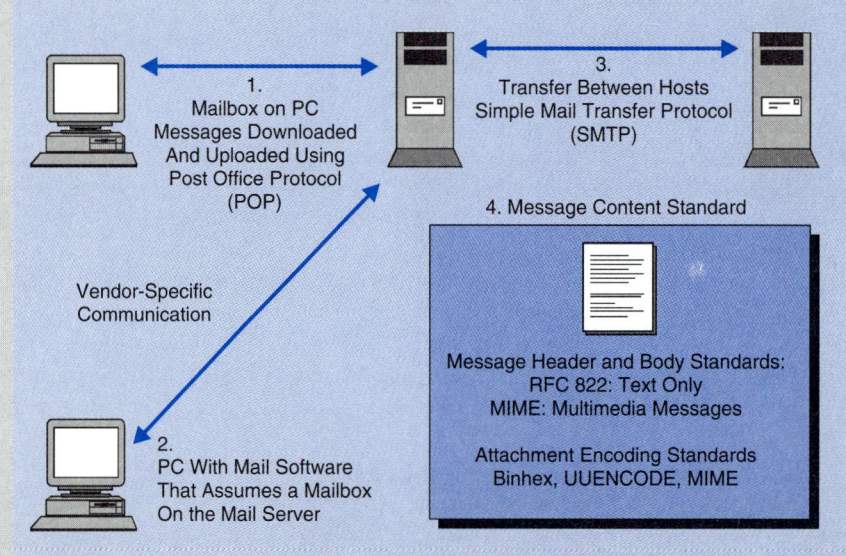

on the server. Then the user could log into the host from anywhere to read and maintain his or her messages.

To handle mail, the subscriber must remain logged into the mail host. This is fine for LAN connections, but it is expensive if the person must pay for connection time to the mail host. (People who are traveling often have to do so.) For such needs, the IETF defined the **Post Office Protocol (POP),** shown in Figure 10.4. Here the electronic mail application software is on a notebook computer or desktop computer. POP allows the user to log in, download incoming mail, send outgoing mail prepared ahead of time, and then log out. When the subscriber has finished reading incoming mail offline, he or she can then create replies, log back in, and upload them to the mail server using the POP standard.

Message Structure

SMTP and POP are both concerned with message delivery, but the structure of a message also needs to be standardized. Recall that the structure consists of a header and a body.

Initially, Internet standards specified that both the header and the body had to consist of simple ASCII text.[3] Even today, email is primarily a text delivery service in most organizations.

The IETF, however, has defined ways to put information other than text into both the message body[4] and the message header.[5] Collectively, these standards are referred to as **Multipurpose Internet Mail Extensions (MIME).** The IETF has added many MIME extensions since creating the MIME framework in 1993.

Internet Encoding for Attachments

One peculiarity of the Internet is that it will only transmit seven-bit ASCII (see Chapter 3). PC ASCII, however, has eight bits. So if you transmit a binary attachment, such as a word processing document, it will not travel through the Internet.

The solution is to **encode** your attachment before transmission. Encoding takes your document and converts it into a legitimate seven-bit form that can travel through the network. To give a crude example of encoding, suppose you

---

[3]RFC 822.
[4]RFC 1521.
[5]RFC 1522.

continued

> wish to send the two bytes 11111111 and 00000000. Then encoding might create three bytes. The first would contain 1111111—the first seven bits of the first byte. The next would contain 1000000—the last bit of the first byte and the first six bits of the second byte. The last byte would contain 00—the last two bits of the third byte.[6]
>
> At the other end, the receiving mail system will have to decode the message. It will return it to the original two bytes: 11111111 00000000. The receiver will then be able to read the attachment.
>
> Unfortunately, there are several encoding schemes in common use. Macintosh users have traditionally used *Binhex*. UNIX users have traditionally used *UUENCODE*. IBM PC compatible users find themselves using both. *MIME* (discussed previously) is also becoming common. Each encodes the attachment differently, so both sides must use the same encoding scheme.
>
> ---
> [6]Although this example gives the spirit of attachment encoding, it would not work in practice.

While X.400 standards have not had as extensive use as Internet mail standards, they are favored by transmission carriers in many countries, particularly in Europe.

Message Handling System (MHS)

In PC networking, the earliest widely used standard was the Message Handling System (MHS) standard promoted by Novell. It is still widely supported, although it has tended to give way to SMTP.

Mail Directory Standards

When organizations have dozens or even hundreds of mail servers, senders must have some way to locate the addresses of intended recipients. First, they must identify what mail system the receiver uses. Then they must learn the receiver's **formal address** on that system. These formal addresses often are highly cryptic, for instance:

$$D54 + bpg$$

Figure 10.5 shows that **directory servers** allow you to give a person's name and other information. The directory then tells you the person's formal address, which you can type into the To: field of your message. For instance, if you ask "Who is Brown?" you will get a list of people named "Brown." You then select the mail server and the formal address of the Brown you are seeking.

Many organizations now have multiple mail servers—often one at each site. As we saw in Chapters 8 and 9, it is possible to synchronize directory servers. As a first step, many companies have installed synchronized mail servers. This allows someone to send email to anyone else in the firm without having to know their mail server. They simply send the message to the person's email address, and the directory servers handle the details. As a second step, many companies are now using synchronized general directory servers that maintain information about a broad spectrum of network resources, as discussed in Chapters 8 and 9. In these generalized directory servers, electronic mail addresses are simply one more set of resources to manage.

Mail Etiquette and Legality

Although electronic mail is widely used, it is still a very young medium of communication. Informal standards of behavior called **etiquette** and legal standards of behavior are still being worked out or are not understood by many users. *For more on electronic mail behavior, see the box "Electronic Mail Etiquette and Laws."*

Email at VeriFone and NuPools

As noted in Chapters 1 and 2, VeriFone makes extensive use of electronic mail. Email is so deeply embedded in the corporate culture that anyone declining to use electronic mail would soon find themselves with another job. Even responding slowly to messages is a major faux pas. In this highly distributed organization, electronic mail is absolutely essential.

VeriFone also uses electronic mail extensively to keep the company's dis-

FIGURE 10.5

Directory Server

If you need to know someone's local electronic mail address, you tell the *directory server* the person's name and other information. The directory server returns a list showing subscribers who match the request. If there are multiple directories, they must be synchronized, so that a user can query a single server for information about any user on the network.

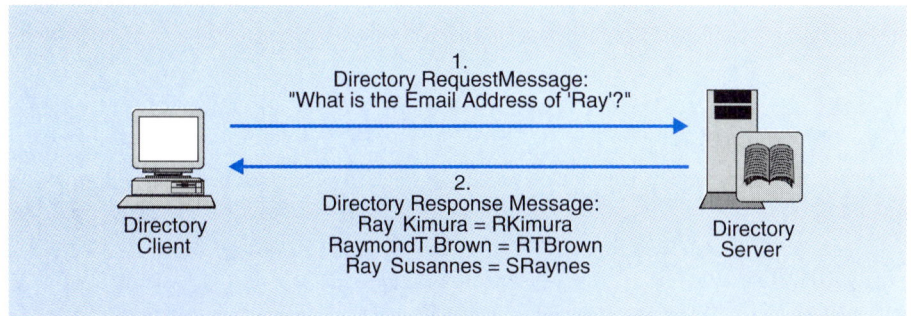

Electronic Mail Etiquette and Laws

Unfamiliarity sometimes leads to rather unpleasant surprises. As a result, many organizations have published rules governing the behavior of electronic mail users. Often these rules are informal and are classified as *etiquette*. In some cases, however, these rules represent corporate policy and can result in sanctions against those who break them. In yet other cases, the rules are set by outside legal institutions.

Flaming

The most frequently discussed behavioral problem is **flaming**. In flaming, a sender sends a harsh emotional message to another user. Often this message is fired off in the heat of the moment after receiving a message from another person. For better or worse, email allows us to dash off replies before we have had a chance to cool down. There is disagreement over how serious flaming is in real corporations. It may well be that flaming in electronic mail is correlated with flaming in other types of media and in meetings in an organization [Davidson, 1995]. Some corporations have styles that are blunt or even rude. Even in these firms, however, flaming in permanent email form may be frowned upon. Some email systems allow you to "take back" a message before the other party has read it, but most do not have this function, and there is no provision for it in standards for mail delivery from one system to another.

Spamming

In the television series, *Monty Python*, there was a skit in which almost everyone is running around yelling "SPAM, SPAM, SPAM." This drowns out the people who are trying to talk.

On the Internet, **spamming** involves copying everybody on every message you send and sending many messages, on the mistaken belief that people want to hear your opinion about everything.[7] Informal feedback from disgruntled receivers is usually sufficient to discourage spamming.

---

[7]There is also a second meaning for spamming on the Internet. It involves deluging others with hundreds or even thousands of email messages. This essentially fills their inboxes, making it impossible for them to find legitimate messages. If the volume is great enough, the user's mailbox size limit will prevent incoming legitimate messages. This *inspamming* is especially bad if the receivers are charged for each message received by their mail service providers.

Forwarding

If you receive a message, you can forward it to someone else. For instance, if you receive a request for a copy of something, you might forward it to your secretary.

Unfortunately, if you send someone a private message, that person can forward it to others—perhaps even to a distribution list. This unexpected forwarding can be done either maliciously or simply with lack of forethought to consequences. Two simple rules are never to send anything that others should not see and never forward something without permission.

Replying

When you get a message, you can send back an immediate reply. This is dangerous if the original message went to multiple people. Some reply functions automatically reply to everyone who received the original mail. If you do not realize this, you are likely to use reply to send a message to the author of the message. Yet everyone who got the original message will also get the reply. This can be acutely embarrassing.

Spoofing

It is not very hard in most mail systems to send a message that seems to be from someone other than yourself. This is **spoofing**. In one company, a message supposedly from the president announced that there would be layoffs the next week. The message really was from a prankster. It took weeks to make this clear to the employees, however, and even then, doubts remained.

Corporate Privacy

Many employees believe that their electronic mail is **private** and that only they can read it. In fact, it appears that companies in most legal jurisdictions have extensive rights to read electronic mail sent or received by their employees. This has resulted in terminations and other sanctions. Unless companies have clear policies stating that email is private, it is best to follow the simple rule that you should not state anything that you would not want your boss or another employee to read.

continued

> ### Legal Discovery
>
> Even if a company decides to make electronic mail private, this has no bearing on how courts view electronic mail. When there is a lawsuit against a firm, there is a **discovery** process in which the plaintiff can require the defendant firm to produce all relevant documents, including electronic mail messages, that bear on a case. Informal messages meant for internal consumption are often very influential with juries. A good rule is never send anything that you would not want a jury to read.

persed workforce "in the know" about the company's activities and performance. Will Pape and other senior managers regularly broadcast digests of key information to everyone in the organization.

At NuPools, email is used primarily by the field site managers, Kevin Brown and Tod Kimura, to communicate with one another and with Luz Martinez, the contracting manager, in the corporate headquarters. Calvin Whyte, the owner, often uses electronic mail when he is on trips, as does Kristi Stone, the purchasing agent, when she travels to learn about new supplies.

GROUPWARE

Perhaps no term in information systems today has been more abused by marketers than *groupware*. Almost any product that does not run on a stand-alone PC has been labeled "groupware" by corporate advertising staffs. Although definitions vary, for our purposes in this book we will define **groupware** as information technology aimed at supporting work processes in managerial or professional groups, where *group* means a workgroup of limited size (say 2 to 20 members) whose members are fairly interdependent.

Note the stress on managerial and professional work. In general these products are not designed for *production systems*, that is, high-volume systems used by clerical workers or other production workers. We will look at production systems in the next section.

Note also the limitation on group size. Most groupware tools are well-suited to the support of workgroups that are relatively flat, that is, do not have two or more levels of hierarchical structure with sharp divisions of labor among subgroups. This may (and should) change in the future, of course.

Another reason for limiting group size is that intensely interacting workgroups in corporations tend to be rather small. In a sample of 165 project teams, only 12 percent were larger than 16 people [Kinney and Panko, 1996]. Subjects were allowed to select any project they had worked on in the last six months, so the data may not be representative, but the trend is impressive. In addition, in a study that examined data on 45 conference room meetings from 14 managers [Panko and Kinney, 1995], the average conference room meeting size was seven people, and extremely large meetings were rare.

Categories of Group Processes

Figure 10.6 shows a taxonomy for group processes. The two dimensions are due to Johansen [1988] and DeSanctis and Gallupe [1987].

Same Time/Different Time ■

A key distinction is whether the people are working together at the *same time* or at *different times*. Same-time work can include face-to-face meetings, telephone calls, desktop videoconferencing, and a number of other types of synchronous communication. Managers spend about 60 percent of their days in same-time communication [Panko, 1992]. In general, same-time communication is highly interactive, with extensive give and take. Ideas fly freely, and the result is a "group thinking" process in which it is very difficult to establish who produced what ideas and why particular decisions were made.

In contrast, different-time processes tend to use text and other permanent artifacts to a greater degree. Because people have more time to compose messages, there tends to be more rational organization to the communication. At the same time, work takes longer to perform.

FIGURE 10.6

Taxonomy of Groupware Dimensions

There are two basic dimensions to groupware. One is whether the participants work at the *same time* (*synchronously*) or at *different times* (*asynchronously*). Another is whether participants work in the *same place* or in *different places*.

	Same Time (Synchronous)	Different Time (Asynchronous)
Same Place	Electronic Meeting Room	Departmental Electronic Document Management (EDM)
Different Place	Videoconferencing Desktop Conferencing	Email Computer Conferencing Electronic Document Management Project Management

Same Place/Different Place ■

The second dimension is whether the participants are in the same place or in different places. Same-place interactions can involve meetings at deskside or in conference rooms. Research has shown that managers spend 55 percent of their working days (not just communication time) in same-place face-to-face communication [Panko, 1992]. Obviously, supporting same-place communication is necessary if we wish to have a large impact on managerial work.

At the same time, it is often inconvenient to bring participants together for a meeting. In a sample of 105 project teams selected by the respondents [Kinney and Panko, 1996], 31 percent of the project team members worked at a different site than the respondent. Over half of the project teams had at least one member from another site, and a quarter had half or more of their members working at a different site than the respondent.

Different-place systems are also important because managers and professionals spend a quarter of their days dealing with written communication [Panko, 1992].

KEYBOARD-BASED GROUPWARE

To many people, the term *groupware* is almost synonymous with Lotus Notes. However, Lotus Notes is only one example of a family of applications called **desktop groupware**. In desktop groupware, the user works at a client PC or some other desktop machine. He or she has one or more keyboard-based applications to help the user participate in group efforts.

Normally, desktop groupware is a different time/different place tool. Other members of the team work at different computers, often at different sites. When they work, furthermore, they work at different times, having access to what others have done previously.

Services

Generally, desktop groupware has a number of related tools.

Verbal Communication ■

The most basic element in any keyboard-based groupware system is **verbal communication,** that is, communication using *words*. Keyboard-based groupware systems tend to use electronic mail as their communication tool. In fact, one early study of groupware indicated that most systems were used primarily for electronic mail [Bullen and Bennett, 1990].

Group Composition ■

In stand-alone computing, word processing is the most widely used type of application software. Groups also need to compose word processing documents, as well as spreadsheet models and other group artifacts. **Group composition** tools extend word processing to groups.

Often one member creates a draft. Others add to it, approve it, or make modifications after the initial author has finished. There may be many iterations in this *different-time* process. In addition, one person may draft the whole document, or different people might draft different parts.

Group Memory ■

Groups need to maintain records of what they have done and why they have done them. They need to maintain a common set of notes, documents, analyses, and other outputs.

Prose documents present special problems, which are addressed by **electronic document management (EDM)** tools. These systems store prose documents in electronic form. This allows rapid retrieval through **full-text searching** or through the searching of fixed-format **catalog entries** for each document. EDM is now being extended to include multimedia documents, including the ability to search for specific graphic elements.

Electronic document management also brings needed control to group memory. EDM can track **versions** of a document that is created over a long period of time. This will allow searchers to find the most recent version of a document. It will also allow the team to go back and look at the changes made over time. EDM also allows a team to create a *retention strategy* to dictate how long different group outputs will be maintained. Not only is storage expensive, but as you increase the number of items stored, you also increase the time and difficulty of locating any individual document.

In some cases, it is critical to retain knowledge of why certain decisions were made. This is especially true in design. **Design rationale** systems not only retain information about what design decision was made. They also retain information on the alternatives considered and the pros and cons of each. This allows the team to revisit its decision later. It also allows the team working on the next-generation design to understand why its predecessor made certain decisions that may at first seem counter-intuitive.

Coordination ■

Group members must coordinate their individual efforts over time. In many cases, such as in approving a document, processes must be handled in a certain sequence, often by multiple team members. As shown in Figure 10.7, **workflow** systems define who should do what, in what order. When one person finishes a task, the workflow system passes the job to the next person. The

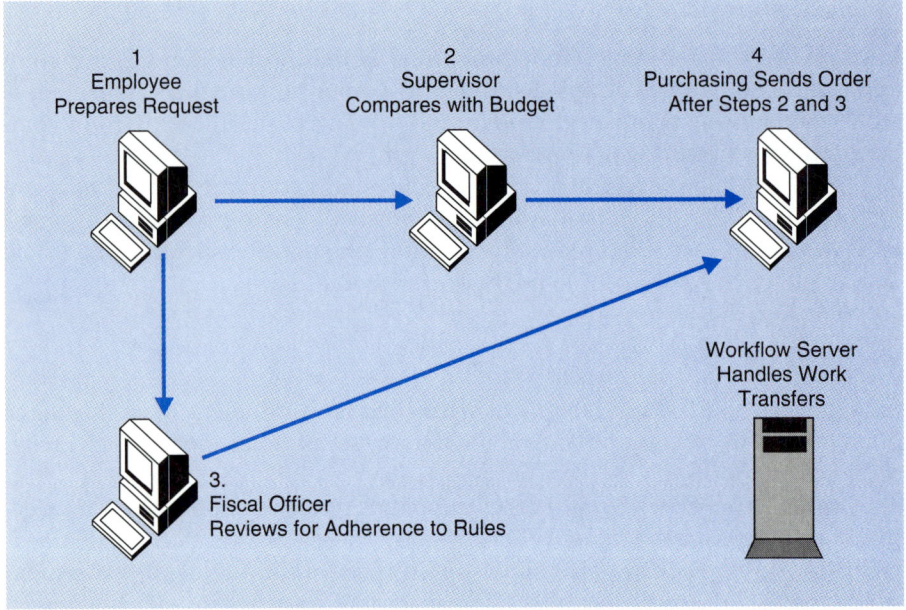

FIGURE 10.7

Workflow Automation

In *workflow automation*, there are many tasks that must be completed. The arrows show that some tasks must be finished before others. The workflow system may provide support for individual tasks. When a task is finished, it routes the work to the person responsible for the next task. The workflow system also provides tools for managing the workflow task.

workflow system may even provide specific tools to each person to help him or her do the assigned work.

Workflow systems also provide tools to *manage* the workflow process. For instance, the administrator is always able to find out who is currently working on the job, in case it seems to have become "lost" in the shuffle of work. The administrator can then contact the responsible person to expedite the work. At the end of each job, the administrator can also see how long the process took. If the process is repetitive, the workflow system can even compute average times for the whole job or for various tasks to help evaluate and improve the workflow process.

Finally, groups need to plan their work. Figure 10.8 represents a **project management system**. In such systems, various tasks can be planned ahead of time. As each task is finished, it is checked off. If a task is taking longer than expected, the project manager can see the impact this will have on other tasks. This may seem like a workflow system, but there is a critical difference. Workflow systems actually *implement* a process, while project management systems are for *planning* future work. The management tasks certainly overlap workflow and project management, and in general it is desirable to have integrated project management and workflow.

Control

Although managers and professionals sometimes chafe at the notion, it is important to control their group processes. Managers and professionals are highly paid. Their work time must be minimized to reduce costs. In addition, time is often of the essence when they work. Managers and professionals often take too long to finish crucial tasks.

We have noted control aspects of several systems already. First, it is important to locate bottlenecks during actual work. Second, it is important to conduct a "post mortem" on each project, to see how long it took and how much it cost. Third, for statistical quality control, it is important to build a database on performance across multiple projects, so that cost and time can be anticipated for future projects, and so that chronic problems can be identified for systematic attention.

Application Development

We have discussed a number of key components found in various groupware systems. Any particular project may require several. Figure 10.9 shows that **ap-**

FIGURE 10.8
Project Management System

While workflow systems implement multitask projects, *project management systems* are designed for planning these projects and managing them.

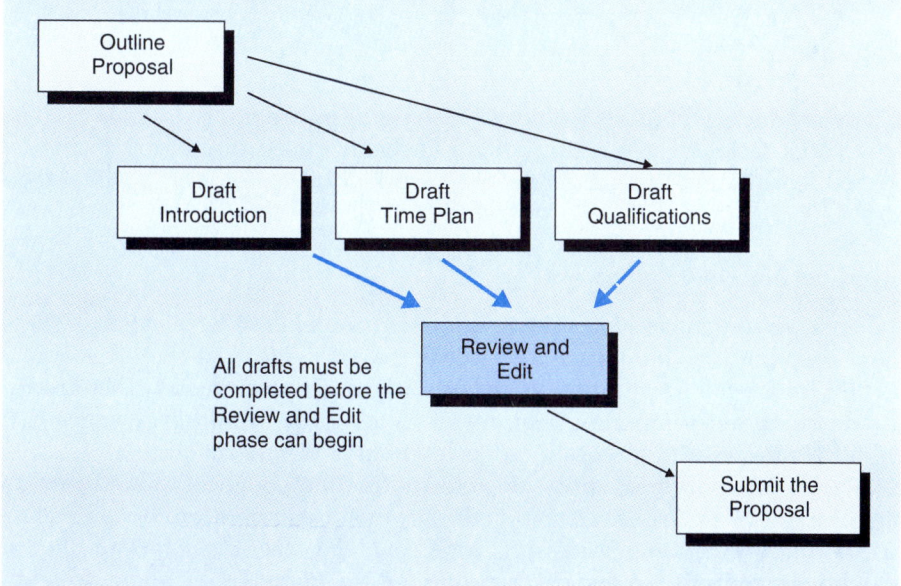

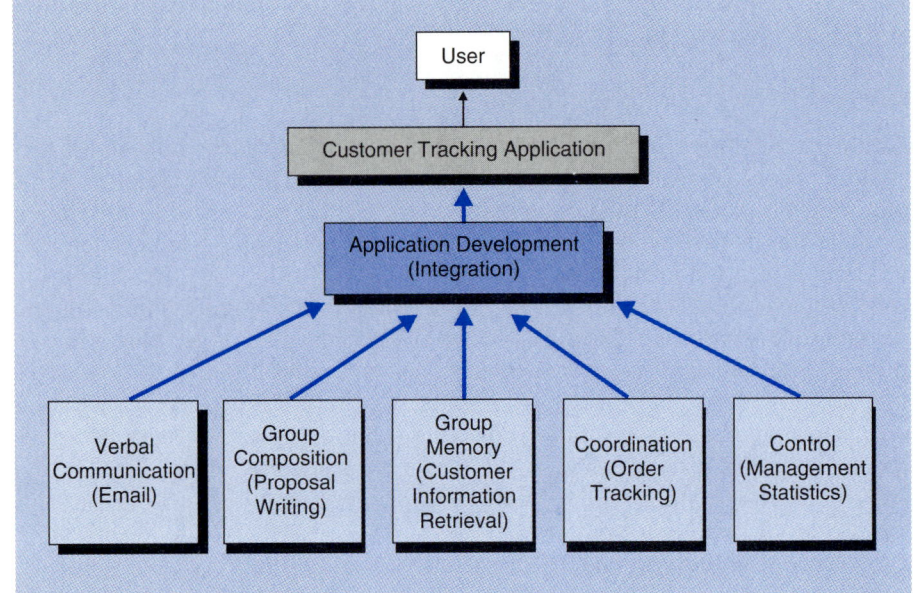

FIGURE 10.9

Application Development

In *application development*, there are tools for integrating multiple modules into what looks to the user like a single coherent application, such as customer tracking.

plication development tools allow the group leader to select a number of modules and link them together into a single application with a single interface. For instance, an application might allow sales force users to track their customers. To group members, it is as if they had a custom-designed group support system.

Lotus Notes

If any product symbolizes keyboard-based groupware in corporations today, it is **Lotus Notes**. Notes is primarily a keyboard-based system originally designed for different-time, different-place work. Most Notes users work at their desks.

Application Development ■

When a person enters Notes, he or she sees a number of applications. These have been put together with Notes' application development tool. This ability to develop custom applications for sales staffs, project teams, and other groups is one of the most important features of Notes. Some would even argue that Notes is primarily an application development environment.

Notes offers a broad array of modules that may be incorporated into an application. Notes has electronic mail, document composition, document retrieval, electronic forms processing, workflow, and other tools. It is adding capabilities constantly, so our description here is certainly out of date. In addition, many third-party companies are selling add-in modules that may be incorporated in applications. This is greatly strengthening the importance of Notes.

Replication ■

What if a company has many Lotus Notes users? This will require supporting a number of Notes servers, as shown in Figure 10.10. Traditionally, Notes servers were limited to about a hundred users, but this can now be increased by moving the server software and data to a RISC workstation server and/or to symmetric multiprocessing, in which the PC or workstation server has several microprocessors instead of just one. This ability to grow to more powerful servers as usage grows without having to change the software and data is called **scalability**. Even with SMP on RISC workstations, however, a company may still need multiple Notes servers. *For more on SMP, see Module I.*

If you have multiple servers, the information on each server must be **replicated** (copied) on other servers. In database applications, such as airline reservation systems, replication across multiple databases must be almost instantaneous. The replication of groupware information, in contrast, usually can be done more slowly. Often replication is done overnight among Lotus Notes servers. This replication is rather automatic and can be tuned by the organization. Along with application development, replication is one of the two key strengths of Lotus Notes.

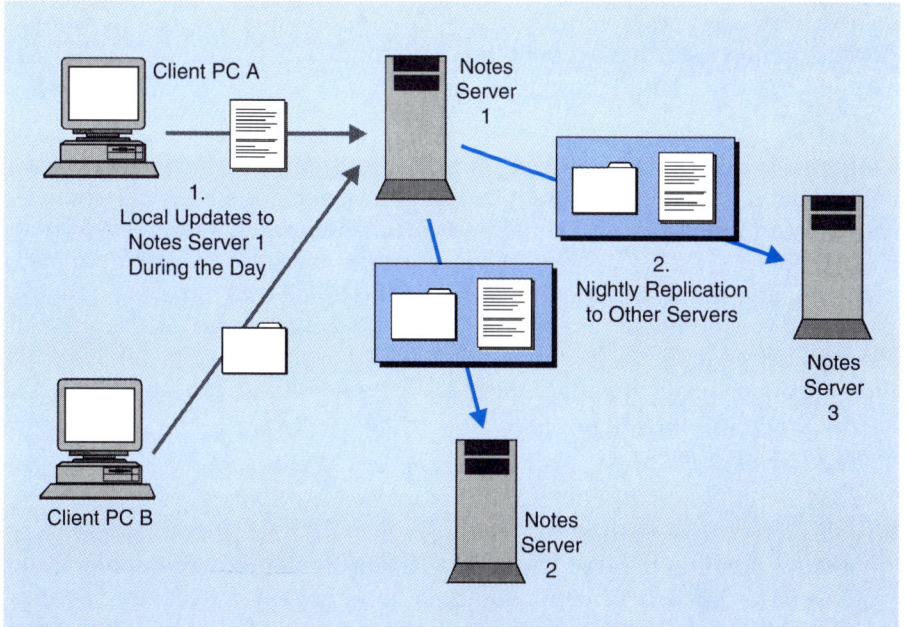

FIGURE 10.10

Replication in Lotus Notes

Information in individual Notes servers must be *replicated* (copied) to other Notes servers. Typically, this can be done overnight because the information does not change rapidly.

Internet Groupware

The growth of the Internet, including the World Wide Web, have created new possibilities for groupware that are just beginning to be explored. Using standard browsers, these systems bring many of the benefits of traditional systems while adding the features of the World Wide Web, including CGI processing on the webserver and downloading applications to the user's PC via the Java language. One pioneering effort was the **TCBworks** system at the University of Georgia. It provides group discussions and voting—tools often found in computer conferencing systems. TCBworks uses CGI to handle user inputs. *For more on the World Wide Web, including CGI and Java, see Module C.*

VIDEOCONFERENCING

When groups have geographically distant members, meetings can be difficult and costly to arrange. Particularly among groups that meet very often and find travel burdensome [Noll, 1976] it may make sense to hold at least some meetings via videoconferencing.

Room-to-Room Videoconferencing

Figure 10.11 illustrates room-to-room videoconferencing in which the equipment is in a conference room. The users in the two rooms can see and hear one another, almost as if they were in the same room.

Costs ■

Videoconferencing has been used since the 1960s, but two things have held it back in the past. One, of course, was cost. However, costs have fallen steeply and will continue to do so. A minimal videoconferencing system costs under $50,000 per site. This is still expensive, of course, but it is acceptable for many purposes, and the cost of a room will continue to fall over time.

In addition to the cost of a room, there is transmission cost. Although room-to-room videoconferencing can be done at 64 kbps, the results are jerky pictures and poor image quality. Fractional T1 leased lines, T1 leased lines, and Frame Relay are needed for good service. *For more on T1, Fractional T1, and Frame Relay, see Chapter 7.*

Standards ■

Another problem in the past has been a lack of standards. Videoconferencing systems from different vendors would not work together. Recently, however, the ITU-T has developed a fairly comprehensive set of standards for video-

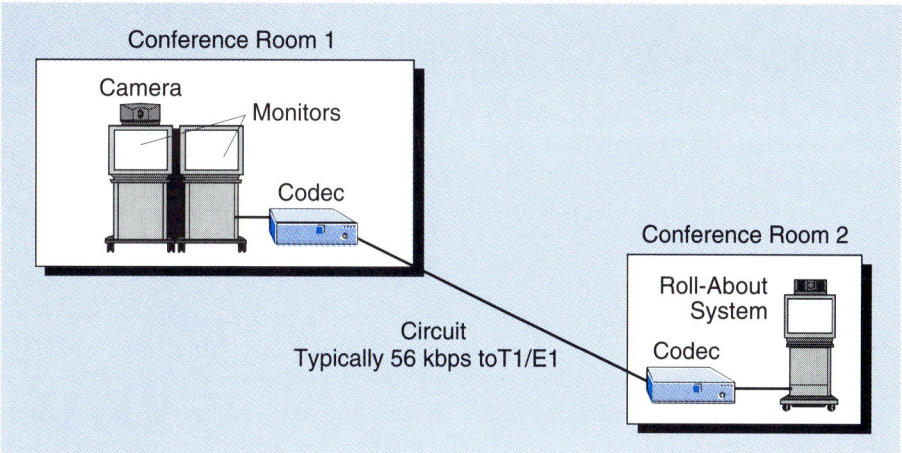

FIGURE 10.11
Room-to-Room Videoconferencing

In *room-to-room videoconferencing*, you need cameras, monitors, microphones, and a control console at each end. You also need a *codec* to convert analog television signals for transmission over digital transmission lines. Monitors typically are computer monitors, which can show high-resolution computer images as well as video images.

conferencing. These are the **H.320 standards**. This family of standards already covers most key aspects of videoconferencing, and ITU-T is constantly adding new standards and enhancing older members of the family. Today interoperability among the systems of different vendors is quite good.

Compression ■

Video signals are analog. High-speed transmission lines are digital. As discussed in Chapter 3, we need a **codec** to translate analog video source signals into digital transmitted signals.

In videoconferencing, codec standards also cover compression. Digitizing a video signal generates a 96 Mbps data stream. By reducing quality and by compressing redundancy out of the signal, videoconferencing can use speeds as low as 64 kbps, at the cost of poorer audio and video quality. *For more on video compression, see Module J.*

Cameras and Monitors ■

The growth of consumer video has slashed the costs of both cameras and displays. Figure 10.11 shows a **roll-about** conferencing system that can be moved from room to room. In 1994, the Business Research Group found that this was the most common type of room-to-room conferencing system [*Data Communications*, 1994]. More expensive units are built into the room's walls and furniture.

Typically, there are two monitors. One might be used to see the remote site, while another might show what the camera is transmitting from this room.

In addition, there usually are two cameras. One might show all of the people in the room, while another might point at the person speaking. This allows rapid switching between views.

At the desk facing the cameras and monitors, there are microphones. These tend to be mounted almost flush with the surface of the desk.

Graphics

Graphics has always been a need in conferences. Many early systems had cameras pointing down from the ceiling, so that a participant could draw on a small pad. This worked, but television has very low resolution.

Today most videoconferencing monitors are computer monitors with at least VGA screen resolution.[2] It is possible to link the display in one room with a computer in the other room. This will allow both sides to see documents and computerized "slide shows" in high resolution.

Control Consoles

The moderator at each end has a **control console**. This allows him or her to set up the conference and to control such things as sound volume and where cameras are pointing. In some cases, the moderator can even remotely control the camera in another room, just as in a real meeting you can decide where to look.

Multipoint Conferences

Traditionally, videoconferences only linked two sites. But there are other possibilities. Figure 10.12 shows a multipoint videoconference. This requires a **multipoint control unit (MCU)**. These are standardized under the H.320 standards set, in **T.120**.

Benefits

Videoconferences are less expensive than travel, although not by as much as one might think. When people travel, they usually attend several meetings.

In general, travel benefits are more subtle than straight one-for-one trade-offs between face-to-face meetings and videoconferences might indicate. Many groups meet periodically face-to-face, then meet by video in between. Having video may allow them to meet more than they would if they had to travel. This adds to cost but improves communication.

Video has many other benefits. If travel is necessary, then most firms limit the number of participants. However, with video meetings, it is common to have additional people present at each end. This brings more people into the

---

[2]VGA screens have 480 horizontal lines with 640 pixels (dots) per line. As the box "Video Compression" discusses, a video picture has only half this horizontal and vertical resolution. If you display a computer image on an ordinary television, it will be very blurry.

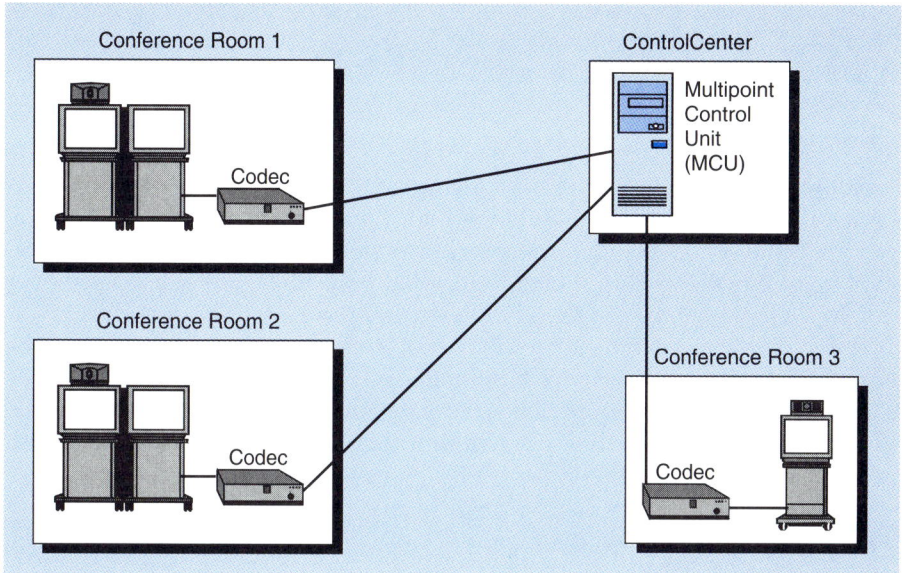

FIGURE 10.12

Multipoint Videoconferencing

It is possible to link multiple sites in a videoconference. This requires a *multipoint control unit (MCU)*.

communication. If additional experts are needed, furthermore, they might be just down the hall.

Video meetings tend to be shorter than face-to-face meetings. Perhaps because of the cost involved or perhaps because video is less emotionally warm, video meetings tend to be more formal and businesslike. As long as there are adequate opportunities for getting to know other members of the team, this briefness can be very beneficial, given the number of people involved in many meetings, and given their salaries.

Videoconferencing at VeriFone and NuPools ■

Although VeriFone is noted for its extensive use of electronic mail, to build and maintain links among people, its chief information officer, Will Pape is concerned about the limitations of electronic mail for human contact. The company is now exploring videoconferencing in an effort to bring more effective interactions when its people communicate over long distances.

NuPools has only used videoconferencing once. Kristi Stone, the purchasing manager, used it to communicate with a supplier. Kristi drove two miles to a copy center that had a videoconferencing system. There she chatted with a supplier, who had an in-house videoconferencing system. The supplier demonstrated several products. Kristi was quite impressed by the presentation.

Desktop Conferencing

Standards for videoconferencing were first created for room-to-room conferencing. However, **desktop videoconferencing** is now becoming attractive, especially on LANs, where there is high transmission capacity. The growth of ISDN also allows fairly rapid transmission between desktop machines. *For more on ISDN, see Chapter 7 and Module G.*

Equipment ■

Unfortunately, room-to-room conferencing standards are not yet ideal for desktop use. As a result, users often have to buy proprietary systems, such as Intel's ProShare desktop videoconferencing system. These systems usually require at least ISDN speeds, which means that you either cannot use them over ordinary telephone lines or that, if you do, the results will be very poor.

Intel's ProShare is the state of the art in desktop videoconferencing at the time of this writing. It has a voice channel operating at 16 kbps. The rest of its 128 kbps ISDN capacity is split between data and video, with data having priority and with video having a maximum rate of 96 kbps. This gives decent quality video, although the video freezes when data are transferred.

Document Conferencing ■

ProShare and competing systems also offer **document conferencing**. This allows the communicating partners to see the same document. In general, they can also mark up the document. This is a valuable feature because many telephone calls are made to discuss a document, spreadsheet, schedule, or some other form of visual information. In general, document conferencing uses VGA resolution, rather than the lower video resolution of the conferencing system. It can do this because it does not have to transmit many video images per second. It only has to transmit a single VGA screen image. *For more on document conferencing, see Module J.*

Telephony

In the past, groupware was seen as standing in opposition to telephony. However, managers spend 5 percent of their days on the telephone, and voice is an extremely important medium of communication. In the market, competing Microsoft and Novell telephony API standards are now trying to link groupware with telephony.

Ideally, in the future there will be no distinction between email messages and voice mail messages. Both will arrive via a common in-box. One will even be able to copy voice messages into document, as annotations.

Electronic Meeting Rooms

Videoconferencing links people at different sites, but traditional conference room meetings, in which everyone is present physically, also need support.

Managers spend a quarter of their days in conference room meetings [Panko and Kinney, 1995]. When they leave their desks, which are loaded with computer hardware and network connections, they often enter a world where the highest technology is an overhead projector. They need better support for the same time/same place conference room meetings that consume so much of their lives.

Hardware ■

Figure 10.13 shows the technology of an **electronic meeting room (EMR)**. Each participant has a PC. In addition, there is a *moderator* who has a special PC that controls the functioning of the EMR. There is also an EMR server to hold EMR application programs and data. These machines are linked together by a LAN.

In addition, there is a **public screen** (sometimes several) large enough for everyone to see. A video projector paints shared images on this screen. This video projector can cost $10,000 or more and tends to be one of the two main costs in the EMR. The other is application software.

FIGURE 10.13
Electronic Meeting Room

Each meeting participant in an *electronic meeting room (EMR)* has a PC. The moderator also has a PC to control interactions. Application software is stored on the *EMR server*. A LAN links the PCs together. There is a public screen for showing common information.

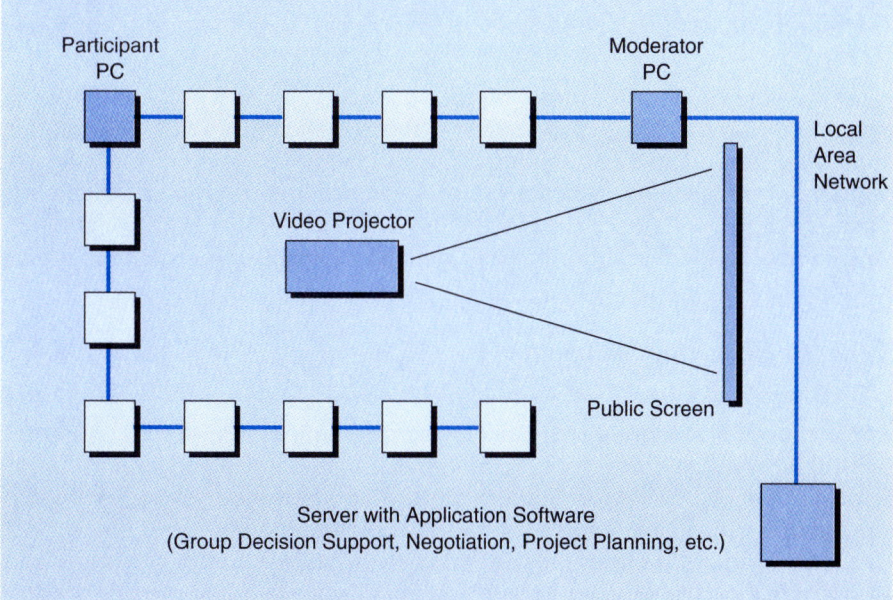

Application Software ▪

As in any system, application software is the real key to productivity and performance. Typically, EMR application software is organized around an **agenda system,** which specifies how the meeting will flow.

For each agenda item, the moderator will either have a traditional live discussion or will select an application software tool. For instance, suppose that an agenda item is a report. The moderator might call up PowerPoint or some other **presentation tool**. The presenter would then take over the presentation.

The next item might be an exploration of possible uses for a new product. A **brainstorming** tool would allow everyone to type in ideas. Because participants can type in parallel, electronic brainstorming groups almost always produce many more ideas than face-to-face groups [Gallupe et al., 1992]. In addition, contributions can be **anonymous,** so others will react to the quality of the suggestion rather than to the status (high or low) of the suggester.

When it is time for a decision, the moderator might call up a **voting tool**. The tool will not only handle voting, but it will also show any patterns in voting. For instance, if a ranking method is used, items that receive very different rankings from various people may be understood differently by the participants. An initial vote may turn up problems in understanding.

In a project meeting, the moderator might bring up the **project management system** and review the status of various tasks.

Another application with apparent potential is **group composition**. Groups can meet in an electronic meeting room to do joint writing or to review and edit a document produced before the meeting.

PRODUCTION SYSTEMS

So far, we have looked at groupware tools for managers and professionals. Organizations also have large **production systems,** which deal with high-volume operational work, usually involving large numbers of people who are often in multiple departments. A typical example is an order entry system, which drives internal processes until the product is delivered and paid for. Several network technologies (in addition to basic data processing) are promising for production systems.

Production Imaging Systems ▪

Many organizations have production systems that are drowning in paper. For example, even a moderate-size bank will process more than a million checks each night. Mailing these back to customers at the end of the month is extremely expensive. So is storing them physically. Many other firms, such as insurance firms, also have enormous paper storage problems.

One solution is **imaging,** that is, taking a digital picture of the paper and storing only the digital image. Production imaging systems are capable of enormously high volumes. The images they produce, although large in terms of bits, can be stored in a fraction of the room needed for paper documents.

When a document is stored, a *header* containing searchable fields that describe aspects of the document is stored with the electronic document image. This allows rapid retrieval whenever needed.

Production Workflow Systems ■

Many business documents in production systems have to go from person to person and from department to department as they are processed. This tends to result in long response times to customer orders and other business needs.

We discussed *workflow* earlier in this chapter. It automates the process of moving the work from one person to another. It also provides expediting information on specific tasks and general management statistics to help assess the performance of the process. Production workflow systems tend to be very large and to cut across multiple departments.

One advantage of workflow systems over conventional programming comes when there are changes in **business rules,** for instance, how credit limits are determined. In conventional programming, you would have to rewrite at least part of a program's code. This is difficult and time-consuming, and it tends to create errors in other parts of the code. In a workflow system, you merely rewrite the business rule.

Production Electronic Document Management (EDM) Systems ■

As in the case of workflow, *electronic document management (EDM)* takes on a different meaning in production systems than in groupware systems. Groupware *docubases* may hold only several hundred documents for a particular team. In contrast, production EDM systems may contain several hundred thousand documents. These documents might be quite similar (for instance, insurance contracts), so finding a specific document can be quite difficult. This requires extremely sophisticated retrieval software.

CONCLUSION

In this chapter, we have looked at networked communication applications. We began discussing the concepts of *middleware* and *application program interfaces (APIs)*. We can expect to see these concepts increase in use as electronic mail, workflow, and electronic document management become underlying systems as well as applications with which the user works directly.

We looked at *electronic mail* and saw that it was no longer simply a tool for delivering brief messages. It is such an important tool that its usefulness is being extended in many ways.

Under the heading of *groupware*, we saw that there are many types of groupware. Functionality needs vary along at least two major dimensions: whether the participants work in the same place or in different places and whether they work at the same time or at different times. We saw that groupware tools of different types tend to embrace a number of common functions, including verbal communication (communication using words), group composition, group memory, coordination, control, and application development. We looked more closely at a few tools, including Lotus Notes, videoconferencing, desktop teleconferencing, and some emerging areas. We contrasted groupware tools with *production* tools.

In the next chapter, we will continue to look at applications, switching our attention to *networked database applications*. We will also look at *reengineering*, which is essential if an organization wishes to really benefit from networked applications.

CORE REVIEW QUESTIONS

1. What is middleware? What is gained by making electronic mail a middleware application?
2. List some of the advantages of electronic mail.
3. List some advantages of computer conferencing. How do you receive new postings in USENET? In LISTSERV? What is a moderated conference?
4. Name the three standards for exchanges between electronic mail hosts. Which standard is the most widely used? Why do you need directory standards?
5. Define *groupware* using the book's definition.
6. What are the two dimensions of groupware? Where does desktop videoconferencing fit in this taxonomy? What about electronic meeting rooms? Lotus Notes? Workflow automation? Project management?
7. List the services commonly found in keyboard-based groupware tools. Why is application development an important service?
8. In Lotus Notes, explain replication and scalability.
9. What is the major set of standards for videoconferencing? What two things do videoconferencing codecs do?
10. In desktop conferencing, explain the difference between videoconferencing and document conferencing.
11. Explain the difference between groupware and production systems.

DETAILED REVIEW QUESTIONS

1. What is a call? What is an API? In middleware, what program gives an API? What program receives it?
2. Explain the following electronic mail terms: header, body, signature, attachment, MAPI, disposition phase, distribution list.
3. What are the advantages of filtering rules?
4. From the box, "Internet Electronic Mail Standards," what are MIME and POP, and why is each important? Why do you have to encode attachments on the Internet, and what are the three main encoding formats?
5. From the box "Electronic Mail Etiquette and Laws," distinguish between flaming and spamming. Discuss the privacy of electronic mail content in organizations.
6. List what you must have for room-to-room videoconferencing.
7. What is an MCU? What does it allow you to do?
8. List the hardware and software in an electronic meeting room.
9. What are the advantages of combining groupware with telephony?

THOUGHT QUESTIONS

1. At what layer in our Basic Communication Model in Figure 1.1 would you find electronic mail standards?
2. Why do you think electronic mail has grown so explosively while more sophisticated groupware products have lagged? Do not limit yourself to what was said in the text.
3. You want to send someone an attachment. What information will you need from that person?
4. What groupware services would you like to have for a team designing a new software product? Explain.

For online exercises, please visit this book's website at: http://www.prenhall.com/panko

REFERENCES

BULLEN, V. V., and BENNETT, J. L. (1990, October). "Learning from User Experience with Groupware," *CSCW'90: Proceedings of the Conference on Computer-supported Cooperative Work*, Los Angeles, pp. 291–302.

Data Communications (1994, May 21). "Videoconferencing Still a Group Activity," p. 15.

DAVIDSON, E. (1995). University of Hawaii, personal communication with the author.

DESANCTIS, J., and GALLUPE, B. (1987). "A Foundation for the Study of Group Decision Support Systems," *Management Science*, 33(12), 1589–1609.

GALLUPE, R. B., DENNIS, A. R., COOPER, W. H., VALACICH, J. S., BASTIANUTTI, L. M., and NUNAMAKER, J. F. JR. (1992). "Electronic Brainstorming and Group Size," *Academy of Management Journal*, 35(2), 350–369.

HILTZ, S. R., and TUROFF, M. (1993). *The Networked Nation*, revised edition, Cambridge, MA: MIT Press.

JOHANSEN, R. (1988). *Groupware: Computer Support for Business Teams*. New York: The Free Press.

KINNEY, S. T., and PANKO, R. R. (1996, January). "Project Teams: Profiles and Member Perceptions: Implications for Group Support System Research and Products," *Proceedings of the Twenty-Ninth Hawaii International Conference on System Sciences*, Kihei, Maui. Los Alamitos, CA: IEEE Computer Society Press.

NOLL, A. M. (1976, November). "Teleconferencing Communications Activities," *IEEE Communications Society*, 8–14.

PANKO, R. R. (1992). "Patterns of Managerial Communication," *Journal of Organizational Computing* (2:1), 95–122.

PANKO, R. R., and KINNEY, S. T. (1995, January). "Meeting Profile: Size, Duration, and Location," *Proceedings of the Twenty-Eighth Hawaii International Conference on System Sciences, Vol. 4*, Kihei, Maui. Los Alamitos, CA: IEEE Computer Society Press, pp. 1001–1012.

CHAPTER 11

Networked Database Applications and Reengineering

INTRODUCTION

In the last chapter, we looked at networked communication applications. In this chapter, we will look at networked database applications. We will also look at reengineering, which is critical in obtaining payoffs from networked applications.

DATA APPLICATIONS

A great deal of the information in a firm consists of highly structured information. In the 1970s, there were about as many pages of business forms sent in organizations as pages of prose memos, letters, and reports [Panko and Panko, 1977]. Data processing has now automated many forms-handling processes. Data processing stores this highly structured information in **databases**.

Client/Server Databases

As discussed in Chapters 2 and 8, database systems often use client/server processing. As Figure 11.1 shows, the database processing is done on a powerful server. The *client program*, typically running on a PC on the user's desk, sends a *request message* to the *server program* on the *database server*. The database server in turn handles the request and sends back the desired information in a *response message*.

Appropriate Load on the Client Machine ■

Client/server database processing allows the two machines to be specialized for their respective roles. The database server can be optimized for database queries. This requires a powerful server computer. The client machine, in turn, can focus on two things. First, it provides a graphical user interface to the user—

FIGURE 11.1

Client/Server Database Processing

In client/server database processing, the client PC sits on the user's desk. The client program sends a request to the database server, asking for information from the database. The database program on the server typically is a relational database program. It searches through the database, finding the required information. It then sends back a response message that includes the requested information. The client program analyzes the data and presents results to the user. Compared to file server program access, client/server database processing reduces the load on the client machine, reduces the load on the network, provides data safety, allows a single client to collect data from multiple database servers and combine the results, and allows rapid application development through the creation of small programs that provide immediate payoffs.

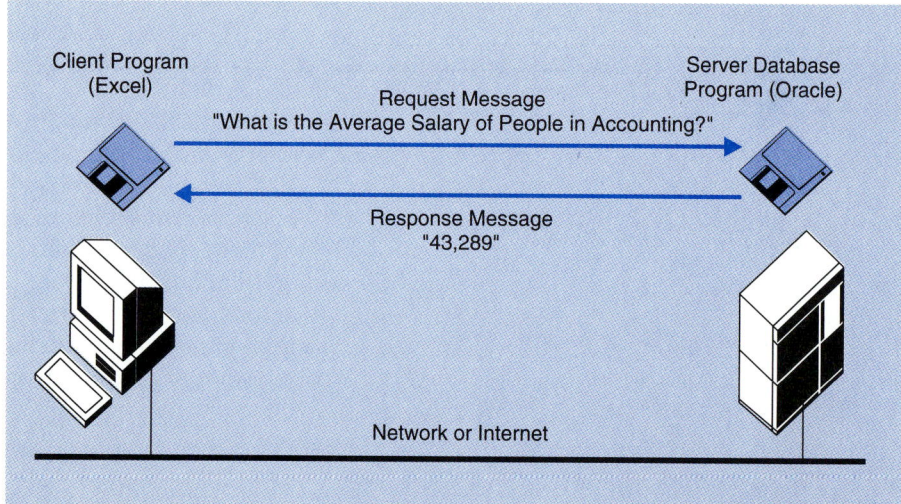

a task at which desktop PCs excel. Second, it analyzes the data received from the server.

Compared to file server program access (see Chapters 2 and 8), client/server program access reduces the processing load on the client PC. This means that a firm may be able to use 80 percent to 90 percent of its existing PCs as clients for many database applications.

Of course, another alternative is to place all of the processing on a minicomputer or mainframe and to give the user a simple terminal. This might be inexpensive, but terminals tend to have limited user interfaces. Graphical user interfaces require very fast communication between the microprocessor and the display. They cannot be handled adequately by a remote processor, even at LAN speeds.

Reduced Network Load

In file server program access, both the database program and a substantial part of the database must be downloaded from the file server to the client PC. For heavy database tasks, this could overwhelm even a fairly fast LAN.

In contrast, in client/server database processing, the response message only contains the information requested. Typically, this information is very small compared to the size of the database. The database program on the server does not have to be downloaded at all. So the load on the network in client/server processing is much lower than in file server program access.

Data Safety

Database programs that use file server program access, such as Microsoft Access, have to be small to execute on client PCs. As a result, they lack many of the features possible in client/server database processing. If there is a network breakdown or the client program fails, there may be lost information, duplicated information, or simply confusion about what information has been entered into the database. Client/server database programs are much more robust, offering many of the **data safety** features of traditional mainframe database systems.

Compiling Data from Multiple Sources

Figure 11.2 shows an even more important reason to use client/server processing. The information that a manager or clerical worker needs may be scattered across multiple databases on multiple database servers.

Almost all client/server database servers today are **relational database** systems. They deliver their responses as tables called *relations*. It is (relatively) easy for the client to join relations from different databases.

Being able to join data from multiple sources means that managers and other workers can find whatever information they need fairly rapidly. There is no need to write complex programs to extract data from multiple incompatible databases, as there was in the days of mainframe database processing.

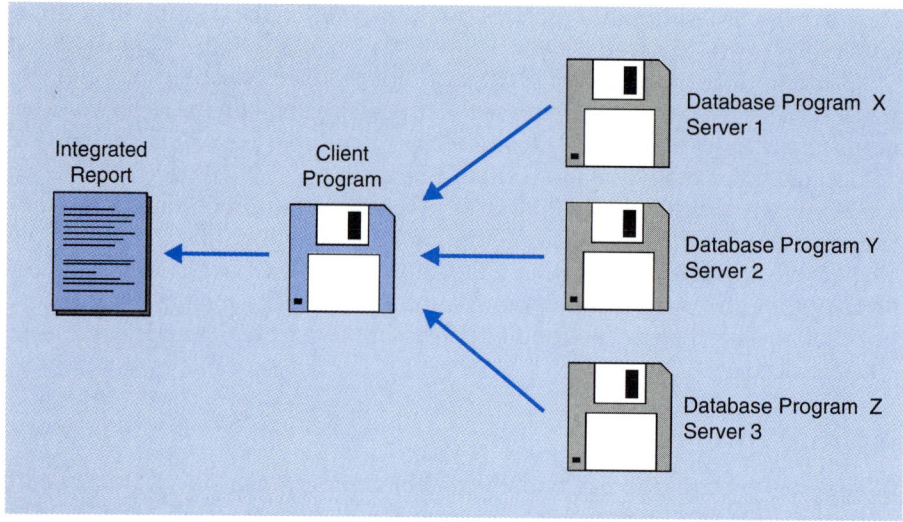

FIGURE 11.2
Access to Multiple Databases from a Single Client

A single client PC can access data on multiple client/server databases by sending request messages to each. The responses will be in comparable form because all client/server databases are relational. This allows the client PC to combine the responses.

Rapid Incremental Application Development ■

Perhaps the most important advantage of client/server computing is **rapid application development**. Traditionally, if you wanted to write a program for a business process, such as order entry, you would have to write an extremely large and complex program to handle all major functions. This might take two or three years. During that time, you would receive no benefits at all.

Figure 11.3 shows incremental application development in client/server computing. Once the database structure is built, the first increment is a program for entering data. In many cases, this can be built in two or three days. Once this is done, you can add increments as needed. You can add a clerical function to check for the purchaser's credit rating. You can add another to give management statistics. This process continues indefinitely.

With each increment, you get immediate benefits. The rapid payoff allows you to recoup initial investments in server hardware and database server software fairly rapidly.

Initial increments might be built with simple programming languages, such as Visual Basic, which allow rapid development but are not highly efficient. Later, increments that are used extensively can be reprogrammed for efficiency in a more sophisticated language, such as COBOL or C++. This allows you to enjoy rapid development and ultimate efficiency. In addition, by the time the optimized version is needed, there will be considerable user experience, which may result in the development of a better application.

Software

There are two general classes of software that are needed to do client/server computing. We will look at each briefly.

Server Software ■

Of course, the most critical piece of software is the server database program. This might be Informix, Oracle, SQL Server, Sybase, or some other client/server database server package.

Database is a sophisticated application that requires a sophisticated server operating system. In 1995, UNIX had 81 percent of the market for database servers [Forrester Research, 1995]. UNIX achieved this dominance because it is

FIGURE 11.3
Rapid Application Development

Instead of writing one large program with high functionality, rapid application development with client/server processing can be done by creating many small incremental programs. Each incremental program is developed quickly using a rapid development language such as Visual Basic. This gives immediate payoffs. Later, high-volume incremental programs may be rewritten in a more traditional programming language.

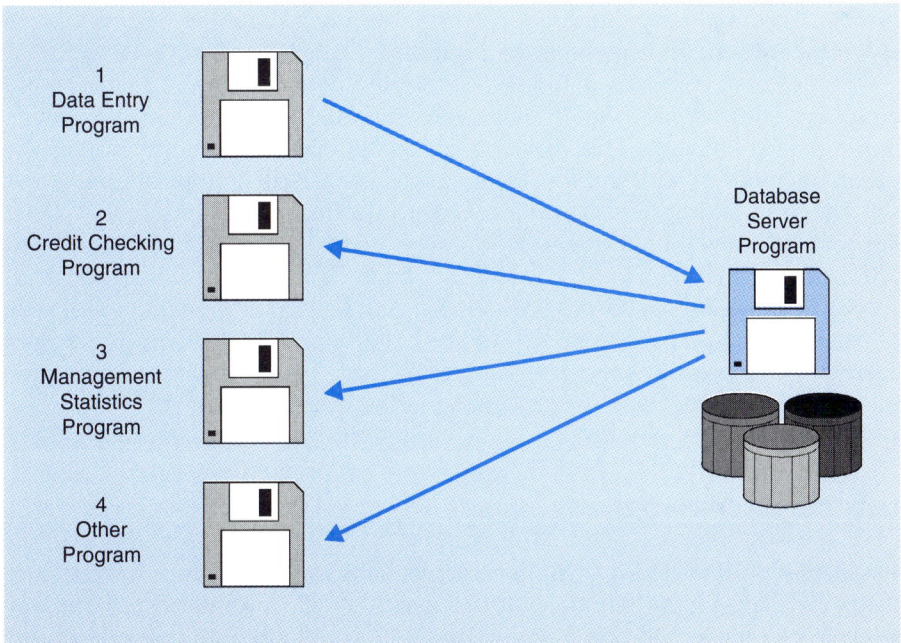

an extremely mature and sophisticated operating system. The rest of the market was split between IBM's OS/2, Microsoft Windows NT Server, and the Macintosh system. (Windows 3.1, Windows 95, and MS-DOS were too limited to support database processing.) Since then, Windows NT Server has cut into UNIX's market share.

One key aspect of both UNIX and Windows NT Server is that both support **symmetric multiprocessing (SMP).** This means that a server can have more than one microprocessor. Most SMP servers at the time of this writing have two to four microprocessors. The microprocessors, furthermore, can be the powerful RISC microprocessors of workstation servers. Database programs written to run on SMP operating systems can serve several times as many users as single-processor database programs. Client/server databases continue to grow in size, and SMP is becoming crucial in large organizations. *For more on symmetric multiprocessing (SMP), see Module I.*

SMP brings **scalability,** which means that you do not have to change your database as its size grows. You can begin by installing the database on a high-end PC server. Then, as usage grows, you move it to a RISC workstation server or SMP PC server. Later you can move the database to an SMP RISC workstation server. Nowhere along the line will you have to redesign your database or applications.

Developing Client Software ■

There are three general choices for developing client software for client/server database applications. The first is to use a **general fourth-generation programming language,** such as *Visual Basic* or *C++*. Visual Basic applications can be developed very rapidly. C++ applications take longer to develop but are more efficient and can be more complex.

The second approach is to use a **dedicated database client development language,** such as *PowerBuilder*. These languages have special tools for developing database clients. This makes them unsuitable for developing other applications but very efficient for developing database client applications.

The third approach is to turn an **existing application program** into a client. For instance, several spreadsheet programs, such as Excel, can act as clients. They can send requests to the database server and place the results in a spreadsheet for immediate analysis. With embedded programming languages, such as Microsoft's *Visual Basic for Applications*, the spreadsheet program may not even look like a spreadsheet program to the user. This approach is valuable if there will be a large need for processing in the client program.

Standards and Middleware

The biggest problem today with client/server database processing is limited standards for client/server requests and responses. Without standards, a company tends to become locked into a single database vendor's proprietary protocols.

SQL

All client/server database programs send request messages in **Structured Query Language (SQL)** format. However, SQL merely specifies the query. It does not specify the full format of the request and response messages. It is not enough for request messages. Request protocols must add the specifications needed to manage request/response exchanges. They do it, furthermore, in different proprietary ways.

Open Data Base Connectivity (ODBC)

Microsoft has promoted the **Open Data Base Connectivity (ODBC)** standard for client/server interactions. Most database server programs will work with ODBC. On the client side, however, many client programs on Macintoshes and UNIX workstations cannot communicate via ODBC. As shown in Figure 11.4, ODBC is an application-layer standard.

Database Middleware

Another limitation of ODBC is that it only works with relational database management systems. There are many older database systems that use other database organizational structures. There are also many old programs written in COBOL that create individual files rather than full databases. Older **legacy** applications cannot be ignored because they hold critical corporate data. Even if you are going to replace a legacy system, furthermore, your new database server program will have to work with the legacy system during the transition period for data transfer.

In the last chapter, we discussed the idea of middleware. Figure 11.5 shows how this concept applies to database processing. Note that the **database middleware** software sits between the application and transport layers. The client

FIGURE 11.4

ODBC Is an Application-Layer Protocol

Open Data Base Connectivity (ODBC) defines how the client application program and the server database application programs communicate. It is an application-layer standard.

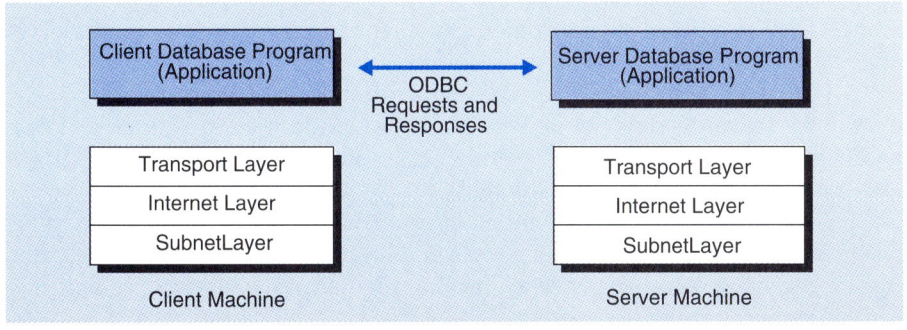

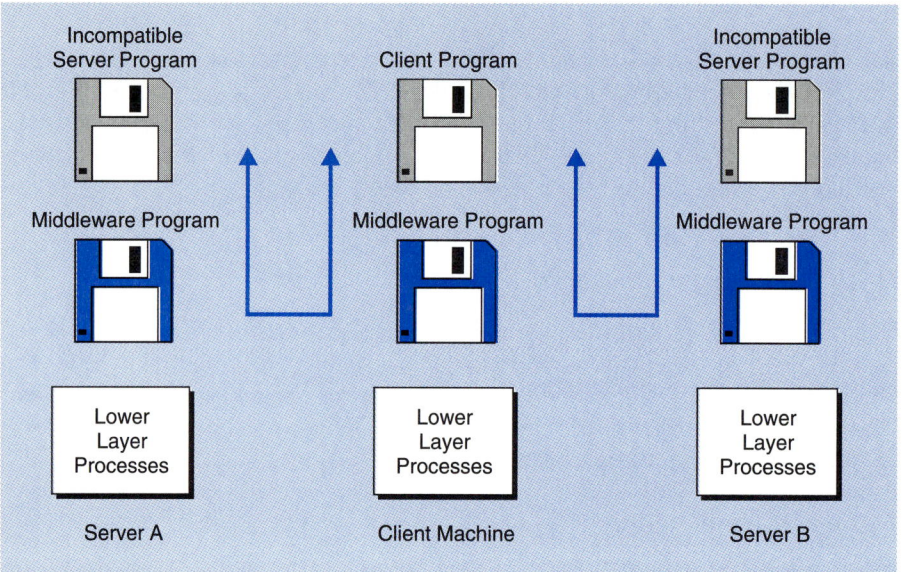

FIGURE 11.5

Database Middleware

Database middleware programs add a layer of software between the application and the transport layers. The client program sends its usual request message. The database middleware program passes the request to the database middleware program on the server machine. The middleware program on the server puts the request in whatever format is needed to get the desired data. Database middleware programs translate formats. They also handle lower-layer transmission details, such as transport- and transmission-layer concerns. This again makes life easy on the programmer. Database middleware also links database servers to traditional legacy database programs.

program merely sends a request message in some supported format. It passes this request to the middleware program via an interface call. The database middleware program grabs the request message, then passes it to the middleware program on the server or mainframe. The database middleware program on the server or mainframe translates the request into whatever format is needed to get the desired data. In other words, database middleware translates formats. The middleware program then passes the request message to the server application software.

Middleware also handles interactions with the transport and network layers for the software developer. This limits the amount of computer platform and networking expertise the developer will need. By handling such details, it also increases programmer productivity.

Finally, database middleware links modern relational database servers with older legacy systems. This supports data transfers during the transition from a legacy system to a new system. It also allows data transfer with enduring legacy systems.

CHAPTER 11 Networked Database Applications and Reengineering

Still, middleware is not perfect. Database middleware products are very expensive. There are also several forms of middleware, so selecting middleware software is difficult. And at the present time, there are no standards for database middleware APIs. *For more on database middleware, see Module J.*

Data Warehouses for Decision Support

If a manager needs information for decision support, he or she may have to perform several complex queries. Even a single complex query can place a considerable burden on the database server program. This will slow response times for production users, such as clerical workers, who need extremely rapid response to their commands.

Data Warehouses ■

Figure 11.6 shows that many organizations address this problem by creating **data warehouses**. The figure shows that the data warehouse extracts decision

FIGURE 11.6
Data Warehouse

In data warehouses, the client program does not send requests to production database servers. Instead it sends them to a data warehouse server. The data warehouse server collects data from production servers, cleans them up, and permits client queries against its data extracts. This protects production servers from performance degradation during queries and from unauthorized changes by client users. It also creates clean information and makes it easier for users to query information from multiple databases.

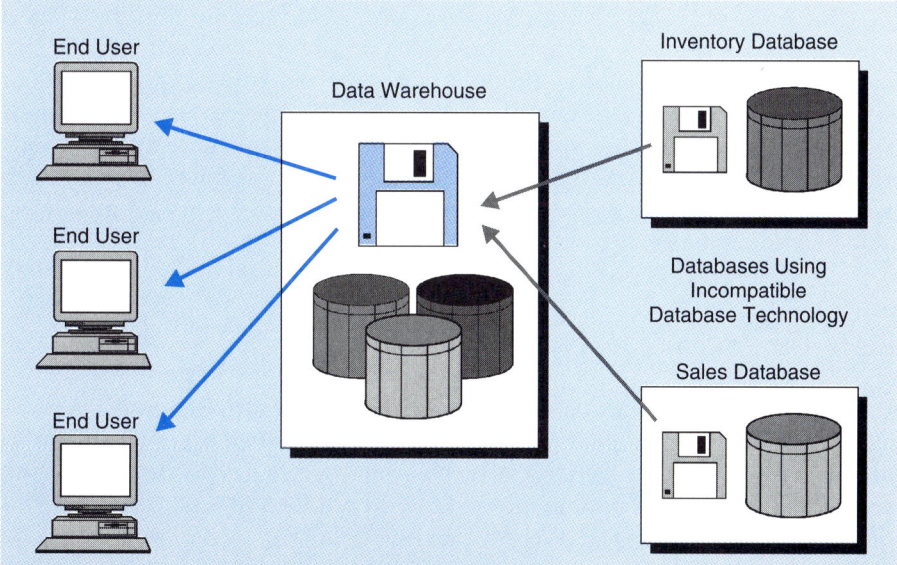

data from multiple databases and individual data files. A manager then executes queries against the extracted data on the warehouse server, rather than against production data on the production database server.

Data warehouses have several advantages besides reducing the load on production servers. First, because they extract data from multiple sources, they reduce the amount of work that the manager has to do.

Second, they can offer **clean data**. In many cases, different production servers store information for the same variable in different ways. For example, there may be variations in whether a person's first name is spelled out or whether an abbreviation is used. This will make it difficult to match information in different databases for a given person [Leinfuss, 1994]. Doing a query under these conditions can produce misleading results. The data warehouse administrator removes such problems in the data.

Third, there is no danger of managers accidentally modifying production data without permission. At worst, they can damage the warehouse copy. To reduce even this possibility, the data warehouse software may only offer read-only connections to managers.

The downside of data warehouses is cost. The data warehouse software itself is expensive. So is setting up and administering the data warehouse.

Online Analytical Processing (OLAP) ∎

To analyze data in a warehouse (or in separate databases), one option is to do all the work on the client PC after retrieval. Many data warehouse programs, however, offer tools that execute on the warehouse server itself. These tools are known collectively as **online analytical processing (OLAP).** They are tailored to database information. They allow the user to do forecasts and to look at the data in a wide variety of ways.

One problem in OLAP is that a user may wish to cut through the data in various ways. He or she might want to look at sales by region. Later the user might want to look at sales by product. These sorts of requests can be difficult with relational databases. As a result, some OLAP systems use **multidimensional databases**. Suppose you have a database on sales. There might be individual cells holding individual transactions by region, product type, customer, and other information. Multidimensional databases provide quick aggregation along any dimension.

Three-Tier Processing for Production Data

Data warehouses are for decision support, but it may also make sense to add a third tier to production data processing. **Production data processing** consists of the real-time (usually clerical) interactions needed in day-to-day accounting, payroll, billing, and other activities. Although production applications are not glamorous, they are critical to corporate functioning.

Just as we are seeing a shift from pure client/server processing to three-tier processing in decision support, we are seeing a three-tier trend in production data processing as well. In this case, the middle tier consists of the processing needed to execute complex business rules, such as determining whether a customer's credit is sufficient for a transaction. As these business rules have grown more complex, processing them on a client machine has become increasingly difficult. Many applications now require "fat clients" that are as powerful as many servers. In a three-tier approach to production data processing, the database server retrieves data, the business rules server does the processing needed for complex business rules, and the client program presents results to the user and also does simple analysis.

Distributed Databases

Even with RISC processing and symmetric multiprocessing, a single server may not have sufficient power to serve all users. Even a mainframe acting as a server might not have enough power.

In addition, if an organization is geographically decentralized, having a single database will result in the heavy use of long-distance transmission facilities. All queries will have to travel to the central source.

Figure 11.7 shows that one solution is to create a **distributed database** in which there are multiple copies of the database (or at least parts of the database) rather than just one copy. Then most users will be able to use a nearby database server. There will be less network traffic, and you will be able to sup-

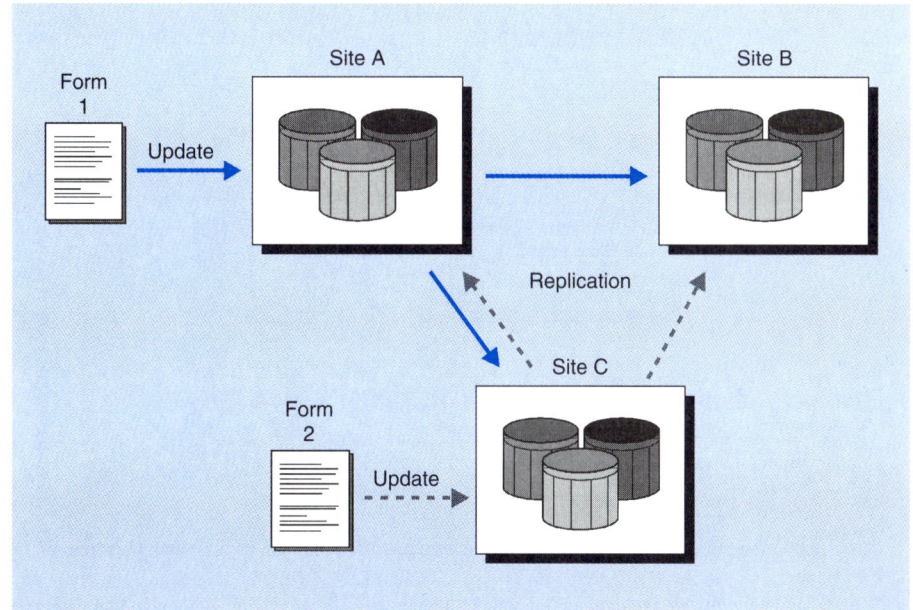

FIGURE 11.7

Distributed Database

In a distributed database system, there are multiple copies of the database. Changes on one must be replicated (copied) to other copies. This has performance advantages, but it is extremely difficult to do.

port more simultaneous users. In addition, if there is a natural disaster, such as a flood, the disruption will be minimized.

Distribution is not too difficult to do if you have a database that is primarily read and seldom updated. Then you can create read-only copies at multiple sites. In this case, you must decide how often to replicate (copy) central data to secondary sites.

Unfortunately, many production databases are updated constantly. If you allow updates at individual sites, furthermore, then replication will have to go in multiple directions. There will also be problems if two sales agents at different sites sell the same part without knowing that the other has already sold it. In general, distributed databases represent the bleeding edge of database technology.

Multimedia Databases

In the past, databases were limited to alphanumeric information—text and numbers. Now, however, it is increasingly common to see fields that contain graphics information, such as photographs (see Figure 11.8). It is even becoming common to see fields containing voice, video, and other multimedia data.

In addition, many multimedia applications create large numbers of data files that must be organized. This is another application for multimedia databases.

A major issue in multimedia databases is how to locate specific multimedia objects, such as graphics images. For alphanumeric databases, it is possible to execute "find" commands that look for text strings. For multimedia information,

FIGURE 11.8

Multimedia Database

Older databases were alphanumeric, including only textual and numerical information. Newer multimedia databases include photographs, voice recordings, and even video.

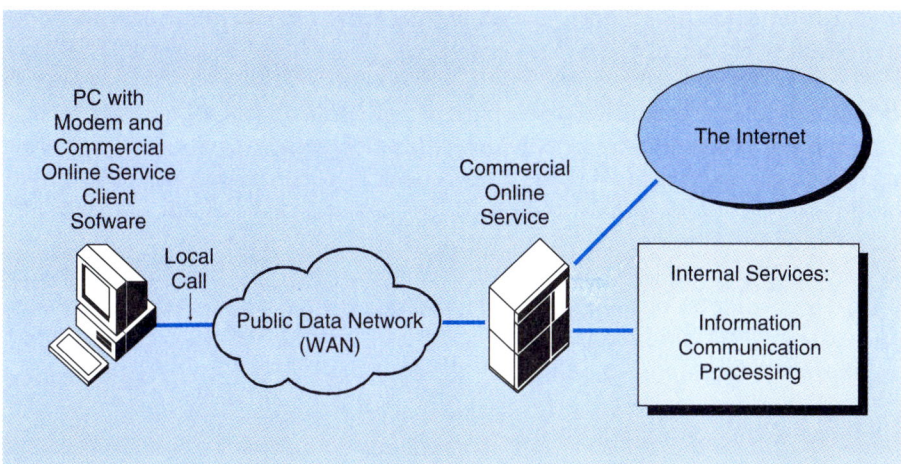

FIGURE 11.9

Commercial Online Service

The commercial online service vendor must have computing capacity, application software, and usually information of commercial interest. The vendor must also have a network to reach customers.

text string searches do not work. We must find ways to search for graphics elements, sound characterizations, and other multimedia characteristics of files.

Commercial Online Services

The Internet offers a wide variety of information services. Because most of them are free, few economic incentives are provided for companies to create complex information and other costly services needed in business. Long before there was the Internet, a number of companies offered **commercial online services** that used networks to deliver services to users for a fee.

Elements of a Commercial Online Service ∎

Figure 11.9 shows the elements of a commercial online service. First, the company offering the service must purchase or lease computer capacity. Second, the company must create the application software needed to provide the service. Third, in most cases, the company must collect information of value to others.

Fourth, the company must have a way to reach users via a computer network. Typically, the company will use a commercial network, such as TYMNET or Telenet. In some cases, the company will construct its own network if it is large enough.

To use the server, customers must first make billing and payment arrangements. After that, they merely dial into the service.

Consumer-Oriented Services ∎

The most widely known commercial online services are **consumer-oriented services**. These include America Online, CompuServe, Delphi, the Microsoft Network (MSN), and Prodigy.

Consumer-oriented services appeal primarily to residential consumers, although they have many business customers as well. Because of their residential target market, their prices have to be kept low. A dollar or two an hour is typical for people who use these services a great deal.

Another characteristic of consumer-oriented services is that they offer a wide variety of services. They try to be one-stop services offering news, entertainment, interactive games, online ordering, and special services for people with common interests.

Although consumer-oriented services are widely known, they represent only a tiny fraction of the commercial online services market. Businesses can and do spend far more money for information and processing services that add value to their firms.

Business Processing Services ■

Business processing services move data processing outside the firm, onto the commercial online service's computers. The biggest business processing service is *payroll processing*. Many businesses find it much easier to let an outside company handle their payroll. They merely enter employee and work data, and the payroll processing service does the rest—including automatically depositing payments in employee bank accounts. Payroll processing is a specialized function involving complex and rapidly changing legal rules. Payroll processing firms maintain the expertise required by these rules. NuPools uses a commercial online service for its payroll processing.

Clearinghouses are also important in the business processing market. In medicine, for instance, there are many insurance agencies that hospitals must work with to receive payments. Medical payment clearinghouses receive claims from hospitals. They collect all claims for a given insurance agency and provide them in a single package—in the insurance agency's preferred format.

Electronic Data Interchange (EDI) ■

One processing service that merits special attention is **electronic data interchange (EDI)**. When firms do business with one another, they have to exchange many business documents—orders, bills of lading, invoices, and so forth. In the past, one firm would print output from a computer, write it onto electronic forms, and mail the forms to the other firm. The other firm would rekey the information into its own computer.

Obviously, it would be better to move electronic documents directly from one firm's computer to the other firm's computer. This saves time, reduces labor costs, and reduces errors, which can be extremely costly.

If two firms are to communicate, they must agree on standards. There are two basic sets of *EDI standards—X.12* and *EDIFACT*. Within each, there are many options. If a firm has many business partners, dealing with each can be very tedious and expensive. As a result, **EDI clearinghouses** allow firms to submit business documents in their own preferred format. The EDI clearinghouse changes the format into one that the receiver prefers and then delivers the document.

Although EDI is probably the best-known business processing service, it is very small compared to other business processing services. However, it is crucial as a first step toward the interorganizational systems that we will discuss later.

Business Information Services

The other major component of the business sector is **business information services** in which the subscriber to the online service is seeking information rather than processing power per se.

Credit checking is the dominant application in business information services. When you apply for a loan, a bank will contact one or more credit-checking companies to get a comprehensive report on any financial problems. They also do this for their commercial customers. On a smaller scale for individual transactions, whenever you go through a supermarket and run your credit card through a VeriFone scanner, the system does a quick check on any problems with your credit card.

Investors need *financial information* about companies. Financial information services provide this information, using required reports that firms must file with the government and using other information. Competitors too find this information valuable.

Electronic Funds Transfer (EFT)

Banks and other financial institutions have to move trillions of dollars each day. A number of highly secure **electronic funds transfer (EFT)** networks are in place. This is a very specialized niche, which is often controlled by bank consortia rather than by an independent commercial online service.

Electronic Commerce on the Internet

The cutting edge of electronic commerce today is the Internet. A surprisingly large number of firms now buy and sell over the Internet. At the time of this writing, however, Internet security is still uneven, but this has not proven to be a complete impediment because, for moderate increases in risks, credit card companies simply charge more per order to make up for occasional losses. Several organizations have proposed standards for secure Internet electronic purchasing. By the time you read this book, reasonably good fixes to the technical security problem will probably exist. See the book's website for updated information.

Internet electronic commerce is exciting because the Internet is worldwide. You can literally reach an audience of millions of people. Even niche markets are enormous if you can figure out how to reach them. You can provide extensive data on your products, route email technical questions to experts for next-day response, and sell your product while the buyer is still logged into the computer. You can completely redefine sales and distribution channels.

Database Applications at VeriFone and NuPools

Although new database technologies are receiving a great deal of attention, many firms continue to do most of their work on mainframes. Mainframes have good management tools, good database tools, and good security.

VeriFone is a classic example of a company that has continued to rely on mainframe database applications. This continues the VeriFone philosophy of using older tools in innovative ways in order to focus on innovations rather than technology. However, a growing number of people in the firm now feel that these old legacy systems have worn out their usefulness and need to be replaced by client/server applications. Within Verifone, there is considerable discussion about the merits of staying with mainframes or moving to client/server processing and other innovations.

Nor does small size mean an inability to use a client/server database system. We saw in Chapter 8 that NuPools has a client/server application that it purchased as packaged (prewritten) software. NuPools is rather small to have customized applications built for it. In addition, as noted earlier, NuPools uses a commercial online service to handle its payroll processing. It is considering doing credit checking online for its larger construction projects.

Reengineering

In the early 1990s, Hammer and Champy's [1993] *Reengineering the Corporation* became a best-seller. The ideas in the book were not really new. Hammer, for instance, had been promoting the basic ideas in the book for over a decade. But in the 1990s, faced with growing competition and shrinking profit margins, managers read the book in huge numbers.

Quite simply, the book argued that if you continue to work the same way you always have, you will get few benefits if you add networking and other information technology. In the Industrial Revolution, after all, the steam engine produced few gains until it was coupled with a revolution in work called the factory.

In the same way, the benefits will be small if you try to apply information technology to **reengineer** (redesign) current work practices. For instance, Hammer and Champy discussed the Ford Motor Company's accounts payable system (pp. 39ff.). They noted that the company expected to get only a 20 percent productivity gain by automating its current processes. By rethinking and reengineering the process, however, Ford was able to slash the number of jobs from 500 to 125.

Obliterate

One theme of reengineering is to completely **eliminate** as much work as possible. Look at the **business rules**—the written (and unwritten) assumptions that guide how people work. We have already seen business rules in the context of three-tier production systems.

Next consider how things would improve if some rules were changed to eliminate whole processes. For instance, Ford decided to pay suppliers when a part was actually used rather than when it was delivered. This caused the parts companies to become expert in delivering parts on time, freeing Ford of the job and placing the job where the real expertise existed in the first place. Individual invoices and a tedious reconciliation step were completely eliminated in the new process, allowing a considerable reduction in staffing needs.

Many business rules that once made sense (or seemed to) no longer make sense today. They often impose high costs to reduce small risks. In many cases, they were present only because crude paper-based technology limited communication and forced the work to be spread over several departments. Many business rules, in other words, were created to cope with an inefficient manual system.

Unfortunately, many obsolete business rules are so integrated into business thinking that their potential inappropriateness is difficult to recognize, much less challenge. In addition, changing business rules always create some real risks.

Consolidate

Another principle is consolidation. When you call a mail-order firm today, a single person can take your name, check your credit rating, determine if what you want is in stock, take your order, and sometimes even begin the automatic stock-picking process. In the past, multiple people in multiple departments would have to have done all of this in sequence, and it would have taken days rather than seconds. This is an example of **consolidating** functions into the hands of a single person. Hammer and Champy [1993] call this person a **case worker**. Database technology and other networking technologies have made it both possible and profitable.

Obviously, the first step in workflow automation is to study the current system. Reorganization to allow the process to be handled by case workers or by small *case teams* may radically reduce the complexity of workflow processes.

Teams

Another area in which companies are struggling to improve is in teamwork. Where tasks cannot be relegated to individuals, they must be handled in teams.

Long ago, Bennis and Slater [1968] foresaw an age in which companies would become **adhocracies,** consisting entirely of teams that would be pulled together as needed and then quickly disbanded so that members could work on other teams. More recently, authors such as Davidow and Malone [1992] have called such organizations **virtual corporations**. In many cases, participants in projects will not even be employees of the firm but members of other firms or consultants brought in for specific expertise.

Although adhocracies and virtual corporations might sound futuristic, as noted earlier, managers and professionals already spend a great deal of their time working on teams. As the pace of teamwork increases, electronic support will be ever more necessary.

Interorganizational Systems

In electronic data interchange, we saw earlier that two firms exchange business documents to increase the speed of transactions and to reduce costs. But this may be only the beginning of cooperation between buyers and sellers. Marketers have long argued that manufacturers, wholesalers, and retailers form **marketing channels** that must be able to act almost as single firms to reach customers. Under pressure from manufacturers selling directly to customers via the Internet, there seems to be growing interest in vertical interorganizational systems that consist of two or more firms working closely together to improve a marketing channel.

EDI allows firms to exchange business documents. The next step is to give the other party more intimate access to a firm's data and processing resources. For instance, a manufacturer might give a *trusted customer* access to the manufacturer's inventory and pricing databases. This would allow the client to check if a product is on hand, look up its price, and place an order.

Or a manufacturer might give a *trusted supplier* access to its production scheduling and inventory databases. The supplier would determine when more goods were needed and deliver them *just in time* for their use in manufacturing. This reduces inventory levels and the chances of components becoming obsolete while sitting in inventory waiting to be used.

Although interorganizational systems are exciting, they produce few benefits unless you redesign your relationships with your buyers and customers. There must be higher levels of trust, and new business arrangements must be negotiated carefully. You are not merely adding technology. You are creating fundamental changes in the ways that firms work together in vertical distribution channels.

Outsourcing

Traditional interorganizational systems are *vertical*, involving interactions among separate firms in the marketing channel from manufacturer to customer. In turn, **outsourcing** is creating *horizontal* interorganizational systems. A firm retains only its **core business functions** in-house. These are functions in which the firm has specialized expertise that allows it to add value. It **outsources** all other business functions to specialists, as in the case of payroll processing. Many firms are also outsourcing much of their information systems function.

In the past, outsourcing has involved arm's-length relationships on marginal activities. In the future, however, as outsourcing increases, the horizontal partners will have to coordinate their activities more tightly. This is especially true when the business partner provides consulting expertise and other valuable services. Tomorrow's virtual corporations, in fact, may have vague boundaries, with team members coming from other firms as needed.

New Channels

It is difficult to believe that the cable television shopping channels did not even exist a few years ago. Cable provided capacity, and a few companies capitalized on this new resource to become billion-dollar firms in a few years.

The Internet and other networking environments are beginning to provide similar opportunities. Marketing directly to customers through the Internet has the potential to destroy many traditional marketing channels. Internet marketers can sell cheaper, provide more data to their customers, and react to questions by consulting central experts. In addition, while physical space limits what retailers can stock, companies that sell directly to users from central locations can stock a wider variety of goods and can stock more of each item.

Telecommuting

Many people already spend parts of their working time at home instead of in an office building. In some cases, people work from their home as consultants, doing what in-house office workers did in the past.

Many social pundits tend to look at **telecommuting** as a way to reduce travel time and congestion. However, many questions remain. Who will fund the expensive equipment and network connections? Will telecommuting harm communication, since managers now spend 55 percent of their days talking face-to-face? Is telecommuting good for only some jobs, and then only for part of the working week?

CONCLUSION

In this chapter, we have continued to look at applications. We saw how networking is revolutionizing *database* processing. In addition to bringing client/server processing, networks have encouraged better tools for managerial decision support applications and for production systems. Firms are also looking outside their borders to get processing and information services from commercial online services.

Finally, under *reengineering*, we saw how firms are likely to reorganize work in the future in order to respond to the opportunities created by networking.

CORE REVIEW QUESTIONS

1. List the advantages of client/server databases and explain why each exists.
2. Explain the need for scalability. How is scalability created in client/server computing?
3. Compare SQL and ODBC.
4. What benefits does database middleware attempt to bring? What are legacy systems, and how do they create special problems?
5. What benefits do data warehouses bring?
6. Which is larger—commercial online services oriented to residential consumers, such as CompuServe, or commercial online services oriented to business?
7. Distinguish between processing and information services. Give one example of each in the commercial domain. Where does EDI fit into this distinction?
8. Describe security for electronic commerce on the Internet. Why are many firms pursuing Internet electronic commerce anyway?
9. What is reengineering? Why is it important?
10. Explain why obliterating and consolidating are central to reengineering.
11. Why is team support important?
12. What is an interorganizational system? Distinguish between vertical and horizontal interorganizational systems.

DETAILED REVIEW QUESTIONS

1. Distinguish between request and response messages.
2. Explain the server operating system picture for database servers.
3. What is symmetric multiprocessing (SMP)? What is necessary for SMP?
4. What are OLAP, multidimensional databases, and three-tier processing?
5. Why do some firms do distributed database processing? Why do most avoid it?
6. What is a marketing channel?
7. What is outsourcing? What determines what business functions should be outsourced?

THOUGHT QUESTIONS

1. Explain why business rules are central to almost all forms of reengineering.
2. Why is updating in distributed processing so difficult to do?
3. Why do you think there is so much interest in reengineering today?
4. Name one business rule in universities that should be changed. Explain how changing it would reduce costs or improve educational delivery. Explain the risks that changing it would create.
5. Think of a business opportunity in which using the Internet to reach customers directly could produce major financial growth. What problems do you foresee?
6. What do you foresee as the benefits and problems of telecommuting?

PROJECTS

1. Collect current data on the online services market. A good place to begin may be *Standard & Poor's Industry Surveys*.
2. VeriFone continues to use mainframe databases. Prepare a memorandum stating the problems that VeriFone may face if it continues to do so. The memo should also list the steps that VeriFone should take in the next year if it implements its first client/server database application to replace an existing application.

For online exercises, please visit this book's website at: http://www.prenhall.com/panko

REFERENCES

BENNIS, W. G., and SLATER, P. E. (1968). *The Temporary Society.* New York: Harper & Row.

DAVIDOW, W. H., and MALONE, M. S. (1992). *The Virtual Corporation.* New York: Harper.

FORRESTER RESEARCH (1995). Cited in Mayer, J. H. (1995, February). "Unix Vendors Prep for Slugfest with NT," *Client/Server Computing*, pp. 35–42.

HAMMER, M., and CHAMPY, J. (1993). *Reengineering the Corporation: A Manifesto for Business Revolution,* New York: Harper Business.

LEINFUSS, E. (1994, July). "Rub a Dub Dub: Give That Data a Scrub," *Client/Server Computing*, pp. 62–67.

PANKO, R. R., and PANKO, R. U. (December 5–7, 1977). "An Introduction to Computers for Human Communication," Paper 21:1, *NTC'77 Conference Record*, Los Angeles, CA: National Telecommunications Conference.

MODULE A

The OSI Architecture

INTRODUCTION

This module is designed to be read after you have completed Chapter 2, although most of it can be read before you have finished Chapter 1. You will understand even more if you have finished the section on synchronous framing in Chapter 4.

This module discusses the *OSI architecture*. The OSI architecture is used by all architectures in the subnet layer of Figure 1.2. So it is important for you to understand the lowest layers of OSI. At upper layers, OSI standards have been less popular, but OSI application-layer standards and network management standards do see some use, especially in Europe and in the networks of transmission carriers.

THE OSI ARCHITECTURE

As Chapter 1 discussed, **OSI** is an abbreviation for **Reference Model of Open Systems Interconnection.** The OSI architecture is the joint creation of ISO and ITU-T. **ISO** is the International Organization for Standardization (easily confused with OSI). ISO sets international standards in many industrial areas, including computing. **ITU-T,** in turn, sets international standards in telecommunications.

In the 1970s, ISO and ITU-T realized that they would need to work together because data communications lies at the intersection of computing and telecommunications. Their first step was to create an architecture or overall design framework for setting standards. (They used the term *reference model* instead of *architecture*.) This architecture specified the types of standards to be developed.

As shown in Figure A.1, OSI divided the three technical layers of our Basic Communication Model in Figure 1.1 into seven layers. In this section, we will discuss the OSI layers in terms of our three basic technical layers, beginning with OSI's widely used subnet-layer standards.

The OSI Transmission Layers

OSI divided the transmission layer into three distinct layers: the physical layer (OSI Layer 1), the data link layer (OSI Layer 2), and the network layer (OSI Layer 3). Each is designed to standardize a specific set of processes. Taken together, they form the transmission layer in our Basic Communication Model, at least for subnets (see Figure 1.2).

The Physical Layer (OSI Layer 1) ■

Figure A.2 illustrates the **physical layer (OSI Layer 1).** As the name suggests, this layer is concerned primarily with physical matters. The figure shows a

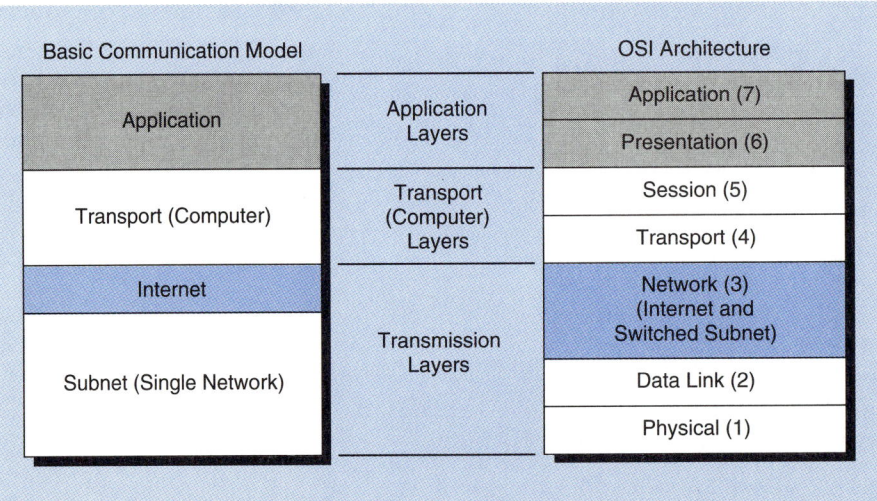

FIGURE A.1

The Basic Communication Model and OSI

OSI divides the three technical layers in the Basic Communication Model into additional layers. The OSI architecture has seven layers instead of three.

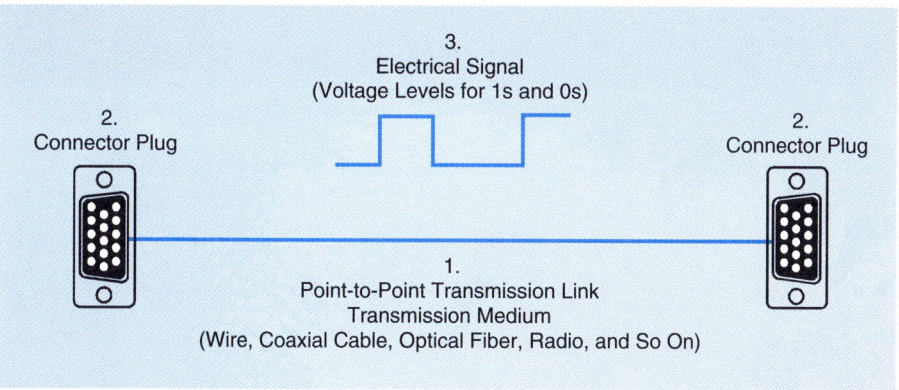

FIGURE A.2
The Physical Layer (OSI Layer 1)

The OSI physical layer defines transmission over a point-to-point connection. It defines the physical transmission medium, the physical connectors at the two ends, and electrical signaling.

point-to-point physical connection. It consists of a transmission medium with connectors at the two ends.

Chapter 4 discusses transmission media, including the most popular transmission medium, unshielded twisted pair (UTP). Another important transmission medium is optical fiber. OSI physical-layer standards define such things as the thickness of the medium and the maximum length of a point-to-point segment. They also define electrical characteristics, such as impedance.

There are many types of connectors. You can see several if you look at the back of the PC. Some are D-shaped, while others are round. Most have a number of connector pins. OSI Layer 1 standards define the physical layout of the connector and the pins. They also define the electrical characteristics of the connector.

Finally, OSI physical-layer standards define how 1s and 0s are represented as voltage levels or in other ways. For instance, in the popular Ethernet (802.3) LAN standard discussed in Chapter 4, a 1 is 0 volts and a 0 is −2.05 volts.

If organizations purchase physical media, connectors, and transmission equipment that follows OSI physical-layer standards, the transmission over the point-to-point connection should be extremely reliable. Few 1s or 0s will be lost or reversed.

Data Link Layer (OSI Layer 2) ■

Figure A.3 illustrates the **data link layer (OSI Layer 2).** The figure shows that the data link layer is also concerned only with point-to-point connections. However, while the physical layer is concerned with physical and electrical matters, the data link layer is concerned with packaging 1s and 0s for transmission.

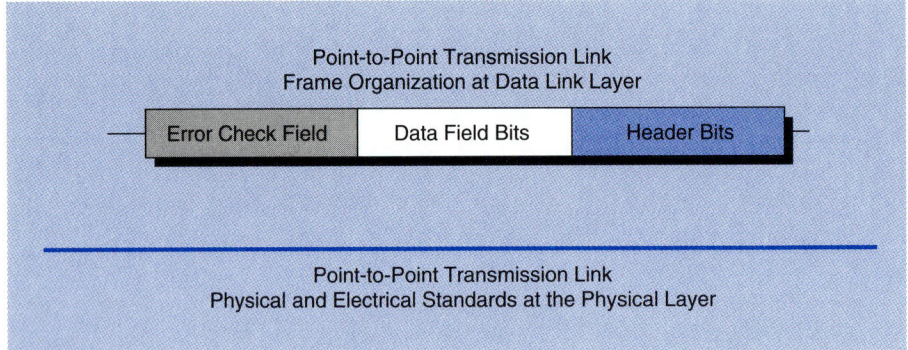

FIGURE A.3
Data Link Layer (OSI Layer 2)

The OSI data link layer packages bits for delivery. Instead of traveling alone, bits are organized into groups called frames. This allows error checking and other data quality controls. In terms of concepts in Chapter 4, a frame is a synchronous unit at OSI Layer 2.

The figure shows that bits do not travel alone down the transmission line. The sending process organizes them into groups called frames. If you have already read Chapter 4, these groups are synchronous frames.

Why package outgoing messages into frames? Chapter 4 goes into considerable detail on the advantages of this approach. For now, we will merely list one advantage—error detection.

The frame in Figure A.3 has an error-check field. This field contains a number that is calculated on the basis of all other bits in the frame. The sender performs the calculation and places the binary number in the error-check field.

At the other end, the receiver repeats the calculation. If there have been no errors during transmission, the receiver's calculation should give the number in the error-check field. If its calculation does not match the number in the frame, it knows that an error has occurred. It throws away the frame and asks the other side to retransmit the frame.

The OSI Layer 2 standard specifies the exact makeup of the frame. In addition to specifying an error check field, it specifies a number of other fields that help control the transmission.

Figure A.4 shows a process called **multiplexing** in which a single point-to-point transmission segment can contain multiple devices. The devices take turn transmitting. While this may result in devices having to wait to transmit, sharing a single transmission segment reduces costs and is widely used in LANs today.

The Network Layer (OSI Layer 3) ■

Figure A.5 shows a **packet switched** data network. It shows Station A transmitting a message (packet) over an access line to a switching node on the net-

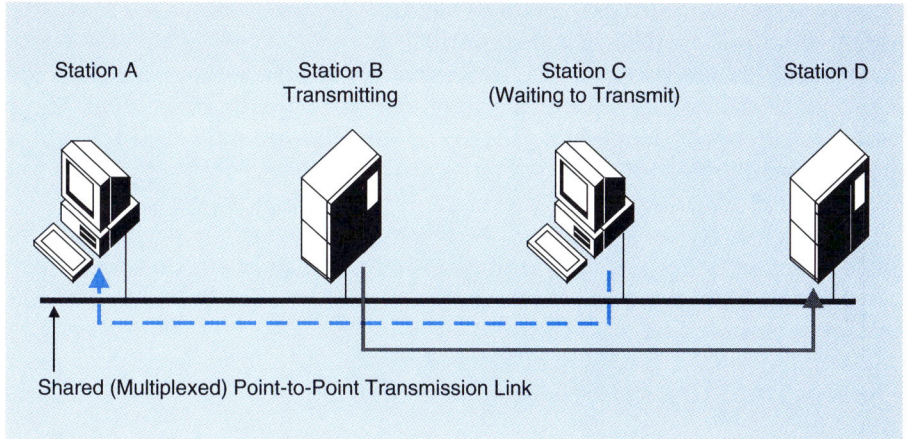

FIGURE A.4

Multiplexing

Multiple stations can share a single point-to-point transmission segment. This is called multiplexing. It reduces costs. However, the stations must take turns transmitting.

work. This message is called a packet. While the data grouping at Layer 2 is called a frame, the synchronous data grouping at Layer 3 is called a **packet**.

When the packet reaches the first switch, the switch has a decision to make. It has two outgoing point-to-point segments leading to other switches. It must pass the packet along to one switch or the other. It makes the choice based on

FIGURE A.5

Packet Switched Network

Station A transmits a message called a packet. This packet passes through a number of switches until it reaches Station B. The job of the network layer (OSI Layer 3) is to manage the routing of the packet from one end of the switched network to another.

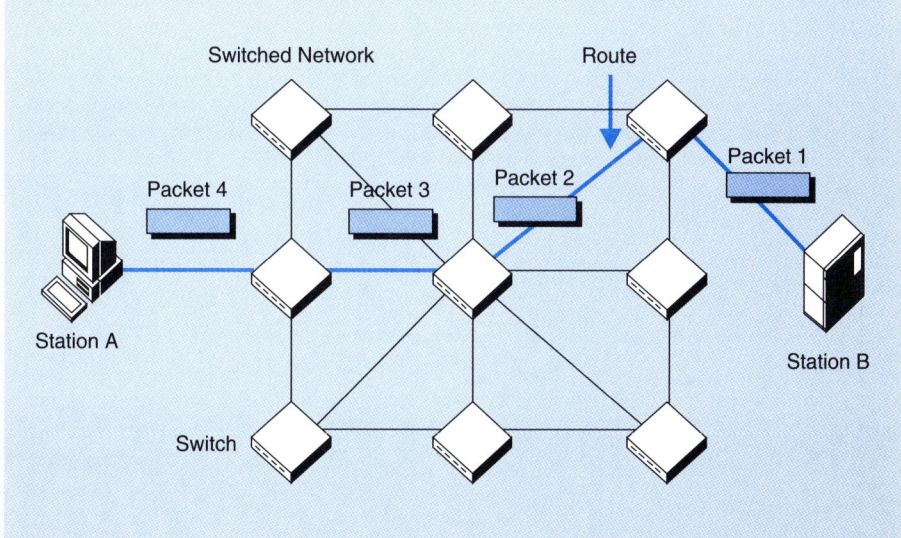

its general knowledge of the network's conditions (congestion, failed lines, segments with the least delays, and so forth).

The end-to-end path that the packet travels is called a **route**. If there are many switches, there is an almost infinite number of alternative routes. Some are obviously better than others. The job of the **network layer (OSI Layer 3)** is to create standards so that the source host and switches can route the packet to the right destination station, picking an optimal route if possible.

Not all networks use switches. LANs, for instance, often consist of a single segment without switches. In this case, there is no alternative routing. There is only one possible route from one station to another. (To see that this is true, consider Figure A.4.)

Protocol Data Units (PDUs) at the Subnet Layer ■

Recall from Chapter 2 that at each OSI layer (except for the physical layer), peer processes on different devices exchange organized messages called **protocol data units (PDUs).** In terms of concepts in Chapter 4, each PDU is a synchronous data package. Figure A.6 shows this process at the lowest layers of the OSI architecture. These layers govern transmission through a single network (subnet in an internet).

ON STATION A. First, Station A's network-layer program creates a network-layer PDU. PDUs at OSI Layer 3 are called **packets**.

Second, the network-layer program on Station A passes the packet to the data link layer program. This program places the packet into a data link layer PDU. In OSI Layer 2, PDUs are called **frames**.

FIGURE A.6
Protocol Data Units

Peer processes on two devices exchange messages in the form of protocol data units (PDUs). In transmission, each data link layer process passes its PDU (frame) to the data link layer process on the receiving device. In reception, each data link layer process passes the network layer PDU (packet) to its device's network layer process. The network layer process on switches determines the best way to send the packet back out. It then passes the packet to the data link layer process on the chosen data link to another switch for delivery over that data link.

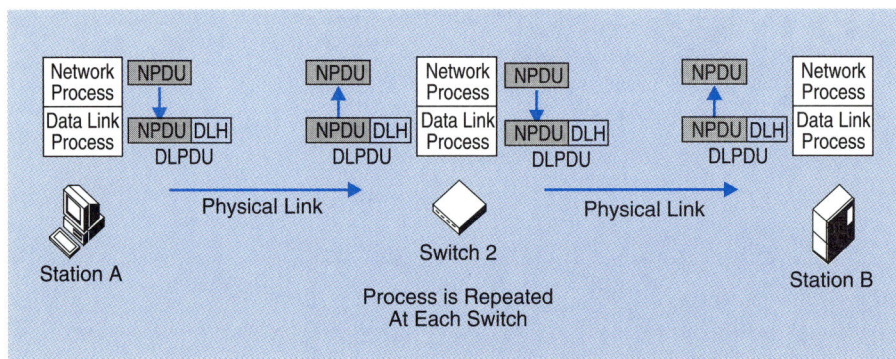

Third, the data link layer program on Station A transmits the frame to Switch 1, using the OSI physical-layer connection.

On Switch 1. Switch 1's data link layer process receives the frame from Station A. It removes the packet from the frame and passes the packet to the network-layer process on Switch 1. The network-layer process decides how to route the packet. It selects Switch 2 as the next switch in the route.

The network-layer process on Switch 1 then passes the packet back to its data link layer process. That OSI Layer 2 process then places the packet in a new frame. It transmits this frame to Switch 2, using the OSI Layer 1 connection between Switch 1 and Switch 2.

On Switch 2. Switch 2's data link layer process receives the frame from Switch 1. It removes the packet from the frame and passes it to the network-layer process. This time, the network-layer process decides that it can deliver the packet to Station B.

The network-layer process passes the packet back down to Switch 2's data link layer process. That OSI Layer 2 process then creates yet another new frame and places the packet inside that frame for delivery. It passes the frame to Station B, using the physical-layer connection that connects them.

On Station B. When the data link layer process on Station B receives the frame, it removes the packet from the frame. The OSI Layer 2 process then passes the packet to the OSI Layer 3 process on Station B.

That network-layer process then deconstructs the packet. It removes the transport-layer PDU from the packet and passes it to the OSI Layer 4 process on Station B.

Notes. Note that only one packet was created in this sequence of steps. Station A created it. Station B received it. This allows unified routing across the network, thus fulfilling the role of OSI Layer 3, the network layer.

However, there were three data links. These links connected Station A with Switch 1, Switch 1 with Switch 2, and Switch 2 with Station B. For each data link between devices on the network, there was a separate OSI physical-layer connection and data link layer connection. There were three OSI Layer 1 connections. In addition, three OSI Layer 2 protocol data units (frames) were created.

OSI Internetting

When the OSI framework was considered, ISO and ITU-T did not envision complex internetting. Consequently, they did not create a separate internetting layer for this complex process. Nor could they easily add a layer between the network and transport layers because they had numbered the layers consecutively. As a compromise, they divided the OSI network layer into three sublayers, as Figure A.7 indicates.

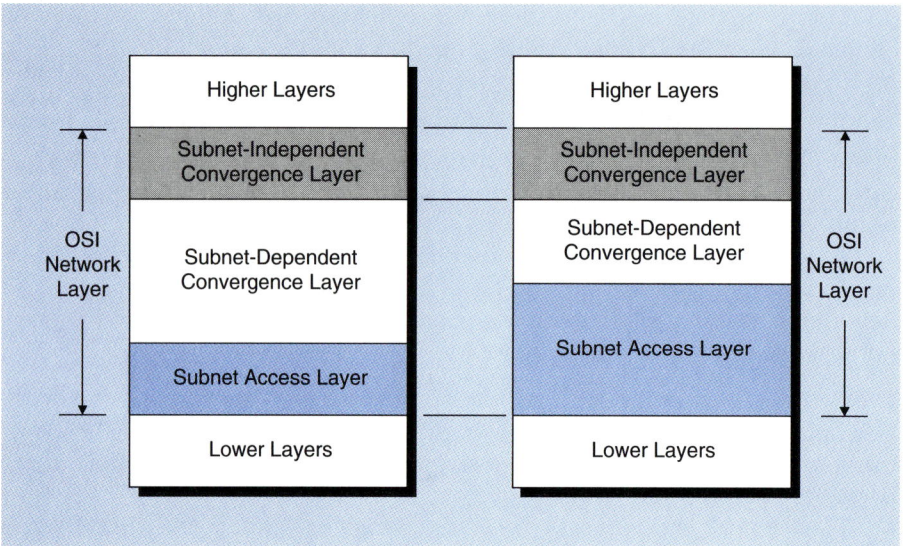

FIGURE A.7
OSI Internetting

For internetting, OSI sublayered the OSI network layer. The *subnet access layer* is the original single-network layer. The subnet-independent convergence layer handles the internetting. The subnet-dependent convergence layer adds functionality to the subnet access layer, so that the subnet-independent convergence layer can assume a given level of functionality below it.

The Subnet Access Layer ■

OSI calls the bottom sublayer of the network layer the **subnet access layer**. Recall from Chapter 1 that a **subnet** is a single network in an internet. This is merely the old network layer. It defines standards for transmission through a single network. If there is no internetting, the higher sublayers are not used.

The Subnet-Independent Convergence Layer ■

The top sublayer, in turn, is the **subnet-independent convergence layer**. This is the layer that actually handles internetting. The subnet-independent convergence layer, as its name suggests, does not care what subnet access layer is used. In other words, all single-networks (subnets, in TCP/IP terminology) can use the same internetting layer standard.

The Subnet-Dependent Convergence Layer ■

Not all subnet access layer standards have the same functionality, so OSI defined the **subnet-dependent convergence layer**. This sublayer adds functionality to the subnet access layer, so that the top sublayer can be independent of the functionality of the lowest sublayer.

The subnet-dependent convergence layer standard, as its name suggests, is specific to the subnet access layer standard. The designers of the subnet-dependent converge layer know the deficiencies in functionality in their particular subnet access layer standard, so they can easily add the functionality needed to get the functionality to the level the subnet-independent convergence layer requires.

OSI Transport-Layer Standards

The role of the transport layer is to allow two stations to communicate even if they have different hardware and operating systems. In fact, there is no way for one station to tell if the other station is an IBM mainframe, a DEC minicomputer, a Hewlett-Packard UNIX workstation, a Windows PC, or a Macintosh. Even a vending machine can be a station on the network. It too will be able to interact with other stations as an equal.

The Transport Layer (OSI Layer 4) ■

OSI has a layer called the transport layer (OSI Layer 4). We will see, however, that it only does some of what a transport protocol needs to accomplish. Its next layer, the OSI session layer (OSI Layer 5) does the rest of the work of the transport layer in our basic communication model.

Figure A.8 shows the OSI transport layer at work. It shows Computer A transmitting a transport-layer protocol data unit (TPDU) to Computer B. This TPDU is delivered inside a network or internet packet, but this is irrelevant.

Note that the computers have different architectures and operating systems. However, because they both communicate using the standards of the transport layer, they can exchange messages effectively.

Because the computers are so different in terms of capacity, the transport-layer standard must do many things to regulate communication. For instance, if one machine is much larger than the other, there must be a way for the slower machine to ask the faster machine to slow down or to stop transmitting temporarily. This is **flow control**.

Transport-layer protocols also have to deal with *host crashes* [Tanenbaum, 1989, pp. 409ff]. Computers crash with some frequency. After a host restarts, it must use OSI Layer 4 protocols to find where it was before crashing. For instance, if it was in the middle of receiving a large file, it must have a way after restarting to check with other hosts to see if it had outstanding transactions. It must then continue the file transfer where it left off.

A third major function of OSI Layer 4 is to make up for limitations in the underlying network. The network is often owned by a transmission carrier, so the host administrator has no control over its quality of service [Tanenbaum, 1989, pp. 371ff.]. If delivery is done over an internet, furthermore, service will almost certainly be unreliable. It is common to handle error correction and other quality of service matters at the OSI transport layer.

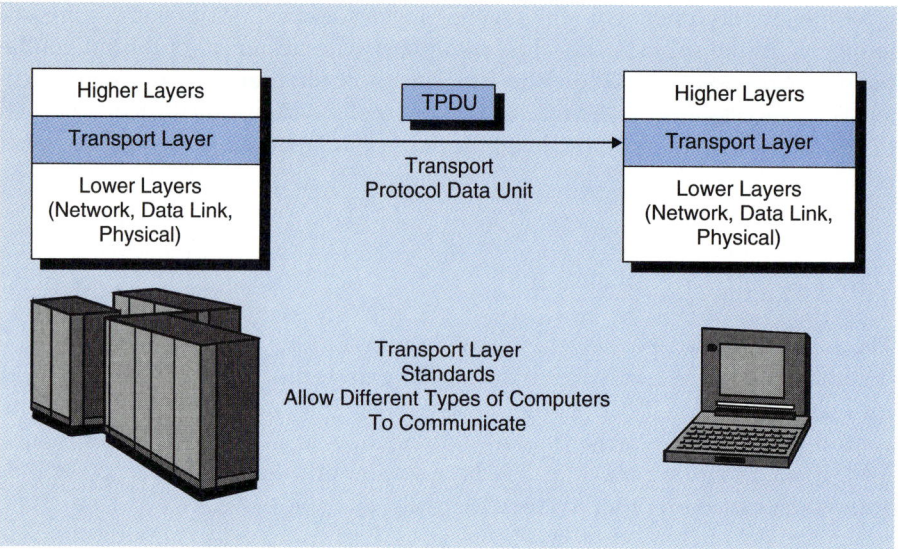

FIGURE A.8
Transport Layer (OSI Layer 4)

The transport-layer processes on the two stations exchange transport-layer PDUs despite differences in hardware architecture, operating systems, capacity, and so forth. This allows a PC to communicate with a mainframe as easily as it can communicate with another PC.

Session Layer (OSI Layer 5) ■

The transport layer can deliver messages (TPDUs) to a host computer. But as Figure A.9 shows, a host computer may have multiple application programs running. There must be a way to specify which applications should receive the enclosed information. This is the main function of the session layer (OSI Layer 5).

The session-layer protocol also manages transmissions between the two application programs. For instance, if the transport-layer process breaks down, the session layer usually remembers what happened up to the last "synchronization point."

In some cases, in addition, the OSI Layer 5 protocol will require one application to wait until the other has transmitted a message. In other words, the session layer can regulate transmissions between the application programs.

Application-Layer Protocols

OSI also subdivides the application layer in our basic communication model. Its two application layers are the presentation layer (OSI Layer 6) and the application layer (OSI Layer 7).

Presentation Layer (OSI Layer 6) ∎

The **presentation layer (OSI Layer 6)** has several purposes. One is to address the problem of differences in the ways different machines store numbers and other information. For instance, one machine might have 32-bit words (basic units of processing) and might call the first bit the left-most bit. Another machine may have 24-bit words and might consider the first bit to be the right-most bit. If information is to be transferred effectively, there has to be translation between the two formats.

Figure A.10 shows how the OSI presentation layer addresses this problem. OSI has created a generalized way of representing many types of information. This is **Abstract Syntax Notation 1,** better known as **ASN.1**. When Application A transmits, its presentation-layer process translates the application message into ASN.1. On the other computer, the presentation-layer process on Station B's machine translates the message from ASN.1 to that machine's native code. It then passes the message to Application B.

OSI Layer 6 also has another purpose. Many ways of representing information are used in multiple applications. For instance, a facsimile image-scanning standard might be used in graphics applications. To prevent each application from creating an incompatible standard for the same purpose, a joint standard is created at the presentation layer. This is, in fact, the layer at which facsimile image-scanning standards are created.

FIGURE A.9

Session Layer (OSI Layer 5)

The session-layer processes on the two stations connect a specific application program on one multitasking station with another OSI Layer 5 process on another multitasking station. The session-layer processes on the two stations also regulate interactions between the application programs, for instance, determining when each may transmit and restarting communication after a breakdown.

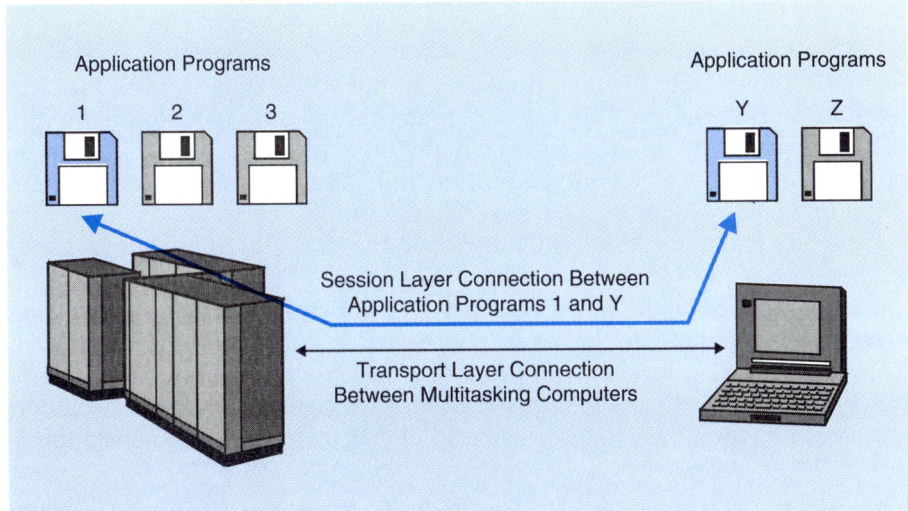

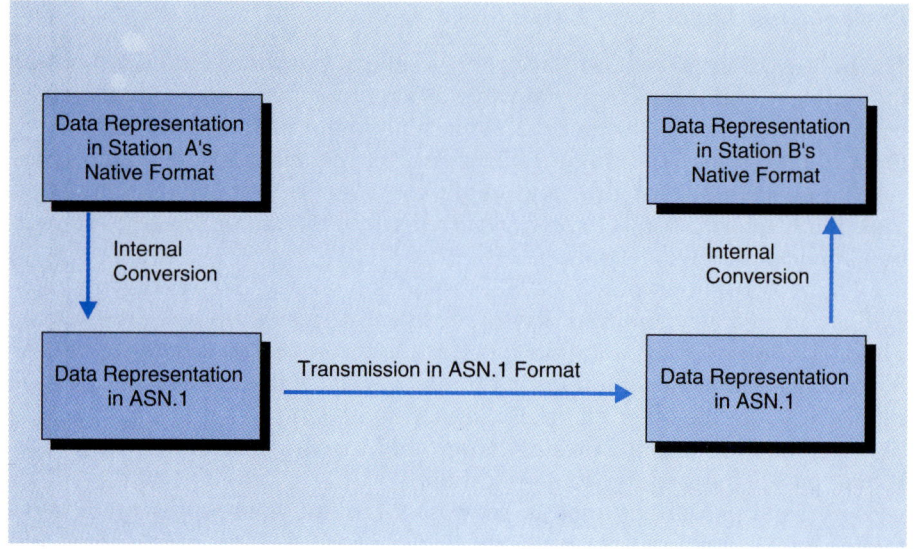

FIGURE A.10

Abstract Syntax Notation 1 (ASN.1) at the Presentation Layer

Different computers have radically different ways to store and represent information. At the presentation layer, ASN.1 provides a common representation language for data. The sending machine translates the information into ASN.1 and transmits the information. The receiving machine translates the information from ASN.1 into its own native format.

Application Layer (OSI Layer 7) ■

The role of the **OSI Application Layer** is probably the easiest to understand of all layers. Its role is to link the application processes on the two stations. Its PDUs allow electronic mail programs from two vendors to work together (see Chapters 2 and 10). **OSI Layer 7** standards also provide interoperability between many types of browsers and many types of World Wide Web servers on the Internet (see Chapter 2 and Module C). Of increasing importance, application-layer standards also govern connections between the directory servers (see Chapters 8 and 9) that provide the glue that holds networks together.

Over time, OSI's designers gradually came to realize that the application layer is an extremely complex layer. They found that for many standards, such as the X.400 electronic mail standards, they had to divide it into sublayers of functionality. Unfortunately, they subdivided the OSI application layer differently in different applications, so we cannot talk in general about OSI sublayering.

Common Functions

Some functions are found at multiple layers:

- **Error detection and correction**. There can be errors or breakdowns at any layer. Many layer standards will check for errors and ask for the retransmission of incorrect PDUs. Many, in addition, will maintain a connection despite a brief breakdown in the next lower-layer process.
- **Multiplexing**. If a given layer is N, it may multiplex PDU streams from several layer $N + 1$ processes. For instance, a single transport-layer connection between two stations can multiplex traffic from multiple session-layer connections between multiple application program pairs on two multitasking stations. On a LAN, multiple data link layer connections between pairs of stations can be multiplexed onto a single physical-layer segment.
- **Security**. Although we did not discuss it, security is likely to be implemented at multiple layers. It is not uncommon to encrypt data streams at multiple layers (see Chapter 9 and Module J).

CONCLUSION

In this module, we saw how OSI was created. We then discussed its seven layers, working our way up from the lowest layer to the highest.

CORE REVIEW QUESTIONS

1. Briefly describe the name and purpose of each of the OSI layers. Given an OSI layer number or name, be able to give the other. Given a function, be able to tell into which layer it should be put.

2. Two stations communicate via a switched network. The route that messages take passes through ten switches. How many network layer PDUs will there be in the transfer? How many data link layer PDUs will there be in the transfer? How many physical layer connections will there be in the transfer? How many routing decisions will there be in the transfer?

3. What do we call an OSI Layer 2 PDU? An OSI Layer 3 PDU?

4. What functions tend to be implemented at multiple layers?

DETAILED REVIEW QUESTIONS

1. At what layer do we standardize voltage levels used to represent 1s and 0s?
2. If you have a point-to-point connection—say, when you dial into an Internet service provider from home—what transmission layers of OSI will be involved?
3. What is an alternative route?
4. Why don't many LANs have a network layer?
5. Explain how the transport and session layers work together to link application programs on different stations. Distinguish between their roles.
6. What is ASN.1? Why did OSI create a distinct presentation layer?
7. Why is the application layer often subdivided in standards?

THOUGHT QUESTIONS

1. Given the perspective of time, suggest two ways in which the OSI architecture could have been designed differently.
2. In the preceding Core Review Question 2, how many transport layer PDUs will be created? How many session-layer PDUs?

For online exercises, please visit this book's website at: http://www.prenhall.com/panko

REFERENCES

TANENBAUM, A. S. (1989). *Computer Networks*, Second Edition, Upper Saddle River, NJ: Prentice-Hall.

MODULE B

The Internet

Introduction

This module assumes that you have read Chapters 1 and 2. It can be read front to back, like a book chapter. Module C is a continuation of this module, focusing on the World Wide Web and browsers.

No aspect of data communications has been more in the public news than the Internet. As noted in Chapter 1, the Internet is a worldwide transmission network. Its millions of host computers serve tens of millions of users. Its size and diversity make it fertile soil for the sprouting of hundreds of new services. Some of these services are brilliant, others banal. Still others, including access to pornography, incite many people to anger.

The Internet is already too widespread and diverse for any module (or book, for that matter) to describe it. Our goal in this module is to discuss how the Internet evolved, how it manages worldwide connectivity, and what its major services (electronic mail, File Transfer Protocol, and so on) offer today. Module C focuses on a particular service, the World Wide Web.

Basic Concepts

Figure B.1 shows that the Internet consists of many different networks in different parts of the world. Special switches called **routers** (see Chapters 6 and 7) link the networks together. As a result, any Internet host can send messages (called IP datagrams) to any other, as if they were on a single network.

The key to the Internet's ability to link its millions of hosts together is a universal system of **host addresses** called **IP numbers.** Each host on the Internet has a unique IP number, which acts like a telephone number. When one host wants to send an IP datagram to another, it merely puts the target host's IP number in the destination field of the datagram. The sender then transmits the message to a router. If the router cannot deliver the message itself, it passes the message to a router closer to the target host's location. Eventually, a router delivers the datagram to the target host.

FIGURE B.1

The Internet

The Internet consists of many networks in many parts of the world. Switches called routers link these networks together. Every host on the Internet has a unique host address called an IP number. When one host wants to send a message (called an IP datagram) to another host, it merely puts the target host's IP number in the destination field of the datagram and sends this datagram to a router.

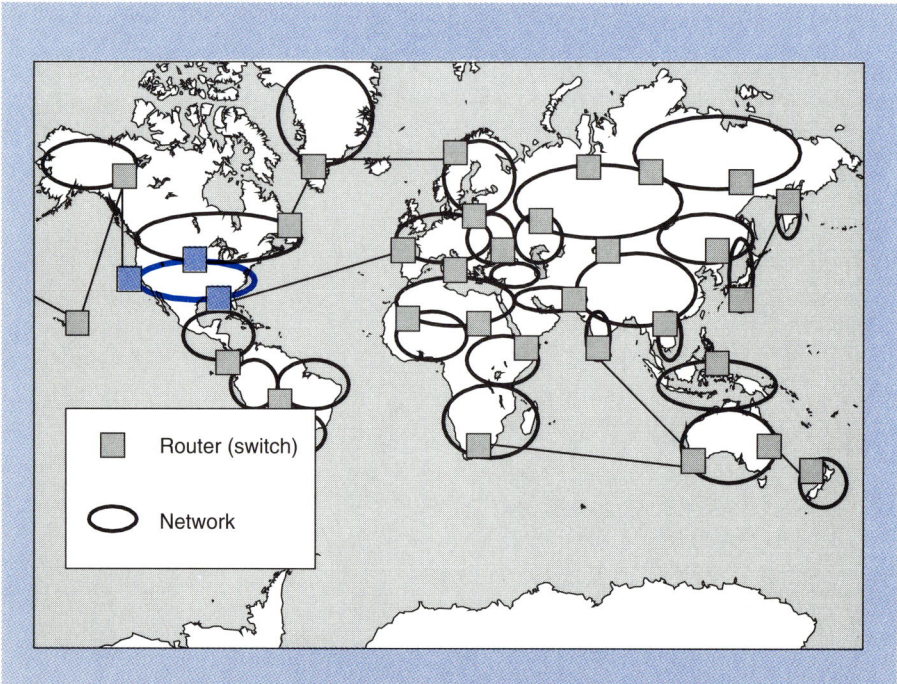

Internet host addresses actually come in two forms. One is the **IP number,** which consists of four numbers separated by periods. A possible host number is "128.171.17.13". Sometimes this is followed by a port address, to refer to a specific connection within the host. To continue our example, if the port address is 50, the host number would be "234.17.2.156:50".

Although IP numbers always work, they are difficult to remember. So most hosts also have **IP names**. The author's email host, for instance, has the host name *DScience.cba.hawaii.edu*. This host name has four parts, and it is tempting to believe that these correspond to the four parts of the IP number, but this is not generally the case. Capitalization, by the way, is not important.

Special hosts on the Internet called **domain name system hosts** keep lists of IP numbers and IP names. As Module H discusses, if a host knows a host's IP name but not the host's IP number, it can contact a Domain Name System host for the number to put into the IP datagram's destination host field (see Chapter 7). *For more on IP addresses, see Module H.*

ORIGINS

The Internet did not reach its current size through some conscious plan. Rather, the Internet has grown chaotically over time. While there are some principles that seem to characterize the Internet's evolution, they are rather vague and are more cultural norms than regulations.

The ARPANET and Telnet

During the 1960s, the **Defense Advanced Research Projects Agency (DARPA)** of the U.S. government funded a broad range of advanced computer technology projects. By the late 1960s, DARPA was acutely aware of a growing problem. In many cases, the software it commissioned could only run on a single computer. Researchers at other sites would have to travel to that computer if they needed to use the software. Or they would have to connect their terminals to the distant host via expensive, slow, and error-prone telephone lines.

Figure B.2 shows the solution that DARPA created. It built a nationwide computer network that would link the major research centers receiving DARPA funding. This wide area network (WAN) would give everyone reliable terminal access to software at other sites. This was the **ARPANET**.

Figure B.2 shows that a user logged into a local host computer, in this case named *Host1*, via a terminal. Then the user executed a program on Host1 called **Telnet**. This program then connected the user to a second host computer, *Host2*. The user then logged into that second host. Afterward, the user would not even be aware that the first host existed. Telnet would make it seem as if the user's terminal connected directly to the second host.

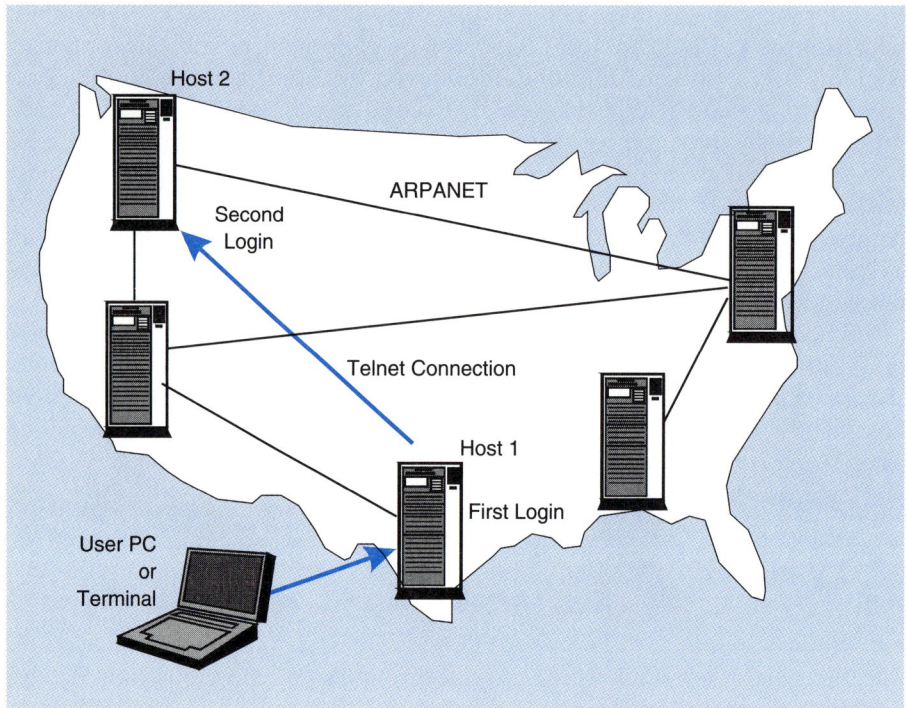

FIGURE B.2

Telnet via the ARPANET

The user works at a terminal or a PC emulating a terminal. The user logs into his or her normal host, Host 1. He or she then runs the Telnet program on Host 1. This connects the user to Host 2, where he or she again logs in. From this point on, the user is effectively logged directly into Host 2. Telnet allows people to use applications that are not present on their normal host computers but are available on other hosts. The user does need an account name and a password he or she can use on Host 2, however.

Figure B.3 shows a simple Telnet session. This session takes place today, but it could as easily have taken place in the early 1970s. Telnet, as a product, is largely frozen in time.

First, the user logs into his or her host computer. This host gives the user its prompt, in this case, *C:>*. At the prompt, the user types *Telnet* to begin the session. The user then asks Telnet to open a connection to the second host.

When the second host responds, the user must type an appropriate account name and password. Note that this account name and password are needed to get access to the *second* host. The user previously typed a different account name and password to get access to the first host.

After working with the other host a while, the user wants to stop working on the second host. The user then types the escape character, which is usually *control-]*. This brings up the Telnet prompt. In this example, the user merely

Quits out of Telnet. This closes the connection to the second machine. The user then gets the *C:>* prompt again. He or she is now talking to the operating system of the first host.

ARPANET: Electronic Mail

Although the ARPANET was built for Telnet service, it did not take researchers long to figure out how to add electronic mail to the network. Soon electronic mail became the dominant service on the ARPANET. Even today electronic mail dominates the number of messages transmitted on the Internet. At first, the host mail systems used an extremely crude protocol. Later they upgraded to what we now call the **Simple Mail Transfer Protocol. SMTP** is not sophisticated, but it is quite serviceable. *For more on Internet electronic mail, see Chapter 10 and Module J.*

Connecting Networks

After a while, the military wanted a network of its own. So DARPA oversaw the creation of a sister network, MILNET. Of course, people on the ARPANET and MILNET soon found themselves wanting to share information. So DARPA ordered the creation of a protocol to transfer data across networks. The eventual result was the *Internet Protocol* that we discuss in Chapters 5 and Module H.

In the early 1980s, the military decided to standardize on the Internet's architecture, TCP/IP, for all of its data communications. At first, vendors objected. However, the military's insistence on TCP/IP was firm, and vendors soon responded with a flood of TCP/IP offerings. *For more on the TCP/IP architecture, see Chapter 1 and Module H.*

1.	C:>**Telnet**	Enter the Telnet program at your host's prompt.
2.	Telnet>**open** *host address*	Open a connection to the other host.
3.	Trying *host address* . . . Connecting to host address Escape character is '^]'.	Connecting to the other host.
4.	Host>*Login* . . .	Log into other host.
5.	Host>*commands*	Give commands to the other host.
5.	. . .	Give more commands to the other host.
6.	Host>^]	Gives the escape character (control-]). From this point on, you will be giving commands to Telnet instead of to the other host.
7.	Telnet>**quit**	Quits the Telnet program.
8.	C:>	Back to your original host's prompt.

Note: What the user types is in boldface.

FIGURE B.3

A Telnet Session

The Internet

Soon other networks wanted to connect to the ARPANET. In computer science, there was CSNET. Among business schools and some other academic units, there was BITNET (Because It's Time Network). Members of one network often needed to deal with people and resources on other networks.

DARPA began to connect these other networks to the ARPANET and so to one another. Because of the military's use of TCI/IP, the software needed for networks to connect to the ARPANET was readily available and reasonably priced.

In time, this growing internetwork began to link networks in other countries as well. The result was the creation of a new entity, the **Internet,** which consisted of all of the individual subnets.

At first, DARPA owned most of the transmission lines. Eventually, however, NSF's NSFNET took over as the main transmission backbone in the United States. Later, the U.S. backbone link was privatized. IBM, Merit, and MCI created a new company, *Advanced Network Services*. NSF gave it a contract to upgrade and maintain the U.S. backbone under the name ANSNET [Comer, 1995].

In other parts of the world, the backbones used in individual networks vary widely in ownership. For instance, in Europe, the EBONE backbone is comparable to ANSNET [Comer, 1995].

Managing the Internet

Faced with this increasingly complex environment and wanting to get out of the business of trying to manage it, DARPA created the **Internet Society**. This nonprofit organization produces standards through its **Internet Engineering Task Force** suborganization.

Control of the Internet ■

It would be a mistake to say that the Internet Society really *controls* the Internet. The various networks that participate in the Internet communicate as peers, and they are owned by different organizations. Although the Internet Society has some moral power because of its heritage, the Internet Society and the Internet Engineering Task Force cannot force compliance to technical standards or standards of conduct on the Internet.

Partly as a result of this situation, and partly as a result of the Internet's traditional culture of laissez-faire computing, the Internet today is a rather chaotic environment. There are few real rules to constrain you from doing creative things on the Internet, and there are few recourses if someone abuses access to the Internet to harm others. The Internet is today's Wild West.

Controlling Host Numbers ■

The one unquestioned "power" of the Internet Society is its control of IP numbers and IP names. Everyone agrees that there needs to be a central clearinghouse for addresses. Otherwise, there would be no way to ensure that host IP numbers and names were unique.

However, even here, the Internet's control is diffused. The Internet Society allocates blocks of IP numbers to individual organizations, which then allocates IP numbers to individual hosts. This preserves uniqueness in IP numbers.

For IP names, there must be a central worldwide organization to register names. This is **InterNIC**. *For more on name assignment, see Module H.*

Human Communication Services

Chapters 2 and 10 discuss electronic mail. Although electronic mail is still one of the biggest uses of the Internet, the Internet today offers a variety of human communication services beyond electronic mail.

LISTSERV

One innovation came out of the BITNET community. This was the LISTSERV program. **LISTSERV** is a way of providing many of the benefits of computer conferencing through electronic mail.

A LISTSERV server maintains mailing lists. When someone posts a message to one of these mailing lists, the LISTSERV program retransmits the message to the list's subscribers. These messages arrive in the subscriber's email in-basket, just like ordinary messages. In this way, the user does not have to make an extra effort to check for new postings.

Subscribing ■

To get on this mailing list, you send an email message to the LISTSERV program on a particular host. This message has only the following line:

```
SUBSCRIBE <list-name> <yourname>
```

Figure B.4 illustrates this situation. For example, suppose you want to join the GLOBAL-L list. You would send your subscribe message to LISTSERV@puka.org. Your message would say "subscribe GLOBAL-L <yourname>". Here <yourname> is your personal name. The author, for instance, would type *subscribe GLOBAL-L Ray Panko*. There is no need to type your electronic mail address in the body of the message because the LISTSERV program already knows your address through the *From:* field in your message.

Submitting Comments ■

After subscribing, however, you would not send comments to LISTSERV@puka.org., which is the manager for multiple lists. Instead, as Fig-

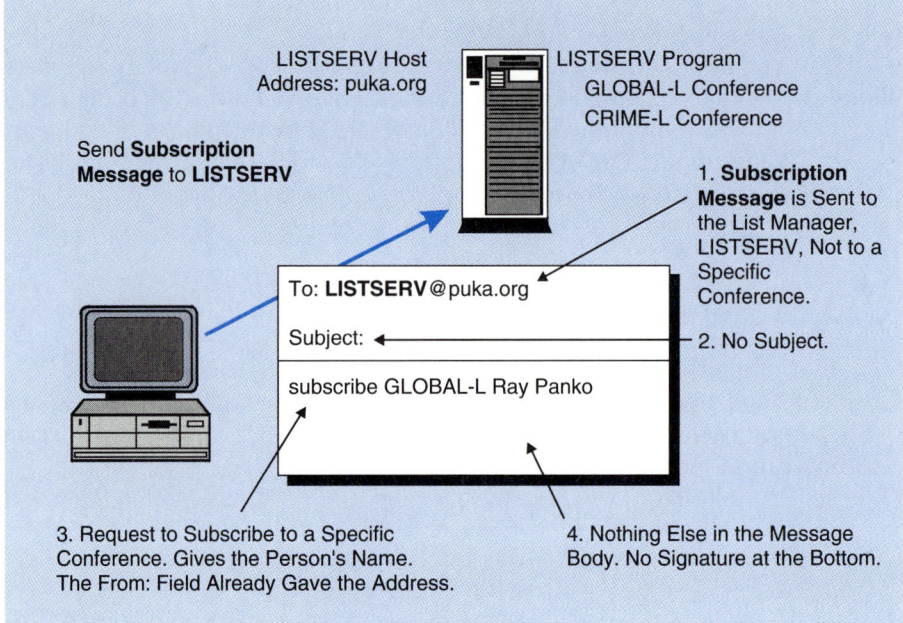

FIGURE B.4
Subscribing to a LISTSERV List

To subscribe to a LISTSERV list, a person must send a message to the LISTSERV management program on the LISTSERV host. The LISTSERV management program may manage several lists. On this host, the LISTSERV program manages two lists: GLOBAL-L and CRIME-L. To subscribe to GLOBAL-L, you send a message to LISTSERV@puka.org. The message has no subject field. Its body has a single line, "subscribe GLOBAL-L Ray Panko". This tells LISTSERV that the sender wants to subscribe to a list, that the list to which he wishes to subscribe is GLOBAL-L, and that the sender's name is Ray Panko. There is no need to include the email address in the body because it is contained in the *From:* field of the message.

ure B.5 illustrates, you would mail them *to the name of a specific list*. In this case, you would address the message to GLOBAL-L. This tends to cause confusion. Note that you *subscribe* with a message *to the LISTSERV program*, but you then *post comments to the list name*.

Unsubscribing from a List ■

If the list does not prove worthwhile, you can send a second message to the LISTSERV program at puka.org:

```
UNSUBSCRIBE <list-name>
```

Again, there should be nothing in the body of the message except this single line.

This second command will take your name off the list. Note again that in the case of GLOBAL-L, you would send the UNSUBSCRIBE command to LISTSERV, the overall LISTSERV program, rather than to the GLOBAL-L address to which you have been posting messages.

Moderated Versus Unmoderated Lists ∎

Some LISTSERV lists are **moderated**. This means that incoming messages are not posted automatically to the mailing list. A human moderator reads each message, posting only those that he or she feels are appropriate to the topic and that are in good taste. In contrast, there is no screening in unmoderated lists, and flaming (abusive messages) is all too frequent.

The Attraction of LISTSERVs ∎

LISTSERV lists are attractive because you do not have to log into a separate program to read messages. Postings come directly to your normal email mailbox.

Some people who get many messages each day prefer to handle all of their

FIGURE B.5

Submitting a Comment to a LISTSERV List

After subscribing, you automatically receive messages that others send to the GLOBAL-L list. When you wish to post a message, you send it to the name of the list (GLOBAL-L), not to the name of the LISTSERV. To repeat, you send content messages to a different address (GLOBAL-L) than you send supervisory messages (LISTSERV).

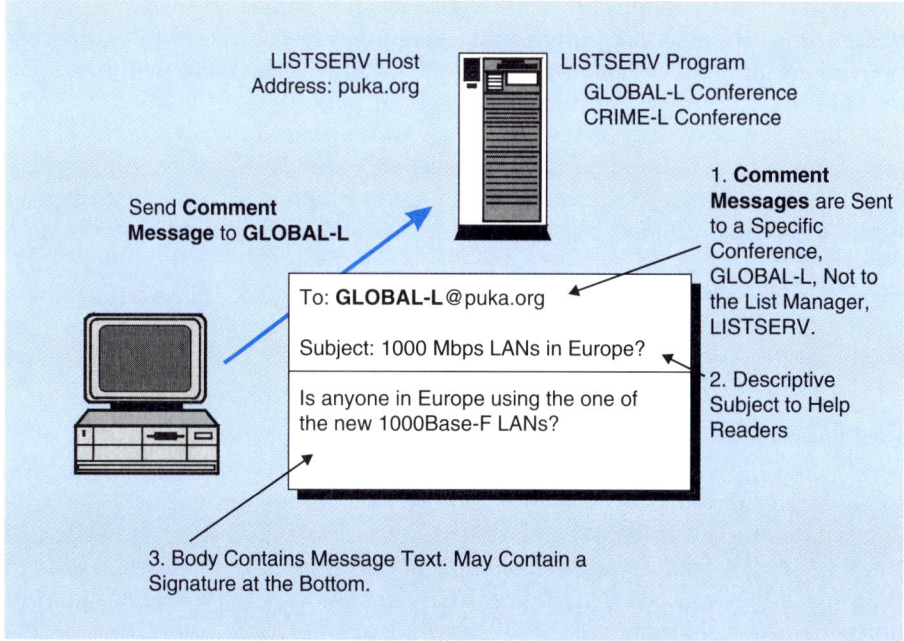

LISTSERV mail separately. This is possible because all messages we have been using come from GLOBAL-L. So if you wish, you can set up your electronic mail filtering rules (see Chapter 10) to file all incoming messages from GLOBAL-L into a file folder with a name like GLOBAL-L. In this way, when you go into your normal in-box, there will be no GLOBAL-L traffic there. You can read the GLOBAL-L postings later.

USENET

Another conferencing system that grew out of the UNIX community is **USENET**. USENET does not use electronic mail, so it does not clutter your in-box with irrelevant messages. Instead you use a **USENET reader** program that allows you to look at traffic on individual USENET conferences. At any moment, you only read the messages for a single conference. When you finish, you can switch to another conference.

USENET, at any particular site, is a huge database with messages for hundreds of conferences called **newsgroups**. USENET newsgroups include everything from forums on particular operating systems to sexually explicit exchanges. Some conferences are vibrant and alive. Others are moribund.

Figure B.6 shows that USENET database programs at different sites exchange new messages. This exchange often happens overnight. This way, in the morning, each USENET database will have roughly the same content. We say "roughly" because things are seldom so smooth in real life. Not every message is updated to every copy of the database. Some sites will only keep messages on a limited number of conferences.

USENET newsgroups typically take on such names as *comp.risk*, which deals with computer risks. In general, there are a few top-level categories for newsgroups. *Comp*, for instance, refers to computers. There you will find such newsgroups as *comp.risk* and *comp.protocols.iso.x400*.

Rec is for recreation. There you will find such newsgroups as *rec.arts.sf.movies* for science fiction movies. *Sci* is for science. *Soc* is for social issues. *Alt* is for alternative matters. Alt groups tend to be based on rather odd and controversial topics. You will find *alt.alien.visitors* and *alt.binary.pictures.misc* for miscellaneous bit-mapped pictures. *Misc* is for miscellaneous topics.

Figure B.6 shows that you need a client USENET reader program to access USENET servers. At a server, you can **subscribe** to several newsgroups that you wish to read.

Real-Time Discussions

Email is asynchronous. The sender and the receiver are not necessarily active at the same time. LISTSERV and USENET are also asynchronous. But sometimes groups of people want to converse in real time. They want to be able to type group messages on their keyboards, and they want to be able to see what other members are typing.

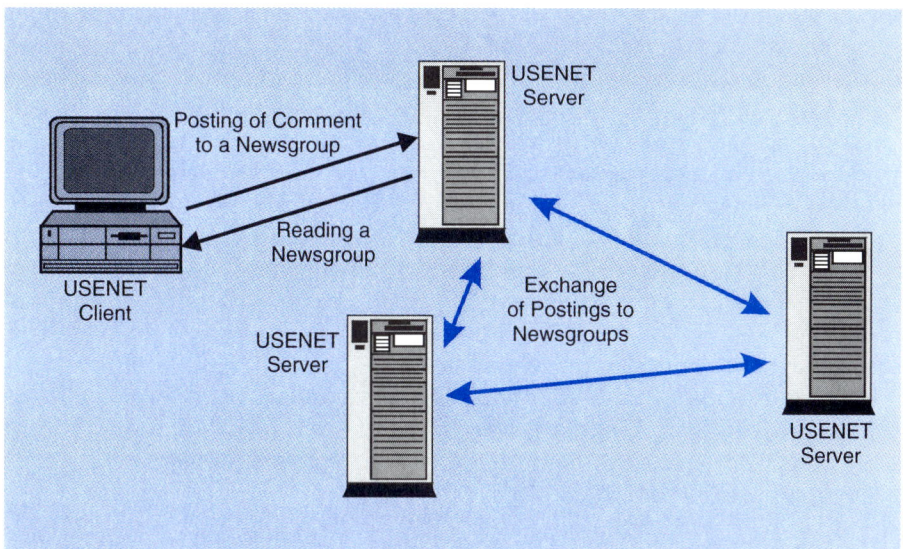

FIGURE B.6
USENET

USENET is a large database of comments in a large number of newsgroups. Different USENET servers exchange new messages, often overnight. In this way, users can link to any of several USENET servers and see roughly the same messages. Users must have client news reader programs to access USENET servers.

Talk ■

Talk allows you to talk to one other user. To initiate a talk session, you type Talk <accountname@hostname>. Usually, the other party has to agree to talk to you. (They may be busy.) Your screen splits in half. One half shows what you are typing. The other half shows what the other person is typing. You can both talk at the same time.

Internet Relay Chat ■

Internet Relay Chat (IRC) allows you to chat online with *many* other people. When you join an IRC *channel*, your screen again splits in half. As in Talk, one half shows incoming messages. The other shows what you are typing. *IRC* discussions tend to be verbal free-for-alls, with no holds barred.

MUDs and MOOs ■

There is now software to allow groups of people to build imaginary worlds called MUDs or MOOs,[1] in which individuals take on identities called **cyber-**

---
[1]These are acronyms for Multi-User Dungeon (showing the Dungeons and Dragons origin of these programs) and MUD, Object Oriented.

selves. For instance, the environment might be a castle, and your cyberself could be a wizard. If you type a command to walk into a room and say something like "look," you will be given a written description of what the room looks like.

You can also create a description of your character. When people *look* at you, they see a description you have written for your character. Objects and players can even "do" things. For instance, if you pick up an odd object, it could explode. You could also throw thunderbolts at others.

Players type messages back and forth, to threaten one another, bargain, or just chat. They can even add nonverbal communication. Suppose your game name is "Flash." If you type "POSE hangs his head in sorrow," the other person would see "Flash hangs his head in sorrow."

In one example of this communication environment, Media MOO, you enter MIT's Media Lab. You can walk around, go into rooms to hear real-time seminars, or chat with people in the corridor. Sometimes, a veteran will even give you a helicopter tour.

Newbies and FAQs

One problem in asynchronous and synchronous conferences is that newcomers often do not understand the existing group's informal rules. For instance, many of these **newbies** ask questions that have already been asked thousands of times before, and they behave in ways that upset veterans.

When you join any activity, you should look for a file that lists **frequently asked questions (FAQs).** The FAQs are questions that are asked so often that veterans are sick of hearing them. Looking through the FAQs first will enable newcomers to avoid asking such queries.

In addition to answering often-asked questions, FAQs usually lay out the ground rules of the group as well. This allows you to understand the group's Do's and Don'ts. At the same time, many rules are informal and not are written anywhere. It is wise to hang back a bit and simply **lurk** (watch) when you join a new group. This will give you a chance to understand how the group really conducts itself.

Videoconferencing

Although Internet links tend to be rather slow, some operate at hundreds of kilobits per second. This allows low-cost (and also low-quality) videoconferencing.

One Internet videoconferencing tool is the CUSEEME program created at Cornell University. To use CUSEEME, all you need is a camera and sound card on your PC, an Internet connection, and CUSEEME software. Because the system uses a software codec (see Chapter 10) instead of hardware, it is very slow, but the cost is so low that many Internet users still use the system.

Another videoconferencing system is the MBONE, which allows one-to-few transmission of video and audio. For instance, the MBONE could be used to teach a class via video. Each student would get an MBONE connection to the class.

Internet Telephone Service

If the Internet can be used for videoconferencing, it is even easier to provide voice telephone service over the Internet. Several standards for **Internet Telephone Service** are now in use. Of course telephone companies strenuously object to "free" long distance calling on the Internet.

INFORMATION RETRIEVAL SERVICES

Chapter 2 divided most applications into human communication applications and information retrieval applications. So far, we have looked at the former services. Now we will turn to the latter.

Anonymous FTP

One important information retrieval service is **Anonymous File Transfer Protocol (FTP),** which allows you to retrieve files from a host computer.

The host authority implements anonymous FTP software. The host authority then makes a directory and subdirectories open to anyone. It loads these directories with files that many people are likely to want, such as reports, news releases, and (in recreational anonymous FTP), games.

Figure B.7 shows a typical anonymous FTP session. You first invoke the FTP program. Then you *open* a connection to a distant host computer.

The host will ask for your password. In anonymous FTP systems, the password is *anonymous*. When asked for your password, in turn, you type your Internet electronic mail address.[2]

After you connect, you move through the directory structure, viewing the contents of directories much as you would in MS-DOS. Many FTP programs support MS-DOS commands, such as DIR and CD. Others require you

---
[2] It is courteous to do so, although very few anonymous FTP sites will refuse your admission if you type something else.

1.	C:>**ftp**	Enter the FTP program.
2.	ftp>**open** *hostname*	Open an FTP connection to the remote host.
3.	200 hostname ftp server ready Name: **anonymous** Password: *youremailaddress* system greeting	Connect to the other host.
4.	ftp>**dir**	To see a list of files in the current directory. On UNIX machines, the command is likely to be "ls".
5.	ftp>**cd outside**	Change the current directory to "outsidem," which is one level down.
6.	ftp **dir**	Do a Dir in the outside directory. You find the file Good.Stf.
7.	ftp>**binary**	Set file transfer type to Binary, to transfer a file that has other than simple ASCII characters. If you will transfer an ASCII file, type "ASCII" instead of "Binary."
8.	ftp>**get Good.Stf**	To start the file transfer.
9.	Transfer complete	FTP program tells you that the transfer is complete.
10.	ftp>**help**	To see a list of FTP commands.
11.	ftp>**help bye**	To get help on the Bye command.
12.	ftp>**bye**	To quit FTP.
13.	C:>	Back to your host prompt.

Note: What the user types is in boldface.

FIGURE B.7

Anonymous File Transfer Protocol (FTP) Session

to know how to move through directories using UNIX commands. For more information on this, see the box "MS-DOS and UNIX Command Line Commands."

When you find the file you want, you need to know if it is an ASCII file, which only contains printable ASCII codes (see Chapter 3 and Module D), or if it is a complex file with non-ASCII codes, such as a program or a word processing document. If it is an ASCII file, you give the command *ASCII*. Otherwise, you give the command *Binary* to tell FTP that you want to transfer a file that is not ASCII. If you try to transfer a binary file as ASCII, you often get a file size of zero, indicating that the file did not transfer.

Finally, you invoke the *Get* command. This causes the transfer to begin. The FTP program on the distant host communicates with your local host's FTP program. This allows them to handle any errors that occur.

The *Get* command, overall, is similar to the MS-DOS copy command, although FTP copies files across networks with full error correction and allows you to browse through directories on the other host before you make transfers. The *put* command, in turn, copies a file from your computer to the remote computer.

MS-DOS AND UNIX COMMAND LINE COMMANDS

When you type commands at the command line, syntax and spelling are crucial.

MS-DOS

When you type MS-DOS commands, capitalization is completely unimportant. Spacing, however, can be critical.

Changing Directories ■

There is always one directory on each drive that is the current directory.

The Current Directory If you type *prompt pg*, the prompt will always show the current directory.

CD . . The *CD. .* command will change the current directory to the parent of the current current directory.

CD childname The CD *childname* command will change the current directory to the specified child of the current current directory. If the current directory has the subdirectory Wilkins, the command would be CD Wilkins.

Viewing the Files in a Current Directory ■

The DIR command lists all of the files and subdirectories in the current directory.

DIR If you want to see all of the files and subdirectories under the current directory, give the command DIR, which is short for Directory. It will show each file or subdirectory on a separate line. For each file, you will see the file's size (in bytes) and the date and time it was last modified. For directories, you will only see the name.

DIR /P If you have many files, you can give the command DIR /P. The P means pause or page. After a screen of files and directories has appeared, the system will pause and wait for you to press any key to continue.

DIR /W For a more compact view of your files, type DIR /W. This will show you file names and extensions and directory names in wide view. You will see files and/or directories on each line. DIR /W /P will pause after every screenful.

continued

Seeing the Contents of a Text File ■

If you have a file consisting of nothing but text, you may give the command *Type filename.ext* and MS-DOS will show it to you on-screen.

If the file is too long to fit on one screen, Type *filename.ext / more* and MS-DOS will pause after every 23 lines. Hit Enter to continue.

UNIX

When you type UNIX commands, capitalization is extremely important. Type your commands in lowercase, but when you type the name of a specific file or directory, be sure that your capitalization is correct.

Changing Directories ■

There is always one directory on each drive that is the current directory.

The Current Directory To see the current directory, type *pwd* (print working directory), and UNIX will show the name of the current directory on-screen.

CD . . As in MS-DOS, the *CD . .* command will change the current directory to the parent of the current current directory.

CD childname As in MS-DOS, the CD *childname* will change the current directory to the specified child of the current current directory. If the current directory has the subdirectory Wilkins, the command would be CD Wilkins.

Viewing the Files in a Current Directory ■

The *ls* (list) command lists all of the files and subdirectories in the current directory.

ls The command *ls* will list all of the files and subdirectories of the current directory. It will only show their names.

ls −l The minus L says to show a long listing, which gives detailed information about each file.

Seeing the Contents of a Text File ■

If you have a file consisting of nothing but text, you may give the command *cat filename* and UNIX will show it to you on-screen.

If the file is too long to fit on one screen, *pg filename* or *more filename* and UNIX will pause after every 23 lines. Hit Enter to continue. Some UNIX systems support *pg*, others *more*, and others both.

ARCHIE

There are literally hundreds of anonymous FTP sites around the country. Suppose you want to get a copy of the program, *ABC.DEF*. How would you go about finding which anonymous FTP site has the file and in what directory the file is stored? This is obviously a difficult proposition.

The **ARCHIE** program makes your life easier when you look for specific files in anonymous FTP archives. **ARCHIE servers** dig through all known anonymous FTP sites. They make a list of the directories and the files they find there. They place all this information in a large database.

To use ARCHIE, you must first log into an ARCHIE program. If your site has an ARCHIE program, you merely log into it there. Otherwise, you must Telnet to another site that hosts an ARCHIE program.

After you start the ARCHIE program, you can type the name of the file, ABC.DEF, or only part of it, such as "BC". ARCHIE will then give you a list of all files and directories matching your search. For each, it will show the name of the host, the name of the directory on the host, and the specific name of the file. For files, it shows the size, last update date, and other characteristics.

Gopher

FTP is a product of the early 1970s. Its crude command line interface has not changed much since it was first created. ARCHIE added something new: an improved way to search for information. One of the biggest recent trends in the Internet has been a series of important improvements in ways to seek information.

The first retrieval service to go beyond the command line interface and cryptic file names was **Gopher**. Gopher is a **client/server** system. To use it, you must log into a **Gopher client,** hopefully on your own host. Then you must connect to one of several hundred **Gopher servers**. These may be on different machines, or they may be on the same machine.

Figure B.8 shows that Gopher is menu oriented. You select a top-level menu item that looks interesting, then see a second-level menu. Simply by following menu selections, you eventually find the information you want.

Gopher can show you text information on-screen. But it can do other things as well. For instance, it can download a file to your client program if the last item in the menu is a file transfer.

By today's standards, Gopher has a crude user interface. In addition, it only shows bare ASCII text in a single font. But its menu-driven user interface was a real breakthrough when it occurred.

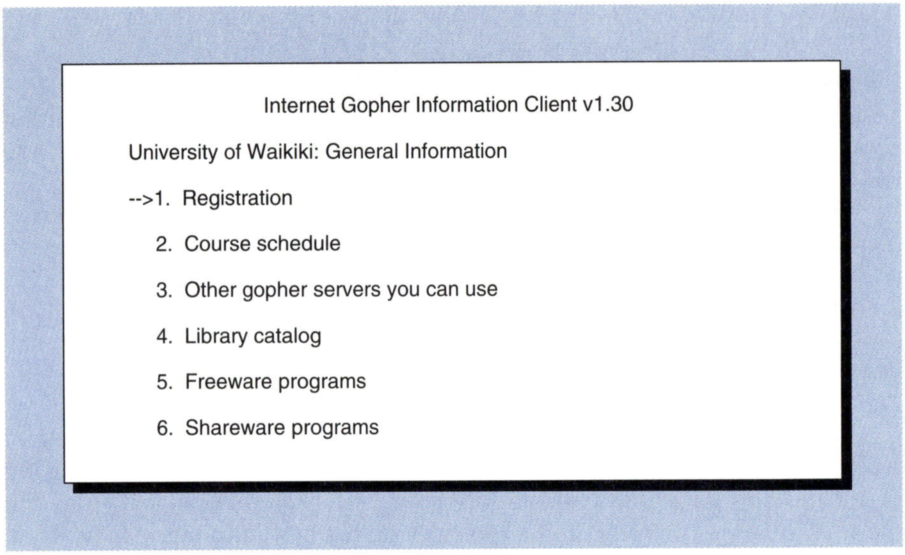

```
Internet Gopher Information Client v1.30

University of Waikiki: General Information

-->1.  Registration

   2.  Course schedule

   3.  Other gopher servers you can use

   4.  Library catalog

   5.  Freeware programs

   6.  Shareware programs
```

FIGURE B.8

Gopher Menu

FTP requires you to memorize a complex series of commands to transfer files. In contrast, Gopher is menu based. If you select *Freeware programs*, you will see another menu listing these programs. You can select one, and Gopher will transfer it to you. Gopher does more than file transfers. For instance, it will allow you to check the *Course schedule*, check on the *Registration* process, or search through the *Library catalog*. In every case, you will merely select choices from a menu.

VERONICA

Just as anonymous FTP was followed by ARCHIE, Gopher was followed by **VERONICA**.[3]

A problem with the Gopher menu approach is that each site has its own specific menu organization. Because there are several hundred Gopher servers in **Gopherspace,** finding information is still difficult.

VERONICA servers analyze the menu structures for many Gopher servers. They index the words in these menus. If you give a keyword, VERONICA will tell you what menu items on what Gopher servers have these keywords.

WAIS

The services that we have looked at so far have been limited to keywords appearing in file names and menus. We would also like to be able to search entire

---
[3]Very Easy Rodent-Oriented Netwide Index to Computerized Archives.

docubases—*collections of full-text documents*. We would like to be able to do **full-text searches** on the basis of words appearing anywhere in the text. We would like these searches to be fairly powerful. The **WAIS** (Wide-Area Information Server) service does exactly this. Unfortunately, most WAIS servers have limited collections of documents, and there are relatively few WAIS servers in operation.

Accessing the Internet

Although there are many ways to get access to the Internet, we will discuss only three of the most common ways: connecting via an online service, using a CSLIP or PPP connection, and using an Internet gateway on a corporate LAN.

Online Services

Figure B.9 shows a connection via an **online service**, such as America Online, CompuServe, Delphi, Microsoft Network, or Prodigy. These online services once offered one-stop shopping for a variety of services. They offered a menu of services, and you accessed these services via the online service. Users were limited to the services that the online service provided.

Recently, online services have added links to the Internet. The service provides a browser (see Module C) for access to various Internet services. The only drawback is that you may be limited to the online service's proprietary browser. In many cases these are good browsers, but they may not be as good as the

FIGURE B.9

Connecting to the Internet via a Commercial Online Service

One way to connect to the Internet is to subscribe to a commercial online service, such as CompuServe. You will need that service's shell program on your PC. In addition to providing its own service, it will provide access to the Internet, usually in the forms of electronic mail and a proprietary browser. This is the easiest way to link to the Internet, but you are often limited by the service's proprietary browser.

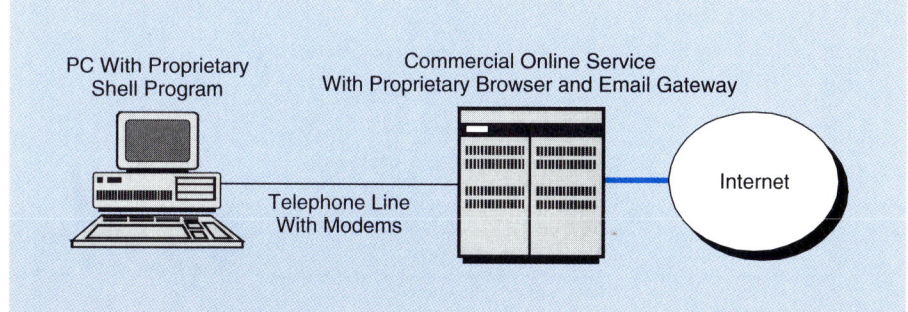

best independent browsers, and they will require relearning for people who are already used to another browser.

CSLIP and PPP Access

The second approach is similar to the first. This is the use of an Internet service provider (ISP) via a CSLIP or PPP connection. Chapter 2 ended with a discussion of this alternative. This is usually less expensive than going through an online service, because the Internet service provider usually offers few services (although electronic mail is common). It primarily exists to link you to the Internet.

Access via the Corporate PC Network Internet Gateway Server

Figure B.10 shows a third way to reach the Internet—via a PC network in your corporation. To support this, your organization needs an **Internet gateway** server. This is a communication server (see Chapter 8) that links you to the Internet.

PC network Internet gateways are attractive because they can operate at very high speeds. With a 64 kbps or T1 connection (see Chapter 6), even World Wide Web service will be fairly fast until you do heavy graphics and multimedia transfers.

FIGURE B.10

Connecting to the Internet Via a Corporate PC Network with a Firewall

The third way to get access to the Internet is via a corporate PC network. An Internet gateway server links you to the Internet, but the network administrator hides most details from users. This alternative may give you higher-speed access to Internet services. Many firms install firewalls between the Internet gateway and the PC network to reduce the chances of unauthorized entry to the PC net.

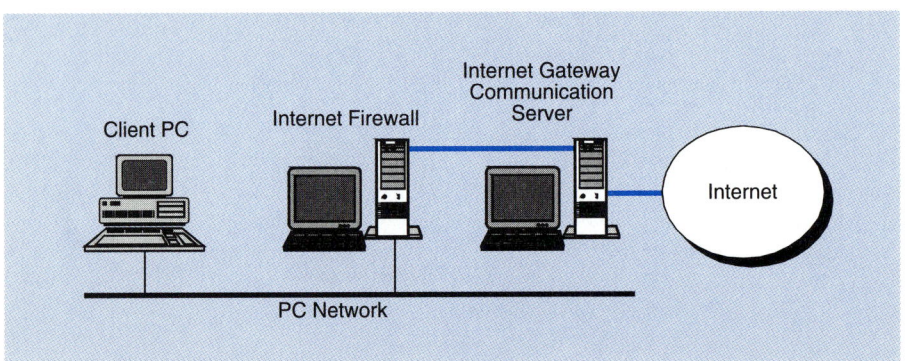

Firewalls

While internet gateways let your people out onto the Internet, they also potentially allow millions of Internet users to get into your corporate network. This can be rather dangerous.

Figure B.10 shows that many firms add an **Internet firewall** to their network access. This is usually a separate computer that places stringent limitations on who can get access to the corporate network from the Internet gateway.

The simplest firewalls work at the internetting layer. They inspect the source Internet address of each incoming message. If the address is on an approved list, the firewall lets the message through. Otherwise, it blocks it.

Unfortunately, it is very easy to **spoof** an Internet address, that is, place a false source address in a message. If the spoofer chooses an approved address, the message will go through. More sophisticated firewalls work at the application layer. They require checking at the level of application programs to ensure that the source is who it says it is and to block application messages from the wrong people. This is more effective, but it is also more expensive.

Tomorrow's Internet

There has recently been a great deal of public debate over the future of the Internet.

Funding

The Internet is a cooperative system. Different organizations own different parts of the overall system. At the same time, it is clear that some of the Internet's transmission capacity is subsidized by governments. Some government officials want to completely privatize the Internet, that is, force hosts to use private lines to link to one another. Others fear that this could kill the Internet. In some countries, such as the United States, this privatization has already come to a great extent.

Crime and Netiquette

Laws to govern behavior on a network are unclear. And even when they are clear, furthermore, they are very difficult to enforce. In the case of antipornography legislation, for instance, how could you enforce the law if a child in Billings, Montana gets pornographic pictures from an anonymous FTP server in Amsterdam?

The Internet community has always had a culture of openness. While this sometimes led to misbehavior, most instances of poor behavior have merely been examples of poor "netiquette." Saying nasty things and sending too many messages were in poor taste but hardly criminal.

Today, however, a growing number of true criminals are taking advantage of this open culture. Con artists offer goods for sale at very low prices, get credit card numbers from buyers, and then disappear with dozens or hundreds of stolen credit card numbers.

There have also been more complex cases. Some Internet users report being sexually harassed by Internet stalkers who keep sending unwanted messages asking for dates or sending messages with sexually explicit contents. Others report getting long series of nasty or even threatening messages from people who got angry with them during a mail or LISTSERV exchange.

For children, there are the issues of pornography and child molesters. Pornography is indeed available on the Internet, even through USENET. And child molesters do stalk discussion groups frequented by children. *For more on email netiquette, see Module J.*

Speed

One thing that does seem clear is that today's typical Internet speeds are no longer sufficient. LANs now run at 10 Mbps and even at 100 Mbps (see Chapter 5.) Even Frame Relay will run at 1 Mbps over long distances, and ATM will run at 100 Mbps or more (see Chapter 5). When many users have access to services offering such speeds at work, they will begin to demand higher Internet access speeds.

Although the path to higher-speed service is unclear, it is likely that the Internet in the year 2010 will be a very different network than it is today in terms of transfer speeds.

CONCLUSION

In this module, we have looked at the Internet. We noted first that technologically, the Internet is a worldwide network of networks integrated by a universal addressing system that assigns each host computer a unique IP number and IP name. To send a message (IP datagram) to a host, you merely put its IP number in the destination field.

We briefly looked at the history of the Internet. We noted that while the Internet Society has oversight over the entire Internet, and while the Internet Engineering Task Force sets standards, neither has any enforcement powers.

We looked at *asynchronous* communication tools, including electronic mail, LISTSERV mailing list conferences, and USENET newsgroup conferences.

We also looked briefly at *synchronous* communication services, including Talk, Internet Relay Chat, MUDs, and MOOs. There is even low-cost (and low-quality) videoconferencing, CUSEEME, and one-to-several videoconferencing (the MBONE).

For information retrieval, anonymous FTP sites open their files to the public. ARCHIE servers allow you to identify which anonymous FTP sites have the file you seek. Gopher servers, in contrast, organize information in a hierarchy of menus. VERONICA servers allow you to identify Gopher sites with menu items dealing with specific keywords. WAIS servers provide full-text keyword searching on documents.

There are several ways to get access to the Internet. The three most popular today are going through an online service, using an Internet Service Provider, and using an Internet gateway on a corporate LAN. Companies that use Internet gateway servers on their networks often build firewalls to prevent access to their internal networks by unauthorized Internet users.

For the future, there are many pressing policy issues surrounding the Internet. Among these are who will pay for Internet transmission lines in the future, how crime and simply rude behavior can be curbed, and how to increase access speeds.

In the next module, we will look at the most important information retrieval service on the Internet today. This is the World Wide Web. We will also look at browsers.

CORE REVIEW QUESTIONS

1. When would you use Telnet?
2. Who controls the Internet? What control mechanisms does it have? What benefits does this loose form of control bring? What problems does this create?
3. Distinguish between LISTSERV and USENET.
4. Distinguish among Talk, IRC, and MUDs and MOOs.
5. What is the purpose of anonymous FTP?
6. What is an FAQ? Why is it important?
7. Distinguish among FTP, ARCHIE, Gopher, VERONICA, and the World Wide Web.
8. What are the three ways listed in the module for getting access to the Internet?
9. In what ways may the Internet change in the future?

DETAILED REVIEW QUESTIONS

1. Describe why host numbering is important on the Internet. Distinguish between IP numbers and IP names.
2. What do domain name system (DNS) hosts do?

3. List the main historical events in the evolution of the Internet.
4. In Telnet, what do you do when you want to stop using the second host?
5. What is the Internet standard for electronic mail exchanges between email hosts?
6. What is the main transmission backbone for the Internet in the United States? In Europe?
7. What is the role of the Internet Society? The Internet Engineering Task Force?
8. There is a host puka2.org. It has a LISTSERV conference, MIS-L. To what address should you send a request to subscribe? What should be its body? To what address should you send a comment after subscribing? To what address should you send a message to unsubscribe from a conference? What should be the body of the message?
9. Distinguish between moderated and unmoderated LISTSERV conferences.
10. What command would you give to Talk to a person on host Puka.Org if the person has the login account CLEVENGE?
11. What is a newbie?
12. If you want to use an anonymous FTP site, why isn't it necessary for you to arrange a login account and a password in advance?
13. In FTP, explain the distinction between the Binary and ASCII commands. What happens if you try to transmit a binary file as an ASCII file.
14. In FTP, what is the command to retrieve a file? To send a file?
15. What would you do if you wanted to get a file pukapuka.exe via anonymous FTP but did not know which FTP server had it?
16. Why is Gopher more user-friendly than anonymous FTP?
17. ARCHIE is to anonymous FTP as _____ is to Gopher.
18. Why is a firewall needed? What are the two types of firewalls? Which is better? Which is easier to implement?

THOUGHT QUESTIONS

1. Why is it necessary to log into the second host in Telnet if you have already identified yourself to the first?
2. What are the major benefits and problems created by the Internet's loose management? How do you think things will change in the future?
3. Which would you prefer to join—a LISTSERV list or a USENET newsgroup? Explain your choice.
4. Do you see any practical business value in IRC, MUDs, or MOOs? Explain.

5. List the three main ways of connecting to the Internet. List the main advantage of each. List the main disadvantage of each.
6. What do you think of government bans on Internet telephone service?
7. Compare and contrast ownership and control on the Internet and the Worldwide Telephone System.

PROJECTS

1. Send an email message to someone on the Internet.
2. Use Telnet to log into a remote host. Run a program there.
3. Retrieve a specific target file via anonymous FTP. Use FTP commands rather than a browser.
4. Use ARCHIE to identify anonymous FTP servers that contain a specific file.
5. Go to a Gopher site and search the menu for information on a specific topic. Use Gopher commands rather than a browser.
6. Produce an update on current standards for commercial security on the World Wide Web.

For online exercises, please visit this book's website at: http://www.prenhall.com/panko

REFERENCES

COMER, D. E. (1995). *The Internet*. Upper Saddle River, NJ: Prentice-Hall.

MODULE C

The World Wide Web and Browsers

Introduction

This module assumes that you have read Chapters 1 and 2 and Module B. Chapter 1 discusses standards, basic networking concepts, internetting, and the Internet. Chapter 2 introduces the two most important services on the Internet today: electronic mail and the World Wide Web. Module B introduces the Internet broadly.

The World Wide Web

Although all of the information retrieval tools in Module B are important, one retrieval tool has risen to the top. This is the **World Wide Web (WWW).** Although electronic mail is the most widely used Internet service in terms of *number of messages* sent, **The Web** is the most important Internet service in terms of *number of bits* sent.

Hypertext

In the early 1960s, Doug Engelbart and Ted Nelson developed a new way of storing information—**hypertext.** As shown in Figure C.1, hypertext information consists of a large number of individual pages (documents). What makes hypertext unique is the way these pages are organized. Pages may contain **links** that point to other pages in the **hyperbase**.

If the user clicks on a link, he or she immediately jumps to the referenced page. The system keeps track of where users have been, so they can jump back to earlier pages they have visited.

Hypertext is a very "democratic" way of organizing information. While it can be used to create a hierarchical or linear organization for a group of pages, it can also be used to organize information in a "flat" way, with all pages being at the same level.

Hypertext mirrors the associative way that people think. We often read something and desire more information on a specific point before going on. Hypertext allows us to do that.

The World Wide Web

Hypertext came into its own when Tim Berners-Lee CERN (the European Particle Physics Laboratory) in Geneva created the World Wide Web standards to

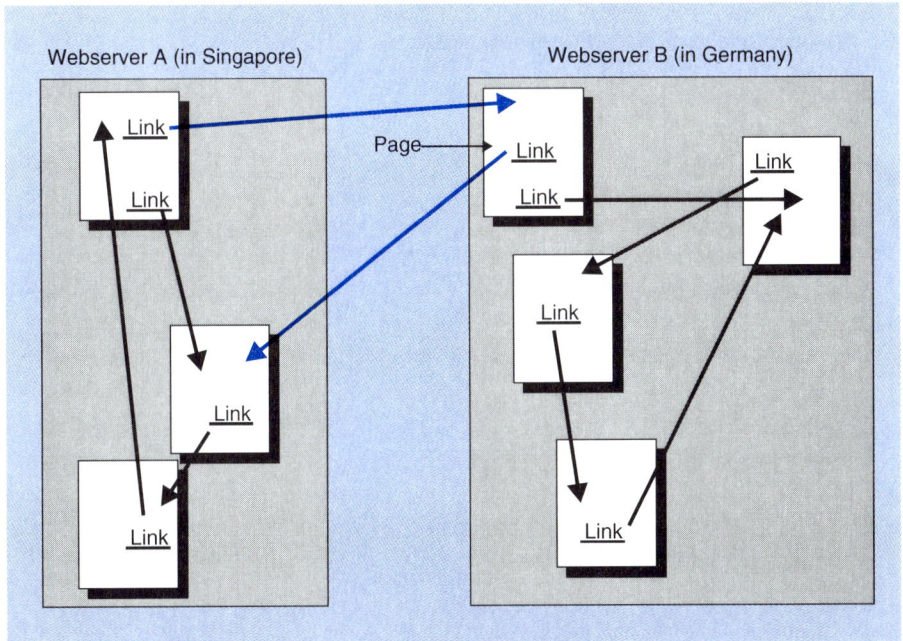

FIGURE C.1

Hypertext Collection of Pages

A hyperbase consists of a large collection of pages (documents). Pages contain links. If the user selects a link, the user jumps to the referenced page. The system keeps track of pages that the user has visited, allowing the user to jump back to an earlier page.

put the hypertext concept into practice. Today the World Wide Web Consortium (W3C) is primarily responsible for enhancing WWW standards.

Client/Server Operation ■

Figure C.2 shows that WWW service, like many other Internet services, uses client/server computing. **Webservers** manage collections of **webpages.**

Some of these pages are designated as **home pages.** Home pages are like any other pages, except that they are designed to be normal entry points for users who are interested in subsets of the webserver's pages. For instance, in a university's webserver, each college might have a home page. People interested in the college would go there first. This college home page would point to webpages for individual faculty members.

Each faculty member might have a home page and associated pages. Note that to the college home page, a faculty member's home page would be a subsidiary page.

NuPools has a home page and 25 subsidiary pages on the commercial webserver it uses. This allows users to check its store hours, see what merchandise

FIGURE C.2

Client/Server Operation in the World Wide Web

Webpages are stored on webservers. Some webpages, called home pages, are created as entry points to collections of related documents. For instance, NuPools has a home page on its webserver. This points to other NuPools pages. Users must have client programs on their PCs. Common client programs are Netscape Navigator and MOSAIC. Client programs communicate with webservers via the HTTP protocol. Clients can jump from one webserver to another.

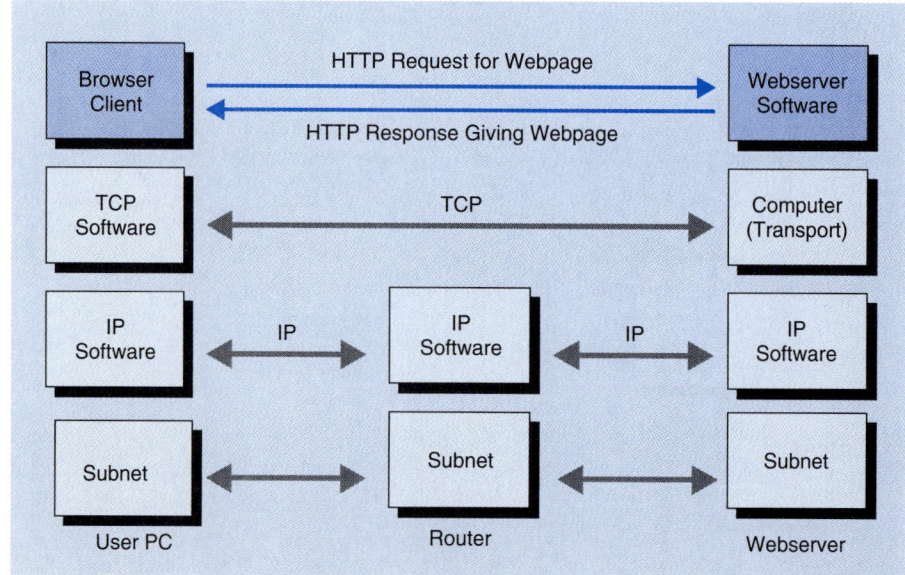

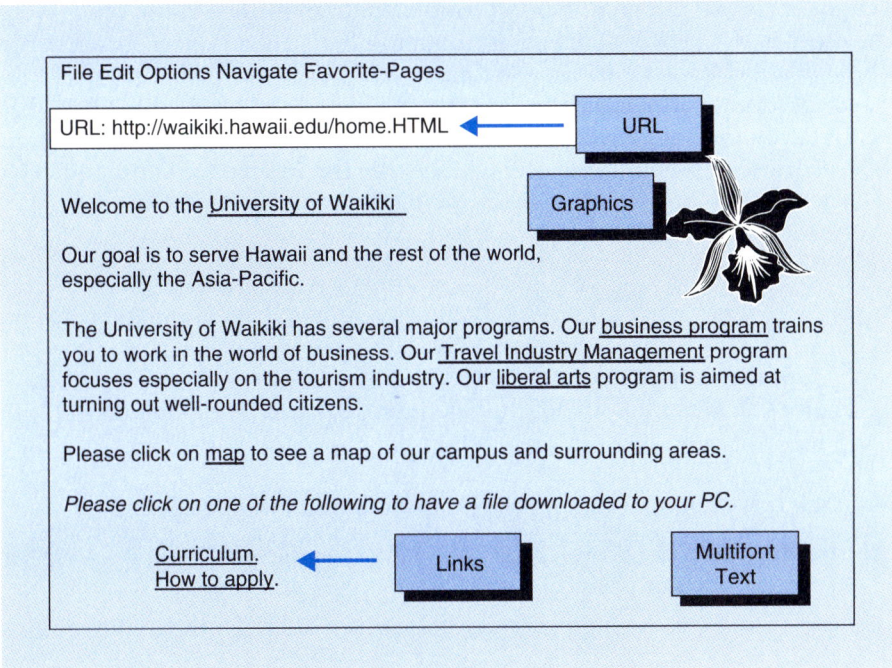

FIGURE C.3

Browser Screen Showing a World Wide Web Webpage

The browser client program displays a webpage. Note multifont text, graphics, and links (underlined blue text). Webpages can also have links to audio and video clips. If the user selects a link, the browser will jump to that webpage. The uniform resource locator (URL) points to the current webpage.

NuPools carries, look at detailed information on important products, and (as discussed later) even place online orders for products.

The user must have a **WWW client program** to use the World Wide Web. The most common client programs are *Netscape Navigator*, *MOSAIC*, and Microsoft's Internet Explorer, but there are many others. Any WWW client program can work with any webserver. In general, these client programs are called **browsers** for reasons we will discuss in a future section in this module.

Figure C.3 illustrates what the user sees on-screen. This figure shows that the WWW supports multifont text, graphics, and links (blue underlined text) on the webpage. Webpages can even have links pointing to audio and video clips.

HTTP ■

Interactions between browsers and webservers are governed by the **Hypertext Transfer Protocol (HTTP).** By standardizing client/server interactions, HTTP provides a common language for any WWW client to send requests to any webserver. HTTP is one of the two core standards in the World Wide Web. The other is the HTML standard for defining the structure of webpages. We will look at HTML later in this module.

HTTP even makes it possible to jump among servers. It is quite likely that the link in one webpage will point to a webpage on an entirely different server. If the user selects that link, the user will see the referenced page appear quickly. The user may be unaware that he or she has switched webservers.

URLs

Of course, every page on the World Wide Web must have a unique name so that links may refer to it unambiguously. This unique name is called the webpage's **uniform resource locator (URL)**.

Figure C.3 illustrates a URL. In this figure, the URL is the following:

```
http://puka.org/Biz/Intro.htm
```

Although this pattern is complex, it is not difficult to understand.

- The *http://* tells the browser that you want to retrieve something using the hypertext transfer protocol.
- Next *puka.org* is the name of the Internet host at which the resource is located.
- Following this, you specify a path to a particular directory. In this case, you start at the root directory, then go down to the *Biz* directory.
- Finally, *Intro.htm* is the file holding the page you seek.

Some hosts do not care about capitalization in URLs. Many hosts, however, require exact capitalization. The latter include most UNIX hosts.

Surfing the Net, Crawling the Web

If you have an Internet host computer, it takes only a few hundred dollars to add webserver software. In fact, there are a number of shareware webserver programs. No one knows how many webservers there are on the Internet, but the number is certainly in the hundreds of thousands and is growing explosively.

Under these conditions, finding the webpages you want can be daunting. Fortunately, there are a number of ways to *surf the net* or, to use the terminology specific to the WWW, *crawl the web*.

Home Pages

As noted previously, some webpages are designated as *home pages*. These are starting points that organizations create for themselves. So the key to finding information is to find the right home pages. Many companies advertise their home-page URLs, and a growing number of individuals put their personal home-page URLs on their business cards.

Jumping Links

Once you have found one home page, you tend to jump around in its internal links to subsidiary pages for a while, learning about the individual or organization.

In addition, many webpages add links to other interesting home pages. For instance, individuals often include links to professional societies in their webpages. Some individuals include links to webpages that they personally find interesting and think that someone who looks them up may also find interesting. Soon you find yourself jumping links from the initial set of webpages to entirely new home pages.

History Lists

Your browser maintains a list of your most recent webpages. This is your **history list.** You can look at the history list and select a previous webpage directly. This allows you to start in one place, and then pursue a tangential interest for a while. When you are finished, you can return to the main point you were considering.

Bookmarks

What if you find a webpage that interests you and that you may want to return to in a later session? Almost all browsers allow you to add this page as a **bookmark.** Later you can call up your **bookmark list** at any time and jump to one of its webpages.

As they crawl the web, most people soon find themselves with dozens of bookmarks. Browsers allow you to create **bookmark categories** in your bookmark list and assign new and existing bookmarks to these categories.

You can even save your bookmark list as a file and treat it as a webpage. (Browsers can read files on your disk drives as well as webpages from webservers.) Many people use their bookmark list as their initial page when they start their browser.

In addition to finding interesting webpages on their own, people find out about relevant webpages by exchanging lists with friends and colleagues. In fact, professional societies and other affinity groups often have their own home pages that list many relevant home pages.

Menu Directories

Another way to seek home pages is to turn to **menu directories.** These are organized lists of home pages, arranged in a hierarchical menu structure.

You first pick a category from a top-level menu. You then get a new menu of more specific choices. For instance, at the top level, you might select *computers*. At the next level, you might select *data communications*. At the next level, you might select *Internet*. Eventually, you make a final choice and get a list of specific home pages. These home pages are shown as links. If you click on the link, you jump to that home page.

The most widely known of these menu directories at the time of this writing is Yahoo (http://www.yahoo.com).

Keyword Search Engines ■

Another way to find information is to use a **keyword search engine.** You type in a search term or combination of terms, and the keyword search engine looks through its database to find qualifying home pages. It then rank orders the pages that it finds according to various criteria. Finally, it presents you with an annotated list of home pages in rank order. Each home page listed is shown as a link so that you can jump to it.

How do keyword search engines get their lists? They have special search programs called **webcrawlers.** When webcrawlers find a webpage, they jump to all of its links, recording information on the pages they find. They then jump to each of the links of those pages. In this way, they soon encounter and record information about a very large number of webpages.

The Mystery of the Lost Home Pages ■

Unfortunately, any list of webpages quickly becomes outdated. Some companies discontinue their home pages entirely. Others move them to other webservers and only leave forwarding addresses (in the form of links at the old home page) for limited periods of time. Even menu directories and keyword search engines vanish without a trace.

For new users, this is not too much of a problem. When they first start building lists, almost all of the webpages they place in their bookmark lists are current. Within a few months, however, a significant fraction of their bookmarks will point to vanished sites.

Commercial Transactions

A webserver can do more than supply information pages to browsers. Pages may include forms that you can fill in. Your browser will then return these forms to the webserver for processing. For instance, a page might contain an order form for pure Kona coffee. You would fill in the order form and hit a link called something like *Place an Order*. Your browser would then send the order form back to the webserver.

The original HTTP standard did not support secure message exchanges. It was easy for a determined thief to intercept your credit card number when you were sending in your order form. A number of extensions to HTTP now provide varying degrees of security to purchasers. However, at the time of this writing no one standard dominates. As a result, the browser and webserver must use the same security protocol for transactions to be secure.

Speed

Many Internet watchers expect the WWW to remain a key Internet service for the next few years. However, if it is to achieve its full stature, it has to overcome speed limitations. Even with a 28.8 kbps modem, page changes are very slow for pages with complex graphics, audio, and video. Even with 64 kbps ISDN links, complex page transfers are far from instantaneous. WWW points to the need for much higher speeds on the Internet in the future.

HTML

We noted earlier that HTTP is the WWW standard governing exchanges between client programs and webservers. We also need a standard to represent formatting on a page (including links). On the WWW, this standard is the **Hypertext Markup Language (HTML)**.

In publishing, a markup language is a set of conventions for annotating plain ASCII text to indicate its formatting. Publishers have long used the *Standardized General Markup Language (SGML)*. HTML is primarily a subset of SGML. There have been several versions of HTML. We will only discuss major formatting elements here.

HTML documents begin as plain ASCII text files. The page developer then adds HTML **tags** to these files. One example of a tag that we will see is the following:

```
<A HREF = "more.htm">Further information</A>
```

This obviously creates a rather messy-looking ASCII file. When a browser reads the file, however, it interprets the tags and presents the browser user with a fairly elegant page with rich on-screen formatting.

It is important, then, to distinguish between two views of the file. First, there is the **tag view** in which all of the tags are visible. The other is the **browser view**, which portrays the formatting that the tags specify.

Creating HTML Pages

There are two primary ways to create HTML pages.

Word Processing Programs ■

First, you can use your standard word processing program. You can simply type in the special HTML tags. This is rather laborious.

To ease your work, you can create an HTML document template containing such basic sections as <HEAD> and <BODY>. You can create macros to add other tags.

You must remember to save your files as simple ASCII text files, not as word processing files. Word processing files have extensive internal formatting codes that browsers will not understand.

If you are on a machine that limits your extensions to three letters, you generally use the extension **htm** for your webpages. On other systems, you generally use the extension **html.**

Most word processors today cannot show pages in browser mode. Fortunately, you can use your browser to see how your page will look when it is on the web. Almost all browsers allow you to open a local file instead of entering a URL. They will then present the file in browser view. This will allow you to see how your page will look to others when you post it to the web.

You do not even need a webserver for this. You can store the webpages on your own PC and run your PC's browser against them locally. So you can prepare your webpages completely before moving them to a webserver for presentation to the whole world.

Unfortunately, as you build a webpage, you may have to switch between the word processing program and the browser dozens of times. This is awkward. Some word processing programs now allow you to see your page in browser view, but these are still not common.

HTML Editors ■

There are also dedicated **HTML editors** that make it much easier to create home pages. First, they make it easy to add tags. Second and more important, they allow you to toggle immediately between the tag view and the browser view. You can check each change immediately. A number of simpler HTML editors are available as shareware.

In some ways, the distinction between word processors and HTML editors is growing smaller over time. As noted earlier, some word processing programs now offer features for building tagged pages and for viewing them in browser mode.

However, HTML editors tend to be a generation ahead of word processing programs in tools for developing webpages. In addition, a growing number of HTML editors can go beyond individual pages to help you manage collections of webpages. They may be able to show you a graphical view of your webpages and how they relate to one another in a form similar to Figure C.4. For serious webpage developers, HTML editors are a must.

A Simple Page

Figure C.4 shows a fairly "simple" webpage. It contains only ASCII text, and the formatting is comparatively simple. It is, however, extremely difficult to read because it is littered with HTML **tags,** which are contained within angle brackets < >. In this chapter, we capitalize tags, but capitalization is not important. You should, however, be consistent.

```
<HTML>
<HEAD>
<TITLE>Sample Webpage</TITLE>
</HEAD>
<BODY>
<HR>
<H1>Purposes</H1>
One purpose of this sample webpage is to show the distinction between the webpage in its ASCII format with HTML codes and what the page looks like when read with a client reader program.<P>
The second purpose is to illustrate the most common codes.<P>
<HR><H1>Codes</H1>
Note that most codes bracket the part of the document they describe. So at one end of "Purposes", which is a level-one heading, we have "H1". At the other end, we have "/H1".<P>
The pair "HTML" and "/HTML" identifies this as an HTML document.<P>
The pair "HEAD" and "/HEAD" marks the head portion of the document.<P>
The pair "TITLE" and "/TITLE" marks the title of the webpage.<P>
The pair "BODY" and "/BODY" marks the body portion of the text.<P>
Each paragraph ends with a single "P". Unfortunately, different versions of HTML handle this code differently.<P>
</BODY>
</HTML>
The webpage contains plain ASCII text. Markup codes, which tells the browser how to format the page on-screen, are contained in angle brackets < >.
```

FIGURE C.4

Simple Webpage in ASCII Format with HTML Markup Codes

Paired Codes ■

In general, tags come in pairs. For instance, <HTML> and </HTML> mark the beginning and end of the page, respectively. This marks the page as an HTML document. Many browsers do not require this pair, but using them is still good practice.

Major Sections ■

Next come the mandatory <HEAD> and </HEAD> tags to mark the beginning and end of the heading section, which contains information about the document. Similarly, the mandatory <BODY> and </BODY> tags mark the beginning and end of the body part, which the browser is expected to present on-screen.

Title ■

Within the heading section, the <TITLE> and </TITLE> tags mark the page's title. The browser normally presents the title somewhere on the screen but usually not in the area showing the body text. In Figure C.3, for instance, the title appears above the area showing the body text. Note that the title text does not appear in the text below the bar.

First-Level Heading

In the text, <H1> and </H1> mark a first-level heading. The browser will portray such a heading differently than normal body text. Usually, it will make the text larger and boldface. It is also possible to add second-, third-, and even fourth-level headings to long pages.

If first-level headings are too large, you can even use second- and fourth-level headings and not use first- (and third-) level headings.

Paragraphs

For paragraphs, there is <P> at the beginning of each paragraph and a </P> at the end. Although this is perhaps the most common tag, different versions of HTML specify paragraphs differently. Some put <P> at the back and nothing in the front.

Horizontal Rules

One of several tags that do not need bracketing is <HR>. This represents a horizontal rule—a line running across the page. This is a nice way to separate sections.

Locations of Tags

In general, Figure C.4 shows the tags for a line on the same line as the text they mark up. For instance, it shows the following:

```
<H1>Purposes</H1>
```

The placement of the tags on the same line is optional. We could have gotten the same results with the following:

```
<H1>
Purposes</H1>
```

The Browser Image

Figure C.5 illustrates how a browser will represent the marked-up page in browser image. You can see that although HTML is complex, the browser presents a simple and attractive picture to the user.

One interesting aspect of the World Wide Web is that while it specifies tags, it does not specify *in detail* how first-level headings, paragraphs, rules, and several other types of formatting should be represented by browsers. It is common, as in the case of Figure B.13, to show first-level headings in boldface and in a larger font, but this is not universal. Some browsers even allow users to specify how they want various markup codes to be displayed.

The display of the webpage, in other words, is under the control of the browser, not the webserver or even the page's designer.

> **Purposes**
>
> One purpose of this sample webpage is to show the distinction between the webpage in its ASCII format with HTML codes and what the page looks like when read with a client reader program.
> The second purpose is to illustrate the most common codes.
>
> ---
>
> **Codes**
>
> Note that most codes bracket the part of the document they describe. So at one end of "Purposes", which is a level-one heading, we have "H1". At the other end, we have "/H1".
> The pair "HTML" and "/HTML" identifies this as an HTML document.
> The pair "HEAD" and "/HEAD" marks the head portion of the document.
> The pair "TITLE" and "/TITLE" marks the title of the webpage.
> The pair "BODY" and "/BODY" marks the body portion of the text.
>
> Each paragraph ends with a single "P". Unfortunately, different versions of HTML handle this code differently.
>
> **The browser interprets the markup codes and formats the webpage for on-screen presentation.**

FIGURE C.5

Sample Webpage as Viewed by a Browser

The browser interprets the markup codes and formats the webpage for on-screen presentation.

Adding Links

While the text in Figure C.5 is attractive, it is not hypertext. Nowhere is there a link to another page. We would like to be able to add links.

Simple Link ■

Suppose, as shown in Figure C.6, that we want the page to contain the text link *further information*. If the user clicks on this link, he or she will be taken to another HTML file, *more.htm*. Suppose that *more.htm* is in the same directory as the page into which we are building the link. In this case, where we want *further information* to appear on the page, we would enter this text string:

```
<A HREF = "more.htm">further information</A>
```

Here the bracketing **anchor tags** are <A> and </A>. The "A" indicates that we have an anchor.

The <A> tag is rather complex. Following "A" in the angle brackets is the following code.

```
HREF = "more.htm"
```

Here HREF means "hypertext reference." In turn, *more.htm* is the name of the file that we wish the browser to load if the user selects the link.

If a user selects *further information* with the mouse, he or she will see *more.htm*.

FIGURE C.6
Page Containing a Hypertext Link

> **Page as seen in a browser:**
>
> That's a brief tour of our campus. If you would like to know more about us, click here for <u>further information</u>. Thank you for taking the time to read our webpage.
>
> **Page in HTML format:**
>
> That's a brief tour of our campus. If you would like to know more about us, click here for <A HREF="more.htm">further information</A>. Thank you for taking the time to read our webpage.
>
> **Notes**
>
> <A> ... </A> defines the start and end of the anchor.
> Between them is the text to appear on the page as viewed by the browser (further information).
>
> Within <A> is the hypertext reference (HREF) "more.htm". This is the relative URL of the page to which the link points.

Links to HTML Files in Other Directories

The form of HREF we have just seen works if *more.htm* is in the same directory as the document calling it. But what if *more.htm* is in the parent directory of the directory holding the file containing the *further information* link? In that case, we would write the following:

```
HREF="../more.htm"
```

This may seem odd, but if you have used MS-DOS, this is how MS-DOS represents a path to a directory one level up from the current directory.

Similarly, suppose *more.htm* is one directory level down in the *details* directory. Then HREF would take the following form.

```
HREF="details/more.htm"
```

We call the preceding three forms of HREF **relative addressing** because they express directories relative to the current directory.

Like MS-DOS, the anchor tag can also contain **absolute addressing** in which the user specifies the full path, including the drive and the path from the top (root) directory to the directory holding *more.htm*. The absolute reference might be the following:

```
HREF="/D|/www/global/details/more.htm"
```

This is similar to MS-DOS referencing, but it is not quite the same. First, it begins with a slash, which MS-DOS does not use to begin an absolute path description showing a drive. Second, D (the drive name) is not followed by the usual colon (:). Instead it is followed by a vertical slash (|). This is done because the colon has special meaning in references.

In general, relative addressing is the desirable form to use. In this way, if all of your files have to be moved to a different part of a webserver's directory structure or to another webserver, they can be moved together, keeping their relative addresses intact. Or, if you prepare your group of webpages on a word processor or HTML editor on your local hard disk drive, you can move them easily to a webserver.

Referencing a Specific Location in a Document ■

So far, we have looked at links that refer to a whole document, but what if we want to refer to a specific location in a target document? Perhaps we want to refer to a section in *more.htm* that begins with the first-level heading, **Wiring Concentrator.** Figure C.7 shows how to do so.

First, in *more.htm*, we would find the text reading <H1>Multiport Repeaters</H1>. Then we would change it to something like the following:

```
<H1><A NAME="mprs">Wiring Concentrator</A></H1>
```

Note that we first bracket the text, Multiport Repeaters, with <A> tags, then with <H1> tags. We would not insert the tags in an asymmetrical manner, such as <H1><A> ... </H1></A>.

FIGURE C.7

Linking to a Named Location in a Target Document

Create a name in the target document using the *NAME=" "* parameter in <A>.

<H1><A NAME="mprs:><Wiring Concentrator></A></H1>
<P>Another name for wiring concentrator in Ethernet 10base-T is <em>multiport repeater<em>. First, the wiring concentrator certainly has multiple RJ-45 ports—usually about a dozen. Second, the wiring concentrator is certainly a repeater. If a station transmits in on one port, the signal is repeated (broadcast) out of every other port.

This will not change the way the information is presented on the page.

Wiring Concentrator

Another name for wiring concentrator in Ethernet 10base-T is *multiport repeater.* First, the wiring concentrator certainly has multiple RJ-45 ports—usually about a dozen. Second, the wiring concentrator is certainly a repeater. If a station transmits in on one port, the signal is repeated (broadcast) out of every other port.

In the source document, create a link to this page and named location.

The most visible elements in a 10Base-T network are its <A HREF="more.htm#mprs"> wiring concentrators</A>. These concentrators serve a number of functions.

Notes

You first create a name in the target document.
Then you create a hypertext reference that includes the name.

What we are doing here is creating a **name anchor** in the target document. *NAME="mprs"* anchors a name with the location multiport repeaters.

Second, we use this name in a link in the source document. For instance, we would change our link in the original document to the following:

```
<A HREF="more.htm#mprs">wiring concentrators</A>
```

The pound sign after the file name tells the browser to go to the named location *#mpr* once it reaches the target page, *more.htm*.

Referencing Pages on Other Machines ■

So far, we have talked about links that point to other pages on the same machine. But it is no problem to point to pages on other webservers. You simply need to add the information about the webserver's host and directory structure to the link. Your link, for instance, might look like the following:

```
<A HREF="http://www.puka.org/files/more.htm">further information</A>
```

If the person selects *further information*, the browser will call the webserver *www.puka.org*, using the hypertext transfer protocol. It would ask the webserver to go one level down to the *files* directory, then retrieve *more.htm* and deliver it to the browser.

Note that the information specified by HREF is the file's *URL*. Actually, this *is the general form of HREF*. If a target page is on the same host as the page containing the link, however, we can ignore the host information and just specify the directory and file. If the target page is in the same directory, furthermore, we just specify the file.

Adding Images

So far, we have only looked at text. Obviously, one of the attractions of the World Wide Web is that it can handle images, sound, and video as well as text.

Inline Images ■

There are two types of images. **Inline images** appear on the screen when you view the webpage with a browser. Figure C.8 shows an inline image. Note that the inline image appears right in the text in place of the tagged information. Inline images allow us to mix text and graphics on a single page.

To insert an inline image, you add a tag that has this form:

```
<IMG SRC="router.gif">
```

This tells the browser that it should show an image (IMG) at this place in the document. The source (SRC) of this image is the file *router.gif*, which is in the popular .GIF file format. The browser will go back to the webserver, retrieve this image, and place it where indicated in the text.

In this example, *router.gif* is in the same directory as the webpage. But what if it was one level down in a subdirectory called *images*? You would do the

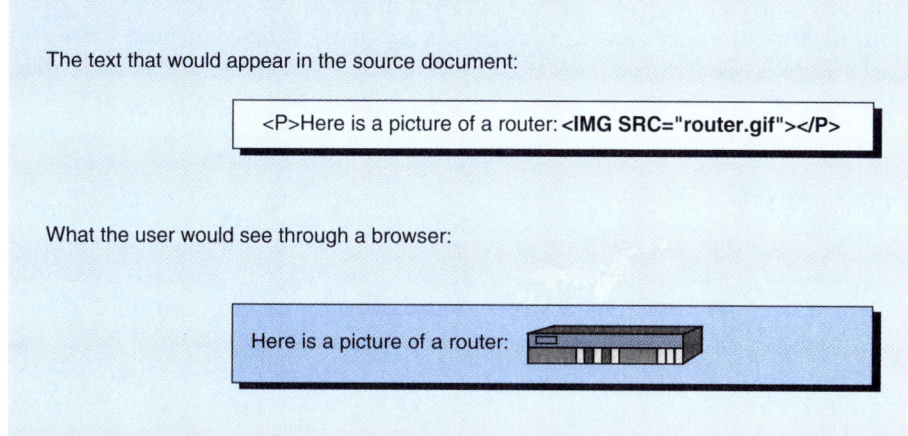

FIGURE C.8
Inline Image

same thing that we saw for HTML pages when links point to other directories. In this case, we would write the following:

    <IMG SRC="images/router.gif">

External Images ■

Images take up a great deal of time to download from the server. This is especially true for large images. For large images, you can create a link that will show the image only if someone selects the link. For instance, suppose you have the link *router* in the text. You want the user to jump to a new page showing just one thing—the picture of the router, as shown in Figure C.9.

In effect, instead of jumping to another hypertext page (.HTML), you wish them to jump to a page showing a graphics image, say a .GIF file. This is an **external image**.

While inline images use the <IMG> tag, external images use the familiar <A> tag. If the image file, *router.gif*, is in the same directory as the page calling it, you would insert the following in your document:

    <A HREF="router.gif">router</A>

Here *<A HREF="router.gif">* is the initial tag. It tells the browser where to look for the image. The *"router"* is what will appear on the page as a text link. Third, *<A>* completes the tag.

Note that this is almost exactly the same syntax as the link to another hypertext (.HTML) page in Figure C.6. The only difference is in the content between the quotes in the Hypertext Reference (HREF) field. Instead of being "more.htm" as it was with a link to a hypertext document, it is "router.gif". In other words, HREF is a general way to jump to another document file, whether that file is an HTML file or another type of file.

However, the browser must know how to display the other type of file. Of course all browsers can display .HTML files. In addition, .GIF is universally supported. Most browsers can support a number of other file types. In addition, if your browser cannot support a certain file type, you generally can find a **helper application**—a program that will allow the browser to display the file.

Images in Links

We have seen that in anchor links, what the user sees in the source document is what we place between the <A> and </A> tags. Up to now, we have only seen text between the tags. In Figures C.6 and C.7, the text between the tags was "further information". It appeared in the source document as *further information*. In Figure C.9, the text was "router". It appeared in the source document as *router*.

However, text is somewhat boring. Perhaps we would like *an image of a router* to appear in the text instead of *router*. Then the user could click on the image of the router to be taken to a page showing more information about the router. Suppose that that page is *details.htm*.

Actually, this is very easy to do, as Figure C.10 shows. Instead of placing text between the <A> and <A> anchors, we place an inline image. In this case, we would have the following.

```
<A HREF="details.htm"><IMG SRC="router.gif"></A>
```

Again, when the browser shows the source document, it merely displays an image of the router. When the user clicks on the image, the browser jumps to *details.htm*.

FIGURE C.9

External Image

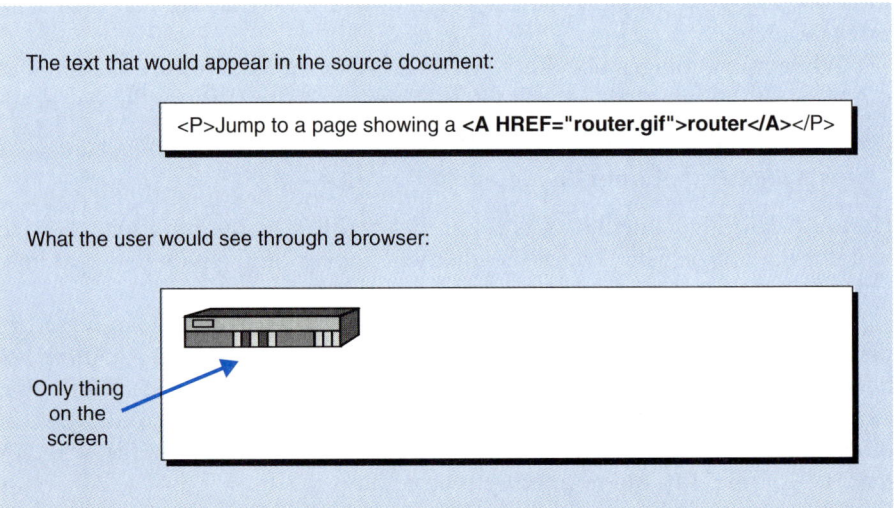

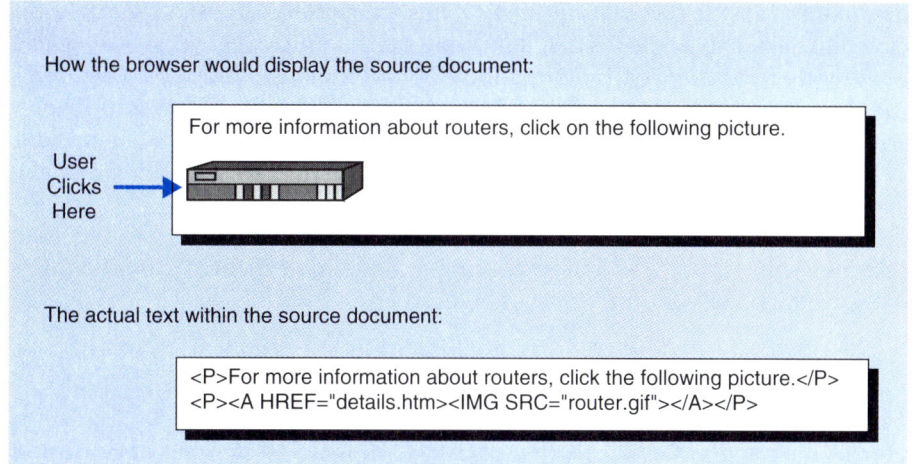

FIGURE C.10
Image in Link

Other Markup Codes

We have looked at only a few of the many markup tags defined in HTML. The appendix, "HTML Codes" provides a list of the hypertext tags in common use. The most recent version of HTML at the time of this writing is **HTML 3.2.** Some older browsers cannot read the most recent HTML codes.

Sometimes Netscape defines codes not in the formal HTML specifications. It is good to stay away from them because other browsers cannot read them.

Representing Text ■

There are two types of tags for adding properties to text. The first is explicit formatting. The second is logical formatting.

EXPLICIT FORMATTING. Explicit formatting is what you are used to in word processing. If you want to control how the text will appear, you specify it exactly. For instance, if you want boldface, you place text between the <B> and </B> tags. To have boldface **router** appear on-screen when the user views the page in a browser, you would put <B>router</B> in the text of the HTML source document.

LOGICAL FORMATTING. In the HTML philosophy, however, it should be up to the browser to determine how to represent text in terms of color, size, and other attributes. This allows readers to customize their view of the documents they read.

As a result, HTML offers logical formatting, which indicates to the browser the type of text it is but leaves the browser free to format the text as it sees fit. For instance, you would put <STRONG>router</STRONG> in your HTML document. Most browsers will display this as boldface, but the user may decide that he or she wishes to display strong information differently—say in red.

Another form of logical formatting is headings. For instance, you might use the second-level heading tags, <H2> and </H2>. Browsers will display second-level headings less distinctly than first-level headings and more distinctly than third-level headings, but the specifics are up to the browser and how the user has configured the browser.

This idea of allowing the users to decide how things will appear on their screens seems rather odd and unnerving at first, but it is the preferred way of working with HTML on the World Wide Web.

Special Characters

One of the most annoying things about HTML is that a number of characters have special meaning. One is the left angle bracket, <, which is also the "less than" symbol. The appendix to this module contains other common reserved characters.

If you need to type one of these special characters, you replace it by a text string. The text string begins with an ampersand (&). Following this is a letter code. (For "<" the letter code is "lt".) Finally, there is a semicolon. So instead of typing "<", you type "<".

Lists

Often you wish to display images in a list. For instance, you might wish to display a list of your favorite courses in an HTML document. Suppose that you wish to display it in a list with each **list item** numbered. This is an **ordered list** in HTML terminology.

Figure C.11 shows how the text will appear in your HTML source document. Note that the tags <OL> and </OL> mark the beginning and end of

FIGURE C.11
Ordered List

Here is the text that appears in the HTML source document:

<P>My favorite courses:</P>
<OL>
<LI>Data communications
<LI>Drama
<LI>Conversational Spanish
</OL>

Here is the text that appears on the browser screen:

My favorite courses:
1. Data communications
2. Drama
3. Conversational Spanish

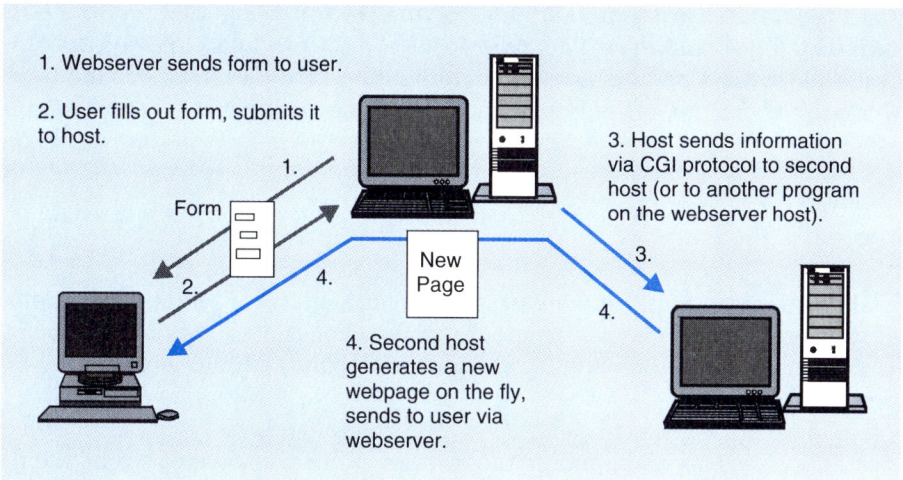

FIGURE C.12
Form and CGI

The webserver sends a form to the browser. The user fills out the form at the browser and submits it back to the webserver. The webserver passes information in the form to another program via the common gateway interface (CGI) standard. That other program processes the information, creates a new webpage on the fly, and sends it back to the browser.

the ordered list. If you had wished to have an **unordered list,** which shows bullets instead of numbers before each point, you would have used <UL> and </UL> to delimit the list.

Note also that before each list item is the tag <LI>. There is no </LI> after each list item, however.

Figure C.11 also shows how the ordered list appears on the browser screen. Note that there is no blank line between each item in the list.

Forms and CGI

Initially, the World Wide Web was a system for retrieving preformatted webpages. Now, however, the Web is becoming an interactive tool. Figure C.12 shows that a webpage can be a form. The user can fill in the form in his or her browser, then submit the form.

When the webserver receives the form, it passes the information in its fields to another program. The interface for doing this is the **Common Gateway Interface (CGI)**.

That other program processes the information in the form. It then creates a new webpage, formatting it on the fly. The webserver then sends this webpage back to the browser.

There are many programming languages for using CGI information, including C++ and even Visual Basic. On UNIX machines and to some extent on Windows and Macintosh machines, there are also *scripting* languages that are simpler than full programming languages. The most widely used is PERL.

Java

CGI programs execute on the webserver. This is good for many applications. However, it is impossible to create rapid changes on the user's screen. Animation, for instance, would take too much bandwidth to work on the Internet.

Sun Microsystems created a new approach called Java. Figure C.13 shows that Java programs are stored on the webserver but are downloaded on the fly to the browser. Once on the browser, they then execute locally.

Java programs are called **applets.** Because applets execute on the client PC, they can do work that requires instantaneous screen refreshes. For instance, they will allow animation and instant sound responses.

Potentially, even larger application programs—such as word processing programs and clients for client/server applications—could be downloaded as needed by extensions to Java technology. This could greatly simplify electronic software distribution.

FIGURE C.13

Java Applet

The Java applet is stored on the webserver but is downloaded to the browser. Executing on the client PC, it can do animation and other work requiring instantaneous screen refreshes. Potentially, all application software could be stored on the server and downloaded by client PCs as needed. This could simplify electronic software distribution.

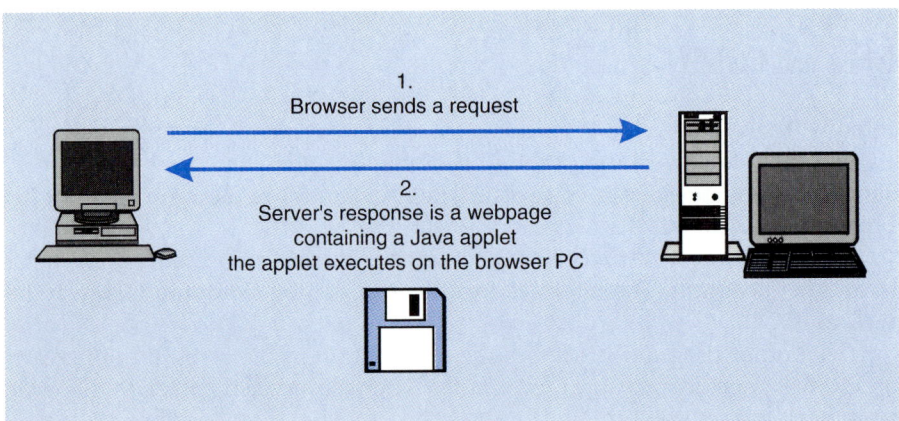

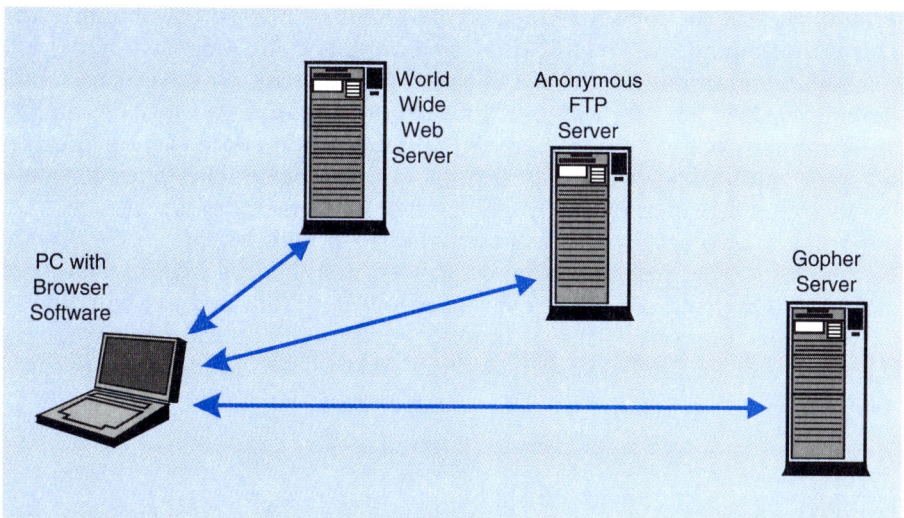

FIGURE C.14
Browsers Can Access Multiple Types of Servers

Browsers are not limited to accessing World Wide Web servers through the HTTP protocol. They can also access other types of servers, such as anonymous FTP and Gopher servers, via appropriate protocols. Thanks to standards, almost any browser (Netscape Navigator, MOSAIC, and others) can access almost any server on the Internet.

BROWSERS

Module B showed the complexities of using command line statements to use Telnet and FTP. Even Gopher, ARCHIE, and VERONICA are relatively crude by today's user interface standards. Having to memorize odd sequences of steps is a great deterrent to use.

Browsers are beginning to change all this. We have already seen browsers in one context, namely the World Wide Web. By using URLs beginning with "http://", the user is telling the browser to send a request to a webserver via the HTTP protocol.

Browsers, however, are not limited to working with World Wide Web servers. As Figure C.14 shows, a browser can work with a number of other types of servers, including anonymous FTP and Gopher servers. To access one of these servers, the user merely gives the appropriate URL.

For instance, they also give you access to anonymous FTP servers. In this case, you would open a connection to an anonymous FTP server by giving a URL such as the following:

```
ftp://ftp.puka.org
```

This command will log you into the anonymous FTP server on the host ftp.puka.org rather than to a webserver. As shown in Figure C.15, you will see a list of files in the top-level directory, together with a list of subdirectories. This is the equivalent of a DIR (directory) command in DOS.

Both the files and subdirectories will appear as links. If you select a subdirectory, you will jump to that directory. You will again see a list of files in that directory, plus the names of lower-level directories. In addition, your browser will give you the choice of jumping back up to the next-higher directory level. In this way, you can move easily among the FTP server's directories.

You can also select files. What happens then will depend on the type of file you select. For instance, suppose you select a text file, which consists of plain ASCII text and usually has the extension *TXT*. Your browser will display the text file on-screen. This is useful because FTP directories often have *readme.txt* files that give details on the files and directory names that appear at that level.

FIGURE C.15

Anonymous FTP as Seen by a Browser User

The contents of an anonymous FTP directory appear to the browser user as links. If you select a link that is a directory, the browser will take you to that directory and show you its contents. If you select a link that is a text file (.txt), the browser will show the file on-screen. If you select a file that is another type of file, such as a program file or graphics file, the browser will let you download it to your disk drive. Note that the browser handles the details of logging in and deciding whether to download the file as a text file or as a binary file. Note that the URL begins with ftp://, indicating that communication will follow the FTP protocol.

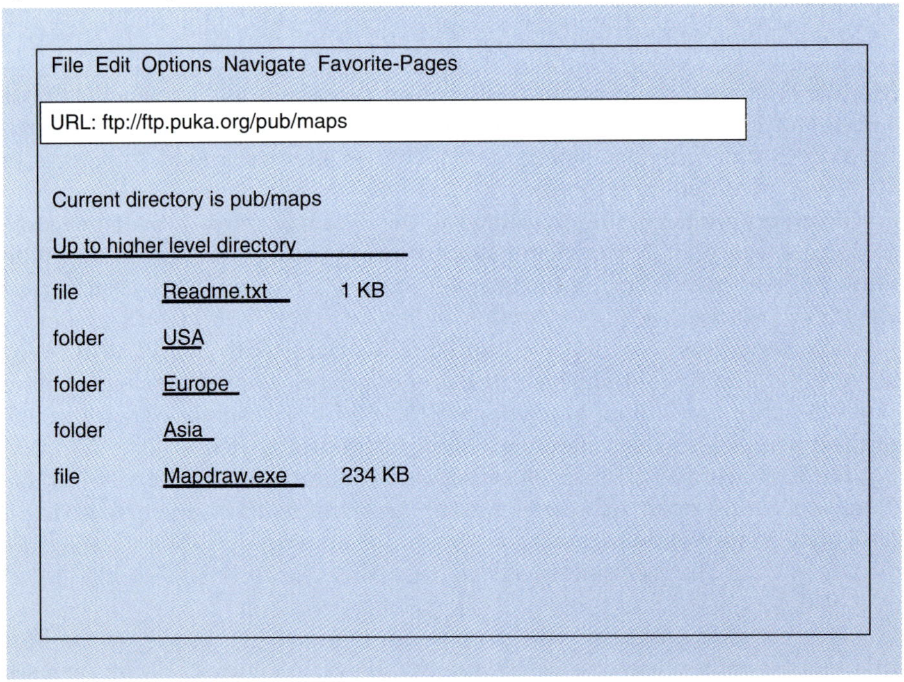

```
File  Edit  Options  Navigate  Favorite-Pages

URL: ftp://ftp.puka.org/pub/maps

Current directory is pub/maps

Up to higher level directory

file      Readme.txt      1 KB

folder    USA

folder    Europe

folder    Asia

file      Mapdraw.exe     234 KB
```

Suppose, instead, that you select a nontext file, such as a program file, a graphics file, or a spreadsheet. In that case, your browser usually will inform you that you do not have the ability to view the file on-screen. It will ask if you want to save it. If you say "yes," your browser will give you a chance to store the file in one of your directories.

Note that we have almost the functionality of command-line FTP without the details. Your browser logs you into the anonymous FTP server automatically, giving your email address as your password. The browser also decides whether to download files as text or as binary files, without your intervention. In general, browsers make command-line services much more user-friendly and attractive.

Figure C.15 shows you that browsers also connect to Gopher servers. To reach a Gopher server, you specify a URL beginning with "gopher://". The browser then logs you into the Gopher server that you designate as the host name. Now you see the traditional Gopher menu items, but they appear as links that you can select to go down to a subdirectory or back up to the next higher directory. If the menu item is a file, your browser will download it for you. Browsers typically provide simplified access to a number of other Internet services, including electronic mail, USENET newsgroups, and Telnet.

Browsers gradually are becoming single programs for users to learn in order to get access to a wide variety of information retrieval and communication services. In fact, Netscape now refers to Navigator as a client program rather than as a browser. With programming languages such as Java, browsers are moving beyond information retrieval and communication to become essentially platform-independent operating systems.

CONCLUSION

Module B introduced the Internet and most of its services. This module concentrated on two specific topics of broad importance. The first was the World Wide Web. The second was browsers.

The World Wide Web has already become the dominant information retrieval tool on the Internet. An author creates a page with the crude but fairly powerful Hypertext Markup Language (HTML). The user then reads the page with a browser. The browser formats the page according to user preferences.

Initially, HTML was limited to the retrieval of pages prepared previously by an author. With forms and CGI, however, a user can interactively fill out information, send the form to the webserver, and receive a new webpage created in response to what the user has said. With Java, it is even possible to download programs (applets) to the browser for execution on the user's PC. In general, the World Wide Web has become a general focus for innovation on the Internet.

To many people, browsers are merely ways of viewing webpages. But browsers are not limited to working with webservers via the HTTP protocol. They can interact with other types of servers, using different protocols. For instance, they can interact with FTP servers using the FTP protocol. In general, instead of beginning the URL with http://, the user begins the URL with ftp:// or some other protocol.

CORE REVIEW QUESTIONS

1. What is hypertext, and how is it used in the World Wide Web?
2. What is HTTP? What is HTML? What is an HTML tag?
3. The World Wide Web is a client/server system. What is the client program, and on what machine does it execute? What is the server program, and on what machine does it execute?
4. Distinguish between webpages in general and home pages.
5. What is the protocol governing client requests and server responses in the WWW?
6. For what words does the abbreviation URL stand? What is a URL? What are its major elements for webpages? Is capitalization important when you type URLs?
7. What function do both menu directories and keyword search engines serve? How are they different? Can they be combined?
8. What is the WWW standard for formatting webpages? Are webpages plain ASCII files?
9. What are tags? Why are they needed? Is capitalization important when you type tags?
10. Are browsers limited to the World Wide Web? Explain.
11. How can you get a browser to work with an FTP server?

DETAILED REVIEW QUESTIONS

1. What is a bookmark list? Why is it useful?
2. Create a home page with the title *First Page*. It should have a first-level heading giving your name. Below it should be paragraphs showing your telephone number and Internet email address. (For privacy, you may create a fictitious telephone number.) Print the webpage, showing all tags. View the page with a browser and print the browser image.
3. Create another webpage to be located one directory level down from First

Page in the directory *misc*. This webpage should have the title *My Address*. It should have your postal address (may be fictitious). Make a link in it pointing to *First Page*. Add a link to *First Page* pointing to your address page. Make both links relative instead of absolute. Print both pages showing tags and as seen by a browser. Verify that you can jump back and forth between them via the links.

4. Add an inline graphic to *First Page*. The file is *little.gif*. It is one level down, in the pixes directory. Add an external graphic to your *First Page*. It is *big.gif*. It is in the *newpix* directory, which is on the same level as the directory holding first.htm. Print *First Page* and the *big.gif* page showing tags and in browser format. Use relative addressing.

5. Add character formatting to *First Page*. Print the page as in Detailed Review Question 2.

6. Add an ordered list and an unordered list to *First Page*. Print the pages as Detailed Review Question 2.

7. If you want the readers to see "<A>" when they look at a page with a browser, what do you actually type in the HTML source document? (*Hint:* "<" and ">" are reserved characters.)

8. Who controls how information appears on the webpage—the webpage author or the browser user?

9. Explain why CGI is important. Explain why Java is important. Explain how they differ.

THOUGHT QUESTIONS

1. Why is the World Wide Web attractive to users?
2. Do we really need the concept of a home page?
3. Do you think the problem of lost home pages will get better or worse?
4. Why do you think hypertext took so long to catch on? (The concept was created in the 1960s.) Why do you think the World Wide Web is exploding today?
5. How can browsers work with multiple types of servers? (Not how a user uses another type of server, but how, technically, it is done.)

PROJECTS

1. Given a specific URL, use a browser to read that home page. Print it. Add it to your bookmark list. Jump to related links on that webserver. Write

down the URLs you encounter and use this information to see how the HTML webpages are organized internally. Jump to a link pointing to another webserver. Note how long it takes. Note any problems you encounter. Write down the new URL.

2. Look up home pages in a menu directory. Look up home pages in a keyword directory.

3. Use a browser to retrieve files via anonymous FTP. Use a browser to look up information on a Gopher server.

4. Produce an update on current standards for commercial security on the World Wide Web.

For online exercises, please visit this book's website at: http://www.prenhall.com/panko

APPENDIX TO MODULE C: HTML CODES

General Document Structure

Note	Tag	Meaning	Description
	<HTML> </HTML>	Boundary of document	Not always required by browsers but good practice and correct.
	<HEAD> </HEAD>	Header	First part of every document. Required.
	<TITLE> </TITLE>	Title	Required part of every header. Gives name of document for the browser to display as the name of the page. Also used for web searches to identify the page.
	<BODY> </BODY>	Body	The main body section. The heart of the document.
	<P> </P>	Paragraph	Marks beginning and ending of paragraph. Browser puts a blank line after </P>. May only need a single <P> at the end of each paragraph, but the two are required in later versions of HTML.
	
	Line Break	Breaks line. Unlike <P>, browser does not insert a blank line. Placed at end of line at break point. No </BR> tag.
	<ADDRESS> </ADDRESS>	Address	Gives information about the webpage's creator or maintainer. Usually ends the body. Not required but good practice. Used by websearch engines to list the source of webpages.
	<!--x -->	Comment	For documenting the webpage. Not displayed by the browser. Not required, but for good practice.

Text Formatting: Explicit Formatting (Appearance Specified by Page Creator)

Note	Tag	Meaning	Description
	<B> </B>	Boldface	Makes text boldface.

	<I> </I>	Italic	Makes text italic.
3.0	<U> </U>	Underlined	Makes text underlined.
	<TT> </TT>	Typewriter	Uses monospaced font.
3.0	<S> </S>	Strikeout	Draws line through letters. Indicates something deleted from a document. Good for legal documents, comparing versions.
N	<BLINK> </BLINK>	Blinking	Text blinks. Annoying. Considered poor practice.

Text Formatting: Logical Formatting (Appearance Specified by Browser)

Note	Tag	Meaning	Description
	<Hi> </Hi>	Heading	Section heading. The i can be 1 through 6.
	<STRONG> </STRONG>	Strong	Usually displayed as boldface.
	<EM> </EM>	Emphasis	Usually displayed as italics.
	<PRE> </PRE>	Preformatted	Displayed in monospace font. For preformatted information, such as a typewritten document. Also an older way to show tabular information.
	<XMP> </XMP>		Essentially PRE, but embedded tags are ignored and shown as text.
	<CITE> </CITE>	Citation	Citation of a source. Usually displayed in italics.
N	<FONT SIZE=n> </FONT>	Font Size	Size of font. The n ranges from 1 to 7.

Parameters for PRE Tag:
   WIDTH=n, where n is the width of a line in characters.

Alignment of Text

Note	Tag	Meaning	Description
	<BLOCKQUOTE> </BLOCKQUOTE>	Block Indent	Indents paragraph from left and right.
N	<CENTER> </CENTER>	Center Center	Centers text on line.

Parameters for <P>
   ALIGN=x, where x may be LEFT, RIGHT, or Center. HTML 3.0
Parameters for <Hi>
   ALIGN=x, where x may be LEFT, RIGHT, or Center. HTML 3.0

Divider Line (Separates Adjacent Vertical Information Visually)

Note	Tag	Meaning	Description
	<HR>	Horizontal Rule	Displays a horizontal rule line across the page.

Parameters for Dividing Line
   SIZE=n, where n is the thickness of the rule line in number of pixels. N.

Lists

Note	Tag	Meaning	Description
	<UL> </UL>	Unnumbered List	Indents each line. Places bullet before paragraph. May be nested. Each level is indented further.
	<OL> </OL>	Ordered List	Indents each line. Places number before paragraph. May be nested. Each level is indented further.
	<LI>	List Element	Marks an element in a list. No </LI> at the end of item.
	<DL> </DL>	Definition List	For definitions. Has subparts <DT> and <DD>.
	<DT> </DT>	Definition Term	In a Definition List, the term to be defined.
	<DD> </DD>	Definition	In a Definition List, the definition for the preceding Definition Term>.

Parameters for UL
  TYPE=x, where x is the bullet type. May be DISK, CIRCLE, or SQUARE. N.
Parameters for OL
  TYPE=x, where x is the marking type. May be 1, A, a, I, or i. N.

Special Characters lt;

Note	Special Character	Replaced by
	<	&lt;
	>	&gt;
	"	&quot;
	&	&amp;
	ASCII Code	&#n; Here n is a three-digit ASCII Code.
	Registered	&#174;
	Copyright	&#169;
N	Registered TM	&reg;
N	Copyright	&copy;

Backgrounds and Colors (Netscape)

Note	Tag	Meaning	Description
N	<BODY parameter> <BODY>	Body Parameter	Several options

Netscape Parameters for BODY
  In all cases, h is a hexadecimal number.
  BGCOLOR=h, sets a color for the background.
  BACKGROUND=URL, uses an image at the URL as a background texture.
  TEXT=h, specifies the color of text.
  LINK=h, specifies the color of links.
  VLINK=h, specifies the color of visited links.
  ALINK=h; specifies the color of the active link.
  Note that colors should be coordinated.

Links

Note	Tag	Meaning	Description
	<A HREF="URL"></A>	Link	See chapter.

<A NAME="x"> </A>	Name a Target		The material enclosed in the link may be referred to by its name (x) in links.
<A HREF="#x"> </A>	Link to Target		For jumping to target on the same page.
<A HREF="URL#x"> </A>	Link to Target		For jumping to target on a different page.

Inline Images

Note	Tag	Meaning	Description
	<IMG SRC="URL">	Inline Image	Locates graphics file at specified URL. Displays at specified point on the page.

Parameters for IMG
  ALT="x", where x is text to be displayed by text-only browsers. After SRC parameter.
  ISMAP, to say that the source file is an image map with selectable areas.
  ALIGN=x, where x may be TOP, BOTTOM or MIDDLE. To align text with image.
  WIDTH=x, where width is the width of the image in pixels. N.
  HEIGHT=x, where x is the height of the image in pixels. N.
  BORDER=x, where x is the width of the border in pixels. N.
  HSPACE=x, where x is the width of horizontal space around the image in pixels.
  VSPACE=x, where x is the width of vertical space around the image in pixels.

Tables (HTML 3.0 Unless Specified as Netscape)

Note	Tag	Meaning	Description
3.0	<TABLE> </TABLE>	Table	Delimits the table.
3.0	<TR> </TR>	Row	Delimits a table row.
3.0	<TD> </TD>	Cell	Delimits a table cell in a row.
3.0	<TH> </TH>	Cell	TD is for data. TH is for heading. Default is bold and centered in the cell.
3.0	<CAPTION> </CAPTION>	Caption	To put a caption on the table.

Parameters for TABLE
  BORDER, to display a border.
  BORDER=n, where n is the width of the boarder in pixels. N.
  CELLSPACING=n, places n pixels between cells.
  CELLPADDING=N. Places N pixels between text and border.
  WIDTH=n, where n is the width of the table in pixels or if followed by %, percent of page. N.

Parameters for TR
  ALIGN=x, where x is LEFT, RIGHT, or CENTER. For horizontal alignment.
  VALIGN=x, where x is TOP, MIDDLE, or BOTTOM. For vertical alignment.

Parameters for TD and TH
  NOWRAP, for no line breaks
  COLSPAN=n, where n is the number of columns for the cell to span.
  ROWSPAN=n, where n is the number of rows for the cell to span.
  ALIGN=x, where x is LEFT, RIGHT, or CENTER. For horizontal alignment.
  VALIGN=x, where x is TOP, MIDDLE, or BOTTOM. For vertical alignment.

WIDTH=n, where n is the width of the cell in pixels or if followed by %, percent of cell width. N.

Parameters for CAPTION
ALIGN=x, where is TOP or BOTTOM. To specify whether caption is on top or bottom of table.

Forms

Define the Form
    <FORM ACTION="URL" METHOD=x> <FORM>
                    Where x=GET (receive) or POST (transmit)

Selection List
    <SELECT parameters> </SELECT>
    NAME="x", where x is the name of the selection list.
    SIZE=n, where n is the number of options from which to select.
    MULTIPLE, indicates that the user may select multiple options.

Option
    <OPTION parameter> is an option that can be selected.
    SELECTED, indicates the selected option. Allows a default value initially.

Input Field
    <INPUT parameters>
    NAME=x, where x is the field name.
    CHECKED, for check boxes and radio boxes.
    SIZE=n, where n is the number of characters.
    MAXLENGTH=n, where n is the maximum length of input strings, in characters.

Text Area
    <TEXTAREA parameters> </TEXTAREA> for the input box
    ROWS=n, where n is the number of rows.
    COLS=n, where n is the number of columns.
    NAME="x", where x is the name of the box.

MODULE D

Advanced Topics in Point-to-Point Transmission

INTRODUCTION

Module D covers a number of topics in point-to-point transmission. Its role is to provide more detailed information for the material in Chapters 3 and 4. It is not meant to be read from beginning to end like an ordinary chapter. Although it follows the basic order of Chapters 3 and 4, there are seldom logical transitions from one major section to the next. It contains information on the following topics:

- The encoding of pixel-mapped graphics and selected topics in advanced text encoding.
- The relationship between speed, noise, and bandwidth.
- Modems.

DIGITAL CODES AND ENCODING

In Chapter 3, we discussed ASCII briefly. Now we will look at digital coding in more depth.

Pixel-Mapped Graphics Codes and Compression

ASCII and other character codes are fine for simple typed text, but they cannot handle graphics.

Pixels ■

Figure D.1 shows that facsimile represents each page by a large number of dots (called **pixels** or *picture elements*). There are many pixels on a page. If facsimile is being sent at a resolution of 200 dots per inch horizontally and vertically, then an 8.5 by 11 inch sheet of paper has 8.5 x 200 x 11 x 200 pixels, or 3.7 Mpixels.

Black-and-White Images ■

In facsimile, each pixel is black or white. So you only need to store one bit per pixel. For instance, you might represent black by 1 and white by 0. We call this **bit-mapped** graphics because we map the pixels on the screen and assign one bit to each pixel on the page map. So 3.7 Mpixels would require 3.7 Mbits of storage.

Color Images ■

Black-and-white images only require you to store one bit per pixel. But if you want color, you must store several bits per pixel. If you store four bits per pixel,

FIGURE D.1

Pixel-Mapped Graphics

In pixel-mapped graphics, each point on the screen is a dot or pixel. It may be black and white or colored. The computer must store 1 to 64 bits for each pixel. Pixel-mapped graphics are fine for photographs and other scanned information, but they tend to consume large amounts of memory.

for instance, you can have 16 (2^4)colors. Storing one byte per pixel will allow you to represent 256 (2^8) colors.

Other common bits per pixel for color are 16, 24, 32, and even 64 bits. Using 24-bit color allows you to store eight bits for each of the three primary colors plus another eight bits for blackness.

Of course, storing multiple bits per pixel causes storage and transmission needs to explode. A 24-color page requires 24 times the storage of a black-and-white page.

Common Pixel-Mapped Standards ■

In the PC world, one common pixel-mapped standard is **GIF,** which is used in CompuServe and in HTML documents for the World Wide Web (see Module B). Another is **PCX,** which was created for PC Paintbrush. Scanners usually encode their images as **TIFF** (Tagged Image File Format) files.

Compression ■

Suppose you are merely sending a facsimile image. Say the resolution is 200 pixels per inch horizontally and vertically. If you have an 8.5 by 11 inch paper, this would mean 3.7 Mpixels on each page. Laser printers have even higher resolution, and if you are transmitting color, there can be millions of bytes of information per page.

Fortunately, image information tends to be **redundant.** This means that if one pixel is white, the next several pixels are also likely to be white. So you can send a brief code to represent many white pixels. This encoding allows your computer to compress the data before transmission.

To give an example, a black-and-white facsimile image has 3.7 Mbits of uncompressed image data. Sending this over a 14.4 kbps fax modem without compression would take over four minutes. With compression, however, transmission times are usually under a minute per page.

For pixel-mapped graphics, the **Joint Photographic Experts Group (JPEG)** compression standard is becoming increasingly important outside of facsimile.

Lossless versus Lossy Algorithms ■

There are too many compression standards for still images to summarize here, but it is important to note a basic distinction between *lossless* and *lossy* standards.

Lossless standards use algorithms that preserve all of the information in the picture. When you decompress the picture, you get the full original picture back.

Lossy algorithms, in turn, lose some information. However, they also provide higher levels of compression, often with only a modest loss of information. Most lossy algorithms allow you to decide what trade-off you will accept between storage space and image quality.

Vector Graphics

In pixel-mapped graphics, images are represented by matrices with squares filled in with various colors. In **vector graphics,** in contrast, each image is made up of a large number of elements—lines, circles, ovals, arcs, and text in various fonts.

Advantages of Vector Graphics ■

Vector graphics has two very important advantages over pixel-mapped graphics. First, the individual elements do not lose their identity after they are added to the image. You can resize them, recolor them, move them in front of or in back of other elements, and so forth. In contrast, once pixel-mapped elements are added to the bit map, they lose their identity, much like paint on a canvas.

Second, in most cases vector graphics produces much smaller files than pixel-mapped graphics. For instance, to store information about a circle element, you only need to store the position of its center, its radius, the thickness and color of its circumference, and the color (if any) of its area. This takes only a few bytes of storage. In contrast, storing all of the pixels for a circle might take up kilobytes of storage. Although very complex vector graphics images are larger than pixel-mapped images, this usually is not the case.

Text and Graphics Documents

Many documents today contain both text and graphics. We have already looked at graphics standards. Now we will turn to text standards and integrated text/graphics standards.

Plain ASCII Text ■

Chapter 3 introduced the simplest form of text encoding. This was *ASCII*—the American Standard Code for Information Interchange. As Chapter 3 discussed, ASCII represents each printing character by a seven-bit code. Table D.1 shows these codes.

As Chapter 3 discussed, there is also 8-bit ASCII, which uses eight bits per character. The use of an eight bit doubles the possible number of codes. Unfortunately, there is no standard for the use of these 128 additional symbols.

Although ASCII can represent characters, it cannot represent any of the rich text formatting we have come to expect from word processing programs. ASCII cannot represent boldface or other types of character formatting. For that, other text standards are needed.

Specifically, we need a way to standardize **fonts,** which are a combination of typeface (Times Roman, Helvetica, and so on), emphasis (*italics*, **boldface**, underlined, and so on), and type size (how large the letter is). Vertical size is measured in points. There are 72 points to an inch. The type size in this book

TABLE D.1

The ASCII Character Set

	000	001	010	011	100	101	110	111
0000	NUL	DLE	SP	0	@	P	`	p
0001	SOH	DC1	!	1	A	Q	a	q
0010	STX	DC2	"	2	B	R	b	r
0011	ETX	DC3	#	3	C	S	c	s
0100	EOT	DC4	$	4	D	T	d	t
0101	ENQ	NAK	%	5	E	U	e	u
0110	ACK	SYN	&	6	F	V	f	v
0111	BEL	ETB	'	7	G	W	g	w
1000	BS	CAN	(	8	H	X	h	x
1001	HT	EN	)	9	I	Y	i	y
1010	LF	SUB	*	:	J	Z	j	z
1011	VT	ESC	+	;	K	[	k	{
1100	FF	FS	,	<	L	\	l	|
1101	CR	GS	-	=	M	]	m	}
1110	SO	RS	.	>	N	^	n	~
1111	SI	US	/	?	O	—	o	DEL

Notes:
  Top row shows bits 7, 6, and 5
  First column shows bits 4, 3, 2, and 1
  So for C, bits 7 through 1 are 1000011

NUL	Null (nothing)	DLE	Data link escape
SOH	Start of header	DC1	Device control 1
STX	Start of text	DC2	Device control 2
ETX	End of text	DC3	Device control 3
EOT	Rnd of transmission	DC4	Device control 4
ENQ	Enquiry (Are you there?)	NAK	Negative acknowledgment
ACK	Acknowledgment	SYN	Synchronizing signal
BEL	Bell (ring bell)	ETB	End of text block
BS	Back space	CAN	Cancel
HT	Horizontal (normal) tab	DM	End of medium (paper out)
LF	Line feed	SUB	Substitute
VT	Vertical tab	ESC	Excape
FF	Form feed at printer	FS	File separator
CR	Carriage return	GS	Group separator
S0	Shift out to other character set	RS	Record separator
SI	Shift in	US	Unit separator
		DEL	Delete

is 12 points. This is also the normal size in letters and reports in organizations. In contrast, 10 point text looks like this, while 14-point looks like this.

Font Standards ■

There are two basic standards for representing fonts. One is the **PostScript** format of Adobe. This was first popularized on the Macintosh but became fairly popular on IBM-compatible PCs.

When Microsoft introduced its **TrueType** format for representing text, however, that format soon became dominant in the PC world. Some PC users buy Adobe Type Manager to allow them to use PostScript fonts in Windows, but these usually display poorly.

Integrated Documents ■

What if you have a **compound document** with integrated text and graphics? In addition to representing individual figures, PostScript can represent whole documents if you convert your text, vector graphics, and pixel-mapped graphics into PostScript format. Often this is as easy as telling your program to print to a file using the PostScript format. Unfortunately, relatively few users can read PostScript files on-screen.

Adobe, which owns PostScript, has recently introduced the Acrobat **Portable Document Format (PDF)** for representing many types of documents. If you have a PDF reader program, you can view the document on-screen and even print it. If you have the proper license, you can send someone a PDF file plus a viewer.

In the computer science world, a popular document format is **LATEX.** This can also represent both multifont text and graphics in an integrated format.

The World Wide Web (see Chapter 2 and Module B) has made the **Hypertext Markup Language (HTML)** a popular way to represent compound documents. HTML is actually a simplified version of the **Standardized General Markup Language (SGML)**.

Bandwidth, Noise, and Speed

In Chapter 4, we noted that telephone channels could only carry data at about 30 kbps. Now we will look at how that number was derived. We will also look at the general relationship among three critical variables: bandwidth, noise, and speed.

As we discussed noise in Chapter 4, noise is random electrical energy on the line. From time to time, noise randomly spikes to unusually high or low values. This can change 0s to 1s or 1s to 0s. It can also result in invalid voltage levels that the receiver cannot interpret.

Noise becomes more of a problem as speed increases. As you increase transmission speed, you are really shortening the duration of each bit. At 300 bps, each bit is about three-thousandth of a second long. At 10 Mbps, in turn, the duration of a single bit is a mere one ten millionth of a second.

Over a long enough period of time, the random fluctuations in line noise average out. But as the duration of a bit decreases, the probability of a spike as large as the signal power level grows. There will be more and more errors.

Shannon [1948] quantified the relationship between transmission speed and noise, as shown in Equation D-1. It shows the maximum transmission speed on a line (W) in bits per second. Here the channel has a bandwidth B

(measured by subtracting the lowest frequency in the channel from the highest frequency in hertz). In addition, the **signal-to-noise ratio (SNR)** is S/N. This is the average signal power divided by the average noise power.

$$\text{Equation D-1: } W = B\text{Log}_2(1 + S/N)$$

This equation makes sense intuitively. Increasing bandwidth is like increasing the diameter of a hose. You can push a lot more water through a wider hose. Noise, in turn, is like resistance. As you increase noise, the need to retransmit damaged messages increases.

What is the maximum speed of a telephone line with a typical 1,000:1 signal-to-noise ratio? Log_2 of 1,000 is 10. Multiplying this by the bandwidth (3.1 kHz) gives a maximum possible speed of about 31 kbps.

Generally, noise is fairly well controlled, so the key factor in attaining more speed is increasing the channel bandwidth. This is why high-speed transmission lines are often called **wideband** lines. A television channel, for instance, has about 2,000 times the bandwidth of a telephone channel (6 MHz versus 3 kHz). It can therefore transmit about 2,000 times as much information per second. You need this because of the many pixels needed to send a television screen.

Note that Equation D-1 only gives the *maximum possible* throughput on a transmission line. The first modems only operated at 110 bps, and although modems are now approaching the Shannon limit for telephone lines, they must use remarkably sophisticated engineering in order to do so.

MODEMS

We saw in Chapter 4 that if you want to send digital information, such as computer data, over analog transmission lines, such as telephone lines, you need a device called a **modem.** When a computer transmits its digital source signal, the modem *modulates* the signal, changing it into an analog transmitted signal that can travel down the telephone wire. At the other end, the other modem *demodulates* the transmitted signal, converting it back into a digital received signal.

Nearly every data communications professional will have to know how to buy modems for many years to come. In this module we will look at some of the factors you need to consider when buying modems.

In most of this section, we will look at data modems, which are designed to carry any type of data. We will, however, look at facsimile modems (fax modems), which are designed to send facsimile images, essentially turning the sending and receiving computers into facsimile machines.

Modem Transmission Speeds

The most important characteristic of any modem is its transmission speed—how rapidly it can send and receive information. We saw in Chapter 1 that we measure transmission speed in bits per second. The modems you can buy in a store today typically operate at 14.4 kbps to 28.8 kbps.

As we discussed in Chapter 4 (and in the previous section), the telephone system has a speed limit of about 30 kbps. The best modems are already very near this limit. So we can expect little or no increase in modem speeds in the future.

Modem Speed Standards ■

Obviously, the modems at the two ends of the telephone line need to be compatible if they are to be able to work together. As we saw in Chapter 1, this requires that they communicate according to the same standard.

All modems now being sold follow speed standards set by the ITU-T. A few older modems still follow AT&T standards, but these are disappearing from the corporate world. Table D.2 shows standards for transmission at var-

TABLE D.2

Modem Standards

Speed Standards

Name	Speed	Origin
212A	1200 bps	AT&T
V.22	1200 bps	ITU-T
V.22 *bis*	2400 bps	ITU-T
V.32	9600 bps	ITU-T
V.32 *bis*	14.4 kbps	ITU-T
V.34	28.8 kbps	ITU-T

Error Correction and Data Compression Standards

Name	Type	Origin
V.42	Error correction	ITU-T
V.42 *bis*	Data compression	ITU-T
MNP (Microcom Network Protocol)	Error correction and data compression	Microcom

Facsimile Modem Standards

Name	Speed	Origin
V.17 *ter*	4800 bps	ITU-T
V.29	9600 bps	ITU-T
V.14	14.4 kbps	ITU-T

Other

Standard	Description
BFT	Binary File Transfer. Allows two facsimile modems to exchange general files as well as page images.
AT	Hayes computer command set. Allows PC to send control messages to data modem.
Class 1 Class 2	Commands set for facsimile modems. Allows PC to send control messages to facsimile modem.

ious speeds. The most common modems on the market today follow the V.34 speed standard (28.8 kbps) and the V.32 *bis* speed standard (14.4 kbps).

Speed standards decree how each modem should modulate its signal when it transmits. In Chapter 4, we saw that this modulation can be extremely complex in modern modems, which phase and amplitude modulation and only use some phase/amplitude combinations.

These speeds, by the way, are somewhat optimistic. They tell you how fast two modems following the same standard can communicate over a very good telephone line. If there is too much noise on the line, the two modems will switch automatically to a lower transmission rate in order to reduce errors. Dividing the actual bits to be sent by the speed of the modem standard will underestimate transmission times.

Hand Shaking for Modem Speed

What if an old modem, which can only communicate via the 2,400 bps V.22 *bis* speed standard, needs to communicate with a 28.8 kbps V.34 modem? The answer is that the faster modem slows down to talk with its slower cousin.

This slowing-down process, furthermore, is completely automatic. Whenever two modems connect, they exchange a series of signals before they begin data transmission. Through these initial **hand-shaking** signals, they learn about the other modem's capabilities and negotiate the best possible transmission parameters. When they are finished, they signal their respective computers that they are ready to begin sending data.

Bits per Second and the Baud Rate

Many modem sellers refer to a modem's speed in bits per second as its **baud rate.** For instance, they will call a 28.8 kbps modem a "28.8 kilobaud" modem.

Chapter 3 explained that when modems transmit, they can represent a multiple bit in each line change (baud). For instance, suppose that you use frequency modulation and have four frequencies. You can use these four frequencies to represent 00, 01, 10, and 11. So this allows you to transmit two bits per line change. Your bit rate would be twice your baud rate.

To give a real example, a V.32 *bis* modem, which transmits at 14.4 kbps, really operates at 2,400 baud [Brown, 1992]. Its bit rate is six times higher than its baud rate. The V.34 standard, in turn, specifies several alternative modulation schemes. These have similar ratios for bit rate to baud rate.

Unfortunately, the confusion of baud rate with bits per second is so widespread in practice that you will see it constantly when you look at modems.

Modem Error Correction

Early modems merely handled modulation and demodulation, as their name implied. However, users began to want more from their modems. Most importantly, telephone lines have a good deal of noise, and this produces errors in the transmitted signal. Users wanted their modems to detect such errors and then correct them without involving either the user or the user's computer. Early modems did not do error detection and correction. Most current modems do perform this valuable service.

In addition, to block-by-block error checking, the newest modem standards, especially V.34, specify that the two modems will occasionally stop sending data and will instead check the transmission line for its error rate by sending test data back and forth. They will then adjust their speed or other parameters to cope with the actual line conditions.

Modem Compression

Another benefit that people would like from their modems is data compression. Even the fastest modems are terribly slow by LAN standards. For instance, if you are transmitting a 20-page document, you must transmit about 300 kilobits. Even with a fast modem, this will require 10 seconds. If you are using a word processing program, waiting this long for a file to load from remote storage would be onerous.

Modem Compression ■

Fortunately, most modems sold today can do data compression. Under the best conditions, a data modem can compress data by a factor of 4:1. Figure D.2 shows how data compression works in practice. Here we assume the 4-to-1 compression ratio. We also assume a V.34 modem, which means speeds up to 28.8 kbps. This modem speed multiplied by four gives 115.2 kbps of data throughput (under ideal conditions).

Note that the computer transmits data at 115.2 kbps to the modem. Not all PCs are capable of doing this. Although all PCs can transmit faster than the official speed limit of 20 kbps in the EIA/TIA-232-E standard, not all can transmit at such high speeds. In addition, many operating systems, such as versions of Microsoft Windows before 3.1, may not be able to handle the serial port at such high speeds even if the serial port is capable of sending and receiving that quickly.

The modem then compresses the data by 4-to-1, down to 28.8 kbps. It then sends the data out over the telephone line at 28.8 kbps. The receiving modem then accepts the data at 28.8 kbps, decompresses the data, and sends the data to the receiving computer at 115.2 kbps.

Figure D.2 shows why you can send data at 115.2 kbps, despite the 30 kbps limit of telephone lines. Only compressed data travels over the telephone line, and it stays within the 30 kbps limit of the telephone system.

While compression is good, computer users often run into problems when they use it. Some computers cannot transmit reliably at speeds approaching 115.2 kbps, and even when the computer's hardware is capable, older versions of Microsoft Windows (through version 3.1) had a difficult time supporting high-speed transmission.

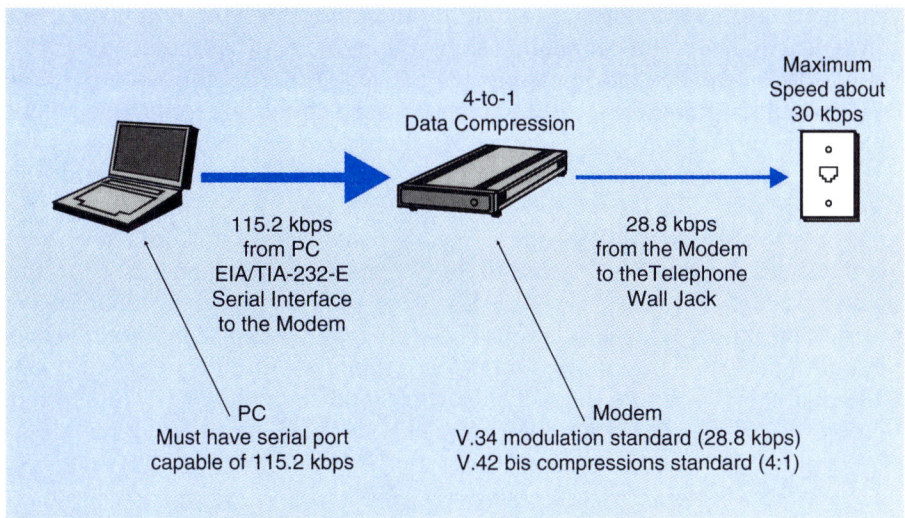

FIGURE D.2

Modem Data Compression

The computer transmits data from its EIA/TIA-232-E serial port to the V.34 modem at 115.2 kbps. The V.42 bis modem compresses the data by 4:1 and then transmits it over the telephone line at 28.8 kbps. Although the computer transmits at 115.2 kbps, the 30 kbps maximum theoretical speed of the telephone line is not violated. The modem places incoming data in a buffer. If the buffer is too full, the modem signals the PC to stop sending. The PC's serial port must be capable of transmitting at 115.2 kbps.

Standards for Error Correction and Compression ■

Table D.2 shows modem standards for error correction and data compression. First, it shows that the ITU-T has *separate* standards for error handling and data compression. For error handling, it has **V.42.** For compression, it has **V.42 bis.** V.42 *bis* is described as offering 4-to-1 compression, but this is a best case. Its real compression ratio is often lower.

When you buy a modem, the box often reads V.42 *bis*, indicating data compression, but it does not mention V.42. This is because data compression only makes sense if you also do error handling. The error rate in an uncorrected compressed data stream would be horrendous. So stating that a modem offers V.42 *bis* implies that it also offers V.42.

Table D.2 shows that there is another standards family, the **Microcom Network Protocol (MNP).** This is a proprietary but widely licensed standard. Microcom has defined a number of versions or *levels* of MNP. MNP Level 5, for instance, offers both error correction and compression, as do higher levels.

Modem Intelligence

All modems can accept data to be transmitted. In addition, most modems can accept *commands* from the computer. For example, the computer might send a command asking the modem to dial a particular telephone number. Or the computer might send a command asking the modem to hang up.

Modems that can accept commands are called **intelligent** modems. They are also called *Hayes-compatible* modems because almost all accept commands in a format developed by the Hayes Computer Corporation.

They are also called **AT** modems because nearly all commands begin with the letters "AT". For instance, "ATDT9,5551765" means "<u>AT</u>tention, <u>D</u>ial using <u>T</u>ones 9, then <u>pause (the comma means pause)</u>, and finally dial <u>5551765</u>." To give another example, the code <u>ATH0</u> means to place the phone back on its hook, that is, to hang up. There are also commands to regulate the loudness of the modem's feedback noises, control how long the modem should let the telephone ring before hanging up, whether the modem should try again if it gets a busy signal, how many times it should try, and so forth. For more information on the AT command set, see Hummel [1993].

Modem Forms

All modems consist of a collection of computer chips and other electronic circuit elements. There are three common ways to package these circuits.

Internal Modems ■

Figure D.3 shows an **internal modem.** It is a printed circuit expansion board that the user places inside a PC's systems unit.

The internal modem is the least expensive modem form. There is no need to package the modem within even a plastic box.

Also an internal modem is "hidden" inside the systems unit and does not add more "mess" to the user's desktop. The only things coming out of the systems unit are two telephone cords: one to the telephone handset and the other to the telephone wall jack.

Adding an expansion board to a PC, however, requires a significant amount of technical knowledge. (See Module I.) Many users lack this knowledge.

External Modems ■

Figure D.4 shows an **external modem.** As the name suggests, an external modem's circuits lie inside a plastic or metal box that sits outside the systems unit.

An external modem has a number of connections. First, it must connect to the PC itself. It usually does this via an EIA/TIA-232-E **serial port.** This is either a 9-pin connector or a 25-pin connector at the back of the PC. A **serial cable** connects the PC and modem serial ports.

Although most serial ports can transmit information at 115.2 kbps in theory, there are often major problems at this speed. So some V.34 modems that use data compression connect to the PC via one of the PC's **parallel ports.** *For more on PC parallel ports, see Module I.*

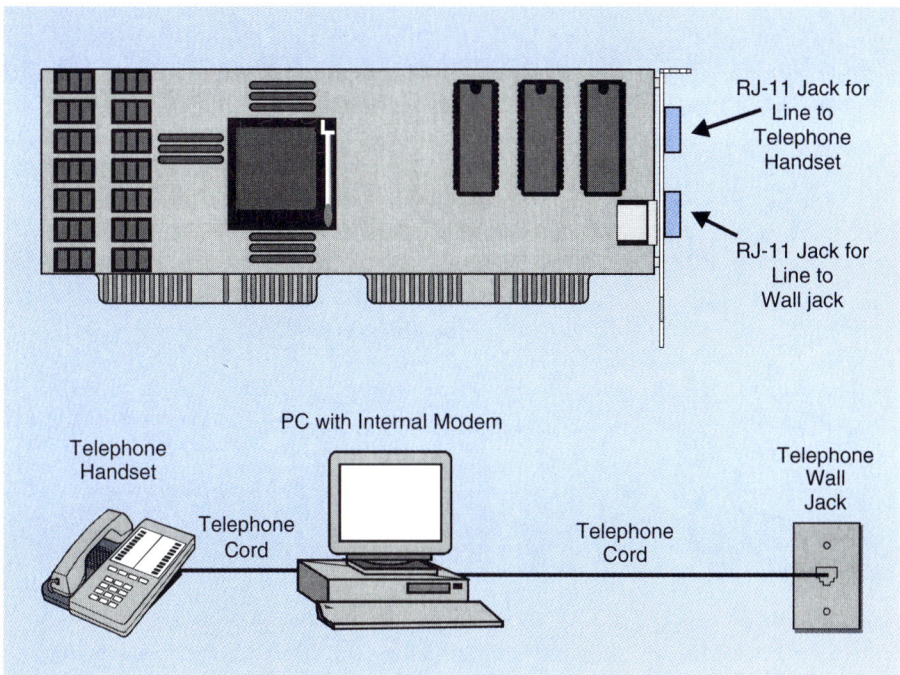

FIGURE D.3
Internal Modem

An internal modem is built onto an expansion board. You must open up the computer to install the expansion board. This can be difficult. However, internal modems are the cheapest modems. As importantly, they do not clutter your desk or need their own power supply. There are two RJ-11 jacks (connectors); one is for a telephone cord to the telephone handset. The other is for a cord to the telephone wall jack.

There also have to be connections using telephone cords. Most external modems have two telephone cord connections. One runs from the telephone handset to the modem. The other runs from the modem to the telephone wall jack. These connections use standard RJ-11 telephone jacks.

Finally, an external modem needs electrical power. Although some have batteries, most need an AC/DC converter connected to wall power. Although external modems are messy in terms of wiring, they are easy to install. For relatively novice computer users, they make sense.

PCMCIA Modems ■

Many notebook computers use a different form of modem, a **PCMCIA modem.** The user inserts the PCMCIA modem into a PCMCIA slot in the notebook.

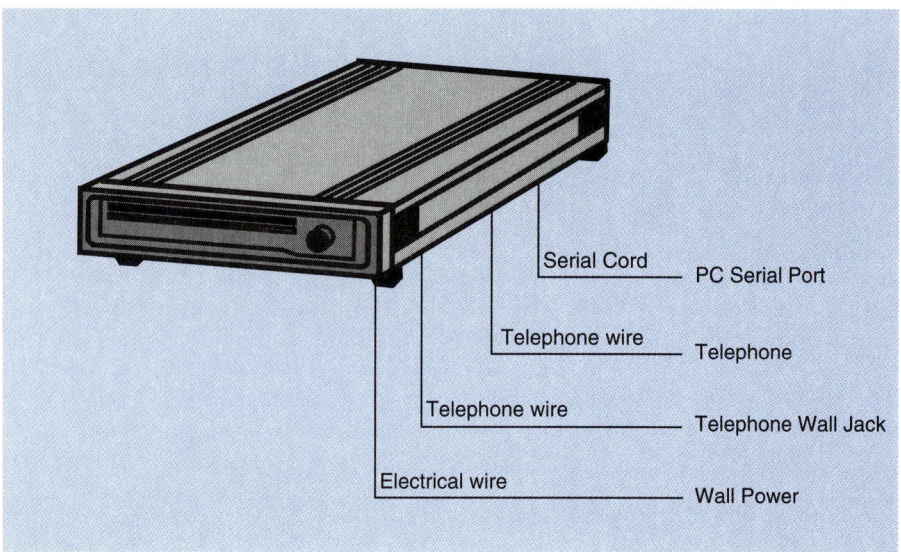

FIGURE D.4
External Modem

External modems are the simplest to install. However, there are many connections. There must be a serial cable between the modem's EIA/TIA-232-E serial port and that of the computer. There must also be a power connection unless the external modem runs on batteries. External modems consume valuable desktop "real estate" and lead to wiring tangles. They are also more expensive than internal modems.

PCMCIA stands for *Personal Computer Memory Card International Association.* As this ungainly name suggests, this plug-in standard was first used to add memory to mobile PCs. Later the standard was expanded to include such things as modems, network interface cards for PC networks, and even removable hard disk drives.

PCMCIA modems are extremely convenient. Even novice users can install them physically and go through simple software setup. Their main drawback is their price, which is considerably higher than the price of external modems. In addition, not many desktop computers have PCMCIA slots.

Facsimile Modems

Up to now, we have been talking about **data modems,** which are designed to move streams of bits from one computer to another. There are also **facsimile modems (faxmodems),** which are designed to transfer scanned images between facsimile machines. Many of the data modems on the market also serve as fax modems, but this is not universal.

Facsimile Speed Standards

As Table D.2 shows, facsimile modems use different standards than data modems. To see if a particular modem will work as a data modem, check the package to see if it follows one of the facsimile modem standards in Table D.2.

Intelligence

Recall that most data modems are intelligent, following the AT command set. Almost all faxmodems are intelligent as well. This allows the communication program to send them commands to control the transmission.

Almost all follow the **Class 1** standard (EIA/TIA-578). Many fax modems follow the more sophisticated **Class 2** standard (EIA/TIA-592). Most facsimile communication programs on PCs support the Class 1 standard and many support Class 2.

Actually, these standards essentially extend the AT command set. Their commands always begin "AT+F", followed by the command. The "+F" designates an extension for facsimile modems. For more information on modem classes, see Hummel [1993].

"Printing"

With the **facsimile software** on the PC, you can send a facsimile message as easily as you can print. For instance, suppose you are in your word processing program and have finished a document. You give the Printer Setup command and you see a choice of printers. One of these "printers" is the fax modem.

If you choose to "print" to the fax modem, a window will pop up on the computer screen. This window allows you to select an addressee from a list. If you select someone on the list, the software fills in that person's name and facsimile telephone number. Otherwise, you type in the receiver's name and number directly. This process creates a cover sheet that lists the name of the receiver, your name, and the number of pages to follow.

You do not have to buy facsimile software from the same company that sold you your facsimile hardware. If you do, however, you may be able to minimize setup problems. Some facsimile software simply will not work at all with some facsimile modems.

Binary File Transfer

Some facsimile programs support the **binary file transfer (BFT)** standard. This allows the fax programs on two connected PCs to talk to one another so that they can transfer whole files, such as word processing files or spreadsheet files. This moves fax modems beyond image delivery to the delivery of editable computer files.

Other Types of Modems

Voice/Data Modems ■

Normally, when you use a modem, your phone is tied up. If you want to talk with the person at the other end, there is no way to do so. For instance, there is no way to conduct document conferencing in which you both see the same document on your screens and talk back and forth to discuss it. As shown in Figure D.5, **voice/data modems,** however, allow you to talk and exchange data at the same time.

Some voice/data modems cut the telephone bandwidth in half, cutting voice quality and data transmission throughput. Others temporarily suspend data transfers when you speak. Neither solution is perfect, but both work reasonably well.

Of course, the modems at the two ends must use a compatible process for mixing voice and data. This creates problems because different voice/modem vendors use different standards for mixing voice and data.

Dial-Back Modems ■

Modem access gives users geographical flexibility, but it creates security problems. Some of these problems are reduced with **dial-back modems.** These modems dial the number, give a user account name and password, and then

FIGURE D.5

Voice/Data Modems

Voice/data modems allow you to send voice and data over the same telephone line. This allows you to talk to the other person as your data is being transferred.

hang up. The central site has a list showing where the user should be and dials that number back.

Managed Modems

Chapter 9 discusses network management. Some (more expensive) modems have built-in network management agents. The central network management program can call a **managed modem** to ask for the status of the modem. In addition, the agents in managed modems can broadcast an alarm if they detect problems with the modem.

CORE REVIEW QUESTIONS

1. Name some major categories for types of data that must be encoded differently (character data, for example, and so on).
2. What is redundancy, and why does it allow compression?
3. If you double the bandwidth, what happens to the maximum possible throughput, given a constant noise level? If you cut noise in half, does throughput increase or decrease? If you double the signal intensity compared to the noise power, does maximum possible throughput increase or decrease? Does the Shannon equation (Equation D.1) give actual throughput or the maximum possible throughput of a channel?
4. The modem box tells you that a modem is a V.34 modem. At what speed can this modem transmit? At what speed will it transmit if the other modem is a V.32 *bis* modem? From this information, can you tell if it does error correction? Compression? Can you tell if it is also a facsimile modem?

DETAILED REVIEW QUESTIONS

1. How many bytes (not bits) would you need to store a black and white image to be printed on a laser printer with a resolution of 300 dots per inch horizontally and vertically? How many bytes would you need to store the image in 24-bit color? Show your work.
2. Facsimile can transmit a page at a resolution of 200 dots per inch resolution horizontally and 100 dots per inch vertically. The transmission takes about a minute at 9,600 bps. What degree of compression does this require? Show your work.
3. Explain the difference between lossy and lossless compression.

4. What is the standard for data modems operating at 14.4 kbps? What is the standard for data modems operating at 28.8 kbps? What does the V.42 *bis* standard govern? The MNP Level 5 standard? The AT command set standard? The V.17 standard? The Class 1 standard?
5. Do all modems do error correction?
6. What is the compression ratio for MNP Level 5? For V.42 *bis*? Are these maximum compression ratios or actual compression ratios?
7. If the maximum possible transmission speed of the telephone system is about 30 kbps, how can a PC using a V.34 modem with V.42 *bis* data compression transmit at 115.2 kbps?
8. Create a table showing the relative advantages and disadvantages of the three modem forms (internal, external, and PCMCIA).
9. Does facsimile have separate standards for speed and compression?
10. What is binary file transfer (BFT)? Is it similar to file transfer protocols (see Chapter 4)? What are voice/data modems, dial-back modems, and managed modems?

THOUGHT QUESTIONS

1. Why do you think the ITU-T developed separate standards for modem speed, error correction, and compression?
2. Your friend is new to computers. She wants to add a modem to her desktop computer. What would you advise in terms of whether she should install an internal modem, an external modem, or a PCMCIA modem? Explain.

For online exercises, please visit this book's website at: http://www.prenhall.com/panko

REFERENCES

BROWN, R. O. (1992). Modems still have a bright future. *Network Management*, 10(11), 74, 76, 78.

HUMMEL, R. L. (1993). *Programmer's technical reference: Data and fax communications.* Emeryville, CA: ZD Press.

SHANNON, C. (1948). A mathematical theory of communication. *Bell System Journal*, 27, 379–423 (July), and 623–656 (October).

MODULE E

Advanced Topics in VT100 Terminal Emulation

INTRODUCTION

Chapter 3 discussed terminal emulation, especially VT100 terminal emulation. Module E presents more information on VT100 terminal emulation, which is also called asynchronous ASCII and dumb terminal emulation for reasons given in Chapter 3.

VT100 TERMINAL EMULATION

VT100 Terminals

As Chapter 3 noted, the most widely supported terminal design today is the **VT100** terminal. This design was created by Digital Equipment Corporation for its internal use. However, the American National Standards Institute (ANSI) defined its **ANSI terminal** standard on the basis of the VT100.[1] Al-

---

[1] The ANSI standard is X3.64. It defines a number of ASCII combinations for full-screen editing. For instance, the 4-byte ASCII sequence "<ESCAPE>[5A" means move the cursor up five rows in the current column.

though Digital Equipment Corporation now has far more sophisticated terminals, the VT100/ANSI terminal design has become a lowest common denominator for host support. Many hosts now support these terminals. In addition, VT100 transmission is simple for PCs to emulate because VT100 terminals use asynchronous ASCII transmission and EIA/TIA-232-E serial operation, which PC serial ports automatically support.

Setting Up the Communication Program

When you install a communication program, you specify some details about your computer and modem.

Serial Port ■

Most importantly, you have to tell the program what serial port your modem is using. PCs normally come with two serial ports, labeled COM1 and COM2. Some have two more serial ports, COM3 and COM4. (Internal modems are frequently configured as COM3 or COM4 because COM1 and COM2 are already on the machine.)[2]

Transmission Speed ■

In addition, you have to specify the speed at which you will transmit. This is a little tricky. Sometimes you specify the modem's speed. With a V.34 modem, for instance, you would specify 28.8 kbps. In other cases, you must specify the serial port's transmission speed. For instance, if you have a V.34 modem that can do V.42 *bis* data compression, you would specify 115.2 kbps because the computer transmits at this speed to the modem. The modem then compresses this data stream and transmits it over the telephone network at 28.8 kbps.

Modem Intelligence ■

You also need to specify the type of intelligence your modem uses. Intelligence allows the computer to send codes to the modem to take a number of actions, such as "hang up" or "dial this number." Normally, the default is *Hayes compatible*. Its command codes are sometimes called the AT command set because most begin with the ASCII codes for "AT," as discussed in Module D.

Telephone Type ■

Usually, there is a default setting for *tone* dialing. If you have an older pulse telephone, you must set the dialing for *pulse*.

---

[2]If the modem uses a parallel port, parallel ports are designated LPT1, LPT2, and so forth. This is short for *line printer*, an obsolete piece of terminology.

Flow Control ■

Recall that in flow control one side asks the other side to pause. We have already seen XON/XOFF software flow control There is also **hardware flow control.** Recall that modems use serial ports. Several of the pins in serial ports are for flow control.

Host Setup

You also need to specify certain parameters for each host computer you wish to use. Every time you want to work with a new host, you need to add parameters for that host.

Telephone Number ■

You will be communicating with the host by modem, so you need to specify the host's telephone number. If you are in the same city as the host, you probably dial the host directly. So the telephone number you enter is the telephone number of the host. If you reach the host via a data network, you would give the telephone number of the network's local access port.

Half or Full Duplex ■

The default in most packages is full-duplex communication. Almost all modern hosts are full duplex.

Stop Bits ■

Almost all modern systems use one stop bit, and this usually is the default.

Speed ■

While there are common defaults for many host parameters, there is none for the transmission bit rate. You must contact each host authority to ask for its speed. This will set the speed at which you can transmit. Sometimes a host can work with terminals communicating more slowly than its best speed, but it cannot work with faster terminals.

Data Characters and Parity ■

You must set the parameters for the data bits and parity. As we saw in Chapter 3, some transmission facilities ignore parity and use all eight bits for data. This allows you to transmit PC data. Some communication programs call this *8 data bits*. Others refer to this as *no parity*. If parity is used, some systems refer to this as *7 data bits*. In this case, the user also must specify whether parity is *odd* or *even*.

Cryptic Forms ■

Sometimes host administrators publish their speed, data bit, and parity parameters in cryptic forms. A typical example is *"9600,7E1"*. The "9600" indicates a speed of 9,600 bps. The last three elements indicate seven data bits, even parity, and one stop bit. Another common example is *"9600,8N1"*. In this case, there are eight data bits, no parity, and one stop bit.

Saving Setup Information

Obviously, it would be desirable to enter all of this setup information once and then save it for later use. Almost all communication programs allow you to do this. Whenever you deal with another host, you merely call up its setup file.

File Transfer Protocols

An important function of communication programs is file transfer. This allows you to transfer whole files with error correction. Your communication software divides the file into blocks. Error checking is done for each block. In a sense, error checking implements synchronous exchanges at the application level. It can do this despite the fact that the underlying transmission system is asynchronous.

There are many standard file transfer protocols for file transfer between PCs and hosts. Some of these are Kermit, XMODEM, XMODEM/CRC, and ZMODEM. Most PC communication programs can transfer files via several protocols. Most hosts offer only one to three. The user's job is to select a file transfer protocol that *both* the host and the PC communication program support.

XMODEM ■

Created for PC–PC file transfers in 1977, **XMODEM** is the lowest common denominator for file transfer. It is not very efficient because after it sends a block, it has to wait for an acknowledgment. In addition, its error detection only uses a single byte and can miss an unacceptable number of errors. XMODEM is widely implemented, but it is primarily for PC–PC file transfers, not for terminal–host transfers.

XMODEM/CRC ■

Increasingly, XMODEM is giving way to **XMODEM/CRC** as the lowest common denominator for PC–PC file transfers. This is XMODEM with a two-byte error-check field. It still must wait after each transmission block to get an acknowledgment from the other side, but at least error checking is more reliable.

ZMODEM

ZMODEM is a much stronger PC–PC file transfer protocol and is achieving good market penetration. ZMODEM is not just an incremental improvement over XMODEM. It is far more sophisticated. ZMODEM does not have to wait after each block for an acknowledgment. It numbers its blocks and sends them without stopping. Later, if it gets a negative acknowledgment on a block, it retransmits just that block. In addition, when two communications programs begin to communicate, they exchange hand-shaking messages that allow them to compare their features, so that they can pick the best possible combination.

Kermit

XMODEM, XMODEM/CRC, and ZMODEM are all PC-oriented file transfer protocols. If you are transferring files to or from a minicomputer or mainframe, you may use the **Kermit** file transfer protocol. Kermit is even more sophisticated than ZMODEM. It also has initial hand shaking and the transmission of blocks without stopping to wait for acknowledgment. In addition, if you divide binary files into eight-bit chunks or send 8-bit ASCII, Kermit will recode the data to ensure that you are not sending one of ASCII's control sequences and that transmission systems that only treat the first seven bits as data will not cause problems.

CORE REVIEW QUESTIONS

1. In VT100 terminal emulation, what do you specify about your computer? About your modem? About the host computer? Explain "9600,7O1".
2. How do you know which file transfer protocol to use with a host computer? Which of the file transfer protocols that the module listed is oriented toward host computers?

THOUGHT QUESTIONS

1. Your friend is new to computers. He or she wants to use a host computer that requires VT100 terminal emulation. What would you tell your friend about what he or she would need to buy? What would you tell your friend about what he or she would need to get from the host authority?

PROJECT

1. Use VT100 terminal emulation to connect to a computer. List the parameters you had to set in the communication software. Explain what each does in terms of data communications concepts. Set up a login script. Download a file to your PC.

For online exercises, please visit this book's website at: http://www/prenhall.com/panko

MODULE F

Advanced Topics in LAN Technology

INTRODUCTION

Module F covers some advanced topics in LAN technology. It is not intended to be read front to back like a chapter. It should be read after Chapter 5. Module F focuses on two basic topics:

- Ethernet physical-layer standards, especially the older but still widely used 10Base5 and 10Base2 standards. It also covers mixing several Ethernet physical-layer standards in the same LAN, for instance, to increase the distance span.
- Further information on 802.5 Token-Ring Networks.

ADDITIONAL ETHERNET STANDARDS

In Chapter 4, we looked at two sets of the Ethernet physical-layer standards from the IEEE 802.3 Working Group. We looked at the

10Base-T physical-layer standard and at the 100Base-X family of physical-layer standards.

In this section, we will look more closely at two 802.3 physical-layer Ethernet standards that predate 10Base-T, namely *10Base5* and *10Base2*. Both, as you can tell by their names, are 10 Mbps baseband standards. Some organizations that networked early still use these two physical-layer standards as their main standards. Others continue to use 10Base5 for longer runs in their 10Base-T networks. As we will see, you can mix 802.3 physical layer standards in a LAN, as long as you separate them by repeaters.

We will close with a brief discussion of the 802.9a *IsoEthernet* standard. This standard adds 6 MHz of isochronous bandwidth to the standard 10 Mbps. This added bandwidth allows real-time voice conversations and low-speed videoconferencing communication.

We will not discuss the 802.3 MAC layer in more detail. All 802.3 physical-layer standards, even those running at 100 Mbps, use the same 802.3 MAC-layer standard.

10Base5

When the 802 Committee began developing 802.3 variants, it developed a naming system. The **10Base5** standard illustrates this system. The "10" stands for the speed of the standard—10 Mbps. The "Base," in turn, stands for baseband transmission.

The "5," finally, means that a **segment** (single unbroken run) of the cable can be up to 500 meters long (1,640 feet). We will see later that several of these rather short segments can be linked to form a larger LAN.

The Trunk Cable ■

The specific type of coaxial cable used in 10Base5's trunk was created especially for the standard. It is 10 millimeters thick (about 0.4 inches). This is much thicker than the coax cables used to bring cable television signals to the home or to link VCRs to television sets.

Because of this thickness, the 10Base5 trunk cable is fairly difficult to bend and install. Because 10Base5 cable is traditionally colored yellow (the standard merely specifies a bright color), installers refer to it in frustration as a "frozen yellow garden hose."

Medium Attachment Unit (MAU): The Transceiver ■

The 10Base-T standard follows Ethernet's fairly complex method for attaching stations to the coaxial trunk cable. Figure F.1 illustrates this attachment method.

First, attached to the thick cable is the **medium attachment unit (MAU)**. One job of the MAU is to provide a physical attachment to the network. Older MAUs have a "vampire clamp" whose two "teeth" dig into the cable, making contact with both the inner and outer conductors in the cable. This rather crude

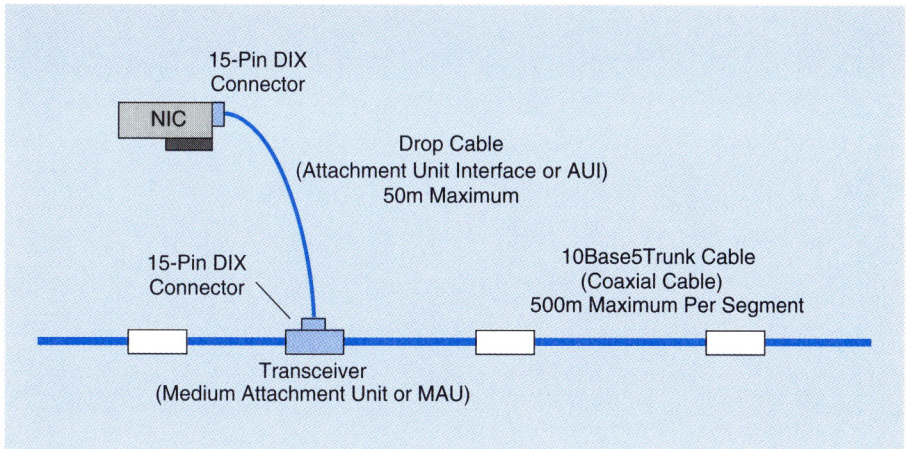

FIGURE 7.1

Connecting a Station to 10Base5 Trunk Cabling

The NIC has a 15-pin AUI connector, which is also called a DIX connector after the names of the companies that created the Ethernet V2.0 standard. A drop cable, also known as an attachment unit interface (AUI), runs from the station to the trunk cable. This is a 15-wire bundle. At the cable, a transceiver, also called a medium attachment unit (MAU), links the drop cable to the trunk cable. The connection between the transceiver and the drop cable is another DIX connector. The transceiver (MAU) attaches to the cable either through a vampire clamp or screw-on N connectors.

attachment method speaks volumes about the antiquity of Ethernet cabling technology. Newer MAUs use screw-on N connectors.

The MAU is also called a **transceiver (transmitter/receiver).** This is the original Ethernet V2.0 terminology. It is a transceiver because it *transmits* signals over the cable and *receives* signals from the cable. It is an active electronic device, not a mere physical connector.

Attachment Unit Interface (AUI): The Drop Cable ∎

The station can be some distance away from the cable. To allow this, an **attachment unit interface (AUI)** runs from the MAU to the station itself. Ethernet V2.0 called this wiring bundle a **drop cable.** This seems like a wiser choice of terminology. The AUI attaches to the station via a 15-pin **AUI connector**.

The AUI can be up to 50 meters long. This is another thick inflexible bundle, but unlike the coax trunk cable, the AUI is a bundle of 15 wires. It is slightly more expensive per foot than the trunk cable.

Segments and Repeaters in 10Base5 ∎

As noted earlier, a single unbroken length of 10Base5 trunk cable is called a **segment.** Figure F.2 shows that MAUs plug into the cable segment at many places. For electrical reasons, the MAUs must be at least 2.5 meters apart. There

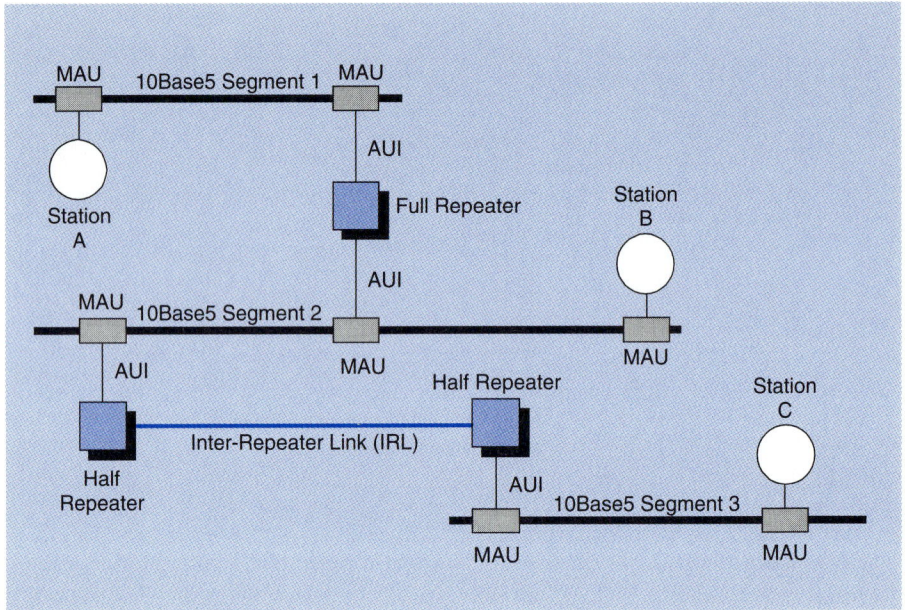

FIGURE F.2

Segments and Repeaters in 10Base5

Cable trunk segments are limited to 500 meters. (Hence, the "5" in 10Base5.) Repeaters can connect segments. Two stations cannot be more than four repeaters or five segments apart. A full repeater attaches to two segments. A half repeater only attaches to one segment. An interrepeater link (IRL) connects half repeaters.

is a maximum of 100 MAUs per segment. Few segments have anywhere near this many MAUs, however.

Figure F.2 also shows that a LAN may have multiple segments. Electronic devices called **repeaters** connect the segments. As the name suggests, repeaters merely repeat everything they hear. If a repeater hears a frame on one of the segments to which it is attached, it copies it without delay to other segments to which it is attached. This is a simple process, so repeaters are very small (about the size of a hard cover book) and very inexpensive.

Figure F.2 shows two types of repeaters. One type, a **full repeater,** connects physically to the two (or more) segments it serves. It has an AUI (drop cable) to each segment. These AUIs terminate in MAUs (transceivers).

The other type of repeater, the **half repeater,** attaches to a *single* segment via an AUI and MAU. It connects to a similar single-network repeater via a point-to-point inter-repeater link (IRL). This may be a section of 10Base5 cabling with no stations on it, or it can be a 10Base-F fiber optic link (see Chapter 4).

Multisegment LANs

In 10Base5, then, the term **LAN** means a group of segments connected by repeaters. The 802.3 terminology also calls this a **collision domain** because if two stations on the LAN transmit at the same time, their signals will collide.

There are two core rules for linking segments with repeaters. The first is that, despite the existence of multiple segments, there is still only a single path between any two stations. If you look at the various stations in Figure F.2, you will see that there is only one possible route between each station pair. In particular, you must avoid loops.

The second core rule is that no two stations may be separated by more than four repeaters. This means that they can be, at most, five segments apart. Because each segment has a maximum length of 500 meters, two stations can be at most 2,500 meters (8,200 feet) apart. In real buildings, two stations can never be quite so far apart.

In the standard, there is actually another limitation. In the five segments, at least two of the segments must be interrepeater links without stations. This is sometimes violated in practice.

Another limitation is that a 10Base5 network is limited to 1,024 stations. However, any network with 1,024 stations would be hopelessly congested. We saw in Chapter 5 that 10 Mbps CSMA/CD networks become congested with 200 or 300 stations.

As a result of distance limitations, 10Base5 LANs are only for small to medium-size buildings. They are not for large buildings, university campuses, or office parks. Still, with a span of 2,500 meters, 10Base5 networks span much larger distances than 10Base-T networks, with spans of only 500 meters. Quite simply, telephone wiring cannot carry signals as far as thick coaxial cable because of higher attenuation.

802.3 Physical-Layer Standard 10Base2

While 10Base5 worked and worked well, it was expensive. The 802.3 subcommittee's second action was to create a standard for a less expensive cabling system. This was **10Base2,** which is also known popularly as *Cheapernet* or *Thinnet*. Like 10Base5, it carries data at 10 Mbps, using baseband signaling. However, its maximum segment length is only 185 meters (607 feet), which is rounded off to "2" in its name. In addition, it can only have 30 stations on a single segment.

Thinner Cable

10Base2 is attractive because it uses RG-58 A/U and RG-58 C/U cable instead of the traditional Ethernet cable. As the standard's popular names suggest, this cable is thinner than the cable used in 10Base5. The 10Base5 cable is 10 mm in diameter. The 10Base2 cable, in contrast, is only half as thick. This thinner cable is cheaper to buy. It is also cheaper to lay because it is much more flexible.

The T-Connector

Another reason why 10Base2 is less expensive than 10Base2 is that Thinnet does not have a separate trunk cable and AUI drop cables. Figure F.3 shows that 10Base2 uses a simpler arrangement. The cable at each station attaches via a simple T-connector. The stem of the T screws into the NIC, via a BNC plug on the NIC. The other connectors on the T-connector are for 10Base2 cable runs to the two adjacent stations. Connections and disconnections can be made in seconds.

In 10Base2, a segment consists of a string (daisy chain) of stations, as shown in Figure F.3. A segment may have up to 30 stations. On the last station, the T-connector not leading to another station must have a terminator installed on the open connector.

Cabling Subtleties

There are only two things that are tricky about 10Base2 cabling. The first is that there must be at least half a meter of cabling between adjacent stations. The second is that the stations at each end of the daisy chain must have a special terminator attached to the T-connections.

Distance in 10Base2

As in the case of 10Base5, you can have up to five 10Base2 segments (four repeaters) between the farthest stations. This gives a maximum span of 925 meters (3,035 feet). As in 10Base5, the standard requires at least two of the segments to be interrepeater links if you are using the maximum five segments.

FIGURE F.3

10Base2 T-Connection

In 10Base2, there are no AUI drop cables. The cable attaches to the station via a T-connector. The run of cable between adjacent pairs of stations, however, is not called a segment. A segment is a daisy chain of stations connected by Thinnet. A segment can have up to 30 stations.

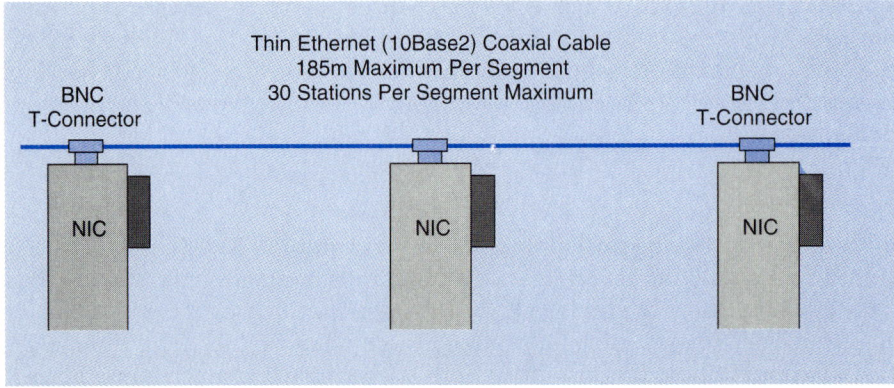

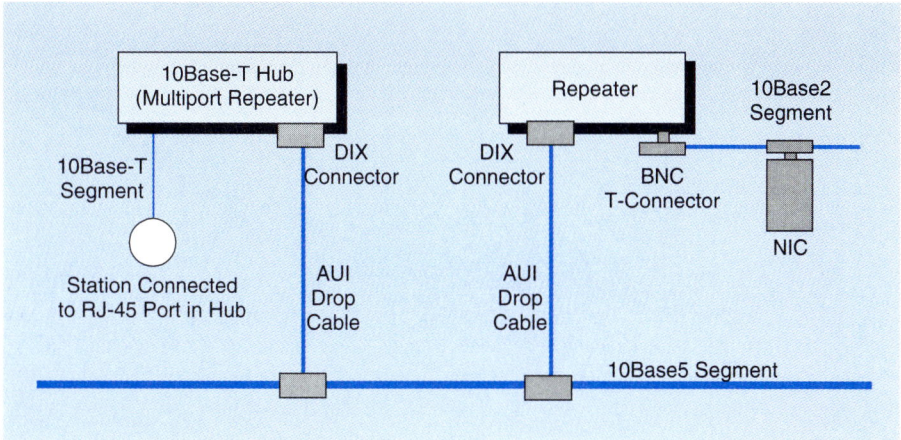

FIGURE F.4
802.3 Network with Multiple Physical-Layer Standards

This network, like all 802.3 networks, consists of multiple segments connected by repeaters. Here 10Base5 and FOIRL are used for longer backbone runs. Runs to the desktop are handled by 10Base2 and 10Base-T cabling. Mixing physical-layer standards causes no problems as long as you respect distance limitations for each segment and use the proper repeaters.

Mixed 802.3 Networks

So far, we have discussed networks that use a single 802.3 physical layer standard. But we would like to mix them together in a network. Figure F.4 shows how to do so.

For longer runs, such a network uses 10Base5 cabling or a 10Base-F inter-repeater link (IRL). This provides a high-speed backbone that can run between offices in a building.

For runs to individual stations, such as network mixes both 10Base-T and 10Base2. There is no reason to mix the two in practice because their distances are not too dissimilar. However, if a firm still has older 10Base2 cabling in place, it can keep older cabling there while replacing some of its 10Base2 cabling with 10Base-T cabling.

Mixing segments with different physical-layer standards requires that you use the proper repeater. Figure F.5 shows a 10Base-T wiring hub, which is properly called a *multiport repeater* in the standard. In addition to its usual 10Base-T RJ-45 connectors, it has a DIX connector, so that it can attach to a 10Base5 segment. Other multiport repeaters come with a BNC connector for connection to 10Base2 segments.

Some repeaters have only two or three ports. These can be any combination of 10Base-T, 10Base5, 10Base2, or 10Base-F. You simply buy the proper multiport repeater for your purpose.

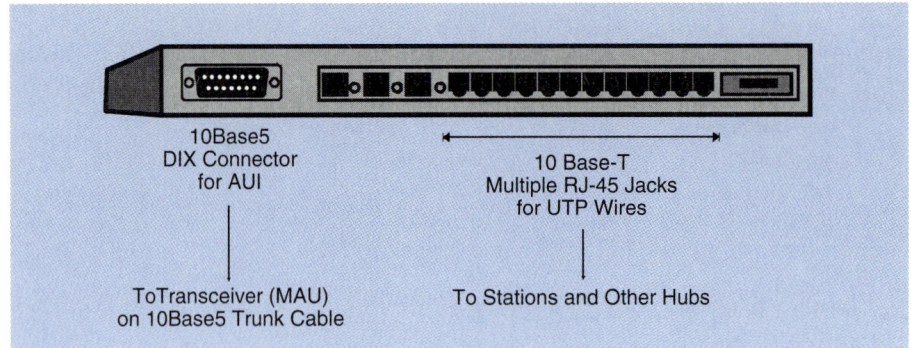

FIGURE F.5

Ethernet Multiport Repeater with 10Base-T and 10Base5 Ports

Note that we have only talked about the physical layer. As Chapter 4 noted, all 802.3 standards use the same MAC-layer standard. Mixing physical-layer standards has absolutely no impact on MAC-layer operation.

The 802.5 Token-Ring Standard

Chapter 5 looked at **token-ring networks (TRNs).** One important point about the TRN standard is that it is strongly driven by IBM. In fact, the actual 802.5 documentation is rather simplified, and most vendors follow IBM TRN standards, which are sometimes incompatible with 802.5 standards for token-ring networks.

Differential Manchester Encoding

In Chapter 4, we saw that to send a signal, you must encode the zeros and ones as combinations of voltage levels. The method that TRN networks use is called *Differential Manchester Encoding*.

Manchester Encoding ■

In Chapter 4, we saw that 802.3 10Base-T uses *Manchester encoding* (as do 10Base5 and 10Base2). In Manchester encoding, there is always a transition in the middle of the bit time. This ensures that the sender will not transmit a long series of bits that all keep the signal high or low. A long voltage transmission without change will not keep the receiver's clock synchronized with that of the sender.

To transmit a 1, Manchester encoding sends a low voltage for half the bit time, followed by a high voltage for the second half of the bit time. A 0, in turn, is a high voltage for the first half of the bit time, followed by a low voltage for the second half. A good way to remember this is that a high ending is a 1 and a low ending is a zero.

There are two possible line changes for each bit transmission. So operating at 10 Mbps, 10Base-T is a 20 Mbaud transmission system.

Differential Manchester Encoding for a 1 ∎

In contrast to simple Manchester encoding, the way that **Differential Manchester Encoding** represents a bit depends on whether the previous bit time ended high or low. To transmit a 1, there is no transition at the start of the bit time. To prevent a long series of 1s from ruining synchronization, there is a transition in the middle of the bit time.

Suppose the previous bit ended high. Then to transmit a 1, you would keep the signal high for the first half of the bit period. Then you would make it low for the second half. The ending state will be different than the starting state.

Or suppose the previous bit ended low. Then to transmit a 1, you would keep the line low for the first half of the bit period and make it high for the second half. The ending state, then, is always different from the starting state when you send a 1.

Differential Manchester Encoding for a 0 ∎

To transmit a 0, in contrast, you *change* the voltage level at the start of the bit period. In other words, you tell a 1 from a 0 by whether or not the line changes at the beginning of the bit period.

Like a 1, a 0 always makes a transition in the middle of the bit period. In this way, you are ensured transitions that keep the receiver synchronized with the sender are ensured. Note that in a 0, there are always two changes. This means that the ending state is always the same as the starting state.

Differential Manchester Encoding for J and K ∎

In addition to representing 1s and 0s, differential Manchester encoding defines two special characters, J and K. Actually, they are called code violations, but as we will see, they are used for signaling.

A J is a 1 without the transition in the middle. Whatever the line state was at the end of the last bit time, a J will continue that state for the entire bit.

A K, in turn, is a 0 without the transition in the middle. Whatever the line state was at the end of the last bit time, a K will change the state at the start of the bit and will hold that changed level for the entire bit. The line state at the end will be the opposite of the line state at the beginning.

Example

Figure F.6 illustrates how 1, 0, J, and K appear on a line. The situation looks more complicated than Manchester encoding, but most of this complexity comes from viewing representations in terms of line changes instead of line values.

Event	Bit to be sent	Initial state of the line	Change at start?	Change at end?	Final state of the line	Remarks
1					High	Starting condition—line high
2	1	High	No	Yes	Low	1 keeps state constant at start, changes in middle. Ends in opposite state.
3	1	Low	No	Yes	High	
4	1	High	No	Yes	Low	
5	1	Low	No	Yes	High	
6	0	High	Yes	Yes	High	0 changes state at start and in middle. Ends in same state.
7	0	High	Yes	Yes	High	
8	0	High	Yes	Yes	High	
9	1	High	No	Yes	Low	
10	0	Low	Yes	Yes	Low	
11	1	Low	No	Yes	High	
12	0	High	Yes	Yes	High	
13	1	High	No	Yes	Low	
14	J	Low	No	No	Low	J keeps state constant at start and in middle. Ends in same state.
15	0	Low	Yes	Yes	Low	
16	J	Low	No	No	Low	
17	1	Low	No	Yes	High	
18	J	High	No	No	High	
19	1	High	No	Yes	Low	
20	K	Low	Yes	No	High	K changes state at start but not in middle. Ends in opposite state.
21	0	High	Yes	Yes	High	
22	K	High	Yes	No	Low	
23*	0	Low	Yes	Yes	Low	
24*	K	Low	Yes	No	High	
25	1					Complete this row.
26	0					Complete this row.
27	J					Complete this row.
28	K					Complete this row.

FIGURE F.6

Line States in Differential Manchester Encoding

For instance, the initial state of the line (Event 1) is high. Event 2 is the transmission of a 1. As the remark notes in condensed form, a 1 keeps line state constant (high) at the start, then changes it (to low) in the middle. At the end of Event 2, the line is in a low state.

The 802.5 MAC Layer Frame: Token Frame

Chapter 4 discussed the 802.3 MAC-layer frame in some detail. We will now do so for the IBM Token-Ring Network MAC-layer frame. We will begin with the token frame itself. Figure F.7 shows that a token frame has three fields: start frame delimiter, access control, and end frame delimiter.

Start Frame Delimiter

The **start frame delimiter** acts like the preamble in 802.3 MAC-layer frames, although it is much shorter. It is only a single byte long. It has the following pattern of bits:

    J K 0 J K 0 0 0

This frame, with two J violations and two K violations in specific locations, is highly unlikely to occur in erroneous transmissions. It uniquely marks the start of the frame.

Access Control

The **access control** field is the heart of the token. This is the field that allows stations to know if the token is free and, if so, whether their priority is high enough for them to take it and transmit. Its eight bits are encoded in the following way:

    P P P T M R R R

PRIORITY. The three P bits give the priority of the token. This gives eight levels, from 0 through 7. Unless a station has priority at least as high as these bits indicate, it must let the token pass, even if it has something to send.

TOKEN BIT. The T bit tells whether this frame is a token or a full frame. In this case, it is a token, so its value is set equal to 1. If it were a full frame, it would be 0.

MONITOR BIT. The M bit is set to 0 by the transmitting station. A special station called the *active monitor* (discussed later) sets this bit to 1. If it sees a 1 bit in an

FIGURE F.7
Token Frame in 802.5

The start frame delimiter acts like an 802.3 MAC-layer frame preamble. The access control field handles the conditions under which a station can take the token. The end frame delimiter terminates the token frame and allows any station to report a transmission error.

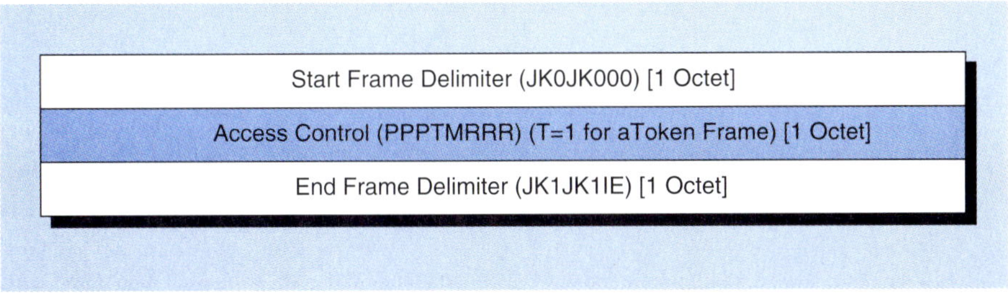

Start Frame Delimiter (JK0JK000) [1 Octet]
Access Control (PPPTMRRR) (T=1 for aToken Frame) [1 Octet]
End Frame Delimiter (JK1JK1IE) [1 Octet]

arriving frame, the active monitor knows that the frame has not been removed by the transmitting station when the frame completed a full run around the ring. The active monitor then removes the frame.

THE TOKEN RESERVATION BIT. The three R bits are the token reservation bits. These come into play when full frames are transmitted, as we will see later.

The End Frame Delimiter ■

The **end frame delimiter** finishes the token. This one-octet frame has the following pattern:

```
J K 1 J K 1 I E
```

Here, again, we start with an unusual pattern of bits containing uncommon J and K violations. But while the start frame delimiter begins with "JK0JK0", the end frame delimiter begins with "JK1JK1".

We will discuss the I and E bits next, in the context of full frames.

The 802.5 MAC Layer Frame: Full Frame

If a station has high enough priority, it takes control of the token. It takes its information and wraps it inside the token, forming a full token-ring network frame. Figure F.8 shows a full frame.

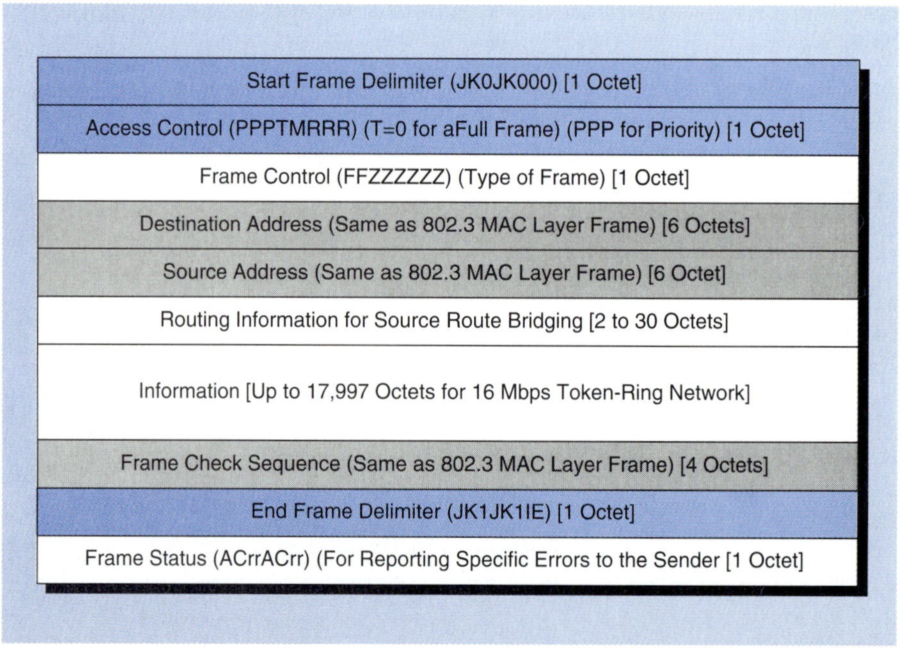

FIGURE F.8

Full 802.5 Token-Ring Network Frame

The 802.5 Token-Ring Network frame has more fields than the 802.3 MAC-layer frame, primarily because token passing is a much more complex protocol than CSMA/CD.

Start Frame Delimiter (JK0JK000) [1 Octet]
Access Control (PPPTMRRR) (T=0 for aFull Frame) (PPP for Priority) [1 Octet]
Frame Control (FFZZZZZZ) (Type of Frame) [1 Octet]
Destination Address (Same as 802.3 MAC Layer Frame) [6 Octets]
Source Address (Same as 802.3 MAC Layer Frame) [6 Octet]
Routing Information for Source Route Bridging [2 to 30 Octets]
Information [Up to 17,997 Octets for 16 Mbps Token-Ring Network]
Frame Check Sequence (Same as 802.3 MAC Layer Frame) [4 Octets]
End Frame Delimiter (JK1JK1IE) [1 Octet]
Frame Status (ACrrACrr) (For Reporting Specific Errors to the Sender [1 Octet]

Access Control

When the station grabs the token, it changes the T bit in the **access control** field from 1 to 0. This signals receivers that they are looking at a full frame instead of just a token.

In addition, when the transmitting station transmits, it changes the three priority reservation bits (RRR) to 000, the lowest priority. If a station wishes to transmit, it looks at the reservation bits. If its priority is higher, it places its priority level in the reservation bits. In this way, by the time the frame has gotten all the way around to the sender, the RRR bits contain the highest-priority level of stations wishing to transmit. When the sending station releases the token, it places this priority level in the main priority bits (PPP) of the token it releases.

Frame Control

The **frame control** field tells what kind of frame this is. It has the following pattern in its eight bits:

```
F F Z Z Z Z Z Z
```

The first two bits tell the stations what type of frame this is. If the F bits are 01, then this frame contains logical link control-layer data. In other, words, its information field contains an 802.2 frame (discussed in Chapter 4).

In contrast, if the F bits are 00, then this frame is a MAC-layer frame only. It is either a token or it is a supervisory frame.

The six remaining Z bits specify particular types of LLC- and MAC-layer frames. We will see some MAC-layer control frames later, when we discuss errors and error recovery.

Destination and Source Addresses

Fortunately, the IBM Token-Ring Network frame (and the official 802.5 MAC-layer frame) uses the same addressing scheme as the 802.3 MAC-layer frame, which we discussed in Chapter 4.

Routing Information

The **routing information** field holds information for routing the frame through a series of source routing bridges. Source routing is the bridging standard created by IBM and allowed by the 802 committee only for 802.5 LANs. This field can be from two to thirty octets in length.

Information

The **information field** holds the data to be delivered. The information field can be very large. In 16 Mbps token-ring networks, frames can be 17,997 octets long. Even in 4 Mbps TRNs, frames can be 4,501 octets.

Frame Check Sequence ■

The **frame check sequence** field is a 32-bit field for error checking. It is the same as the frame check sequence field in 802.3 MAC-layer frames.

End Frame Delimiter ■

In 802.3 MAC-layer frames, the frame check sequence field is the last field. In contrast, TRN frames add two final fields. One of these is the **end frame delimiter.** This one-octet field, as we saw earlier, begins with the unusual pattern "JK1JK1". It ends with two other bits, the I bit and the E bit.

The I bit is used to tell the receiver if this is the last frame in a series. It is set to 1 if it is the last frame. If it is not, the I bit is set to 0.

The E frame is for reporting errors. The transmitting station sets this bit to 0. If *any* station along the ring detects a J or K violation or some other transmission error, it sets this bit to 1. When the transmitting station receives the frame back, it looks at the E bit to check for errors.

Frame Status Field ■

The very last field is the **frame status field.** This octet allows the receiving station to tell the sending station whether or not it has received the frame. The form of this field is the following:

    A   C   r   r   A   C   r   r

Here the r bits are reserved for future use (an unlikely situation at this late date). They are normally set to 0.

The transmitting station sets the A bit to 0. If the receiving station recognizes its address, it resets this address bit to 1. When the frame returns to the sender, the A bits tell if the receiver has seen the frame.

It is possible that a station will recognize its address yet not be able to copy the frame into its NIC's memory. This is the purpose of the C (copy) bit. The sending station sets this bit to 0. The receiver resets it to 1 if it makes a successful copy.

The frame status field comes after the frame check sequence field, so there is no way for the receiver to check for errors in the frame status field. To compensate, the field has two A bits and two C bits. The transmitting station assumes a failure unless it sees 1s on both A bits and on both C bits.

Error Handling

In many ways, a token-ring network is an accident waiting to happen. As we will see, there are two major ways in which a simple failure can disable the network.

While these dangers are very real, the standard contains several ways of minimizing these risks. As a result, TRNs are not at all the fragile networks that they first seem to be.

Breaking the Ring ■

The most obvious problem is that frames have to go all the way around the ring. In addition, the stations do not simply watch the frames go by. They stop each frame, look at it bit by bit, and then regenerate it. As a result, if a single connection breaks, or if a single device fails, no station on the network can transmit.

Losing the Token ■

Another risk in TRNs is losing the token. We know that stations may not transmit until they have the token. What if the token is lost? Then no station may transmit. Again, communication will break down.

Wrapping the Ring Between Access Units ■

How can we prevent the loss of transmission if there is a break in the ring between access units? To allow recovery, 802.5 rings are really double rings, as shown in Figure F.9. Normally, all traffic passes through one of the two rings. The other ring goes unused.

If there is a break between access units, the ring is **wrapped.** As you can see from Figure F.10, the result is still a loop. If you trace the wrapped ring going in one direction, you will still come back to the starting point. This is the essence of a ring network.

Wrapping only works once. If there is a second failure, a wrapping will break the ring into two unconnected pieces. So wrapping is a "first aid" practice. The broken link should be reestablished as quickly as possible.

Within an Access Unit ■

Figure F.11 looks more closely at the access unit. Note the way that devices are connected. Essentially, there is still a ring. The ring comes into the ac-

FIGURE F.9
Double Ring in an 802.5 Token-Ring Network

There are really two rings connecting the access units. Normally, one is used and the other is left idle.

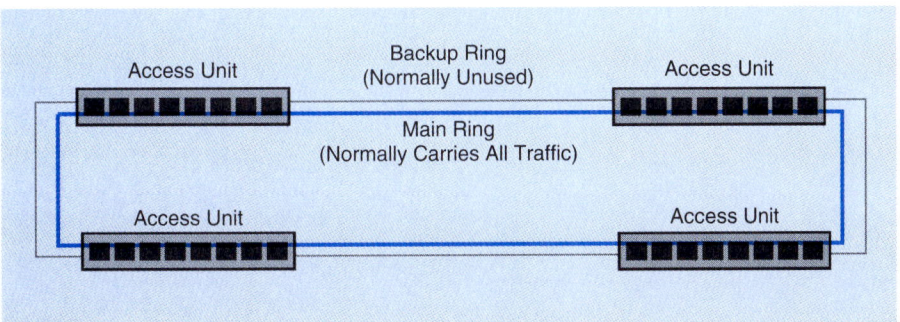

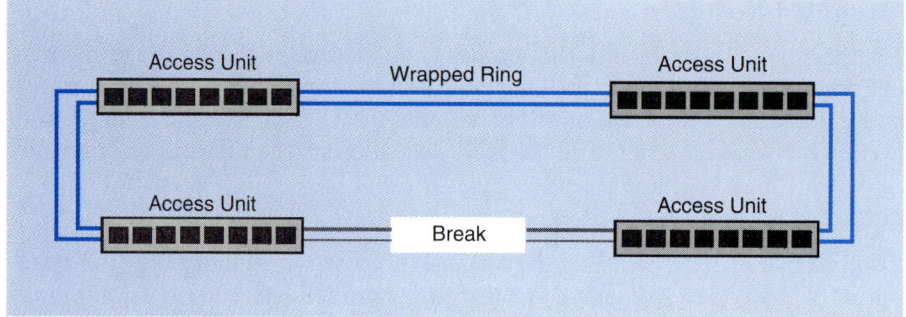

FIGURE F.10

802.5 Token-Ring Network Wrapped Between Access Units

If there is a break in the main ring between two access units, the two access units at the ends of the break can wrap the ring, using the backup ring. Note that there is still a closed loop connecting all the access units, so the ring is preserved. Note too, however, that if there is a second break between access units, the network will be broken into two subrings. Wrapping is a temporary expedient until the break can be fixed.

FIGURE F.11

Connections Within an 802.5 Access Unit

Within an access unit, the stations maintain the ring topology. If you look at the line coming in, you will see that it passes through all of the active stations in a continuous ring. Even the connection to each station is part of the continuous line. Whenever a station powers up, it is automatically connected into the continuous line. When a station powers down, it is automatically bypassed, preserving the ring. With each connection and disconnection, a few frames are lost. These few errors, however, are caught at higher layers.

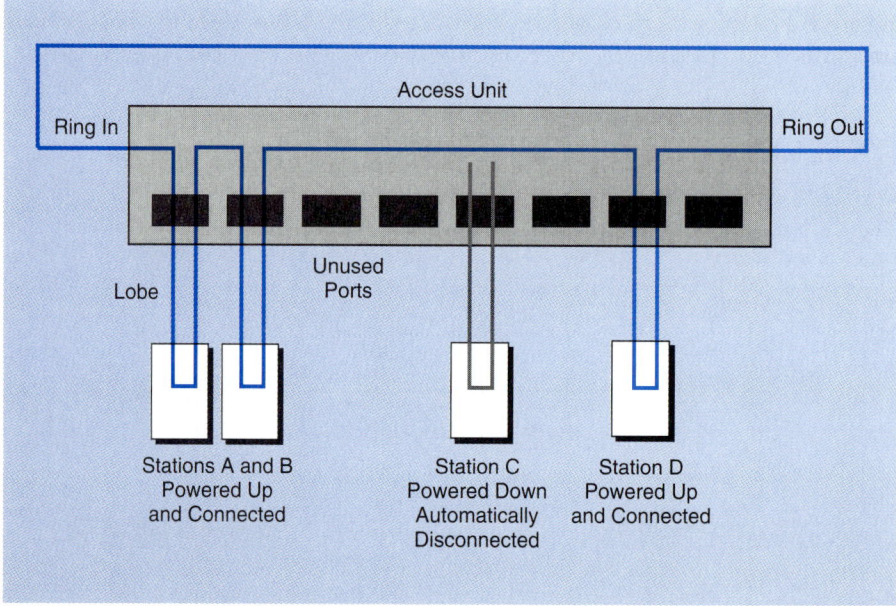

cess unit. It then goes down to the first station and back up to the access unit in an unbroken connection. If you trace the ring from its entry into the access unit to its exit, you will find a single complete path. The ring is unbroken.

The ring extends to the stations themselves. The two lines between the access unit and the station are part of the overall loop of the network. This is why the link to the station is called a **lobe.**

What if one of the stations turns off its power? Figure F.11 shows that when this happens, the ring remains unbroken. Look at station D. It has powered off. The access unit automatically bypasses the device.

Of course, there are a few milliseconds between the time when a device powers off and the access unit redirects the ring to pass through the connection. During this time, packets will be lost. But these errors will be discovered at higher layers, and the packets will be resent.

When you power on a station, there is also a brief interruption as the access unit connects the station to the ring. Again, a few frames will be lost.

The mechanism is remarkably simple. For each port, the access unit has a small solenoid. When the station powers up, it sends power to the solenoid. The solenoid closes, connecting the station's lobe to the main ring. When a station powers down, in turn, even accidentally, the solenoid opens, automatically disconnecting the station from the ring.

Handling Lost Tokens and Other Problems ∎

Handling lost tokens, in contrast, requires a considerable amount of complexity. One station must be designated as the **active monitor.** We saw earlier that the active monitor constantly examines frames that have gone around the ring more than once. It also removes garbled frames with obvious errors.

This station keeps watching for the token. It times the reappearance of the token each time the token passes the station. If that time is too long, it generates a new token. This station also weeds out duplicate tokens.

The active monitor even detects breaks in the ring [Tanenbaum, 1989]. When *any* station fails to hear traffic for a long period of time, it issues a beacon control frame with a frame control field of 00000010. By listening to these beacon frames, the active monitor can estimate where the break has occurred. This can allow automatic wrapping of the ring.

What if the active monitor fails? This failure will be addressed automatically [Tanenbaum, 1989]. Periodically, the active monitor issues an **active monitor present** control frame. Its frame control field is 00000011. If any station notices an absence of this frame for a long period of time, it may issue a **claim token** control frame, whose frame control field is 0000010. If this token gets all the way around the ring, the claiming station assumes the duties of the active monitor.

CORE REVIEW QUESTIONS

1. Name the physical-layer options a company has for Ethernet networks running at 10 Mbps. Can you mix these physical-layer technologies in a single LAN? If possible, why would you wish to do so?
2. Compare and contrast Manchester encoding and differential Manchester encoding.
3. Briefly explain the *main* purpose of the following 802.5 fields: start frame delimiter, access control, frame control, destination address, source address, routing information, information, frame check sequence, end frame delimiter, and frame status. Why is the 802.5 MAC layer frame so much more complex than the 802.3 MAC layer frame?
4. Explain the major risks associated with 802.5 Token-Ring Network and how the standard handles each of them.

DETAILED REVIEW QUESTIONS

1. Explain the physical connections in 10Base5. What are the distance limits of 10Base5?
2. Explain the physical connections in 10Base2. What are the distance limits of 10Base2?
3. In 10 Mbps Ethernet LANs, what are the limits for the number of segments and repeaters linking the two farthest stations? Explain what interrepeater links (IRLs) are and how they relate to maximum distance spans. What is a 10Base-F?
4. Explain how differential Manchester encoding produces 1, 0, J, and K.
5. Why does the 802.5 start frame delimiter have the pattern JK0JK000?
6. Explain the PPP, T, M, and RRR in the 802.5 access control field.
7. In the 802.5 end frame delimiter, JK1JK1IE, what are the purposes of I and E?
8. In the 802.5 frame control field, what types of frames can a station specify in the FF bits?
9. How do the address fields and the frame control sequence fields differ in 802.3 and 802.5, if they differ at all?
10. What is the purpose of the 802.5 routing information field?
11. Describe the ending fields in 802.5.
12. Explain why a break in the ring between access units is dangerous and how 802.5 reduces the danger. Explain why a station suddenly turning off is dangerous and how 802.5 handles it.

13. What is an active monitor? What happens if it fails? How does 802.5 handle this danger?

THOUGHT QUESTIONS

1. Now that you know more about 802.5, why do you think its market penetration has been lower than 802.3? Why do you think that many companies feel that 802.5 is a much better choice anyway?
2. Complete Figure F.6 by filling in the last four lines.

PROJECT

1. You have a building with three office areas in a row. There is a fourth office 150 meters away. Each office is about 40 meters on a side and should have its own 10Base-T hub. Design a LAN using 10Base-T wiring as much as possible.

For online exercises, please visit this book's website at: http://www.prenhall.com/panko

REFERENCES

IBM (1992). *Token-Ring Network: Introduction and Planning Guide*. (Document GA27-3677-05). Research Triangle, North Carolina.

TANENBAUM, A. S. (1988). *Computer Networks*. Englewood Cliffs, NJ: Prentice-Hall.

MODULE G

Telephone Service

POTS

Its official name is **PSTN—Public Switched Telephone Network**—but even telephone professionals call it *POTS* (*plain old telephone service*). Telephony gets little respect and causes even less excitement. It is simply there, always available, like a comfortable old pair of jeans.

However, corporations must take the telephone seriously. Telephony is extremely important in corporate life, and it is also important to data communications professionals.

- While data communications is flashy and new, companies spend far more today on telephony and will for years to come. People talk a lot on the telephone. Managers, for instance, spend 5 to 10 percent of the day on the phone [Panko, 1992].

- In addition, voice transmission generates a great deal of bandwidth. While data tend to come in short bursts, people fill up a telephone channel constantly when they talk.

- Furthermore, the job of managing telephony is exploding in complexity. Technology is creating many new products and a bewildering array of options in each product category.
- Finally, *deregulation*—the relaxation of protectionist regulation—has brought the competition needed to get new products to users.

In one critical respect, even basic voice service has changed profoundly in recent years. In real dollars, interstate long-distance prices fell by 60 percent between 1977 and 1992 [Standard & Poor's, 1992], and they are still falling. International calling charges have fallen even more dramatically.

Because of falling prices, use is growing several times faster than the GNP. We now make even international calls without giving much thought to cost. Compared to even ten years ago, telephony today has a much wider role in organizations, and this role will grow in the future.

THE TRADITIONAL TELEPHONE SYSTEM

A Telephone Call

Figure G.1 shows what happens when you place a telephone call. The call begins in your **customer premises**—your home or office.

Next the signal travels over a transmission line to the first switch of the telephone company. This transmission line is called the **local loop.** It usually consists of one or two pairs of copper wire. In the future, the local loop is likely to switch (at least in part) to optical fiber. Fiber will allow much higher speeds.

The call may travel through several more **switching offices.** The lines between switching offices are called **trunk lines.** Each trunk line usually carries 24 or more voice calls. Some carry hundreds or thousands of simultaneous calls.

At the other end, the final switching office connects you with the party you are calling. Your connection goes over the local loop to the customer premises of the party you are calling.

Carriers

Carriers in the United States

Chapter 7 discusses the transmission carriers who provide services in the traditional telephone system in the United States. That chapter focused on the situation in the United States. Some major points were:

- There is a local service area called the **local access and transport area (LATA).** LATAs are fairly large. There are only 161 in the United States. Even the largest states have fewer than two dozen LATAs.

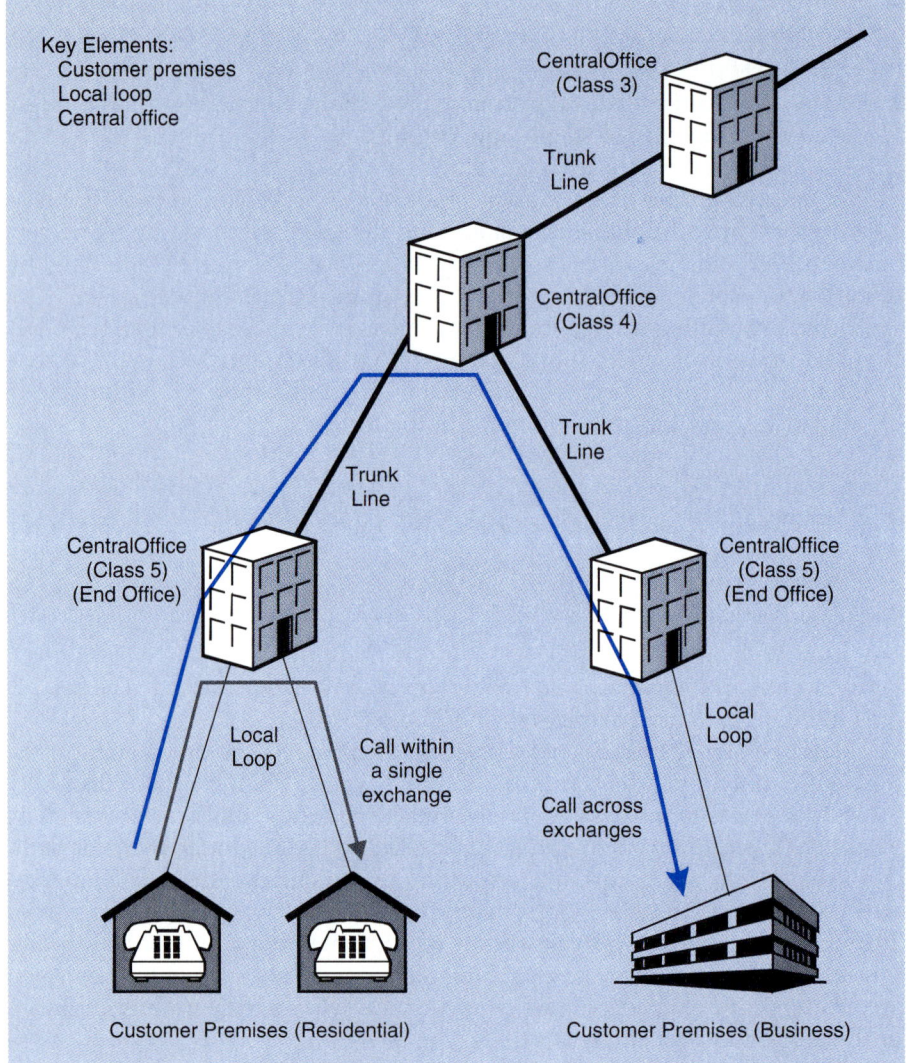

FIGURE G.1

Public Switched Telephone Network (PSTN)

A call goes from your customer premises over the local loop to a switching office. There it may travel to several other switches, over trunk lines connecting the switches. Finally, it goes over the local loop to the customer premises of the party you are calling.

- Traditionally, a LATA was served by a monopoly telephone carrier called the **local exchange carrier (LEC).** It controlled the local loop and all transmissions beyond the local loop.
- Now other companies are beginning to compete with the LEC for service within the LATA. These new entries are called **competitive access providers (CAPs).** These include cellular telephone companies, cable television companies, and companies going head-to-head with the LEC in providing comprehensive transmission service.

- Each LATA has at least one **point of presence (POP).** Located on the premises of the LEC, the POP provides a linking point between the LEC and CAPs, so that the customers of one can call the customers of the other.
- The POP also connects the local carriers to carriers that carry the signal out of the LATA to another LATA or to another country.
- There are two types of carriers that transport signals beyond the LATA. **Interexchange carriers (IXCs)** are domestic companies, meaning that they transport signals only within the United States. They transport signals from one LATA to another. Note the "I" in IXC does *not* stand for *international*.
- For transmission to another country, the POP connects local carriers to **international common carriers (ICCs),** which transport the signal across the U.S. border to other countries.

Carriers in Other Parts of the World

Most other countries do not have a two-tier structure for carriers. Historically, there was a monopoly telephone company that provided all domestic services. This was the **Public Telephone and Telegraph (PTT)** authority. Sometimes there was a second company for international service. For instance, in Japan, the NTT was the domestic PTT. KDD handled international calls.

The **deregulation** (relaxation of competition rules) that brought competition in the form of CAPs and many IXCs in the United States is also strong in many other countries. Most countries now allow competing carriers, although most limit what competitors can offer more so than does the United States.

Regulation

Regulation is relatively complex in telephony, especially in international calling and especially in the United States.

Regulation in the United States

In the United States, there are two tiers of regulation, reflecting the country's political system. At the national level, there is the **Federal Communications Commission (FCC).** The FCC regulates interstate transmission. It also regulates the entire telephone system, for instance, by setting technical standards and basic service conditions that must be met within states.

At the state level, there are **public utilities commissions (PUCs)** that monitor the day-to-day work of telephone companies operating within the state. The PUCs set many rates and set limitations on services.

Historically, the FCC has been more aggressive in deregulating telephone companies than the PUCs. Today, however, PUC resistance to deregulation is crumbling in many states, and competition is increasing at the local level.

Breaking Up Ma Bell

In the United States, there never was a national telephone monopoly. However, **AT&T,** sometimes known as **Ma Bell,** came very close to having a monopoly. It controlled most local telephone companies, and it had a de facto monopoly on long-distance transmission.

In 1984, an antitrust suit by the Justice Department broke up AT&T. One resulting company, which maintained the name AT&T, kept long-distance service and most equipment manufacturing capacity. Local telephone companies, in turn, were divided among seven **Regional Bell Operating Companies (RBOCs).** The seven RBOCs are often called the "Baby Bells." The agreement signed with the Justice Department kept AT&T out of the local markets (intra-LATA service). It also kept the Baby Bells out of long-distance telephony beyond their borders (inter-LATA service).

The next year, for competitive reasons, AT&T voluntarily decided to further split itself into three companies: a carrier that retained the name AT&T, an equipment manufacturing company called Lucent, and a computer equipment and software company, NCR.

That same year, Congress passed the **Telecommunications Act of 1996,** which vastly increased competition in both intra-LATA and inter-LATA service. AT&T can now compete in local telephony, and the RBOCs can compete in long-distance service.

Regulation in Other Countries

In most other countries, there was a national PTT to provide service. There was also a government **Ministry of Telecommunications** to oversee the operation of the PTT. Most ministries are now implementing deregulation to varying degrees. This is bringing in new competitors to the PTT.

International Telephony

The ITU-T (see Chapter 1) sets standards for telephone equipment and transmission lines. It also standardizes many services. However, in most cases, international calling conditions between each pair of countries are largely unregulated. The governments of the two countries engage in bilateral (two-country) negotiations. As a result, long-distance calling rates reflect politics much more than they do cost and distance.

Trends in Regulation

Although countries vary in the degree to which they have deregulated telecommunications, most follow the same rank ordering in what aspects of telecommunications they have deregulated.

First, the most deregulated aspect of telecommunications is usually the *customer premises*. In many countries, firms can do anything they want on their own premises, as long as they obey the technical and legal rules for connecting to public networks.

Next usually comes *long-distance domestic service*. Most countries deregulate long-distance transmission several years before they deregulate local service.

Next usually comes *international calling*. Deregulation here requires bilateral cooperation between the two countries in allowing the other country's international common carriers (ICCs) into their respective markets.

Finally, the least deregulated aspect of telecommunications is almost always *local service*. Regulators often fear that if they allow open competition here, poorer people may be left without telephone service.

CUSTOMER PREMISES EQUIPMENT

Up to this point, we have looked at facilities owned by telephone carriers. Now we will return to the *customer premises*, that is, the building and grounds owned by the customer.

In the past, telephone companies owned all telephone equipment on the customer premises, including all copper wires and even telephone handsets. In most countries, regulators have turned this around completely. The customer now owns everything on the premises, unless it specifically chooses to lease equipment from a telephone company. In effect, companies operate their own internal telephone companies.

PBXs

Many large companies have internal switches called **private branch exchanges (PBX)**. A PBX is somewhat like having your own end office switch. In fact, this is the origin of the term *branch exchange*. (*Exchange* is another word for *switch*.) Figure G.2 illustrates customer premises wiring using a PBX.

In large buildings, a key consideration is where to leave space for the PBX, wiring, and other equipment. PBXs are about two-thirds of a meter deep, allowing them to fit into standard equipment racks. A small PBX that serves "only" a few hundred lines is about 2 meters tall and 1 meter wide. A large PBX that serves 10,000 to 70,000 lines typically is the same height but may be twice as wide. Companies also need room for the mandatory **termination equipment** that telephone carriers require you to place between the PBX and outgoing lines to protect the telephone system.

In addition, most companies have internal telephone operators. These operators need room for an office. It is common to put operators near the central equipment, but it is not necessary. The telephone manager and his or her office staff and technicians need room as well.

Wiring ■

If a PBX serves many users, the company needs to run hundreds or thousands of telephone wires from the PBX to individual telephones. As shown in Figure

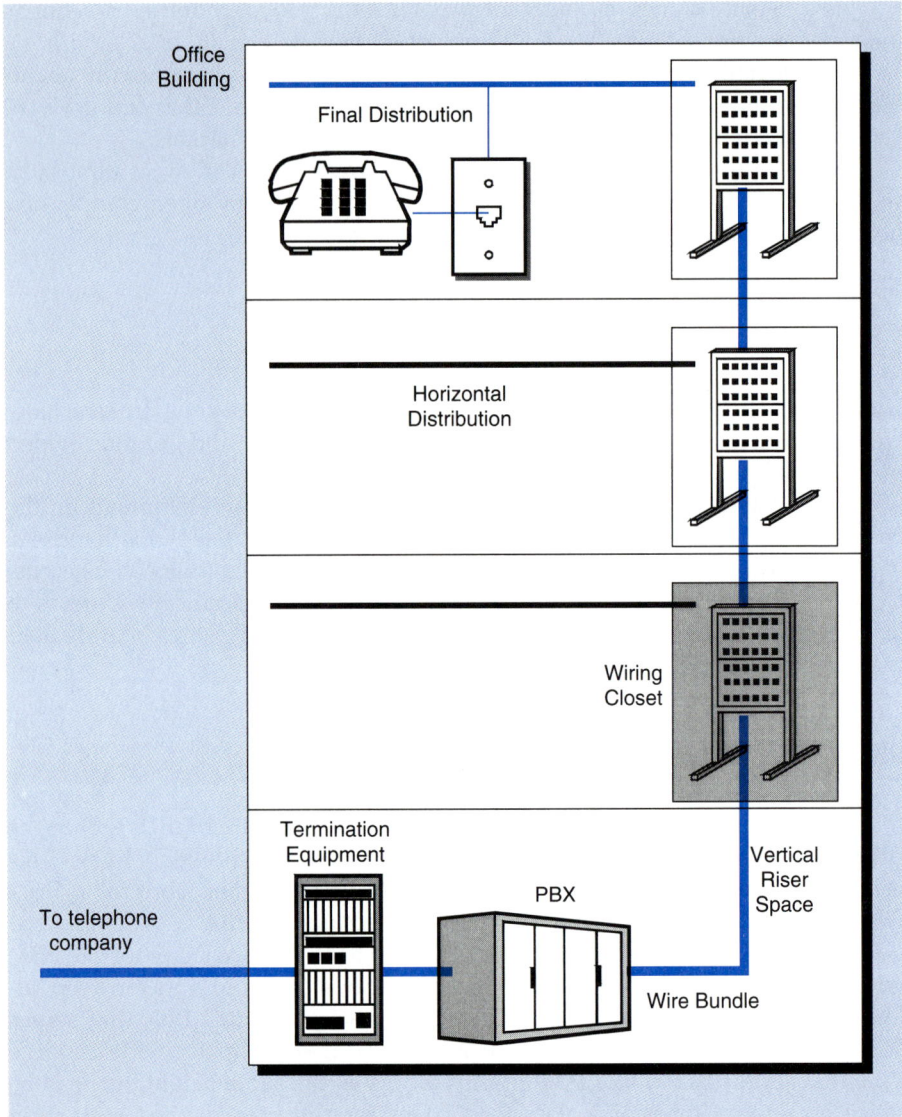

FIGURE G.2

Customer Premises Wiring with a PBX

A PBX is like your own telephone switch. Lines run from the telephone company into termination equipment, which prevents your firm from damaging the telephone company's system with dangerous currents. From the PBX, telephone wires run out through thick wire bundles that contain many pairs of telephone wire. Vertically, these bundles travel upward through the building via vertical riser spaces. At each floor, there is a wiring closet to hold equipment for splitting wire bundles. Some smaller bundles then fan out for horizontal distribution on each floor. In final distribution, a single bundle of two to eight telephone wires goes to a wall jack. Users plug their telephones into this wall jack.

G.2, the wires leave the PBX in thick **cable bundles** that contain hundreds of wire pairs.

The cable bundles move vertically through a building inside **riser** spaces that are 6 inches to 2 feet thick. For new buildings, it is important to leave ample room for riser spaces.

At each floor, the riser terminates in a **wiring closet.** This wiring closet is about the size of a hall closet in a home or apartment. Inside the wiring closet, the thick vertical bundle splits into smaller bundles. Some continue to travel upward to the next floor. Others run out horizontally on that floor of the building. Typically, they run through false ceilings. In other cases, they run inside walls. In improperly built buildings, there are neither false ceilings nor wall space. In that case, the horizontal distribution has to take place through relatively unsightly conduits that look like water pipes.

Finally, the last cable splits into a bundle destined for an individual telephone. As we saw in Chapter 4, this bundle usually contains four pairs (eight wires). It terminates in an RJ-45 connector.

Importance of Customer Premises Wiring ■

There are two reasons we have spent a considerable amount of time talking about customer premises wiring. The first is that the company must manage its own wiring system, so telephone professionals need to be familiar with customer premises wiring.

In addition, several LAN technologies use high-quality telephone wiring, including repeaters that fit into wiring closets. To understand building wiring is to understand many forms of LAN wiring.

PBX Networks

Many large companies have multiple sites. Figure G.3 shows that each site is likely to have a PBX.

The figure shows that the company can link its sites together with leased lines. For instance, a single T1 leased line can handle 24 conversations between two sites. Especially busy connections require T3 lines or even faster leased lines.

If a firm buys all of its PBXs from a single vendor, they will be able to function together like the central offices of the telephone company. A single system of extension numbers will be able to serve everyone in the firm, regardless of their sites. A PBX at some master site should even be able to do remote maintenance on the PBXs at other sites.

For communication outside the firm, the PBXs will have access lines to the local telephone company. The PBXs will also have links to various other transmission carriers other than the local telephone company. This allows the PBXs to select the least-cost line whenever someone makes an outgoing call.

Although the leased lines are likely to be put into place for voice service, they are also good for data transmission. Even if a firm does not completely integrate its voice and data transmission within sites, it usually sends both along the same leased lines between sites.

In effect, the company sets up its own private telephone system. Technically, this is known as a **private telephone network.** By creating private networks, companies can reduce their costs while providing a high level of services.

Many telephone carriers are now trying to get back the business they have lost because of private networks. They now offer **virtual private network service.** In this service, the firm appears to have an exclusive private telephone network. Calls between sites are inexpensive, and so are outside calls. However, the telephone company provides this service using its ordinary switches and trunk lines. Virtual private network service is a matter of pricing, not of technology.

User Services

Figure G.4 shows that, because digital PBXs are essentially computers, they allow vendors to differentiate their products by adding application software to provide a wide range of services.

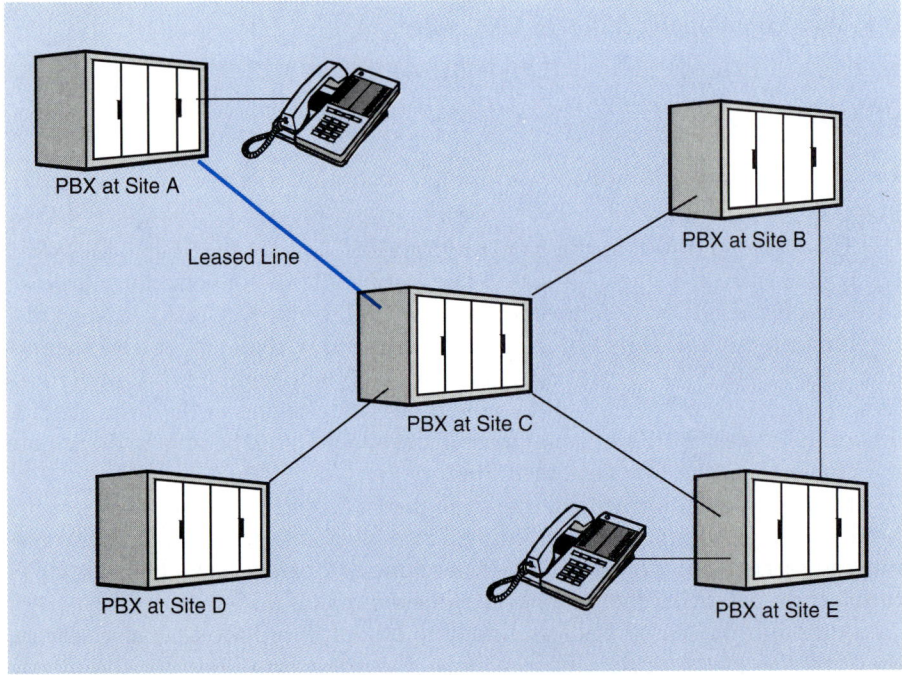

FIGURE G.3

Private Network using PBXs

If you have a PBX at each site, you can connect them with leased lines. They will then act together, so that any telephone at any site can call any other telephone at any other site. The caller may not even be aware that the other party is at another site.

For Users

Speed dialing	Dial a number with a one- or two-digit code.
Last number redial	Redials the last number dialed.
Display of called number	LCD display for number the caller has dialed. Allows caller to see a mistake.
Camp on	If line is busy, hit "camp on" and hang up. When other party is off the line, he or she will be called automatically.
Call waiting	If you are talking to someone, you will be beeped if someone else calls.
Hold	Put someone on hold until they can be talked to.
ANI	Automatic number identification. You can see the number of the party calling you.
Conferencing	Allows three people to speak together.
Call transfer	If you will be away from your desk, calls will be transferred to this number.
Call forwarding	Someone calls you. You connect them to someone else.
Voice mail	Callers can leave messages.

For Attendants

Operator assistance	In-house telephone operators can handle problems.
Automatic call distribution	When someone dials in, the call goes to a specific telephone without operator assistance.
Message center	Allows caller to leave a message with a live operator.
Paging	Operator can page someone anywhere in the building.
Nighttime call handling	Special functions for handling nighttime calls, such as forwarding control to a guard station.
Change requests	Can change extensions and other information from a console.

Voice Response

In interaction, user hits keys for input. System gives prerecorded voice responses. Allows menu operations to allow callers to get access to telephone numbers and other information. Reduces operator time.

For Management

Automatic route selection	Automatically selects the cheapest way of placing long-distance calls.
Call restriction	Prevents certain stations from placing outgoing or long-distance calls.
Call detail reporting	Provides detailed reports on charges by telephone and by department.

FIGURE G.4
PBX Services

CARRIER SERVICES AND PRICING

Having discussed both technology and regulation, we can finally turn to the kinds of transmission services that telecommunications staffs can offer their companies. Figure G.5 shows that users face a variety of transmission services and pricing options.

FIGURE G.5

Telephone Services and Pricing

> **Local Calling**
>   Flat rate
>   Message units
> **Toll Calls**
>   Intra-LATA
>   Inter-LATA
> **Toll Call Pricing**
>   Direct dialing
>   Anytime, anywhere
>   Highest rate
>   800/888 numbers
>   Free to calling party
>   Reduced rate per minute
>   WATS
>   Wide area telephone service
>   For calling out from a site
>   Reduced rate per minute
>   900 numbers
>   Calling party pays
>   Called party charges the calling party a price above transmission costs
> **Personal Telephone Number**
>   Number for a person, not an instrument
>   Can reach the person anywhere

Tariffs

In looking at services in regulated areas, it is important to look at the carrier's tariffs. A **tariff** is a contract with a regulatory agency that specifies what the carrier will offer. There is a separate tariff for each service the carrier offers.

Services

Tariffs specify *two* things. First, they specify the details of the *service* to be provided. It is important to read tariffs carefully because they sometimes leave out things that you assume would be included. They may not make these service features available at all or may do so only for an additional charge.

Prices

Second, tariffs specify how the service will be *priced*. This can be extremely complex. For instance, pricing may depend on usage volume, with larger customers getting price discounts. In addition, there are likely to be initial charges for activation and installation, flat monthly charges, and volume-dependent charges.

Tariffs from competing carriers may have very different pricing structures, making them difficult to compare. You almost always must be able to estimate traffic volume to make comparisons.

Nontariffed Services ▪

Under deregulation, a growing number of services are becoming nontariffed. This allows carriers to compete with one another by negotiating services and pricing with individual customers.

Basic Voice Services

The most important telephone service, of course, is its primary one: allowing two people to talk together. While you get roughly the same service whether you call the next-door office or another country, billing varies widely between local and long-distance calling. Even within these categories, furthermore, there are important pricing variations.

Local Calling ▪

Most telephone calls stay within a LATA. There are several billing schemes for local calling. Some telephone companies offer **flat-rate local service** in which there is a fixed monthly service charge but no separate fee for individual local calls.

In some areas, however, carriers charge **message units** for some or all local calls. The number of message units they charge for a call depends on both the distance and duration of the call. Economists like message units, arguing that message units are more efficient in allocating resources than flat-rate plans. Subscribers, in contrast, dislike message units even if their flat-rate bill would have come out the same.

Toll Calls ▪

While the local situation varies, all long-distance calls are **toll calls.** The cost of the call depends on distance and duration for both intrastate and interstate calls. Because of PUC rate control, some intrastate toll rates are more expensive than interstate calls over longer distances.

800/888 Numbers ▪

Companies that are large enough can receive favorable rates from transmission companies for long-distance calls. In the familiar **800 number** service, anyone can call *into* a company, usually without being charged. To provide free inward dialing, companies pay a carrier a flat monthly fee. Now that 800 numbers have been exhausted, the 888 area code is offering the same service to new customers.

In the past, AT&T dominated 800 calling. Now, however, **800/888 number portability** is allowing AT&T customers to move to competitors without losing their numbers. As expected, this competition is driving down prices. Now even people who operate companies out of their homes or apartments may have 800/888 numbers.

WATS

In contrast to inward 800/888 service, **wide area telephone service (WATS)** allows a company to place *outgoing* long-distance calls for a flat monthly rate. WATS prices depend on the size of the service area. WATS is often available for both intrastate and interstate calling. WATS can also be purchased for part of the country instead of the entire country.

900 Numbers

Related to 800/888 numbers, **900 numbers** allow customers to call into a company. But while 800/888 calls are usually free,[1] callers to 900 numbers pay a fee that is much *higher* than that of a toll call. Some of these charges go to the IXC, but most of them go to the company being called.

This allows companies to charge for information, technical support, and other services. For example, customer calls for technical service might cost $20 to $50 per hour. Charges for 900 numbers usually appear on the customer's regular monthly bill from the local exchange carrier. While the use of 900 numbers for sexually oriented services has given 900 numbers something of a bad name, they are valuable for legitimate business use.

Universal Availability

A potential new service is the **personal telephone number.** In this service, a number would be linked to an individual rather than to a telephone. A call would be able to reach the person anywhere in the country and eventually anywhere in the world. People might even be assigned a personal telephone number at birth.

Advanced Services

While telephony's basic function as a "voice pipe" is important, telephone carriers offer other services to attract customers and to get more revenues from existing customers.

Electronic Switching Services

Earlier we saw that most digital switches are really computers. We also saw the types of applications that vendors now program into their PBXs. Many carriers now offer the same services to home and residential customers.

Unfortunately, different carriers throughout the country tend to offer very different digital switching services. We will see later that one reason for creating the Integrated Services Digital Network is to standardize services. This will allow them to be offered even for calls that span multiple carriers.

---

[1] Some companies actually charge 900 number tolls for 800/888 calls by switching the users once they have called in.

Perhaps the most controversial electronic switching service is **automatic number identification (ANI)** in which the caller's number is displayed in an LCD display on the receiver's telephone.

On the positive side, ANI can allow **service customization.** The caller can automatically be routed to his or her personal service agent or to an agent serving the caller's state or geographical location within a city. ANI could also allow a computer to pull up information on the caller, so that the agent would have full background information for servicing the call.

On the negative side, ANI has serious privacy implications. The telephone has always been an anonymous calling device, and many people are loath to give up their privacy. This is a special problem if the person wants to call up a crisis hot line or some other service for which anonymity is desirable. On the other hand, call receivers rightly complain about prank calls and feel that they should be able to know who is calling their own homes or businesses.

One potential problem with call screening is geographical **red lining.** In red lining, companies refuse service to people living in certain parts of a city. While explicit red lining may be illegal, ANI would provide geographical information that could allow companies to subtly discourage service to red-lined areas.

Because of these problems, many nations are limiting ANI usage. Some permit ANI *call blocking*, in which the caller can block the ANI feature. The receiver, however, can still know that ANI has been blocked in most cases. The receiver can then decide whether to accept an ANI-blocked call.

CELLULAR TELEPHONES

In 1990, there were already 5 million cellular telephones. This number may grow to 40 million by the year 2000 [Standard & Poor's, 1992], driven by continuously falling prices. Paul Kagan Associates forecasts that 40 percent of all people in the United States will be carrying cellular telephones by the year 2000. The OECD [1990], in turn, estimates that 60 percent of all telephone calls in the year 2000 will have at least one cellular partner.

Cellular Concepts

Until the 1980s, mobile telephones were rare. There were only 44 channels in the United States, and other countries had similar limits. Not everyone is on the telephone all the time, so a system can have more subscribers than channels, but in general, only about 20 subscribers can share a radio channel. So early mobile radio could only serve 800 to 900 subscribers, even in the largest cities. This kept prices astronomically high.

During the 1960s, AT&T invented a technology that would bring prices down, making it possible to serve thousands of users rather than a few hundred. It would do this by allowing available channels to be used multiple times in each city, not just once. Figure G.6 shows AT&T's **cellular telephone** concept. The figure shows that cellular service divides a city into a number of geographical regions called **cells.**

Frequency Reuse ■

Cells are the key to **frequency reuse,** that is, using a channel several times in each city. Cellular uses low power, so signals do not travel far. You can use the same channel in nonadjacent cells. If someone talks on Channel 1 in Cell E, someone can be talking on Channel 1 in Cell A.

In practice, you can reuse a channel roughly every seven cells. So with 21 cells, you can reuse channels about three times. Following the 20-to-1 rule, you can support about 60 subscribers with each cellular channel, not just 20.

Low-Power Operation

Because **cellular telephones (cellphones)** only have to transmit over short distances, they can have very low power. A typical hand-held cellphone only uses 0.6 watts of power. This brings small size, relatively long battery life, and most important of all, relatively low cost.

Hand-held cellphones are common sights. In addition, there are **mobile cellphones** designed to stay in cars. They have a bit more power, so they operate in some areas where hand-held units will not. They also cost less because they use the car's electricity. However, they cannot leave the car when the subscriber does.

Spectrum Capacity ■

The third advantage of cellular service, besides channel reuse and low-cost phones, is not intrinsic to cellular operation. This is the fact that national regulators give cellular service much more spectrum capacity than early mobile services. Most cellular systems have about 800 channels. We will see later that this will soon quadruple.

More spectrum is possible because regulators are putting cellular services in high-frequency bands—450 MHz, 800 MHz, 900 MHz, and 1.8–2.2 GHz. These higher-frequency bands are less crowded than traditional mobile bands.

Handoffs ■

What if a subscriber moves from one cell to another? When that happens, the system will automatically execute a **handoff,** passing responsibility for service to the subscriber's new cell site. This happens so rapidly that few people are aware when a handoff occurs.

Handoffs are possible because all of the system's cell sites (antennas plus other equipment) are coordinated from a central point. This is the **mobile telephone switching office (MTSO).**

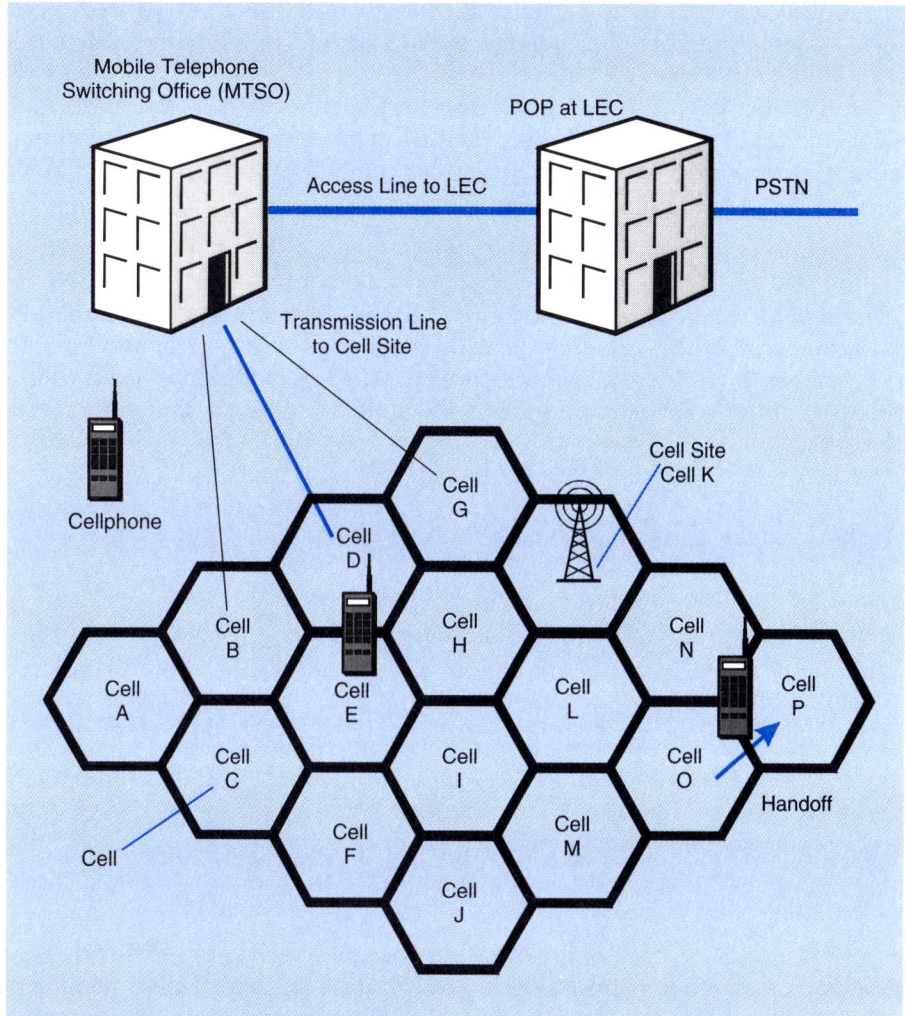

FIGURE G.6
Cellular Telephone System

A cellular system divides its service region into many small areas called cells. In each cell is a cell site that has a transceiver (transmitter/receiver) and an antenna. Users have small cellphones with built-in radios. Because cells are small, the cellphones can have low power, making them relatively inexpensive and small. Dividing the region into cells allows frequency reuse. For instance, suppose a user in Cell E transmits on Channel 1. Adjacent Cell B cannot support Channel 1 because of interference. But nonadjacent Cell A can support a different cellphone transmitting on Channel 1. This multiplies the number of customers that can be served. A central mobile telephone switching office (MTSO) coordinates the cell sites. It also connects the cellular system to the wire-based local exchange carrier. Note that a cellphone is passing from Cell O to Cell P. The MTSO will "hand off" service of the subscriber from the Cell O cell site to the Cell P cell site. Taking your cellphone to an entirely different service territory is roaming.

All cell sites pass their signals to the MTSO. If the other party is also a cellular customer, the MTSO sends the signal back out to the other party's cell site.

If the other party is on a regular (wireline) telephone, the MTSO passes the signal to the local exchange carrier. The LEC handles the rest of the connection to local customers. For long-distance calls, the MTSO connects to an IXC or even an ICC.

Roaming

What if you take your cellphone to another city? You would still like to have the same service. This capability called roaming is becoming common within most countries, but there may still be limitations or inconveniences even within countries. International roaming is very limited, but even this situation is changing.

Analog Cellular Telephones: The First Generation

Figure G.7 shows that today's cellular services are first generation, and that we can foresee two more generations in the near future.

Analog Operation

These first-generation systems are *analog*. When AT&T developed the technology in the late 1960s and early 1970s, digital technology was still prohibitively expensive. This prevented them from receiving the many benefits of digital operation, such as transmission efficiency and low error rates.

Large Cells

A second characteristic of first-generation systems is their use of relatively *large cells* that are one to five miles in diameter. Because the capital cost of a system depends heavily on the number of cell sites that a carrier must construct, and because the demand for cellular service was uncertain, cellular power was set to support fairly large cells. This, of course, limits channel reuse. In a typical city of 1 million people, there usually are about 20 cells, giving a frequency reuse factor of only about 3.

Limited Spectrum

A third characteristic of first-generation cellular systems is *limited spectrum*. A typical spectrum allocation for first-generation systems is 40 to 50 MHz in the 800 MHz band. This is very generous compared to early mobile service, but it still only allows 600 to 800 two-way channels. (The U.S. system has 832 two-way channels.) Simple frequency reuse only boosts this to 1,800 to 2,400 channels. So with the 20-to-1 rule, cellular could only serve 36,000 to 48,000 sub-

First Generation
- Analog service
- Large cells (around 20 per city)
- Around 50 MHz bandwidth
- In the United States: AMPS
- Many other standards in other countries
- Lack of world standard makes roaming difficult
- Dominates the market today

Second Generation
- Digital service
- Same cell size as first generation
- Same bandwidth as first generation
- In the United States, mostly CDPD
- In most of the rest of the world, GSM
- Growing rapidly
- In United States, retrofitting existing first-generation systems

Third Generation
- Called personal communication service (PCS) or personal communication network (PCN)
- Digital services
- Much smaller cells than previous generations
    - Far more channel reuse
    - Lower-power cellphones will cost less
- About three times the bandwidth of earlier generation
- Just starting
- In most of the world, a GSM derivative technology, DCS

LEO Systems
- Low earth orbit satellite systems
- Nearness allows hand-held phones without directional dishes
- Will be able to reach a phone anywhere in the world

FIGURE G.7
Three Generations of Cellular Service

scribers in a typical million-person city. Even with the reuse tricks described earlier, this can only double or quadruple. This is not enough for larger cities, even today.

Lack of Standards ■

A final characteristic of first-generation cellular systems is a *lack of standards*. The U.S. AMPS technology is different from technologies in Europe. In Europe, furthermore, there are several competing standards. Not even the frequency range of cellphones is standardized around the world. There is not even standardization in frequency bands. Most countries use the 800 MHz band, but some use the lower 450 MHz band.

Digital Cellular: The Second Generation

The second-generation systems now being introduced are similar to first-generation systems in some ways. Cell sizes and spectrum allocations tend to be the same or similar, for instance. However, there are two important differences.

Digital Operation

The big change in second-generation systems is the use of digital technology instead of analog technology. This is better for data. It also produces cleaner systems. With multiplexing, it is even possible to have more subscribers for the same amount of bandwidth, although this is not always implemented.

Standards

Although second-generation systems are not perfectly standardized around the world, there have been improvements. There seem to be two competing standards around the world. In the United States and some other countries, there is **cellular digital packet data (CDPD).** Generally, existing analog systems have some of their channels refitted for CDPD service. So CDPD is an incremental addition to existing cellular systems. In Europe and many other parts of the world, in contrast, there is an entirely new system operating in the 900 MHz band. This is **GSM,** the Global System for mobile telephony.

Personal Communication Services: The Third Generation

Although GSM and CDPD will offer many improvements, they still use large cells and have fairly limited bandwidth. Unless forecasts for cellular demand are extremely optimistic, first- and second-generation cellular systems will not be able to meet subscriber requirements in the near future.

Microcells

The most important gains in third-generation digital systems are due to their use of much smaller cells called **microcells.** Instead of being a mile or more in diameter, a PCS microcell may be only a quarter of a mile in diameter or even smaller.

The number of cells increases as the inverse square of the cell size. For instance, decreasing cell size by a factor of 5 increases the potential number of cells by a factor of 25. While the gains in capacity are not completely proportional, they are very large. Having many microcells and, therefore, massive amounts of reuse is the real key to capacity increases in PCS.

In addition, power requirements fall by the *cube* of the cell size. Reduced power requirements will make PCS cellphones very small, light, and inexpensive. It should also reduce concerns about radiation. At the time of this writing, power levels and, therefore, cell sizes have not been set for PCS. So the ratios we have just given are only for purposes of illustration.

Greater Bandwidth

A second key to improved service is increased bandwidth. By assigning third-generation service to the 1.5–3 GHz range, countries can typically allow about 150 MHz of capacity—three times the amount of earlier systems. Thanks to smaller cells and more frequency capacity, third-generation systems can serve ten to thirty times as many users as earlier systems. Most cellular users eventually will be third-generation users.

PCS

In the United States, the third-generation service is called **personal communication service (PCS)** with a generous allocation of 160 MHz of capacity in the 1.8–2.2 GHz band. To create competition, the FCC will authorize several PCS carriers to compete in each market. It has even decided not to standardize the technology in order to provide technological competition.

DCS

Europe, in contrast, is taking a more conservative approach. The third-generation cellular system will be the **distributed communication service (DCS).** DCS will operate at 1710–1785 MHz and 1805–1880 MHz, so it is called DCS-1800. It is a modification of GSM technology, so it should be easy to develop DCS equipment.

Personal Service

The *personal* in PCS signifies that the telephone number will not be linked to a piece of equipment. Rather, it will be linked to an individual person. You may have a credit card sized *access card* that you can plug into whatever cellphone you are near. This credit card will recognize you as the number's owner. You could plug it into your home telephone, your car phone on the way to work, your desk phone at work, and a cordless telephone tied to your organization's wireless PBX.

Even if you use a single phone, it will be able to serve multiple uses. At home, it will act as a cordless telephone working with a low-power unlicensed base unit. The cell in this case will be about the size of your house. On the way to work, the phone would act as a traditional cellphone, using a cellular carrier. At work, it would link to a wireless PBX using unlicensed operation.

Whether you use one phone or several, people will be able to get you anywhere, simply by calling your PCS number. DCS is also likely to see extensive use as a personal communication service.

LEO Services

All of the services we have discussed so far use ground-based transmission facilities. But several companies are now attempting to produce satellite-based systems that will work with low-cost radio transceivers.

All of these systems use low earth orbit **(LEO)** satellites, which circle the earth at orbits of only 100 to 200 miles. This allows inexpensive hand-held cellphones to reach them. In contrast, traditional communication satellites circle the earth at an orbit of 22,300 miles. This provides a geosynchronous orbit, making the satellite appear to hover over one position. This allows you to aim a dish at the satellite. LEOs, in contrast, race across the sky. This makes dishes impossible to use, but the distance to the user is so much less that this inability to use a dish is unimportant.

CONCLUSION

In this module, we have looked at the Public Switched Telephone Network. Corporate telephone use is big, growing, and important; it is likely to be even more so in the next few years. There are now so many options that corporations need fairly large and knowledgeable telephone staffs. This is particularly true for companies that are building or have built private telephone networks.

CORE REVIEW QUESTIONS

1. Define the following terms for the Public Switched Telephone Network: *customer premises, local loop, switching office, trunk line*. Trace the route a call takes from one customer premises to another in a local call.
2. Explain each of the following in the United States: LATA, LEC, CAP, IXC, ICC, RBOC, Baby Bell. Trace what happens in a local call involving a LEC customer and a CAP customer. Trace what happens in a long-distance call between states. Trace what happens in a call from a U.S. customer premises to a party in another country.
3. What is a PTT? Explain the relationship between the PTT and the Ministry of Telecommunications.
4. Explain the two levels of regulation within the United States. Describe the two phases of AT&T's splitting up into smaller firms. What is the significance of the Telecommunications Act of 1996 in the United States?
5. Rank order the following in terms of degree of deregulation: customer

premises operation, local service, and long-distance service. Use the terms *most*, *middle*, and *least*.
6. What is a PBX? Why are PBXs attractive to businesses?
7. Describe the main elements in the vertical and horizontal distribution of telephone wiring. (Be sure to explain the function of the wiring closet.)
8. What is a tariff? What two things does it specify? Are all carrier services tariffed?
9. Compare and contrast 800/888 numbers, 900 numbers, and WATS, in terms of whether the caller or the called party pays and the cost compared to the cost of a directly dialed long-distance call.
10. Explain why cellular telephone systems can serve thousands of simultaneous callers in a large city.
11. Explain the differences between first-generation cellular systems and third-generation (PCS) cellular systems in terms of number of subscribers served. Explain reasons for the difference. Explain why third-generation cellular phones should be less expensive and smaller than first-generation cellphones.

DETAILED REVIEW QUESTIONS

1. Explain the requirements of a telephone system for building space.
2. What is a PBX network? Why do organizations create such networks? Distinguish between private telephone networks and virtual private networks.
3. In PBXs, what is the difference among user services, attendant services, and management functions? List at least three services in each category. Be able to explain all of the PBX services in the module if given their names.
4. Distinguish between voice recognition and voice response systems. What are the relative advantages of each?
5. How do countries decide which ICCs will serve their customers? How are ICC rates set?
6. Describe pricing for local calls and toll calls. What is the advantage of 800/888 numbers for customers? For companies that subscribe to 800/888 number service? What is 800/888 number portability, and why is it important to competition?
7. For what do the initials ANI stand? What does this service do? Why is it good for business? Why is it good for individual people? Why may it be bad for individual people? What is ANI call blocking?
8. Explain the benefits of dividing a service area into cells in cellular telephony.

9. How do cellular systems serve many people (some overlap with the preceding question)?
10. Distinguish between handoffs and roaming.
11. What are the characteristics of first-generation cellular systems? Why are these characteristics limiting? Describe standardization for first-generation services.
12. Distinguish between first-generation and second-generation cellular. Which limitations of first-generation cellular does second-generation cellular relieve? Which does it not relieve?
13. Distinguish between first-generation and third-generation cellular. Which limitations of first-generation cellular does third-generation cellular relieve?
14. LEO transceivers will not use directional antennas. Why will this be possible? What benefits will this bring? What costs will it create?

THOUGHT QUESTIONS

1. Compare building wiring with LAN and local internet wiring in Chapters 4, 5, and 6. Do you think the similarities are accidental? If not, why do you think LAN wiring so closely follows traditional telephone building wiring?
2. How do feel about 900 numbers, ANI, and cellular radiation's effects on the body? Do you think we should have new regulations?
3. Do you think that people should be allowed to use cellular telephones while driving?

PROJECTS

1. From your local cellular company, find how many cells serve your city. If possible, locate the cell sites on a map.
2. Determine the cost of cellular telephony in your area. Determine activation (initial) charges and monthly charges. Most cellular systems provide several alternatives based on monthly calling volume. Compare them.
3. Go to a store that sells cellphones. Compare prices with features, size, and power. Determine if low-price cellphones require you to get an account with a particular cellular provider. If so, determine the activation fee and monthly service charge. See if it is possible to get a contract with a lower

activation fee and/or a lower monthly service charge. See if this raises the price of the cellphone.

For online exercises, please visit this book's website at: http://www.prenhall.com/panko

REFERENCES

OECD, *Communications Outlook 1990*. Paris.

PANKO, R. R. (1992). "Managerial Communication Patterns," *Journal of Organizational Computing*, 2(1), 95–122.

STANDARD & POOR'S (January 23, 1992). "Telecommunications: Basic Analysis." In *Standard & Poor's Industry Surveys*, 160(4), Section 1, New York: Author.

MODULE H

Advanced Topics in Internetting

This module is designed to be read after Chapter 7, although some topics can be read earlier. It is designed to be read front to back like a chapter. Module H gives an overview of basic TCP/IP concepts related to internetting. These topics include layering, IP addressing, the delivery of IP messages (datagrams), TCP and UDP, supervisory protocols such as ICMP, router coordination protocols, and the new Internet Protocol Version 6 (IPv6).

BASIC TCP/IP CONCEPTS

Layering

In Chapter 1, we saw that the TCP/IP architecture is rather simple with only four layers.

- At the top, there is the *application layer* for protocols linking application programs on different machines.
- Below that comes the *transport layer*, which allows computers with different

operating systems and hardware architectures to work together. Here TCP is the most widely used protocol.

- Next comes the *internetting layer*, which governs the routing of packets across multiple networks. IP is the most important protocol at this layer.
- Finally, there is the layer for individual networks (subnets). TCP/IP calls this the *subnetwork access layer*. TCP/IP adopts OSI standards at this layer.

In Chapter 2, we saw that this basic layering appears in the way that messages are transmitted from one host to another. Figure H.1 shows that each layer's processes produce synchronous messages, which we will call, following OSI terminology, **protocol data units (PDUs).** Each layer's PDU, except the last, is encapsulated in the lower layer's PDU, as the information or data field.

IP Numbers

In an internet, the key to connectivity is addressing. Every host computer must have a unique IP number for internet delivery—just as every telephone needs a unique telephone number.

FIGURE H.1
Encapsulation of TCP/IP Protocol Data Units

Using OSI terminology, each layer's process produces a synchronous unit called a protocol data unit or PDU. The application layer's PDU becomes the data field for the transport-layer PDU. This in term becomes the data unit for the internet layer's PDU. The internet-layer PDU, finally, is carried inside the frame or packet of the subnet. As we saw in Chapters 2, 4, and 5, the subnet itself is likely to have a layered PDU structure.

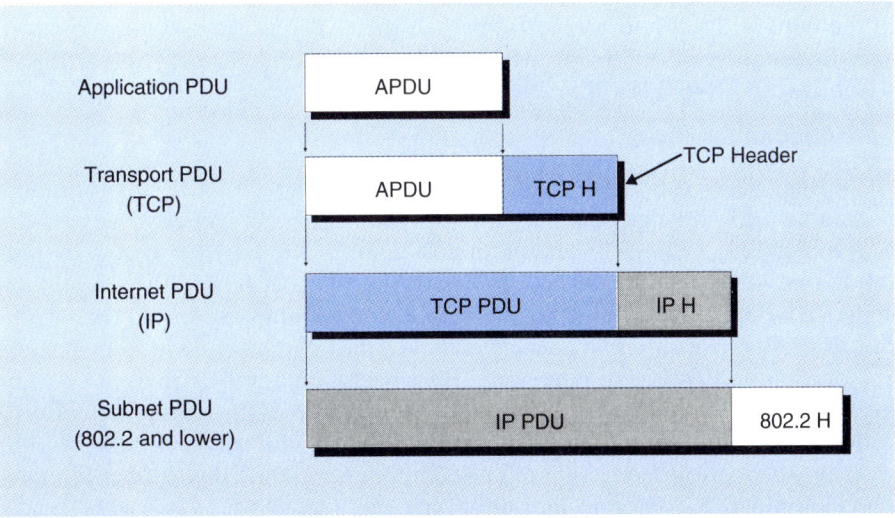

32-Bit IP Addresses

In the IP packet header, shown in Chapter 6, there is a destination address and a source address. Each is 32 bits long. For instance, one host's IP number is the 32-bit number:

    10000000101010110001000100001101

Obviously, this is not very easy to memorize. To make life easier, we normally represent IP numbers in an equivalent but easier-to-remember way. First, we divide the 32-bit number into four bytes, in this case:

    10000000 10101011   00010001   00001101

Second, we convert each byte into its decimal equivalent. For instance, the number 10000000 in binary is 128 in decimal. The binary number 10101011, in turn, is 171 in decimal. The last two bytes are equivalent to 17 and 13.

We then represent the number as these four decimal numbers separated by periods. This gives us the following number IP host address. We pronounce it as "128 dot 171 dot 17 dot 13".

    128.171.17.13

Assigning IP Numbers

To ensure that each host has a different IP number, one solution would be for the Internet Society to assign individual IP numbers to individual hosts. This would be impractical. The Society would be overwhelmed with requests to add, drop, or change millions of 32-bit IP numbers around the world.

Instead, the Internet Society takes a different approach. Instead of assigning individual host IP numbers, it assigns blocks of IP numbers to organizations, which then assign these numbers to their individual hosts.

To implement delegation, the Internet Society divides 32-bit IP numbers into a network part and a host ID part. It then assigns the network part to a specific user organization. For instance, it assigned the following **network part** to the University of Hawaii:

    1000000010101011 or 128.71

It then allowed the University of Hawaii to assign 32-bit IP host numbers beginning with this network part. So, for instance, the university assigned the IP number 128.171.17.13 to one of its host computers. In other words, by giving the university a unique network part, the Internet Society was really assigning the university a block of IP numbers to administer. It was then up to the university to ensure that it assigned no duplicate numbers within this block.

In a sense, this is like telephony. By worldwide agreement, each country gets a country code. It can assign any addresses it wishes after the country code. In the United States, numbers are further subdivided by area code. Within each area code, responsible telephone companies are then given blocks of telephone numbers to assign to individual telephones.

IP Number Classes

One big difference between telephone numbers and IP numbers is that all IP numbers are 32 bits long. In contrast, telephone numbers vary in length from country to country. This fixed size of IP addresses causes problems. If you want to have many network parts to assign to organizations, you have to use a lot of bits in the network part. This will leave you with few bits to represent host addresses. For instance, if the network part of the 32-bit IP address is 16 bits long, then the host part can only be 16 bits long. If the network part is lengthened to 24 bits, then only 8 of the 32 bits would be left for host parts.

The Internet Society addressed this problem by creating several **classes** of networks. Although all classes use 32 bits for the address, some classes use only a few bits for the network part, leaving many bits for host computers. Others use many bits for network parts. This creates a large number of small networks with only a few hosts. Figure H.2 shows how the bits are divided in these classes of networks.

CLASS A NETWORKS In a **Class A** network, the network part begins with a 0 to designate the network as a Class A network. It then uses the next seven bits of the 8-bit Class A network part to designate a specific Class A network. This allows only 2^7 or 128 Class A networks.

The good news is that each Class A network has 24 bits left (32 minus the 8 bits of the network part) for host parts. This allows each Class A network to assign up to 16 million host IP numbers. Few organizations need that many host addresses, which is fortunate because it is impossible to get a Class A network today. The few that were possible all have been assigned or reserved.

FIGURE H.2
Classes of IP Networks

Class	Beginning Bits	Remaining Network Part Bits	Host Part Bits	Approximate Number of Networks in this Class	Approximate Number of Hosts per Network
A	0	7	24	128	16 million
B	10	14	16	16,000	65,000
C	110	21	8	2 million	256
D*	1110				

*Used in multicasting.

Problem: For each of the following IP numbers, give the Class, the network bits, and the host bits if applicable:

10101010111110000101010100000001
11011010111110000101010100000001
01010101111110000101010100000001
11101110111110000101010100000001

CLASS B NETWORKS Next come **Class B** networks. These are the workhorses of the Internet, thanks to their good balance between number of networks and the number of hosts per network. They use 16 bits for the network part of the IP number. The first two bits are 10, to designate the network as a Class B network. This leaves 14 bits to assign to individual network IDs. This is enough for over 16,000 Class B networks.

Using 16 bits for the network part leaves 16 bits for host IDs. This is enough for more than 65,000 hosts on each Class B network. This is sufficient today even for fairly large and computer-intensive organizations, such as universities. The example given previously (128.171) was a Class B network.

CLASS C NETWORKS For **Class C** networks in turn, there are 24 bits for the network part. The network part begins with 110, leaving 21 bits for network IDs. This allows 2 million possible Class C networks.

Unfortunately, each of these networks can only have a few hosts. With only 8 of the 32 address bits left for host IDs, each Class C network can only have 256 possible hosts. This would only be enough for a very small company.

CLASS D ADDRESSES **Class D** addresses are for multicasting. For instance, one multicast address might be assigned to all hosts in a division of a company. When a host sees an incoming packet, it looks not only for its own IP number in the destination address field. It also looks for the multicast addresses to which it has been assigned. If 30 hosts are assigned the same multicast address, then IP messages (datagrams) sent to that Class D address are read by all 30 hosts.

Because of the two-step approach that the Internet uses to assign IP numbers, it is impossible to get anywhere near the maximum four billion number of possible hosts on the Internet. As a result, the Internet Society is running out of network numbers—especially for the high-demand Class B networks. Later we will see that one goal of the new IP standard, IP Version 6, is to avoid the Internet running out of addresses to assign.

Delivery

So far, we have merely said that unique addresses allow the Internet to deliver packets to any destination computer. In this section, we will look at how this is done.

The Host's Role for Local Delivery ∎

Figure H.3 shows the source (sending) host's role in message delivery. In TCP/IP, the source host plays a very active role in message delivery. It does not merely push its message onto the network with an address to which the message should be delivered.

The figure shows that when a source host wishes to transmit a message, its internetting process (program) must have the 32-bit Internet Protocol (IP) number of the destination host.

The IP software on the source host then compares the destination host's

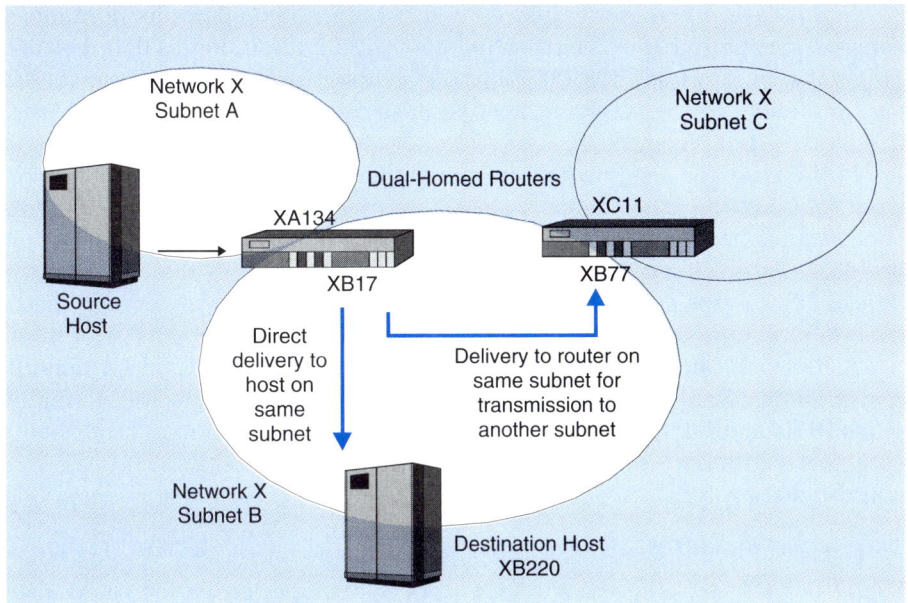

FIGURE H.3

The Source Host's Role in Internet Delivery

The source host's internetting program must have the IP number of the destination host. This 32-bit number contains the destination host's network part and host ID. If the destination host's network part is the same as the sending host's network part, then the sending host knows that the destination host is on its own network. The source host sends the IP datagram directly to the destination host, using the network's subnet delivery protocol. However, if the destination host's network part is different from the source host's network part, then the source host knows that the destination host is on a different network. In this second case, the source host sends the message to a router for delivery.

network part to its own IP address's network part. If these match, then the source and destination hosts must be on the same network. For instance, in Figure H.3, the source host is 128.17.15.12 and the destination host is 128.17.15.20. The network parts (128.17) in these two Class B networks match, so the two hosts are on the same network, Network X.

The sending host therefore transmits the IP packet directly to the receiver, using the subnet's transmission protocol. For instance, suppose the network is an Ethernet LAN. The sending host then places the IP datagram within an 802.2 frame and then within an 802.3 frame. It then sends the MAC-layer frame to the destination host's 48-bit IEEE LAN address (see Chapter 4).

The Host's Role in Internetting ■

In the preceding example, no routers were involved. That changes if the destination host is on a different network than the sending host. Figure H.3 also shows this situation.

In this case, when the sending host (128.17.15.12) compares its IP number network part with that of the receiving host (128.30.13.13), it sees that they are different (128.17 versus 128.30). Now the sending host cannot deliver the IP packet directly to the receiving host because they are on different networks (Network X and Network Y).

In this case, the sending host transmits the IP packet to a *router* on its network for handling. The source host and the router are on the same network. So the source host transmits the IP datagram to the router using the network's protocol. On an Ethernet network, the host again packages the IP datagram into an 802.2 frame and then into an 802.3 frame. This time, however, it sends the 802.3 frame to the router's 48-bit NIC address.

A network can have multiple routers. Each host normally has a **default router,** in this case, router 18.17.93.15. This is the router to which it normally sends IP datagrams for delivery to other networks. The host may learn, however, that a different router is best for a certain destination host. If it does, it will use that router.

The Router's Address ■

Figure H.4 also looks at the situation from the perspective of the router that receives the IP packet. First, it shows that the router attaches to two networks. This allows the router to pass IP datagrams from one network to the other. In TCP/IP terminology, the router is **dual-homed.** Routers that connect to multiple networks in turn are **multihomed** routers.

Suppose the two networks are both 802.3 Ethernet LANs. Then it is obvious that the router must have two NICs, each connecting it to a single LAN. It is also obvious that the two NICs must have different addresses because NIC addresses are unique.

Like hosts, routers have IP numbers. Otherwise, hosts could not send them IP datagrams for delivery. In fact, they have multiple IP numbers—one for each network on which they are homed. In Figure H.4, the router needs *two* IP numbers, one for each network. On Network X, the router is 128.17.93.15. On Network Y, its IP number is 128.30.7.7.

Why does a router need a separate IP number on each network? Recall that an IP address consists of a network part and a host ID. The router cannot have the same IP address on both networks because at least the network part has to be different on the two networks.

The Router's Role for Local Hosts ■

When the router receives an IP packet from Network X, it looks at the destination host's network part and compares it with its own *Network Y* IP number. If the network parts match, the router knows the receiving host is on Network Y. In Figure H.4, for example, if the destination host is 128.30.40.40, the host's network part, 128.30, matches the network part of the router on Network Y. The router knows it can deliver the IP datagram directly to the destination host.

The router looks up the receiving host's 48-bit NIC address and sends it to the IP datagram encapsulated within 802.2 and then 802.3 frames.

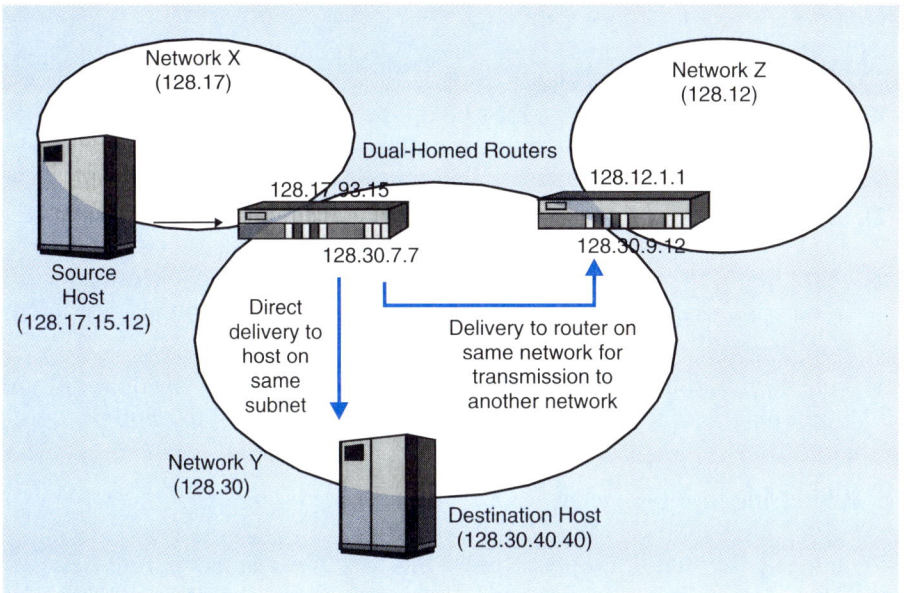

FIGURE H.4
The Router's Role in Internet Delivery

The router is dual-homed, meaning that it attaches to two networks. It has a local subnet address on each network. It also has an IP number on each network. When the router receives an IP datagram from a sending host on Network X, it compares the destination host IP address's network part with the network part of Network Y. First, if they match, the router knows that the destination host is on Network Y. It delivers the IP datagram directly to the destination host, using Network Y's subnet transmission protocol. Second, if the destination host's network part does not match the network part of Network Y, the router cannot deliver the IP packet itself. As the source host did, it forwards the IP datagram to another router on Network Y for delivery to another network.

The Receiving Router's Role in Pass-Through ■

What if the destination host's network part is not the network part of Network Y? Then the router knows that the host is not on Network Y. It cannot handle the delivery itself. Instead it sends the IP datagram to another router on Network Y. That router then takes over responsibility for the message. In Figure H.4, for example, if the network part of the destination IP number is not 128.30, then the first router knows that the destination host is not on Network Y. It passes the IP datagram onto router 128.30.9.12.

Note that if the destination host is not on one of the first router's two networks (X and Y), the router never looks at the host part of the IP number. It merely looks at the network part of the IP number. This makes the router's job slightly easier. A router only has to know the host IP numbers for stations on its two (or more) directly attached subnets, in this case Network X and Network Y.

Recap

Figure H.5 illustrates the four possible conditions schematically.

- First, if the destination host may be on the source host's own network, the source host delivers the IP datagram directly. No router is involved. However, the internet PDU—the IP datagram—is included in the protocol stack.
- Second, if the destination host is not on the source host's network, the source host delivers the datagram to a router on its own network.
- Third, if a router gets an IP datagram, the destination host may be on one of its networks. If so, the router delivers the datagram to the destination host directly.
- Fourth, if a router gets an IP datagram for which the destination host is not on any of its networks, it passes the IP datagram to another router.

IP Addressing in More Detail

We noted earlier that IP numbers are 32 bits long. We also noted that these addresses have two parts: the network part and the host part. We will now look more closely at IP numbers.

IP Numbers

IP addresses can be specified in two ways. First, there are the **IP numbers** we have been discussing. An IP number is a 32-bit binary number. For example, it could be 10000000101010110001000010001101. Because such numbers are difficult to remember and write, we saw earlier that we break it into four bytes (octets). We then convert each octet to a decimal number. In dot format, we then write the address as 128.171.17.13.

FIGURE H.5

Overview of IP Packet Delivery

IP datagram delivery requires the source host and routers to make decisions in a specific sequence. A basic rule is that the first network device—host or router—that can deliver the datagram does deliver the datagram.

Network of Destination Host	How IP Datagram Delivery Is Handled
Same as source host	Source host delivers the datagram directly.
Separated from the source host's network by a single router	Source host delivers the datagram to a router. The router delivers the IP datagram directly.
More than one network away	Source host delivers the datagram to a router on its network. Router delivers the datagram to another router. This router-to-router delivery may happen repeatedly. The final router delivers the IP datagram.

IP Names

This is still difficult to remember. Fortunately, there is a second way to specify IP addresses. This is to give a host an **IP name,** such as *voyager.cha.hawaii.edu*. IP names are easier to remember than IP numbers.

In the case of *voyager.cba.hawaii.edu* we again see four parts separated by decimal points. Although it is tempting to expect the *voyager* to be 255, the *cba* to be 0, the *hawaii* to be 255, and the *edu* to be 2, this is not the case. *There is no relationship at all between the four parts of an IP number and the four parts of many IP names.*

In addition, many IP names do not have four parts. For example, we might see the following host names:

```
www.prenhall.com
microsoft.com
```

Sockets

Sometimes the IP number is followed by a colon and a number. For instance, in the preceding example, the host might be:

```
128.171.17.13:80
```

The number following the colon is called a **port number,** which specifies a particular application program on the destination host. This serves the same function as the OSI session layer (Layer 5). *For more on OSI, see Module A.*

In this case, 80 is the **well-known port number** of World Wide Web server application software. Each major Internet application service has a "well-known" port address. Although hosts can use any port number for any service, most host authorities follow the IETF's port numbering recommendations.

The combination of an IP number and a port number is called a **socket.** A socket uniquely specifies a specific application on a specific host on a specific network on the internet.

Typically, the client software on a client machine will hide the detail of port numbers and sockets from the user.

Subnet Masks

While Class A and Class B networks can support many hosts, the corporate reality is that most organizations do not have a single network. Instead they have many networks, which are typically LANs. So we would like to have a three-tier addressing scheme, consisting of networks, subnets within networks, and hosts within subnets. Each LAN in the network would be a subnet.

The organization might allow individual work units to assign host numbers within their subnets. For instance, at the University of Hawaii, the university assigned a subnet number to the College of Business Administration. It then allowed the college to assign IP numbers to specific hosts.

The initial host numbering system on the Internet did not support subnetting. The "fix" was to create a *second* number for each host on the network. This number is the network's **subnet mask.** Figure H.6 shows how a subnet mask works. Here there is a host numbered 128.171.17.13. In binary, this is 10000000101010110001000100001101. The subnet mask is 255.255.255.0. In binary, this is 11111111111111111111111100000000.

Looking at the host IP number, we see that it begins with the bits "10". From Figure H.2, this means that the network is a Class B network. In a Class B network, the first 16 bits (128.171) represent the network part. All addresses on the network, then, begin with 128.171.

FIGURE H.6
Subnetworking with a Subnet Mask

Givens

Adress	Decimal	Binary (with 2nd and 4th octets underlined)
Host Number:	128.171.17.13	10000000<u>10101011</u>0001000<u>100001101</u>
Subnet Mask:	255.255.255.0	11111111<u>11111111</u>11111111<u>00000000</u>

This is a Class B network, because the IP number begins with "10"
    In Class B networks, the first two bits are **10**
    In Class B networks, the first 16 bits are the network part (10000000**10101011**)
        So the first 16 bits are (128.171)
    The last 16 bits identify the host and subnet (17.13 or **0001000**1**00001101**)

The subnet mask's bits are set to 1 for the network part bits
    This is a class B network, so there are a total of 16 network part bits
    So the first 16 bits of the subnet mask are 11111111**11111111** (255.255)

The subnet mask's final 16 bits are **11111111**00000000.
    The 1 bits mark the subnet bits
    The 0 bits mark host bits

In other words the mask tells us that the *last 16 bits of the IP number* have this form:

Host and Subnet Bits	Subnet Mask	Subnet Bits	Host Bits
0	1	0	
0	1	0	
0	1	0	
1	1	1	
0	1	0	
0	1	0	
0	1	0	
1	1	1	
0	0	(17)	0
0	0		0
0	0		0
0	0		0
1	0		1
1	0		1
0	0		0
1	0		1
(17.13)	(255.0)		(13)

Problem: Give the network class, network bits, subnet bits, and host bits for each of the following:

    IP Number   10101010<u>11110000</u>10101010<u>00000001</u>
    Subnet mask   11111111<u>11111111</u>11111111<u>00000000</u> (255.255.255.0)

    IP Number   01010101<u>11110000</u>10101010<u>00000001</u>
    Subnet mask   11111111<u>11111111</u>11111111<u>00000000</u>

This means that the last 16 bits (17.13) must represent the subnet part and the host part on that subnet. Figure H.6 shows that the subnet mask's 1s represent the subnet bits. So these bits are 00010001 in binary or 17 in decimal.

The 0s in the subnet mask, in turn, designate the bits for the host. So the host part is 00001101 in binary or 13 in decimal.

Putting this together, we know that the IP address is 128.171.17.13 and that the subnet mask is 255.255.255.0. This means that the network part is 128.171, the subnet part on that network is 17, and the host part on that subnet is 13. In other words, if you know the host IP number and the subnet mask, you can determine the network class, the network, the subnet, and the host. Although the method for computing the subnet and host IDs is awkward, inelegant, and difficult for humans to learn, it works well and is easy for computers and routers to implement.

Since you have 8 bits for subnet IDs in this example, you can have 2^8 subnets, or 256. Each of these subnets, in turn, can have 2^8 possible hosts. With this arrangement, the organization might give each work unit a subnet ID. That work unit would then be free to assign IDs to 256 hosts. This would be good if the organization had quite a few work units, each with a moderate number of hosts. In practice, the subnet mask 255.255.255.0 is the default. Most organizations use it.

Another Subnet Example ▪

At the same time, there is nothing magic about the subnet mask 255.255.255.0. For instance, suppose that you wanted to allow each work unit to be able to assign IP addresses to 1,024 hosts, instead of just 256. To represent 1,024 addresses, you need 10 bits. So you would set the last 10 bits in the subnet mask to 0000000000. You would set the first 22 bits of the subnet mask to 1.

Subtracting the 16 bits for the network part, you would have 6 bits for subnet IDs. This would allow 2^6 or 64 possible subnets. In other words, the organization would have subnets that can serve more hosts apiece, but it would have fewer subnet IDs to assign.

In this second example, the subnet mask would be 1111111111111111 1111110000000000. In decimal, this would be 255.255.252.0.

TCP and UDP

So far, we have been looking at the internetting layer. We have been looking specifically at the delivery of IP datagrams. Now we will move up to the next higher layer, the *transport layer*. This layer allows two machines to communicate despite different hardware and operating systems. Although the transport layer serves a different function than the internet layer, the two work together very closely in TCP/IP.

IP Unreliability ▪

The IP protocol is an *unreliable* protocol. Figure H.7 shows that although the hosts and routers make their best effort to deliver IP datagrams, they do not do error correction at the internet layer. If an IP process on a router detects a

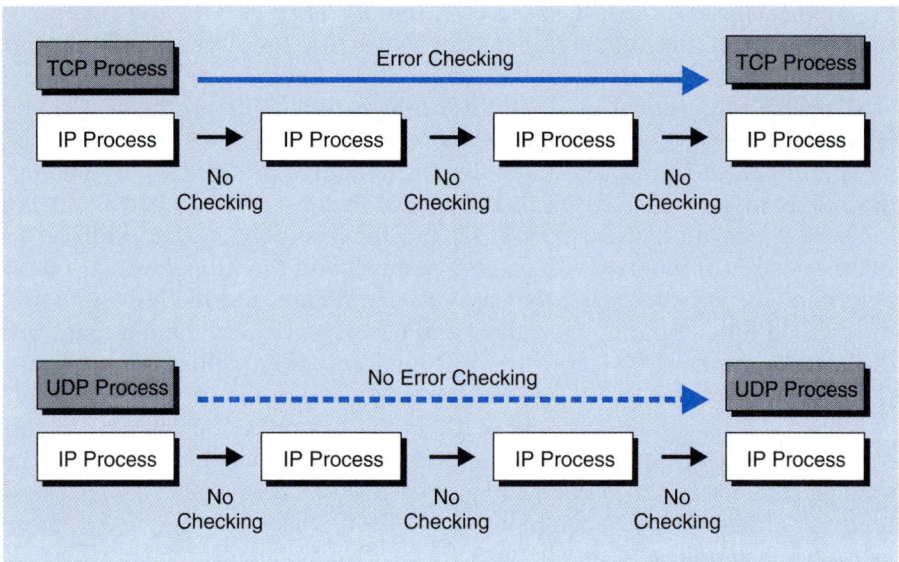

FIGURE H.7
IP, TCP, and UDP

IP is an unreliable service. It makes its best effort to deliver IP datagrams across multiple routers, but if an error occurs, there is no retransmission. At the next higher layer, the two host transport processes may catch errors. A TCP transport process on the destination host does detect errors and asks the peer transport process on the source host to retransmit the missing or damaged TCP PDU (segment). This implements error correction on only two machines (the hosts) instead of on many intermediate routers. Host transport processes that use the unreliable UDP process, in turn, do not do error detection at the transport layer either. UDP is for supervisory messages whose delivery is not urgent and for real-time data, such as voice and video.

header error in transit, it merely drops the IP datagram. In addition, datagrams may arrive out of order, and there can be duplicate datagrams as well.

The main reason for making IP an unreliable service is that extensive error correction over multiple hops between routers on multiple networks would be extremely complex. It would slow IP datagram delivery and reduce throughput. Simply, the internetting layer is not an efficient or effective place to catch errors. At the transport layer, only the two host computers have to do error handling—not all the intervening routers.

The Transmission Control Protocol (TCP) ■

The most common transport layer protocol in TCP/IP is the **Transmission Control Protocol (TCP).** As you might expect from the previous discussion, TCP is a reliable protocol. The sending host stamps each outgoing TCP protocol data unit (called a **segment**) with a sequence number. It also puts an error-check field into the segment. This allows the receiving host to check for errors, lost segments, or duplicate segments. It discards duplicate segments. For damaged or missing segments, it sends a request for retransmission to the sending host, using the sequence number of the missing segment.

While extensive hop-by-hop error checking at the internetting layer would be extremely difficult, TCP only does error checking once—at the final destination. This is quite efficient. Although it places more of a burden on the hosts, this is a good trade-off.

The User Datagram Protocol (UDP) ■

The TCP/IP protocol suite has an alternative to TCP called **User Datagram Protocol (UDP).** UDP does no error checking. In addition, its maximum length is very brief. It is, in every sense, a very stripped-down transport-layer protocol. If errors occur, it is up to the application-layer process to discover the error. In some cases, the error may never be detected at all.

UDP is often used to send nonurgent supervisory messages for which reliable delivery is less important than not burdening the network with traffic. It is also good for real-time applications, such as voice and video, where occasional losses do little harm and where retransmission would introduce unacceptable latency (delays).

Supervisory Protocols

So far, we have talked about basic packet delivery over an internet, but there are also a number of **supervisory protocols** that the hosts and routers use to enable them to work effectively. These supervisory protocols make TCP/IP routing extremely automatic and able to serve millions of host computers. A number of them also optimize routing. Unfortunately, they also add to the cost and complexity of routers, compared to simpler devices, such as bridges and LAN switches.

Address Resolution Protocol (ARP) ■

We noted previously that if the destination host is on the same subnet as the sending host or a router, then the host or router delivers it via the subnet's protocol. For an Ethernet LAN, this would be the 802.3 protocol.

Every host must have two addresses. One is the 32-bit internet-layer IP address. The other is its 48-bit IEEE MAC-layer address on the LAN. Unfortunately, when the host or router wants to deliver an IP datagram to another host on the same subnet, it may only know the host's IP number.

Figure H.8 shows that the **Address Resolution Protocol (ARP)** allows the host to learn another host's MAC-layer address. It broadcasts an *ARP request* message to all stations on the LAN. This ARP packet essentially says, "Hey IP Number X, what is your 48-bit MAC-layer address?" The host with that IP number sends back an *ARP response* message that includes the 48-bit LAN address on its network interface card (NIC).

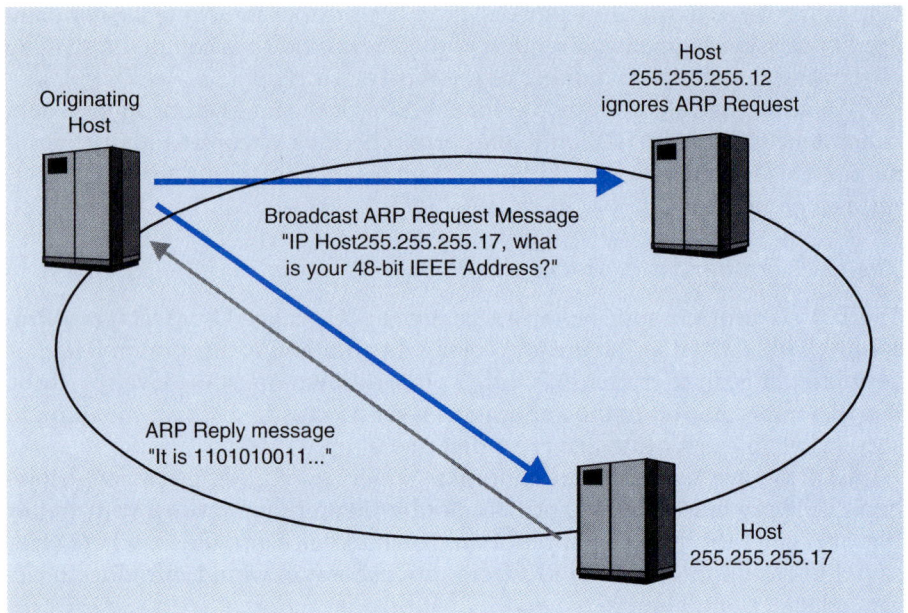

FIGURE H.8
Address Resolution Protocol

In ARP, a host wants to transmit to another host on the same subnet. It knows the receiver's IP number, but not its subnet address (in the case of LANs, its 48-bit IEEE MAC-layer address). The sending host first broadcasts an ARP request to all hosts on the subnet, using the subnet protocol. The ARP request contains the IP number of the target host. If a host recognizes its own IP number in the ARP request, it transmits an ARP response message back to the sending host, again using the subnet protocol. This ARP response message contains the target host's IEEE address.

Autoconfiguration Protocols ■

We have seen earlier that a source host must know its own IP number. Sometimes this address is contained in the machine's configuration files. In other cases, however, the machine must call an autoconfiguration host to learn its IP number. Figure H.9 illustrates this **autoconfiguration** process.

When a host boots up (starts working as an Internet host), it sends an *autoconfiguration request message* to the autoconfiguration host. This message asks for an IP number. The autoconfiguration host sends back an *autoconfiguration response message* with an IP number for the host to use as its source IP address in outgoing IP datagrams.

Autoconfiguration was first envisioned for extremely simple hosts too limited to have their own autoconfiguration files. Today it has a different use. Every time a PC accesses the Internet, it becomes a temporary Internet host. (This is even true if you access the Internet at home.) So autoconfiguration allows each PC, when it wants to access the Internet, to get a temporary IP number. When the PC finishes, it releases the temporary IP number. The next time it uses the Internet, it will get a different temporary IP number.

This allows a company or an Internet service provider to ration its IP numbers. For instance, suppose that a college only has 256 IP numbers for its subnet. Suppose further that it has 400 PCs. By assigning IP addresses "on the fly," the autoconfiguration host allows any PC that wants to use the Internet to do so—as long as there are not more than 256 simultaneous connections.

There are several autoconfiguration protocols. One is the **Reverse Address Resolution Protocol (RARP).** Another is **Bootp.** A third is the **Distributed Host Configuration Protocol (DHCP).** DHCP is becoming popular because it is the protocol supported by Windows 95. Obviously, the host and the autoconfiguration server must use the same protocol.

Domain Name System (DNS)

Recall that the sending host has the task of addressing datagrams to destination hosts. Some users and application programs only know the *IP name* (such as *www.paramount.com*) of the destination host. Because the IP destination field can only hold the 32-bit *IP number*, hosts need a way to determine IP numbers if they only know IP names.

Figure H.10 shows that they can do this by calling a **domain name system (DNS)** host. These DNS hosts maintain tables of host names and host numbers. When the DNS host receives a *DNS request message* containing a host name, it looks up the corresponding IP number. It returns this IP number in a *DNS response message*.

FIGURE H.9

IP Autoconfiguration

The host sends an autoconfiguration request message to an autoconfiguration host. This message asks for an IP number. The autoconfiguration host sends back an autoconfiguration response message that contains an IP number for the host to use.

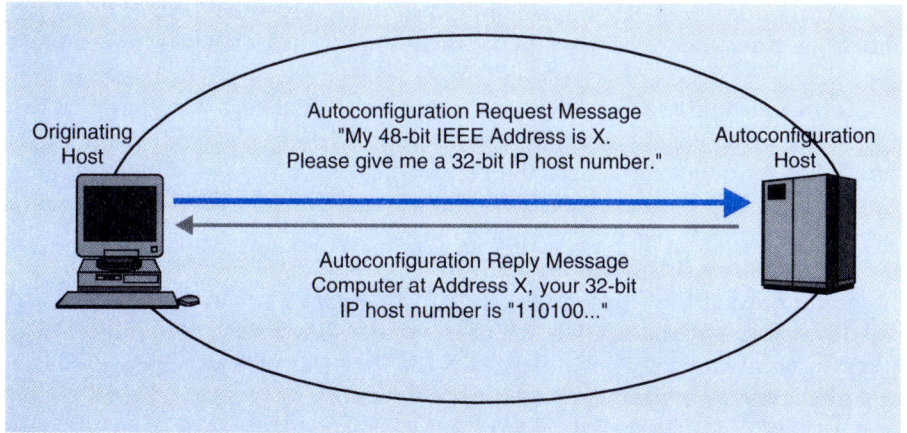

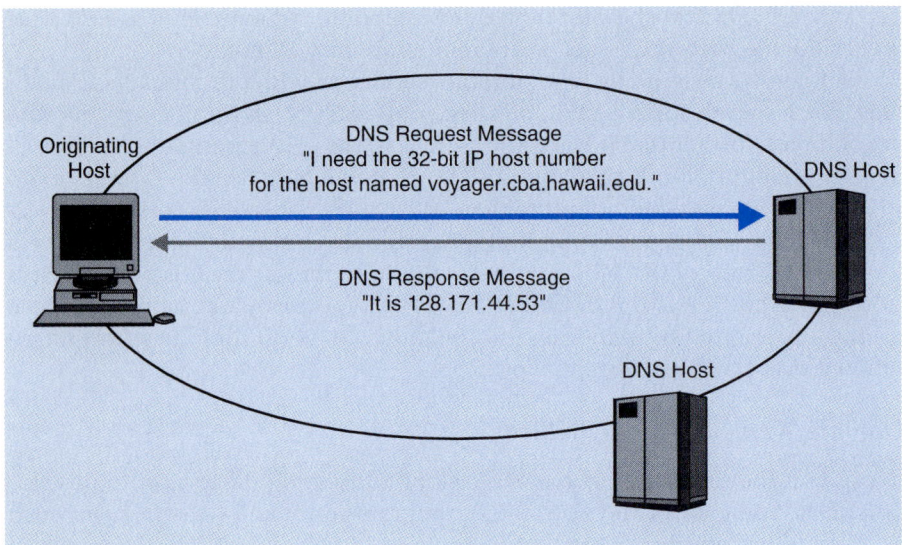

FIGURE H.10

Domain Name System (DNS)

The sending host knows the IP name of the destination host but needs the corresponding 32-bit IP number to place in the destination address field of outgoing datagrams. The sending host transmits a DNS request message to a DNS host. This DNS request message contains the IP name of the destination host. The DNS host looks up the IP number in a table and sends back a DNS response message containing the 32-bit IP number to be placed in the IP destination host field. If a DNS host does not know the IP number, it contacts another DNS host, which responds with the IP number.

If the called DNS host does not know the IP number for the given IP name, it will call another DNS host. That DNS host will reply with the desired IP number. Most DNS hosts maintain relatively small lists of IP names and IP numbers. Special **root DNS hosts** have all (or at least most) host names.

To make things easier, DNS host names are organized into top-level domains. For instance, the author's host address is *hawaii.edu*. Here the *edu* tells that this is an educational organization in the *edu* domain. Another common top-level domain is *com*, which designates a commercial firm. An example of this is *www.prenhall.com*. Other common top-level domains are *net* for networks and *org* for nonprofit organizations.

Some hosts use country codes as their top-level domain. For instance, *au* is Australia, *ca* is Canada, *ch* is Switzerland, *de* is Germany, *fr* is France, *hk* is Hong Kong, *il* is Israel, *jp* is Japan, *nl* is the Netherlands, *nz* is New Zealand, *uk* is the United Kingdom, and *za* is South Africa.

Internet Control Message Protocol (ICMP) ■

In some ways, the main *supervisory* protocol in TCP/IP is the **Internet Control Message Protocol (ICMP).**[1] The ICMP is a workhorse protocol with several important uses.

ERROR MESSAGES First, Figure H.11 shows that hosts (and routers) use ICMP to send error messages. If your application program gives you a "host unreachable" message, it is merely telling you what an ICMP error message has told you. Although routers do not ask for the retransmission of IP datagrams with damaged messages, they typically try to send ICMP error messages back to the sending host to indicate this and other problems.

QUERIES ICMP also allows one host to send queries to another host. The most common of these is handled by a **Ping** program. When a host "pings" another host, it is sending it a message asking the receiver if it is operational. The receiver then sends a reply saying that it is indeed operational. In ICMP terminology, the Ping program sends an ICMP **echo** message.

FIGURE H.11
Internet Control Message Protocol (ICMP)

ICMP is used to transmit error messages from hosts and routers, such as "host unreachable." It is also used to send queries to other hosts. ICMP is also used for flow control between hosts and for a router to tell a host that the route it selected for a packet is not optimal.

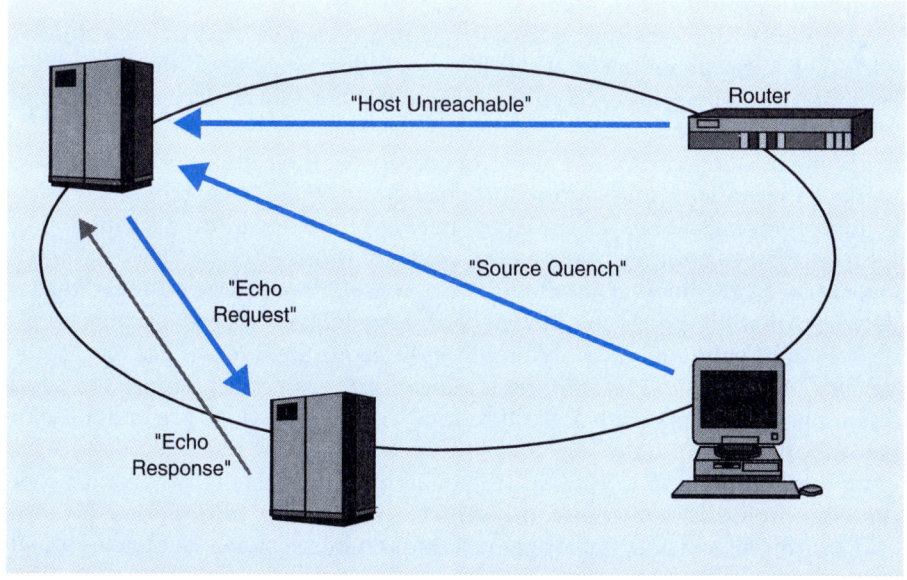

---
[1]RFC 792 [Postel, 1981b].

FLOW CONTROL ICMP also provides flow control. One ICMP message is **source quench.** When a host receives this message from another, it should reduce the rate at which it transmits. It should continue to reduce its transmission rate until it stops receiving source quench messages. When source quench messages finally stop, it should increase its transmission rate slowly. This is a rather weak form of flow control.

ROUTING ICMP is also useful in routing. When a router receives an IP datagram from a host, it may realize that a different router on the network would be better able to handle datagrams to the host. One IP message is a **redirect** message, which advises the sending host to send the message to a different router.

Overall, ICMP offers a rich set of supervisory messages. It supplements the basic IP delivery mechanism. In the introduction to the ICMP RFC, Postel [1981b] notes that "ICMP is actually an integral part of IP, and it must be implemented by every IP module."

Router Coordination Protocols ■

Routing messages across internets requires each router to understand at least its "local neighborhood" of routers, so that it knows where to send IP datagrams that it cannot deliver itself. If routers merely passed on IP datagrams blindly, the datagrams might never be delivered. Every router maintains a file containing information on nearby routers. This file is its **routing table.** Protocols that allow routers to share information in their routing tables are **router coordination protocols.**

Figure H.12 shows that routers exchange messages that tell one another what they know about the local neighborhood. This includes information about how many nearby routers there are, how many hops away each router is from one another, and so forth.

Some router coordination protocols broadcast their information. In others, pairs of routers exchange information using one-to-one messages. The former is good for small networks. In larger networks, however, broadcasting clogs the network with router coordination protocol traffic.

Figure H.13 shows that some networks are **autonomous systems,** that is, networks that are under the control of an organization. For instance, a corporation might have several autonomous systems for its internal use. Each of its subnets, for instance, would be autonomous systems. Linking these autonomous networks together are **autonomous routers,** which are under the organization's control.

Within autonomous systems, organizations may choose any router coordination protocol they wish. For instance, the **Routing Information Protocol (RIP)** is simple and adequate for small autonomous systems, but it uses broadcasting, which can clog large networks. For larger autonomous internets, there is **Open Shortest Path First (OSPF),** which uses one-to-one transmission and can take cost, security, and other route selection matters into consideration.

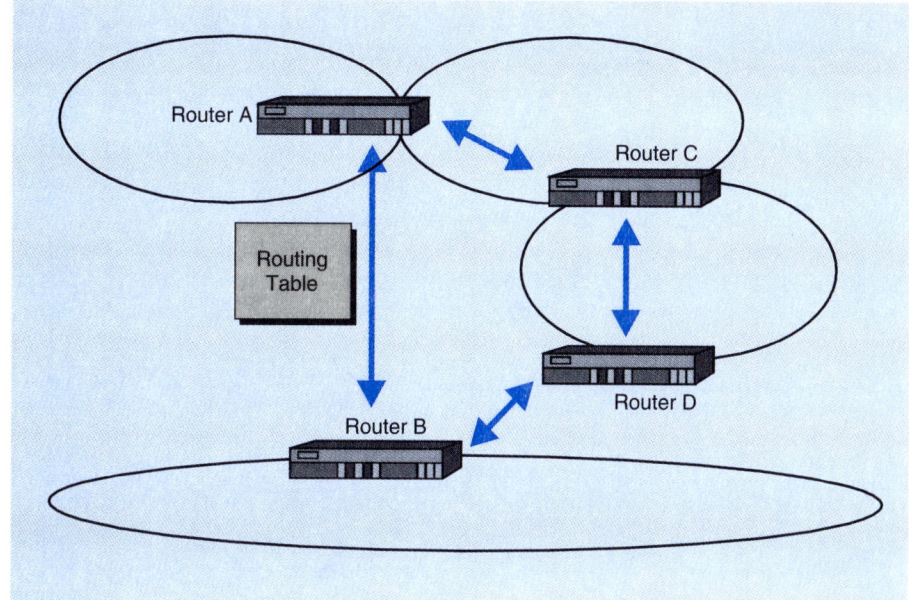

FIGURE H.12

Router Coordination Protocols

Routers must know about nearby routers, so that they can pass on IP datagrams intelligently. This knowledge of nearby routers is held in a routing table. Router coordination protocols allow routers to share information in their routing tables.

When IP datagrams pass out from the organization or into the organizations, they do so through **border routers.** These were previously called *border gateways* because **gateway** is the old term used on the Internet for routers. These border routers must communicate with border routers on other autonomous systems through prescribed standards. On the Internet, the **Border Gateway Protocol (BPG)** is popular.

IP Version 6

As noted earlier, the most widely used version of IP today is IP Version 4 (IPv4). This is the version that has 32 bit addresses. The Internet Engineering Task Force has defined a new version, IP Version 6 **(IPv6).**

128-Bit Addresses ■

IPv4's 32-bit addressing scheme shown in Figure H.6 did not anticipate the enormous growth of the Internet. Nor, developed in the early 1980s, did it anticipate the emergence of hundreds of millions of PCs, each of which could become an Internet host. As a result, the Internet's Network Information Centers are literally running out of the popular and easy-to-use Class B addresses.

IPv6 will expand the address size to 128 bits. This will essentially give an unlimited supply of IPv6 IP numbers, at least for the foreseeable future. It should be sufficient for large numbers of PCs and other computers in organi-

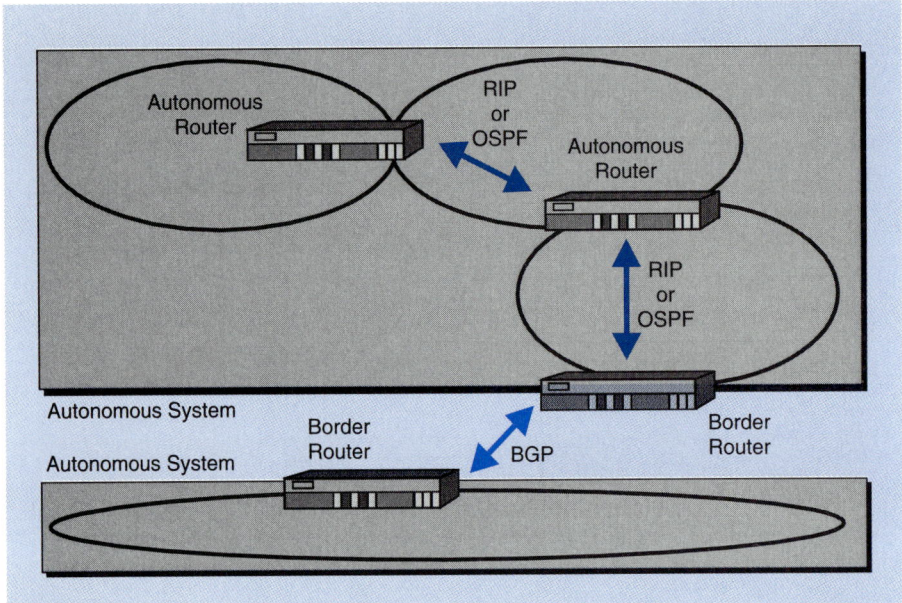

FIGURE H.13

Router Coordination Protocols Within and Between Autonomous Systems (Networks)

Autonomous systems (networks) are under the control of an organization, such as a corporation. Within its autonomous domain, the organization can select its internal router coordination protocols. Typical autonomous router coordination protocols are RIP and OSPF. Other routers link the autonomous systems within the corporation to the outside world. These are border routers (formerly called border gateways). The Internet Society controls border router protocols. A popular border router protocol is the Border Gateway Protocol.

zations. It should even be sufficient if many other types of devices, such as copiers, electric utility meters in homes, and televisions become intelligent enough to need IP numbers.

Authentication, Privacy, and Other Options ■

Although IPv4 can have options, they are difficult to use. In contrast, IPv6 has a well-defined way of handling options through extension headers, which come after the main header.

One of these extension headers provides **authentication.** As Chapter 9 and Module J discuss, this ensures that the source host address really is the address of the source host. Or, at the very least, the source address was not modified during transit. The authentication header is mandatory.

The authentication extension header also provides **integrity** for the entire packet, that is, the assurance that other parts of the packet have not been modified enroute either.

Although integrity ensures that changes have not been made, it does not ensure privacy that others have not read the packet enroute. Another option is **encapsulation security.** This option encrypts the packet, ensuring that the message cannot be read enroute. While the authentication extension header is mandatory, the **privacy extension header** is optional.

Autoconfiguration

Recall that in the Reverse Address Resolution Protocol, a RARP server assigns a host an IP number, usually when the host boots up. IPv6 has extended this capability considerably in its autoconfiguration service. For instance, suppose you are using one Internet service provider and switch to another while you are doing work on the Internet. The autoconfiguration service will switch your computer's IP number immediately to one assigned by the new Internet service provider. (Because IP numbers contain information about the network and different Internet service providers have different networks, the host's IP number cannot remain the same.)

Quality of Service

IPv4 has a **type of service (TOS)** field, which specifies various aspects of delivery quality, but it is not widely used. In contrast, IPv6 has a well-defined **quality of service** field, in which quality of service parameters are assigned to **flows**—series of IPv6 packets with similar quality of service needs. Quality of service parameters might require such things as low latency for voice and video while allowing electronic mail traffic and World Wide Web traffic to be preempted temporarily during periods of high congestion. When an IP datagram arrives at a router, the router looks at its flow number and gives it appropriate priority.

Piecemeal Deployment

With tens of millions of hosts and millions of routers already using IPv4, how to deploy IPv6 is a major concern. The new standard has been defined to allow **piecemeal deployment,** meaning that the new standard is backwardly compatible with IPv4 and can be implemented in various part of the Internet without affecting other parts or cutting off communication between hosts with different IP versions.

CORE REVIEW QUESTIONS

1. Name the four major layers in the TCP/IP architecture.
2. What are the parts in an IP number? Why are there multiple classes of networks in IP addressing? What are the four classes of IP numbers? Which class is used in most universities? How can you tell, by looking at an IP number, the network's class?
3. What is a datagram?

4. What does a host have to know to send an IP datagram to another host? Describe the source host's role in IP delivery, depending on whether the destination host is on the same network or is on a different network.
5. Explain the router's role in IP delivery, depending on whether the destination host is or is not on one of the networks to which the router is directly attached.
6. Distinguish between host IP numbers and host IP names. What is a port number? A socket?
7. What are subnet masks? In subnet masks, what do zeros represent? How can you distinguish between 1s that represent the network part and 1s that represent the subnet part?
8. IP Version 4 addresses are almost exhausted. What is the Internet Engineering Task Force doing about this?
9. Is IP a reliable or unreliable service? How does TCP complement IP delivery? Is UDP reliable? For what kinds of messages is UDP appropriate?
10. How does an organization use autoconfiguration protocols to ration its pool of assigned host ID numbers?
11. What is the Domain Name System (DNS)? Why is it crucial when a host wishes to send an IP datagram?
12. What are the roles of ICMP in internetting?
13. Why do you need router coordination protocols?
14. What is an autonomous system (network)? Name two popular autonomous system router coordination protocols. What is a border router? What other term does TCP/IP use for routers? What is the Border Gateway Protocol?
15. Name the major advances in IPv6.

DETAILED REVIEW QUESTIONS

1. Why do routers have to be at least dual-homed? How many Internet addresses does a dual-homed router have? Why?
2. How many bits are there in an IP number? How are these bits organized in Class B addresses? Why are Class B addresses in especially high demand?
3. You have a Class B network. You want each subnet to have 512 possible hosts. What subnet mask would you create?
4. List at least three major domain names.
5. List some specific types of messages that ICMP can carry.
6. Name three autoconfiguration protocols.
7. How does IP Version 6 enhance autoconfiguration?

THOUGHT QUESTIONS

1. Why do you think the original IP number specification standard failed to provide a sufficient number of addresses?
2. Why do you think the original IP specification failed to anticipate the need for subnet masks?
3. You are setting up Windows 95 for TCP/IP access to the Internet. You must specify a *DNS host* and a *default gateway*. For both, you must include an *IP number*. For the DNS host, you include your own host name (not number). You can select whether you want your *IP number* and *subnet mask* to be *obtained automatically* or whether you will have a *permanent IP number* and subnet mask. Explain the concepts in italics to a person who is not familiar with TCP/IP internetting.
4. Reword Figure H.5 (Overview of Packet Delivery) if subnetting is used. Redo Core Review Questions 3 and 4, replacing "network" with "subnet."

For online exercises, please visit this book's website at: http://www.prenhall.com/panko

REFERENCES

HEDRICK, C. L. (1988, June). RFC 1058, *Routing Information Protocol*.

LOUGHEED, K., and REKHTER, Y. (1991, October). RFC 1267, *A Border Gateway Protocol 3 (BPG-3)*.

MALKIN, G.S. (1993, January). RFC 1388, *RIP Version 2: Carrying Additional Information*.

MOY, J. (1991, July). RFC 1247, *OSPF Version 2*.

POSTEL, J. (1980, August 28). RFC 768, *User Datagram Protocol*.

POSTEL, J., ed., (1981a, September). RFC 791, *Internet Protocol*.

POSTEL, J. (1981b, September). RFC 792, *Internet Control Message Protocol*.

POSTEL, J., ed., (1981c, September). RFC 793, *Transmission Control Protocol*.

MODULE I

Managing a PC Network

INTRODUCTION

Chapter 8 focused on the user's viewpoint in PC networking. In this module, we look at the PC network administrator's job primarily at the level of installing server and client PC hardware and software, establishing user rights in various directories, and doing administrative and user support tasks.

Although this module draws its examples from Novell NetWare, specifically the 3.X series, its focus is fairly general. In terms of servers, it focuses on file server installation, but the installation of other servers is very similar and is usually simpler.

INSTALLING A FILE SERVER

Although a network administrator may only have to install one or two servers a year, these will be among the most critical activities in the network administrator's year. Compared to, say, installing a word processing program in Windows, installing server operating system (SOS) software can be anything but smooth and automatic. In addition, installing a file server requires a great deal of preplanning because once directory structures and other details are established, changing them is difficult. Mistakes made when installing a file server tend to endure for a long time.

Planning the Directory Structure

Before the network administrator installs any software, he or she must first design the directory structure for the file server. While it is desirable to have a simple and clear directory structure on any PC's hard disk, the design of the directory structure is critical for file servers. Dozens or perhaps hundreds of people will share the directory structure. To change it later would result in widespread confusion.

Planning is especially critical in synchronized server networks (see Chapter 8). For such networks, the entire directory structure should be planned before any servers are installed.

Top-Level Directories ■

Figure I.1 shows a model directory structure on a file server. In the figure, each top-level directory has a special purpose.

■ **User** holds private directories for individual users. These private directories correspond to Drive F: in Chapter 8.

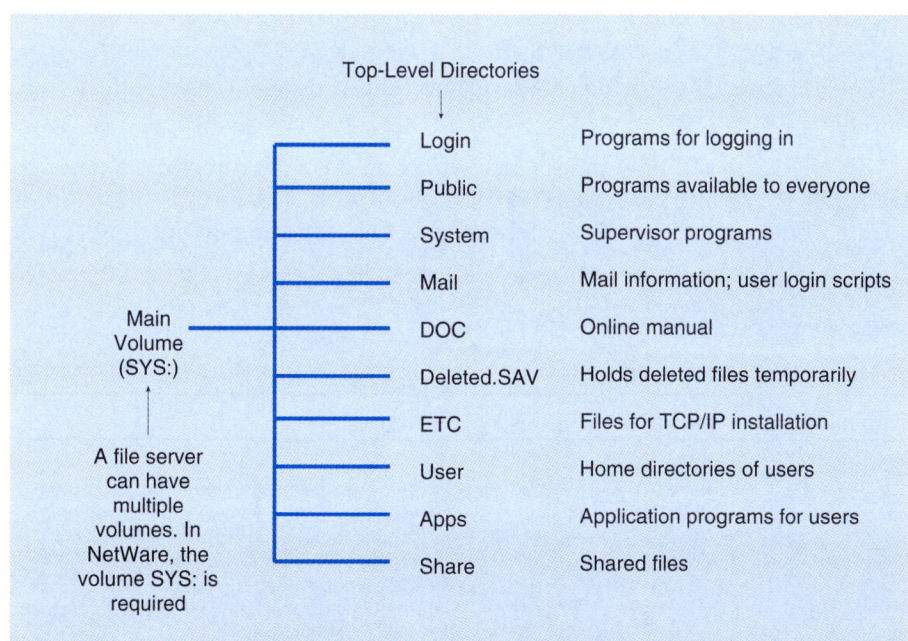

FIGURE I.1

Directory Structure on a File Server

By assigning top-level directories intelligently, the network administrator can manage rights to various directories with a minimum amount of work.

- **Apps** holds application programs that are available to all or most users. Program directories correspond to Drives Y: and Z: in Chapter 8.
- **Share** holds directories of files that will be shared by groups of users. These shared directories correspond to Drive S: in Chapter 8.
- **System** holds files that should only be used by the systems administrator. In Novell NetWare, this corresponds to an account called the *Supervisor*.

Rights

Users must be given specific **rights** in all directories and subdirectories in the system. These rights determine whether or not the user can even see the existence of a directory, whether or not the user can execute programs found there, whether or not the user can create, edit, or delete files in the directory, and so forth. Table 10.1 shows rights in Novell NetWare.

In general, there are three ways to give a user rights.

- The first way is to assign the **user** specific rights in the directory. This allows very precise assignment, but it is extremely time-consuming because there are so many users on the PC network and because there are hundreds or even thousands of directories.
- The second way is to assign specific rights in a directory to a **group** to which the user belongs. One group might be members of the user's department. Another might be members of a project team. A third might be people authorized to use word processing. If a group has rights in a directory, so does an individual member of the group. Assigning rights to a group is easier than assigning rights separately to many individuals, but there are still hundreds or thousands of directories.
- The third way is to let the user or group **inherit rights** in the directory from a parent directory.

Figure I.2 shows how inheritance works. In the *Apps* directory, a user has the rights to execute programs (File Scan and Read). The *Database* directory is a child of the *Apps* directory. By inheritance, the user also inherits rights to ex-

TABLE I.1

Rights in Netware

Symbol	Name	Meaning
S	Supervisory	All rights
R	Read	Can retrieve files on the server's disk drives
W	Write	Can edit files on the server's disk drives
C	Create	Can create new files/directions
E	Erase	Can delete files/directories
M	Modify	Modify file attributes, such as read-only
F	File Scan	Can see files/directories using the Dir command
A	Access Control	Can grant rights to others.

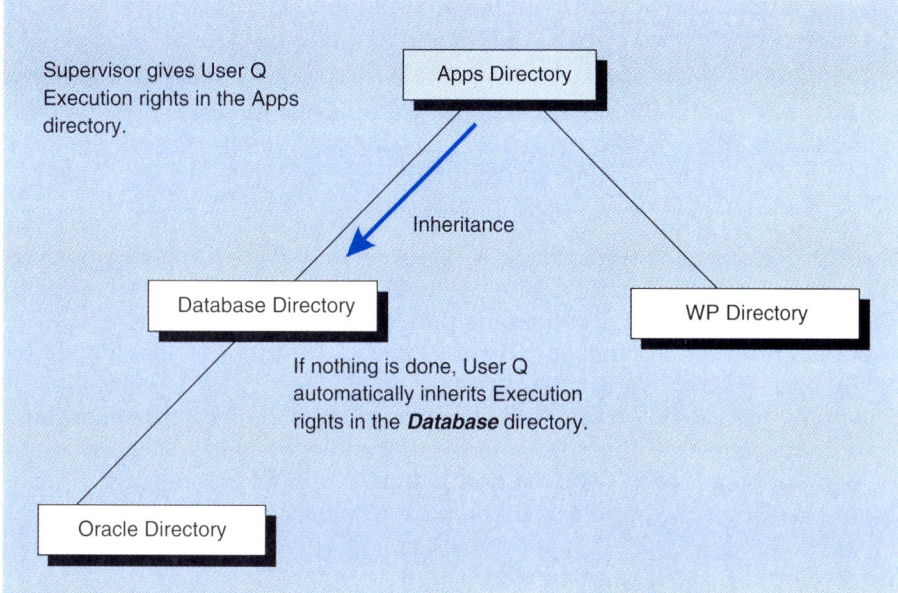

FIGURE I.2

Automatic Inheritance of User Rights

By default, all the rights that a user (or group) has in a directory are inherited automatically in every direct child directory. Of course, inheritance can be blocked or modified.

ecute programs in the *Database* directory. The principle that applies in this and other situations is that by default, all the rights that a user (or group) has in a directory are inherited automatically in every direct child directory.

Of course, it is possible to block default inheritance, so that the user or group inherits no rights in a child directory. It is also possible to modify inheritance so that the user inherits some rights but not others.

Inheritance in Apps ■

It is perhaps easiest to see inheritance in Apps. In general, Apps is for programs to which most users or even all users will have rights. So the network administrator normally assigns execution rights in the Apps directory to a group called Everyone. As you would expect, every user is a member of this group, so every user inherits rights automatically in all child directories. Inheritance also extends to the child directories of these child directories and so on—down to the lowest-level subdirectory.

If there is a directory in which only certain users should have execution rights, then the network administrator can block all inheritance in that directory. The administrator can then assign specific rights to individuals or groups.

While this process is not completely automatic or simple, it is certainly more automatic and simpler than assigning each user rights in each individual directory.

To test your knowledge, what do you think the default rights would be in the User, Shared, and System top-level directories?

Selecting Server Hardware

After designing the directory structure, the next task is to select the hardware for the server. Some file servers will require extensive power. Others can get by with simpler machines.

PC Server or RISC Workstation Server

The first decision is whether to use a PC server or a RISC workstation server. PC servers—even high-end PC servers—cost less than RISC servers. However, if you need extremely high processing power, then you need to choose a RISC workstation server. Fortunately, many server operating systems operate on both PC servers and workstation servers.

If you select a PC server, then you must make a choice between an Intel PC or a Macintosh PC. Intel PC servers can handle Macintosh client PCs and are normally used when there are mostly Intel PCs in an organization. Macintosh servers, however, are ideal if you have mostly Macintosh clients.

Mother Board and Expansion Boards

Figure I.3 shows how printed circuit boards are organized within a PC's systems unit. First, there is a main printed circuit board, the **mother board**. The mother board holds the microprocessor, most or all of the computer's RAM, and other essential circuits. The mother board typically lies on the floor of the systems unit.

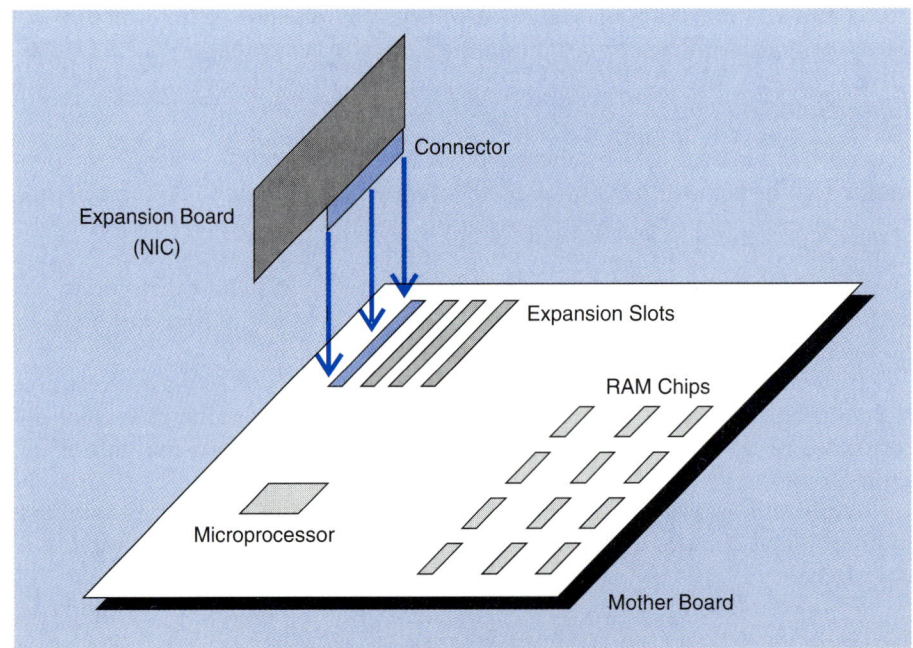

FIGURE I.3

Mother Board and Expansion Boards

The *mother board* holds the computer's essential circuits, such as its microprocessor. *Expansion boards* are for additional electronics, such as NICs and disk drive controllers. Expansion boards plug into *expansion slots* on the mother board.

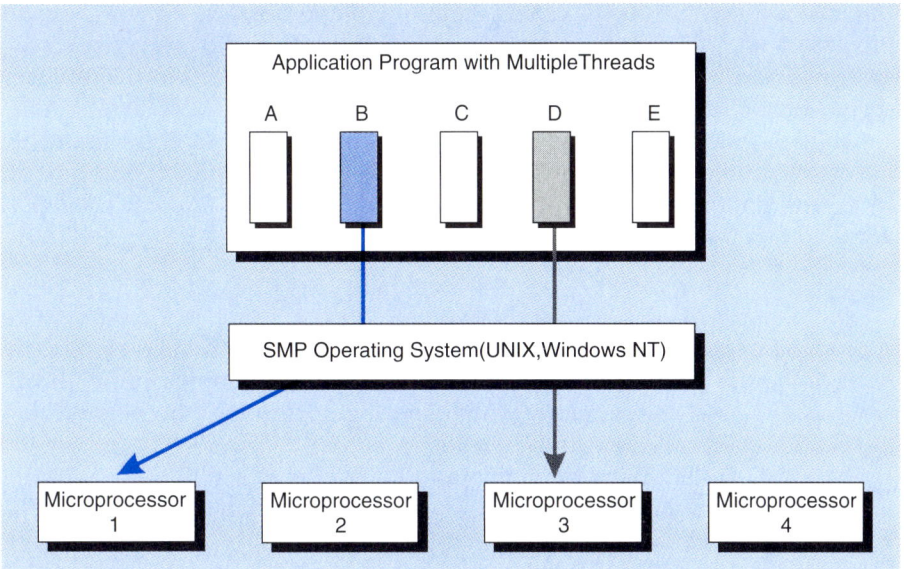

FIGURE I.4
Symmetric Multiprocessing (SMP)

A symmetric multiprocessing computer has more than one microprocessor—generally two to four. An SMP server operating system can work with an SMP application. The SOS assigns threads of the application to microprocessors as they become free. SMP is found on both Intel computers and on RISC workstation servers.

The mother board has **expansion slots** into which you can plug additional printed circuit boards. These additional boards are called expansion boards. NICs usually come as **expansion boards**. So do disk drive controllers. Video cards, in turn, send information to the display.

Microprocessor ■

If you select an Intel platform, then you must decide which Intel processor to use. At the time of this writing, the Pentium is the most widely used microprocessor in high-end servers, but the Pentium Pro is becoming popular. However, the microprocessor situation is changing very rapidly.

Symmetric Multiprocessing (SMP) ■

Most server operating systems can work on **symmetric multiprocessing (SMP) machines.** As shown in Figure I.4, an SMP machine has multiple microprocessors. Most SMP machines have either two or four microprocessors. Some have more, but incremental benefits tend to fall beyond two or three microprocessors.

For SMP to work, you need **SMP applications** written to execute on an SMP server. These applications are divided into many small parts called

threads. Different threads can execute simultaneously on different microprocessors without causing problems. You also need an **SMP server operating system**. These SOSs understand SMP and can assign threads to microprocessors as they become free.

Initially, SMP was only available on RISC workstation servers running the UNIX operating system. However, Windows NT can support SMP on both RISC workstations and on SMP Intel computers. Novell has announced an SMP version of its NetWare SOS at the time of this writing.

Busses ■

Although fast microprocessors are good, many servers are limited by other aspects of the server design. One is the server's bus. As Figure I.5 shows, the microprocessor must communicate with other circuits through a high-speed communication path called the **bus**.

Most PCs today have three busses. One is a **memory bus** for accessing memory chips on the mother board (the main printed circuit board in a PC). The user normally has no access to this bus, which is typically proprietary. Memory chips installed on the mother board are automatically connected to the memory bus. On a server, the mother board should be able to hold at least 128 MB of RAM.

Second, there is the **main bus**. This is a rather slow-speed bus. It typically follows the ISA design created in 1984. This bus only runs about 8 MHz—far below the speed of the microprocessor. However, there are many expansion boards that work with ISA busses and do not require more speed.

FIGURE I.5

Computer Bus

The microprocessor communicates with other circuits via transmission lines called *busses*. Most computers today have three busses. The *memory bus* connects the microprocessor to RAM. The *main bus* is for low-speed expansion boards. The *local bus* is for high-speed expansion boards.

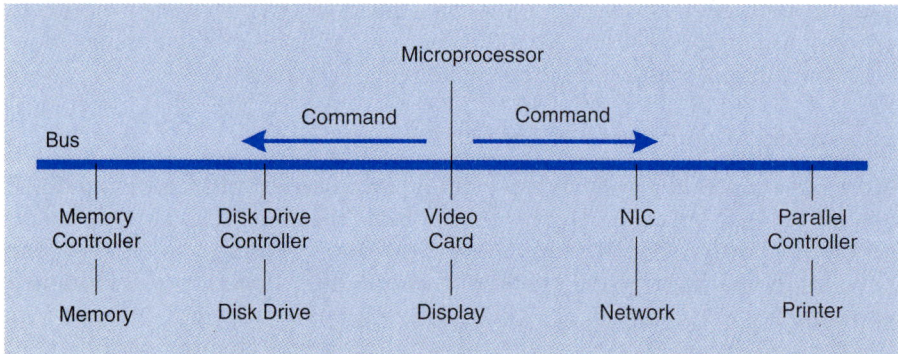

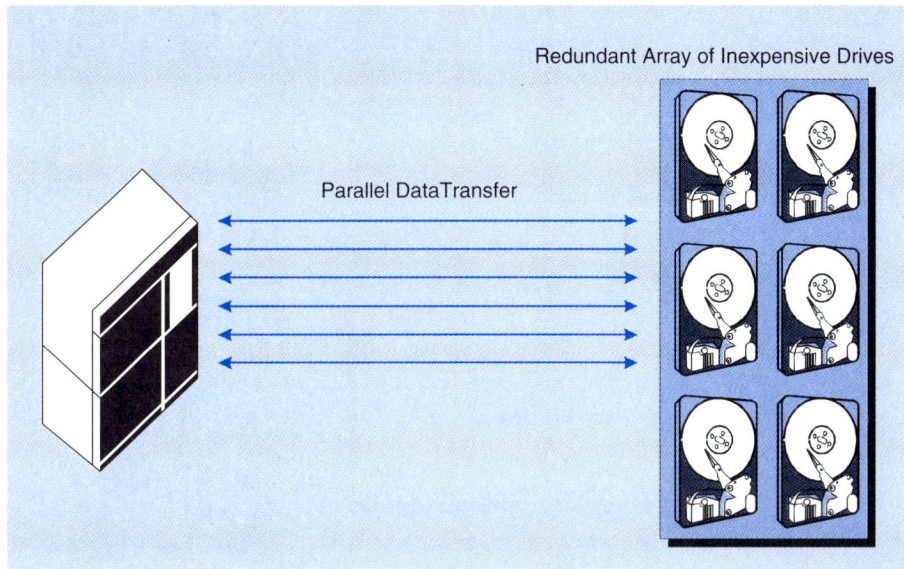

FIGURE I.6

RAID Disk Array

A RAID controller can write bits in parallel to multiple drives. This is faster than serial transmission. In RAID Level 5, information is written redundantly, so that if one drive fails, no information is lost.

Third, there is the **local bus or PCI bus,** which is used for high-speed expansion boards, including disk drive controllers and 100 Mbps NICs. Such expansion boards are expensive, but slower boards would act as bottlenecks, slowing the entire system.

Disk Drive Standard ■

Many servers, especially file servers, tend to do extensive I/O (input/output) to disk drives. This makes the use of high-speed disk drives essential.

Most client PCs today use EIDE (enhanced integrated drive electronics) disk drives. In contrast, most servers use SCSI (small computer systems interface) drives. SCSI drives are faster and can have higher capacity. In addition, a single SCSI controller board can control up to seven hard disk drives and CD-ROM drives.

For high performance and safety, many servers use RAID (redundant array of inexpensive disks) drives. As Figure I.6 shows, a RAID controller controls a group of disk drives.

Controllers that follow the standard write bits to be stored in parallel to the multiple drives. In RAID Level 5, different bits go to different drives, so each byte is scattered over the drives.

From Chapter 4, parallel transmission is faster than serial transmission, so even with inexpensive disk drives, performance is fast. Reads are also done in parallel.

RAID Level 5 stores the information redundantly, so that if one drive fails, no information is lost. Typically, disk drives are the least reliable component in a server, so this is a very valuable feature of RAID drives.

FIGURE I.7

Disk Caching

The server operating system copies frequently used information on the disk drive into RAM into an area called a disk cache. This cache will be tens of megabytes. In this way, information can be accessed from high-speed RAM rather than from slow-speed disk drives.

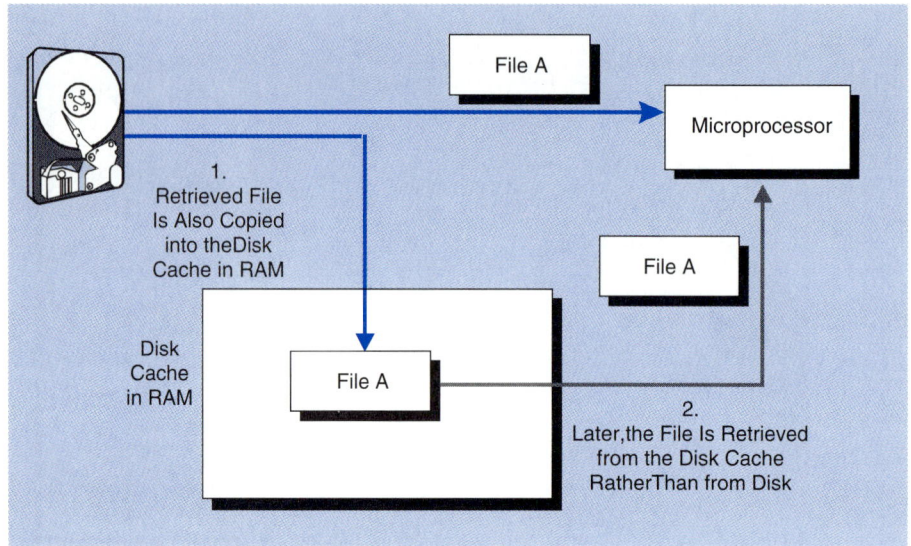

RAM ■

RAM is much faster than disk. As shown in Figure I.7, information that is accessed frequently from disk drives is copied into RAM, into an area called the **disk cache**.[1] These disk caches tend to be tens of megabytes. In fact, most server operating systems use *all* unused RAM as a cache, in order to keep as much information in high-speed RAM as possible. In general, most SOSs are very sophisticated about disk caches.

Disk caches help to reduce the strong tendency of disk drives to cause bottlenecks that limit the performance of the server. In many cases, slow server performance can be enhanced by adding more RAM, which the SOS can use to increase the size of the cache.

Although disk caches are very useful, they are slightly dangerous. If there is a power failure, the information in the disk cache will be lost. If changes have been made to information in the cache and have not been saved to disk, the loss of this information can create extensive problems.

With disk caches, it is critical to have an uninterruptible power supply, so that a loss of power does not flush the disk cache. An uninterruptible power supply gives the administrator time enough to shut down the server gracefully, closing out the cache. Some uninterruptible power supplies can even shut down the server by themselves if the power outage persists.

---

[1] Most disk drives themselves have on-board disk caches, but these usually are only about 256 KB.

Backup ■

For client PCs, normal backup technologies are floppy disks and quarter-inch (QIC) tape drives. However, floppies are too slow for servers, and QIC tape drives are too limited in capacity, storing only about 500 MB. This is far too little for servers. They are also slow.

Tape drives backup systems for servers usually follow higher-quality tape standards. For instance, **DLT** (digital linear tape) can store about 50 GB on a single tape and is much faster than QIC as well. It is also far more expensive.

NIC ■

For high-speed throughput, many servers use **32-bit NICs,** which fit into local bus slots. This is especially important for 100 Mbps NICs. In contrast, 10 Mbps NICs for client PCs normally fit into ISA expansion slots.

Installing the Server Operating System

Users who install modern application programs, such as spreadsheet programs, have a relatively easy time of things. Smooth Install or Setup programs ask the user a few simple questions and then go off by themselves. If you have floppy disks, the program will ask you to insert new disks periodically. If you install from a CD-ROM disk, you will not even have that kind of interruption.

Installing the server operating system in NetWare and many other server operating systems is not as easy. There is a similar phase in which the installation program asks you questions and then works alone, but this is only part of the process. Roughly half of the installation time may fall outside this automated sequence.

Installing Application Software

Unless you have a system with many servers, a typical network administrator will only install a handful of file servers each year. Any network administrator, however, will have to install several pieces of new application software each year. Given the frequency of upgrades, the administrator will have to install several times as many upgrades.

Installing application software on the server can be tricky, and if a firm has many file servers, this server installation can be very time-consuming. A firm may have dozens of application programs to store on servers, and these are upgraded about once a year on the average.

Even if you store the application program on a server, many programs still require you to install at least a little software or other setup files on each client PC. This is called the program's **stub**. This traditionally requires leaving the network control center and making appointments to upgrade individual PCs.

As discussed in Chapter 8, if you have more than one server, it may be possible to install application software on multiple servers through electronic software distribution (ESD). ESD may even be able to install stub programs on client PCs.

Installing NetWare Loadable Modules (NLMs) ■

Figure I.8 shows that a file server stores two types of programs. Some are written to run on the client PC. The file sever merely stores these for delivery to the client PCs. Other programs run on the file server itself, using the server operating system. These programs operate the server, help to manage it, or are client/server server programs. Figure I.8 shows that in Novell NetWare, programs that execute on the server are called **NLMs** (NetWare Loadable Modules).

Note that all application programs have to be written for specific operating systems. When an application program runs, it periodically issues calls (re-

FIGURE I.8

Execution of Programs Stored on the Server's Disk Drives

The file server has two types of programs. Some are written to execute on the client PC. In the case of MS-DOS, these end with the extensions .COM and .EXE. Other programs are designed to run on the file server, using the server operating system. In Novell NetWare, application programs running on the server are called NetWare Loadable Modules (NLMs). They typically have the extension .NLM.

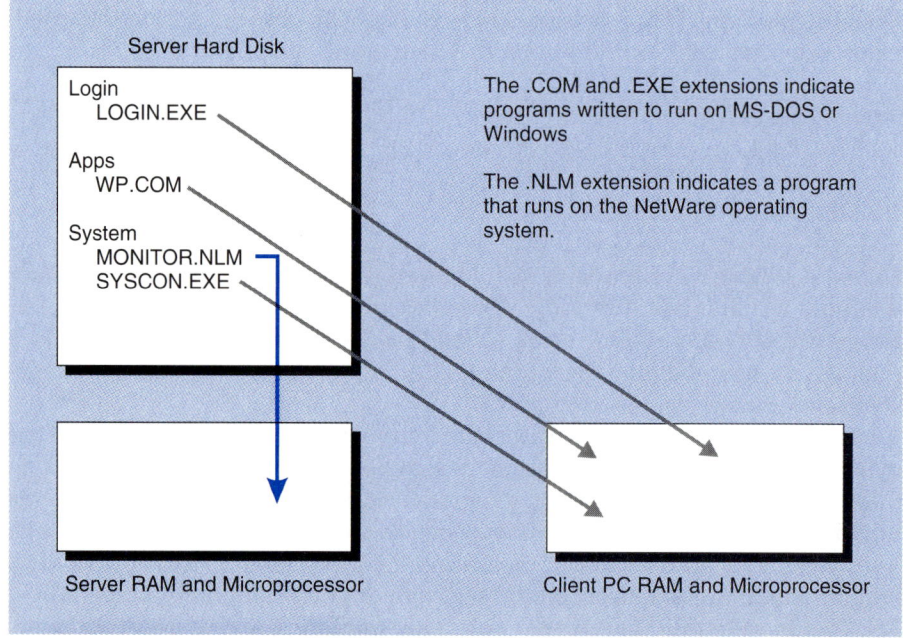

quests for service) to the operating system. These calls are in different formats for different operating systems. So a specific application program can only run on a single operating system. This is why you cannot run a Windows program on an MS-DOS machine that is not running Windows, on a Macintosh PC, or on a UNIX machine. It is also why you cannot run a UNIX or Windows program on a server running the NetWare operating system.

Programs that Execute on the Client PC ■

The file server will store many programs written to execute on the client PC, such as word processing programs and electronic mail programs. For such programs, the file server merely acts as a storage device. For execution, it ships these files to the client PC. We saw in Chapter 8 that this is the essence of file server program access.

All such program files must be executable on the client machines. Application programs that execute on IBM compatible computers must be in MS-DOS or Windows format. (In MS-DOS, most programs have the extensions .COM or .EXE.) Application programs that execute on Macintosh clients, in turn, must be in the Macintosh executable format. In Figure I.8, the word processing application program has the extension .COM.

Besides application programs, most PC networks have administration programs that execute on the client PC. For example, LOGIN.EXE, which the NetWare user executes to log into the network, is stored on the file server but executed on the client PC. In the logging in process, the client shell sends out a call for the nearest server. The server downloads the LOGIN.EXE program to the client PC, where it executes. Many other network management utilities execute on client PCs.

ADDING A USER

Although most administrators only install a few servers each year, they add dozens or even hundreds of users each year. This consists of two steps: (1) installing hardware and software on the user's PC, in order to change it into a client PC, and (2) adding the user's account on the file server. Adding a new user quickly becomes second nature to network administrators. However, the process is still time-consuming, and it is often difficult.

Installing Hardware and Software on a Client PC

Suppose the user begins with a stand-alone PC. The first step is to turn the stand-alone PC into a client PC.

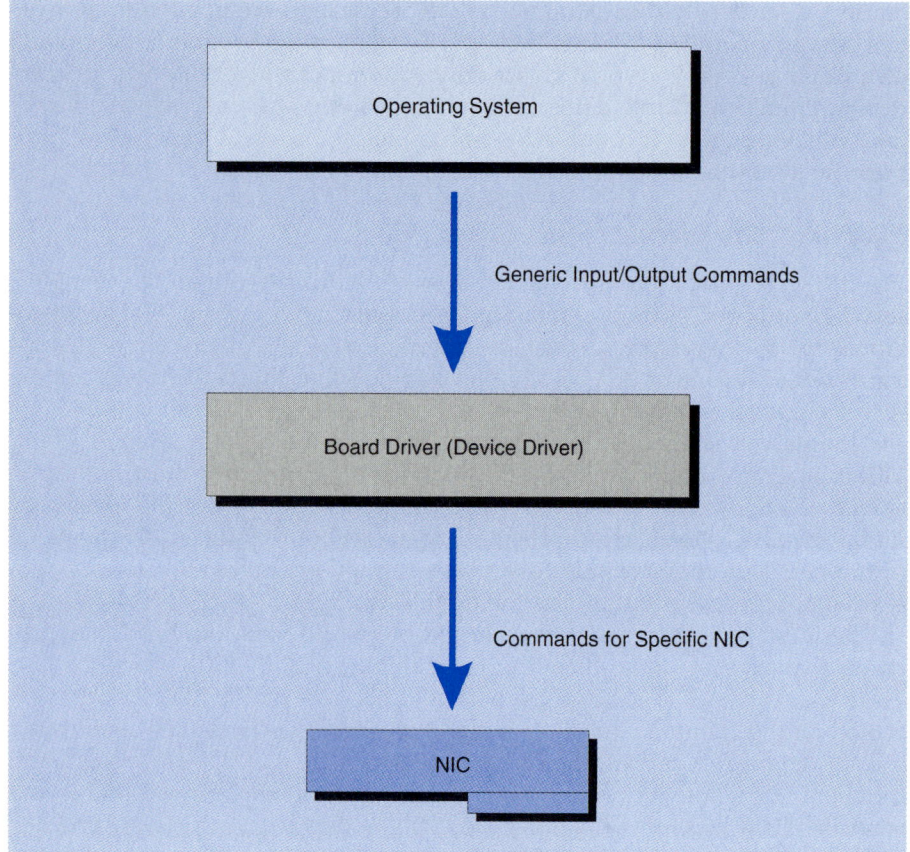

FIGURE I.9

NIC and Board Driver

The board driver software allows the operating system to send commands to the NIC. The board driver is a general class of programs called device drivers.

Adding a NIC ■

The first step in converting a stand-alone PC into a client PC is to install a network interface card (NIC). The NIC is usually an expansion board. In addition, the NIC normally comes with a piece of software called a **board driver**. As shown in Figure I.9, the operating system sends commands to the board driver, which in turn sends them on to the NIC.

Plug-and-Play NIC Installation ■

New PCs use plug-and-play technology that allows you to install the board physically and then install the board driver. When you restart the computer, you will be able to use the NIC immediately.

For plug-and-play installation to work, however, the computer must be capable of plug-and-play operation,[2] you must be using a plug-and-play op-

---

[2]Plug-and-play PCs have ROM BIOS (Basic Input Output System) programs in ROM that are compatible with the plug-and-play standard.

erating system (such as Windows 95 or later versions), and you must be using a plug-and-play NIC. In fact, all of your expansion boards should be plug-and-play boards. Otherwise, there may still be installation difficulties.

Manual NIC Installation ■

Older PCs require a more complex installation process for NICs. As discussed in the box "NIC Address Settings," any expansion board has several addresses that it uses in listening to commands from the microprocessor and in sending requests for service to the microprocessors. If two expansion boards use the same address of any kind, the resulting conflict can cause the computer to crash, losing all information. This need to make sure that the address settings on the NIC are unique can be very time consuming. You must know the settings on all other boards and then select free settings for the NIC. *For more on NIC installation, see the box "NIC Address Settings."*

Network-Capable Client Operating Systems ■

In Chapter 8, we saw that Windows 95 and other new operating systems are network capable. In terms of Figure I.11, this means that they are ready for networking with no software beyond the board driver. The board driver is still necessary because it is linked to the NIC hardware. They are sold as a pair.

Client Shell ■

If the operating system is not network capable, however, additional software is needed. This is called the **client shell** software. Figure I.11 shows that the client shell includes a **redirector program** that wraps itself around the basic operating system. This allows it to intercept application program calls for virtual drives and virtual ports, so that the client PC's basic operating system never has to deal with them.

Figure I.11 also shows **adaption-layer software,** which feeds output from both the redirector and from other programs to the board driver. The redirector is only for file service. Other PC network services allow the client PC software to communicate directly with server software. For instance, in client/server processing (see Chapter 8), the client program will communicate with the server program via the adaption layer and then the board driver.

Customizing the Connection to the File Server ■

Normally, operating systems have ways to customize the client PC's connection to servers. For instance, if you normally log into a particular server, it should be possible for this to be the default server when the user logs in. Most network-capable operating systems and client shells have customization capabilities. In some, you select choices from dialog boxes. In older ones, you create text files containing the information.

NIC Address Settings

The microprocessor must be able to communicate with the NIC. It cannot simply yell, "Hey you!" It must send commands to addresses on the NIC. In the same way, when the NIC needs attention, it cannot simply whistle at the microprocessor. It has to send a signal that contains the NIC's address so that the microprocessor will know which circuit needs attention. In this box, we will look at the addresses used on ISA busses. Figure I.10 illustrates the many addresses that must be used.

It is necessary for each of these addresses to be unique to the board. If another board in the PC uses even one of these addresses, there will be a conflict, which could cause the computer to crash.

FIGURE I.10

Traditional NIC Hardware Settings

On the NIC itself, ISA bus NICs must have four communication settings. The microprocessor will use the board's I/O or *port addresses* to send commands and data to registers on the NIC. The microprocessor will also send data to (and read data from) RAM on the NIC. This will require a *base memory address* for the NIC. For communication in the opposite direction, the NIC will need an interrupt number (IRQ) to send requests for service. Some boards also have DMA numbers for requesting block data transfers.

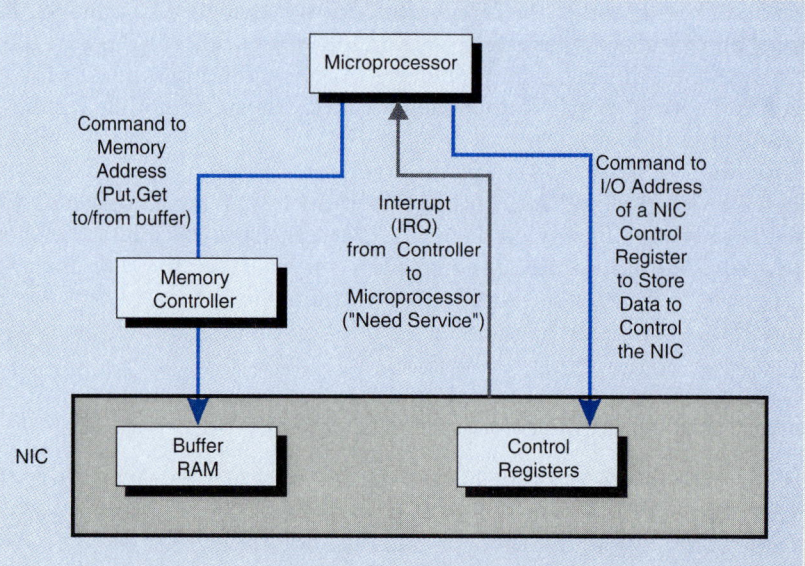

I/O or Port Address

The NIC has multiple **registers** or small holding areas for information. Each may be only a single byte long. When the microprocessor wishes to send the NIC command, it writes information into one or several of these registers. The NIC then acts on the basis of what is in its registers.

Each register has an **I/O (input/output) address,** also called a **port address.** Because there are multiple registers, there must be a block of I/O addresses set on the NIC—one for each register. For instance, a typical block of addresses for a NIC might be 300h to 30Fh. The h indicates that this number is in base 16. There are 16 numbers specified by 300h and 30Fh, corresponding to 16 registers on the card.

Base Memory Address

The NIC has a certain amount of RAM, which it uses as a **buffer** or temporary storage area. When the microprocessor sends data to be sent to the network, the NIC holds the data here until it has packaged the data into frames (see Chapter 5) and until it is free to transmit its frames. Conversely, when new frames arrive, it holds the data in the RAM buffer until it has removed the data from the frame and until the microprocessor can handle the interrupt and remove the data from the buffer.

Every byte in RAM (and ROM too) must have a unique address, or the microprocessor's store and retrieve commands will try to work on two separate bytes of memory simultaneously, with unhappy consequences.

On the NIC, the installer must set the **base memory address,** which specifies the starting address of the RAM buffer. The next byte in the RAM buffer, of course, then has the base memory address plus one.

The installer must set the base memory address so that no byte in the NIC's RAM buffer overlaps with any other location in memory. This can be tricky because RAM buffers on NICs and other devices must be in the narrow **upper memory** area from 640 KB to 1 MB. In addition, the upper memory area holds ROM programs as well.

Interrupt (IRQ) Number

Most of the time, the microprocessor initiates communication by sending commands to the NIC. Sometimes, however, the NIC must send a message to the microprocessor. For instance, when new frames arrive from the network, the NIC must let the microprocessor know, so that the frames can be handled.

To get the microprocessor's attention, the NIC sends an **interrupt request (IRQ)** message. When a microprocessor receives an IRQ message, it interrupts the program currently executing and loads a program that handles the interrupt.

continued

So that the microprocessor will know what circuits sent it the interrupt, each circuit that wishes to send interrupts must be assigned a unique IRQ number.

Computers only have 15 IRQ addresses (IRQ0 through IRQ15). When an installer goes to install a NIC, most of these will already be in use. As a result, the installer must set the NIC's IRQ carefully, or there will be conflicts.

Direct Memory Access Request Number

Recall that an IRQ asks the microprocessor for a priority data transfer. There is a second type of interrupt called a direct memory access (DMA) request. This requests a burst-mode transfer, which transfers a large block of data in a single command.

Some NICs use DMA, while others do not. If a NIC uses DMA, the installer must specify the number of the DMA channel it will use. Fortunately, few boards use DMA transfers, so there usually is a free DMA number for the installer to select.

Most PCs have 15 possible DMA request (DRQ) numbers. The NIC must be set to a particular DRQ number, such as DRQ2, so that the microprocessor will know which circuit sent the DMA request.

Default Values

Hopefully, the **default** (factory-set) values for each of the four settings will match available settings on the PC. If they do, then the installer has no work to do. If one or more of the default values are taken, however, the installer must change the board's default settings to available values.

Manual versus Software Setup

Older boards have switches and other devices on the NIC itself. You must set these physically to specify NIC addresses. This requires you to open the PC and usually to remove the NIC from the computer to make changes.

For newer boards, there are setup programs that configure the board through software.

Configuring the Operating System

We have focused on configuring the board, that is, telling the board what addresses to listen for and what addresses to use when it initiates a communication.

The operating system also has to be configured. It has to be told that the board exists and that it should use the corresponding addresses when talking with it. Typically, this requires running a setup program for the operating system or using the operating system's own configuration utilities.

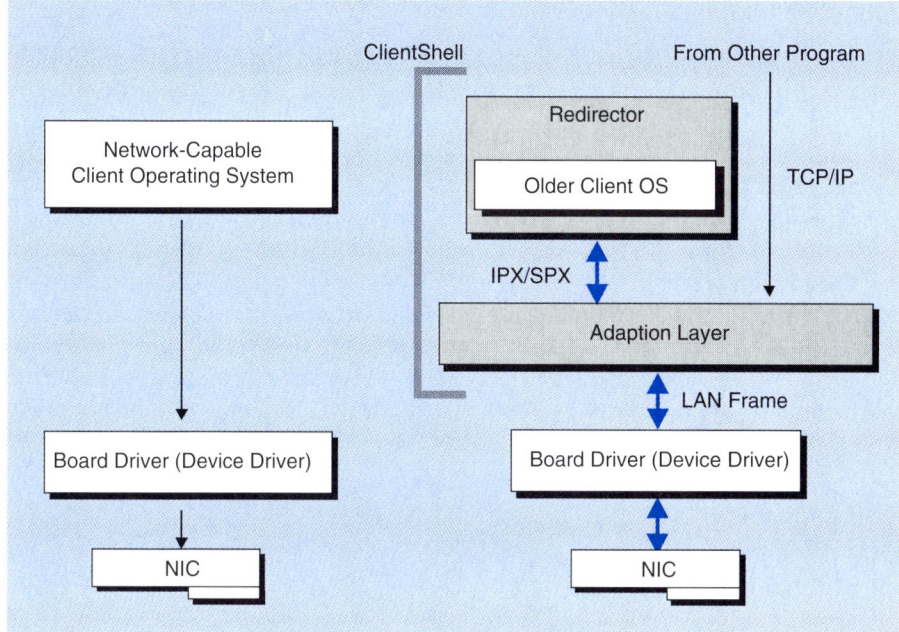

FIGURE I.11

Client Shell

The client shell has two components. The first is the *redirector*, which wraps itself around the operating system to handle calls to virtual drives and ports. The second is the *adaption-layer software*, which allows multiple application programs to use the board driver simultaneously.

Installing a User's Account on the File Server

After you have set up the user's client PC, you must add the user to the file server.

Creating an Account ■

Normally, there is a server utility for adding a user account. This will include such things as the user's account name and password. It may also include restrictions on how users may use the server, for instance, limiting them to working from 6 am to 6 pm.

Assigning the User to Groups ■

The next step is to assign the user to groups. For example, there may be a group for members of the person's department. As we saw earlier in the chapter, this makes it easier to assign rights.

Assigning Rights ■

That same utility will probably allow the installer to assign specific rights to the individual in various directories. At a minimum, the users will get their own private space on a file server. This will be their Drive F: in terms of Chapter 8.

Login Scripts

When the NetWare client shell programs load, they find a server and download a program called something like LOGIN.EXE. This is the program that asks the user for his or her login name and password. Although the program is stored on the server, it executes on the client PC. It passes the information the user enters to the server.

The server then downloads a login script to the user's client PC. This script prepares the client PC for redirection and other matters. For instance, it will tell the network-capable operating system or client shell which drives are virtual drives and which section of the server's hard disk corresponds to that virtual drive. For virtual ports, the script will tell the user which ports are parallel ports and what print queue on the file server corresponds to each virtual port. *For more on login scripts, see the box "NetWare Login Scripts," Containing Figures I.12 and I.13.*

ONGOING MANAGEMENT

Network administrators only install a few servers each year. They install more client PCs, but even these do not dominate their time and attention. Network managers have a great deal of other ongoing work to do, and this work takes many forms.

Monitoring the Server

When you are in the hospital, the nurse constantly reads your "vital signs," such as temperature, blood pressure, and pulse rate. Network administrators also have to collect diagnostic data on an ongoing basis.

Unfortunately, this data collection and monitoring must be done at multiple levels. If performance is lagging, the cause may range from server problems, such as inadequate RAM or too slow a hard disk, to transmission problems, such as a defective wire or a NIC that is sending traffic constantly because it is configured incorrectly.

Because of this multiplicity of problem sources, network administrators must use a number of tools to help them monitor various aspects of the network. We will discuss two tools Novell produces, but there are also excellent third-party tools that do the same tasks.

The MONITOR Program: Ongoing Data

One of these Novell tools is MONITOR. This is an NLM utility program that runs on the server itself instead of on a client PC. MONITOR gives basic information *about the server itself.*

The main MONITOR screen has two parts. The upper screen gives ongoing data about the server: its utilization, the number of connections in use, the number of open files, the number of current disk requests, and several pieces of information about how the server is using its internal memory.

It is easy to see why a network administrator would want to know the number of connections. If this number is growing, it could be a cause of concern because it could lead to server overload. The utilization parameter helps the administrator see if this is the case.

Insufficient RAM ■

It is less obvious why there are so many items that deal with memory. To give an example, the reason is that NetWare uses RAM as much as it possibly can. It divides much of the memory into several types of cache buffers. (We saw disk caching earlier in this module.) The performance of the system depends considerably on the way that these buffers are used.

For example, the server will perform poorly if there is not enough RAM. There will be too little room for the total cache buffers, so many calls to disk will actually have to go all the way out to disk instead of just to a disk cache buffer.

Administrators know that if the free cache buffers fall below 50 percent, they need to plan RAM increases. And if free cache buffers fall below 20 percent, the administrator should take immediate action [Niedermiller-Chaffins, et al., 1994]. Unloading some NLMs from the server and other quick fixes can produce short-term gains, but getting more RAM quickly is the only real remedy. Often a small amount of additional RAM can produce a quantum leap in the performance of a server by reducing disk access.

Slow Disk Drives ■

Another common server problem is disk overload. The number of reads and writes to disk may become too much for the disk drive to handle. The most common symptom of this is the number of *dirty cache buffers.* These are delayed writes to disk from the disk cache. If this number is large, it means that the disk is being pushed to its full capacity much of the time. If so, then the Current Disk Requests number will also be high. A faster disk drive is needed.

Disk Information ■

The lower portion of the main MONITOR screen is a menu. Selecting one of these options will give the PC network administrator closer looks at server-related information.

The *Disk Information* choice, for instance, tells such things as the amount of space left on the disk. If this gets too small, the server is in trouble. Often the administrator can simply broadcast a message to users, telling them to delete some of their old files, but this may only roll back the problem temporarily.

The *Disk Information* choice, in addition, can give hints that the disk is go-

NETWARE LOGIN SCRIPTS

Novell NetWare uses two login scripts. First, there is the *system login script*, which is the same for everybody. Second, there is the *user login script*, which is specific to the individual.

FIGURE I.12

NetWare System Login Script

The system login script is the same for all users. MAP commands tell the client shell what drives are virtual drives. Similarly, CAPTURE commands tell the client shell what ports are virtual ports.

MAP INS S1:=N3:\SYS:PUBLIC

MAP INS S2:=N3:\SYS\PUBLIC\%MACHINE\%OS\%OS_VERSION
    These commands map virtual drives to sections of the server's hard disk.
    S1: is mapped into the public directory on the SYS: volume of server N3.
    If the user calls a program, the client shell will assume the current directory.
    If it does not find the program there, it tries S1:.
    If it fails there, it tries S2:.
    S1: and S2: are search directories. They are like entries in a DOS PATH command.
    The user sees S1: as virtual Drive Z; while S2: become Y.: S3: would be X:.

COMSPEC=S2:COMMAND.COM
    This tells where to find COMMAND.COM.

MAP F:=N3:\SYS:USER\%LOGIN_NAME

MAP S:=N3\SYS:SHARED
    Maps virtual Drive F: to the user's home directory, under the USER Directory.
    Maps virtual Drive S: to the shared directory.

IF MEMBER OF "MARKETING" THEN #CAPTURE L=2 Q=MKTPRINT TI=30
    Captures print jobs to LPT2 (L=2).
    Directs them to print queue MKTPRINT.
    Sets a time out of 30 seconds.

DRIVE F:
    Sets the current drive to F: on the client PC.

System Login Script

Once the user is identified, LOGIN.EXE downloads a **system login script** from the server to the client PC. Figure I.12 shows that this system login script does a number of things.

MAP Commands

Most importantly, the system login script contains a number of **MAP** commands that tell the DOS Requester which drives are virtual drives. Figure I.12 illustrates the MAP commands that do this in a system login script.

Simple MAP Commands

Note that there are two types of **MAP** commands. Commands like MAP F: specify virtual Drive F. They specify that when users refer to F:, they will be pointing to their home directories on the file server. The MAP S: command, in turn, maps virtual Drive S: to the top of the *Shared* directory.

The syntax of a simple MAP command begins the keyword MAP. Next comes the desired user name for the virtual drive, followed by an equal sign. Then comes the directory on the server that will correspond to the virtual drive. To create virtual Drive R:, which will correspond to the *Shared\Project2* directory on the server's SYS: volume, you would create the following command in the system login script:

```
MAP R:=SYS:\Shared\Project2
```

Recall in Chapter 8 that when the client shell (really, the DOS Requester) sees a file call from an application program, it looks at the drive name. From this, it can tell whether the call is for a local drive or a virtual drive. The NET.CFG file specifies the last real drive. The MAP and MAP INS commands, in turn, tell the DOS Requester the sections of the file server's drive to which various virtual drives point.

MAP INS Commands

Other MAP commands start with **MAP INS**. They are followed by drives with such names S1: and S2:. To NetWare, MAP INS is similar to the PATH command in DOS. If the user calls a program, and if the program is not in the user's current directory, NetWare will look for it first in the S1: directory, then in S2:. In effect, MAP INS maps these **search drives** into a search list.

Like the DOS PATH command, search drives only work with programs. If you try to execute a program, it will look successively in the search directories. But if you refer to a data file, the search drives mean nothing. NetWare will not search for your data files in other virtual drives.

continued

Although search drives exist primarily for the system's use, users can also look at the contents of search drives. Of course, MS-DOS cannot accept two-character drive names, such as "S1". So the system maps S1: into Z:, S2 into Y:, S3 into X:, and so forth. To see the contents of the public directory on the Server's SYS: volume, you merely go to Drive Z: and use standard MS-DOS or Windows commands.

Again, there are two ways to map the virtual drive names to locations on the server's hard disk. Direct mapping assigns specific names to virtual drives, such as the F: and S: drives. Search drive mapping, in contrast, automatically assigns drive names and places the search drives on the equivalent of a PATH statement.

CAPTURE Commands ■

Figure I.12, the system login script, also contains CAPTURE commands. What MAP commands do for virtual drives, CAPTURE commands do for virtual ports. They create information in a table that tells the client shell that certain ports are *virtual ports*. The table also tells the client shell what print queue should receive the printing job when the user prints to a virtual port.

There is only one CAPTURE command in the system login script. It tells the client shell that references to LPT2: (L=2) are references to a virtual port. If an application program prints to LPT2:, the client shell will send the output to print queue *MKTPRINT*.

The essential components in a CAPTURE command are the keyword #CAPTURE, the designation of the virtual port name, and the print queue to which the output will be sent. If you want to use LPT2: to represent the *Colorpr* print queue, which serves a color printer, you could give this command.

```
#CAPTURE L=2 Q=Colorpr
```

There are also many optional parameters. NT means no tabs, which is normally useful as a way of overriding the way NetWare historically handled tabs. NFF means no form feed (new page) before each job. NB means no banner (a page that lists your name, so that you can keep your output separate on a shared printer). TI=60 means to time the attempt to print out after 60 seconds if there is no response.

User Login Script

After the LOGIN program finishes executing the *system login script*, it looks for a login script written for that particular user. This is the **user login script**. This login script will execute automatically, when the user logs in, as soon as the system login script finishes. Figure I.13 illustrates a simple user login script.

This script begins with a greeting, which is tailored to the individual. This allows you to greet individuals by name when they log into the system, giving a personal touch to the process.

In addition, the user login script has a MAP command. The MAP commands in the system login script apply to all users. The MAP command in the user login script, in contrast, is specific to that user. So is the CAPTURE command. User login scripts, then, allow you to define virtual drives and ports that are specific to an individual.

The user login script should end with the command *Exit*.

FIGURE I.13

NetWare User Login Script

The user login script executes after the system login script. It adds MAP, CAPTURE, and other commands that are specific to the particular user.

```
WRITE "Good %GREETING_TIME, %LOGIN_NAME
        Provides a personal greeting, by time of day and the person's
        name.
MAP T:=N3\SYS:SHARED\MARKETING
        Creates a drive mapping specific to this person.
        Drive T: is the Marketing directory under Shared.
        N3 is the server, and SYS: is the volume.
        Shared is a directory under SYS:
        Marketing is a directory under Shared.
#CAPTURE L=3 Q=MKTCOLOR TI=30 NFF NB
        Sends print jobs for LPT3: (L=3) to the print queue MKTCOLOR.
        TI=30 times out the printing if there is no response
        after 30 seconds.
        NFF means no form feed after printing.
        NB says not to produce a banner page with the person's name be-
        fore each printing.
Exit
        The user login script (not the system login script) should end with
        the Exit command.
```

ing bad. Immediately after NetWare writes to disk, it reads what it has written. If the two disagree, it knows that a sector on the disk has gone bad. In a *hot fix*, it marks off that sector and writes the results somewhere else. It then notes the problem. If the number of hot fixes is growing steadily, the disk may be starting to fail.

Processor Information ■

The server can also be processor limited. If you load the MONITOR with the command *load monitor p*, another menu choice will show you the load on the processor. This consumes a considerable amount of processor resources itself, so it should not be the default.

LAN Information ■

The *LAN Information* choice notes such things as the number of packets being handled by the server's NIC. If this number grows too large, then the server will be NIC limited in its ability to perform. The administrator can add a second or even a third NIC to the server.

The *LAN Information* choice will also do some basic checking on errors on incoming packets. These data are only rough guides to problems, but they do give the administrator hints to problems in the world outside the server.

Protocol Analyzers

MONITOR looks primarily *at the server*. Its *LAN Information* option provides only a little information about the network outside the server. Most network administrators also use a **protocol analyzer** that looks in detail *at the network*. For protocol analysis, one product that NetWare offers is **LANalyzer for Windows**.

Ongoing Monitoring ■

The main screen of LANalyzer has three large gauges, like gauges on a car's dashboard. In fact, this main screen is called the dashboard. These three gauges monitor packets per second, utilization, and errors. This allows the administrator to see at a glance how the network is doing.

In addition, LANalyzer always looks at the basic transmission patterns that it sees. Using a number of decision rules, it can note a potentially problematic situation. When it does this, it issues an **alarm**. It actually places a little alarm clock in the lower left corner of the screen. It can also sound an auditory alarm. When the administrator clicks on the alarm clock icon, he or she gets more information and even advice on how to interpret the alarm.

Frame Decoding ■

If the administrator suspects a problem, he or she hits *Start* in the *Packet Capture* area of the dashboard. LANalyzer then makes copies of every packet (really, frame) that reaches the NIC. It keeps collecting frames until the administrator tells it to stop and present its results for viewing.

LANalyzer examines each frame in detail. It looks at whether or not the frame is complete, whether its cyclical redundancy check indicates correct transmission or in error, whether the frame is too short or too long, and other diagnostic information.[3]

Combinations of these factors can diagnose errors. For example, if there are many collisions, then the traffic rate will be high, and there will be many short packets that are cut off when the sender detects a collision and stops transmitting before it sends an entire frame. This would lead the administrator to think about segmenting the network with bridges or routers.

If there are many long frames from a single station (this is called *jabber*), then there may be a problem with the software or NIC hardware on that machine. Or suppose that the frames are correct but there are very many from a single client. This may suggest that the client is doing such things as sending an endless series of calls for a nearest server because its software is out of date or was set up improperly.

Server Monitors and Protocol Analyzers ■

Some problems become apparent when you use a program like MONITOR, especially if they are *server* problems. Protocol analyzers, in turn, are good at diagnosing pure *transmission* problems. To get a real understanding of the network, and to diagnose many subtler problems, the network administrator must be able to work with both, collecting subtle clues that collectively point to problem areas. In discussing both programs, we have only scratched the bare surface of what they do.

User Support

In many ways, dealing with the network, client PCs, servers, and application software is the easy part of the job. The hard part is dealing with users. While machines are predictable, users are not. It is tempting to focus on technology, but without excellent user support, the PC network will never achieve its promise.

Training ■

New users must learn to use the network. Of course, they must be trained to log in and give specific commands. More generally, they must understand the basic roles of clients and servers and other concepts, so that they can use intuition to help them generalize from what they have learned.

---

[3] In this section, we will discuss the types of data relevant to Ethernet (802.3) networks. (See Chapter 5.) Different data would be needed for Token-Ring (802.5) networks and other types of networks.

Training also needs to involve network etiquette. In the days of stand-alone PCs, users had broad latitude over their machines. They could set up directory structures any way they liked and work any way they liked. But in a network environment, there needs to be more discipline. Directory structures on client PCs need to be controlled somewhat, or installation and troubleshooting will become nightmares. In addition, in shared directories, people have to be "good roommates." Their actions can affect others considerably.

Training also has to extend to the use of specific application packages. For instance, users have to learn how to give commands in the electronic mail system. They also need to develop a conceptual understanding of what electronic mail is and why it is useful. They also need to understand electronic mail etiquette and even legal matters. Again, broad training is needed if people are to be able to work with technology.

Some network staffs do extensive documentation development. Manuals are often too large for users. *Cheat sheets* give users scripts for common actions. Training guides, furthermore, are useful well after training.

Help Desk Support

Even in small PC networks, the staff will have to deal with several user help calls every day. Large networks generate dozens of help calls daily. Keeping track of these calls is critical or many calls for help will be lost in the heavy workload.

Many PC network staffs address this problem by having a central **help desk** with a single telephone number and email address. Each day one of the staff members staffs the help desk. Each call or email message is logged into a help desk program's database. As the request is cleared, it is checked off in the database.

At any moment, then, the help desk staff member can get a list of uncleared requests. This is helpful in scheduling assistance and in making sure that jobs assigned to individual staff members are cleared promptly.

The help desk software also helps the PC network administrator. It can tell what percentage of jobs fails to clear within the staff's target clearance time. It can also tell if certain types of tasks or certain staff members have particular problems. If users are always sent email questionnaires on their satisfaction with the service they received, feedback is even better. More generally, if there is a growing number of calls of a specific type, this may alert the administrator to a systematic problem.

CONCLUSION

The PC network administrator's job is complex and highly varied. Some aspects of it are technical in the extreme. Other aspects are human behavior and customer service matters. Not surprisingly, courses to train network administrators require several hundred classroom hours.

Network administrators only have to set up a few servers a year, but when they do, the initial work is extremely important. It is critical to choose the proper hardware, so that the server will have the required performance. It is also critical to choose the proper server operating system, as discussed in Chapter 8.

It is necessary to set up a directory structure that users can live with for a long time. In addition, care taken initially can drastically reduce the work that needs to be done later to assign rights to various resources to users. Actually installing the serve software is itself a time-consuming task.

We discussed how to set user rights, which is a major chore for the network administrator. In general, there are three ways to assign rights in a directory: assigning rights to the individual, assigning rights to a group of which the individual is a member, and assigning rights by automatic or modified inheritance from the parent directory.

Installing a new user is an extremely common task. On the client PC, the network specialist must install the network interface card and several pieces of software. The installer must also customize the connection to the server. On the server, the network specialist must assign various rights to the user for various resources. The installer also needs to develop a system login script on the server for all users and a user login script for each individual user.

Finally, we looked at the work of ongoing management. The network administrator must constantly monitor the performance of both the file servers and the transmission network because problems and bottlenecks can appear at either of these two levels. The network administrator also must provide a good deal of user services, such as training and help desk support.

CORE REVIEW QUESTIONS

1. Why is it necessary to create a directory structure carefully before you install a file server?
2. What are the three ways of assigning rights to a user in a directory?
3. What are your choices for servers besides Intel-based PCs?
4. What type of circuits go on the mother board? On expansion boards?
5. What are the three types of busses in a typical PC? What is the main advantage of each?
6. Why do you need a great deal of RAM on a file server? What will happen if you have too little RAM?
7. Your server's hard disk drive will contain programs designed to run on (at least) two different operating systems. On what two operating systems are these programs designed to run? How can you tell on which operating system a program is designed to execute?
8. What are the steps needed to add a new user to a PC network?

9. Distinguish between what MONITOR and protocol analyzers do.
10. What sorts of things should be in network training programs?
11. What does a help desk do?
12. (From the box, "NIC Address Settings") There are four address settings: I/O addresses, base memory addresses, IRQs, and DMA numbers. Which does the microprocessor use in sending commands and information to the NIC? Which does the NIC use to ask for help from the microprocessor?
13. (From the box, "NetWare Login Scripts") What is the purpose of the CAPTURE command? The MAP command? The MAP INS command?

DETAILED REVIEW QUESTIONS

1. Explain symmetric multiprocessing. What do you need for SMP to work? About how many microprocessors would you have in an SMP computer?
2. Why is SCSI better than EIDE? What are the two advantages of RAID drives?
3. Explain why a loss of power can be disastrous when you use a disk cache.
4. If you are using tape for backup, why would you tend to use DLT instead of QIC? Give two reasons.
5. What is a 32-bit NIC? When would you use it?
6. What is the purpose of a board driver?
7. Explain the advantages of plug-and-play NIC installation. What do you need for plug and play technology to work at all? To work well?
8. What are the two parts of the client shell? What does each do?
9. How can you tell if you have insufficient RAM? How can you tell if your hard disk drive is too slow?
10. How can you detect that a network is becoming congested? What is jabber? Why is it bad?
11. (From the box "NIC Address Settings"). Why do you need more than one I/O address for a NIC? Why do you need RAM on a NIC? Where do the addresses of the RAM's memory locations reside in memory? How many IRQs are there? What is DMA? What are default values? Why do you have to configure the operating system as well as the NIC?

(Remaining questions are from the box, "NetWare Login Scripts")

12. What are the main elements of a simple MAP command? Of a CAPTURE command?
13. Create a MAP command to make the *Shared\ProjectX* directory on the file server is SYS: Volume appear to the user to be Drive P:.
14. You already have two MAP INS commands. Add another for the directory

Apps\Utils\Backit on the file Server's Enterprise: Volume. What will the letter of this virtual drive appear to be to the user?

15. Create a CAPTURE command so that when a user prints to LPT1:, the output will go to the *BigLaser* print queue.

THOUGHT QUESTIONS

1. What rights would you give to all users in the User directory? Explain. What rights would you give to all users in the Share directory? Explain. Under the User directory, what rights would you give to users in their own home directory (Drive F: in Chapter 8)?
2. Why won't programs such as SS.EXE and MAIL.COM run on a Novell NetWare file server? Do you think they will run on a UNIX computer? When you log into the file server, you are running a program called LOGIN.EXE. Where does the program execute?
3. You are doing client/server processing (see Chapter 8). The server program's filename is DATAB. It runs on a NetWare file server. What will its extension be?

PROJECTS

1. Your client has seven PCs and wants to add networking. The client has already determined the cost of the network cabling. Using available sources, determine the cost of adding a file server. Include all costs.
2. Get access to a file server as an ordinary (nonsupervisory) user but without a menu system to hide the details of the file server's file structure from you. Using MS-DOS, MS-Windows, or whatever operating system your client PC has, find the directory structure of the server to the extent it is visible to you.
3. Get access again as a supervisor. See how what you can see changes.
4. Install a nonsupervisory user on a server, giving the person appropriate rights, an appropriate user login script, and so forth.
5. Install a file server. First, define the directory structure carefully. Then install the NIC if necessary.
6. Use MONITOR or some other program to study the performance of the server. Report on your results.
7. Use LANalyzer or some other program to study the performance of your transmission network. Report on your results.

For online exercises, please visit this book's website at: http://www.prenhall.com/panko

REFERENCES

Niedermiller-Chaffins, D. R., Heywood, D., Wilhite, C., and Cady, D. (1994). *NetWare Training Guide: Networking Technologies*, Second edition, Indianapolis, IN: New Riders Publishing.

MODULE J

Advanced Topics in Management and Applications

INTRODUCTION

This module is not designed to be read front to back like a book. Rather, it presents a series of advanced topics in network management and applications:

- Encryption and authentication.
- Video compression.
- Document conferencing.
- Database middleware.

ENCRYPTION AND AUTHENTICATION

Chapter 9 discusses encryption and authentication. This section presents more information on these topics.

Encryption

As discussed in the Chapter 9, **Encryption** turns messages into unreadable bit streams,

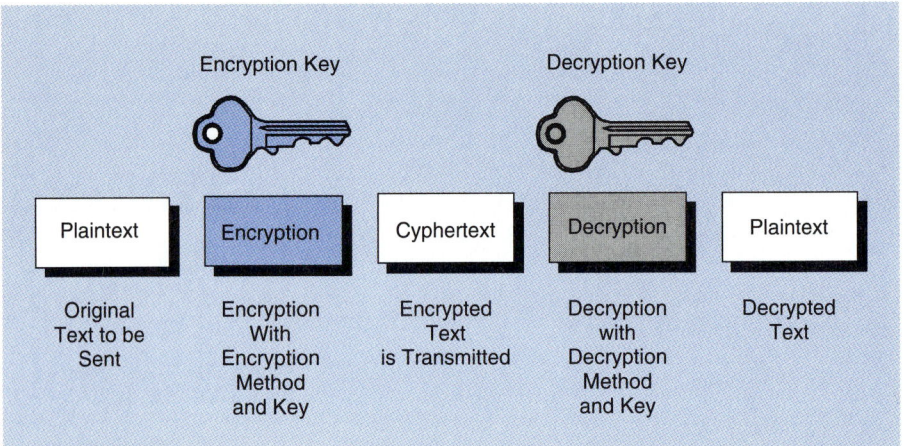

FIGURE J.1

Transmission Using Encryption

The message to be sent is called the *plaintext*. This is *encrypted* into unreadable *cyphertext*. Encryption involves both an *encryption method* and a *key*. At the other end, the receiver *decrypts* the cyphertext, restoring it to plaintext.

so that others cannot make sense of your messages if they succeed in intercepting them. This means that only the people who should read the message can read the message. This virtue is called **privacy** or **confidentiality**.

Basic Concepts

Figure J.1 illustrates the basic elements in the transmission of messages using encryption. The first element in the figure is the original message to be sent. This message is in humanly readable form called **plaintext**.

The next step is **encryption**, which turns plaintext into unreadable **cyphertext**. For instance, suppose you choose a very weak form of encryption in which you number the letters of the English alphabet from 1 through 26 and subtract one from each number to encrypt it. So if you were transmitting the plaintext "IBM," you would encrypt it as "HAL."[1] You would then transmit "HAL" across the network.

Note that encryption involves *two* things. First, it involves an **encryption method**. In this case, the method is "give each letter a number, then to subtract some constant number from each letter."[2]

---

[1] Stanley Kubrick denies that this is how he chose the name of the computer company in *2001: A Space Odyssey*.

[2] Obviously, this is a very weak encryption method. Anyone who has watched the *Wheel of Fortune* television show knows that some letters are much more common than others. To break the encoding, we select the most common number (4). That will probably correspond to "E." Given the frequencies of other letters and of short words, the process would continue. It should not take too long to break the encryption.

Second, there is some **key** value. In our example, it is the constant to be subtracted from the numerical value of each character. In our example, the key was "1." It could have been any other number.

In practice, the encryption method is almost impossible to keep secret, and it should be assumed to be compromised (known by enemies). So in addition to a solid encryption method, you must have keys large enough that the potential opponent cannot guess them economically, even by trying all possible combinations of keys on the encrypted text. The goal, then, is to keep the key secret.

At the other end of the transmission line, the process is reversed. Using a key and knowing the encryption method, the receiver reconverts the cyphertext into plaintext.

DES Encryption ■

In U.S. business data transmission, the most widely used encryption method is **Data Encryption Standard (DES),** which was created by IBM and certified by the National Institute of Standards and Technology in 1977.

This technique uses a key length of 56 bits.[3] This is not a very long key. DES encryption can be broken by guessing the key and applying it to a sample with enough computer time. Mindful of this, business partners often use triple encryption, in which a message is encrypted, decrypted, and encrypted again, using two 56-bit keys [Kaufman, Perlman, and Speciner, 1995]. The total of 112 key bits makes exhaustive search infeasible.

Figure J.2 shows that DES is a **single-key** encryption method. This means that the same key is used both to encrypt and to decrypt the message. Our foregoing simple encryption example was also a single-key methodology. With single-key systems, secure **key distribution**—getting the key to both parties—is critical.

Although DES is considered to be somewhat weak, it is inexpensive to implement in hardware. This allows rapid encryption and decryption in order to avoid latency due to encryption processing. In addition, DES is in the public domain, so anyone can build DES products.

RSA Encryption ■

A much more powerful encryption method is the **RSA** method.[4] This is a **public key encryption system** in which there are two keys rather than one. Figure J.3 shows that a person has two keys. One is a **private key** that only he or she should know. The other is a **public key** that can be released to anyone. In fact, public keys are widely published.

The figure shows that if someone wants to send you a message, he or she encrypts it in *your* public key. At the other end, you decrypt the message using your private key.

---

[3] The actual key if 64 bits, but 8 of the bits are parity bits, giving 56 bits of meaningful key.
[4] Named after the three inventors of the algorithm, Rivest, Shamir, and Adleman.

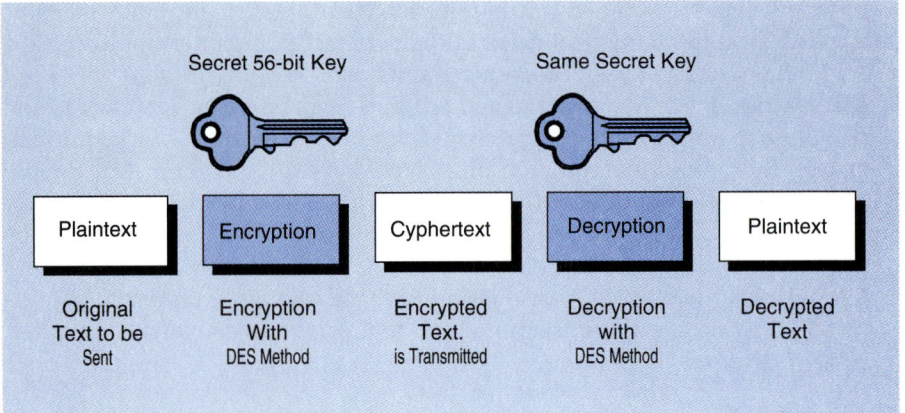

FIGURE J.2

Encryption with the DES Single-Key Method

The *Data Encryption Standard (DES)* is widely used in U.S. business. It can be implemented in hardware for rapid execution at low cost. Unfortunately, its 56-bit keys are too easy to break. Many firms reduce this problem by using two keys for a total of 112 bits. DES is a single-key standard, meaning that both parties use the same key to encrypt and decrypt. *Key distribution* is a potential source of threats.

FIGURE J.3

Public Key Encryption

In *public key encryption systems*, such as *RSA*, there are two keys. You have a *private key*, which you keep secret. You also have a *public key*, which you can give to everyone. When someone wishes to send you an encrypted message, they encrypt it with your public key. You decrypt it with your private key.

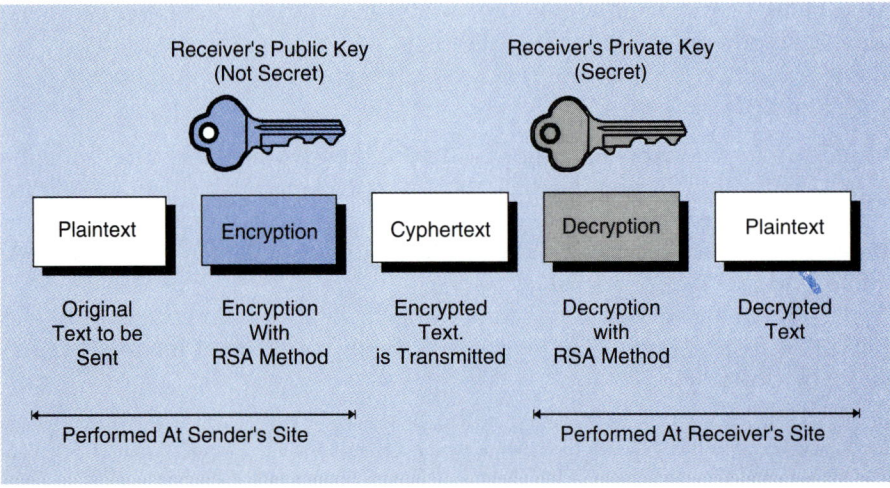

This approach works because some encryption methods make encryption very easy but make decryption difficult. So once the message being sent to you is encrypted, people without your private key will not be able to decrypt and read the message.

Public key encryption is attractive because it eliminates the problem of key distribution, except in the first step of getting the private key to the individual. Anyone can send you an encrypted message by using your widely published public key. There is no need to distribute a different secret key to each communication partner.

Although RSA public key encryption is extremely good and convenient, the algorithm is relatively expensive to implement and slows processing time. Nevertheless, the strength of RSA has made it the dominant approach to new network security technologies.

The other drawback to RSA is that it is a proprietary encryption method. Any vendor wishing to use it must pay license royalties to RSA Data Security, Incorporated. The patents on original RSA algorithms, however, are due to expire in the year 2000 [Kaufman, Perlman, and Speciner, 1995].

Key Escrow ■

What if you lose your private key? Then you will not be able to read messages that others have sent to you. In addition, you may have encrypted many of your general files with your private key. They will be lost to you unless you can get back your private key.

One solution is **key escrow** in which a trusted party keeps your private key in a safe place, as shown in Figure J.4. If you lose your private key, you can get it back from the key escrow company.

In addition, in the United States and some other countries, law enforcement agencies are concerned that with stronger encryption, they will not be

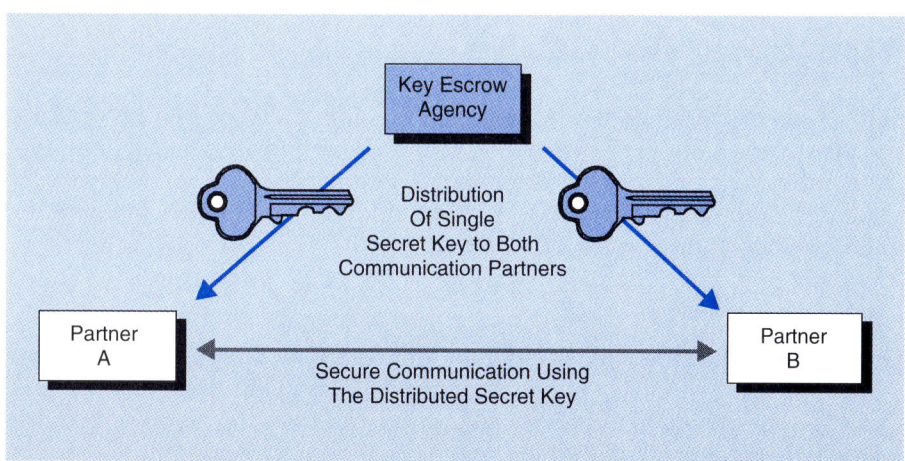

FIGURE J.4

Key Escrow

In *key escrow*, a trusted third party holds your secret key. If you lose your key, the key escrow service can provide it to you again. Law enforcement agencies can get the key to do legal wiretaps.

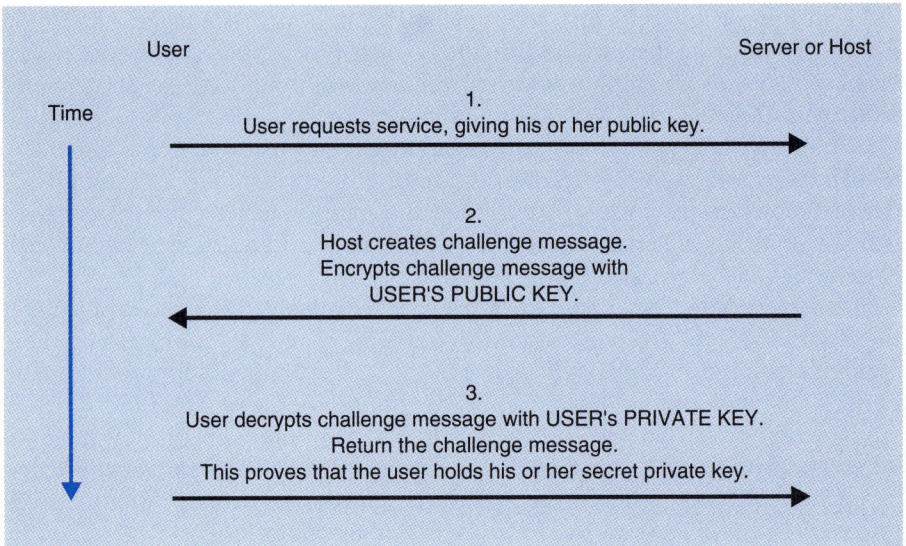

FIGURE J.5

Challenge and Response Authentication

The other party sends a plaintext challenge message. It encrypts it and sends it to you. You decrypt it. You then send back the plaintext to show that you have successfully decrypted it. The other party now knows that you hold the secret key. This establishes your identity.

able to execute legal wiretap warrants issued by the courts. Some proposed legal moves would require organizations to place their private keys (or single keys) in a key escrow account. Then, if the courts permit a wiretap, the key escrow company would surrender the key to a law enforcement agency upon presentation of a warrant.

Authentication

Suppose someone sends you a message. The author claims that he or she is your customer. Can you be sure that your customer really sent the message? **Authentication** is the determination that a message comes from the person or organization claiming to have sent it. Authentication is extremely important in financial transactions.

Challenge and Response ■

One authentication method is **challenge and response**. Suppose you are using public key encryption. The other party needs to establish your identity. In a challenge and response approval, the other party does this by asking you to prove that you in fact hold your secret key. If you can demonstrate that you hold your secret key, you have established your identity with a reasonable degree of certainty.

Figure J.5 illustrates how to do this. The other party creates a plaintext challenge message. It encrypts the challenge message and sends it to you. You decrypt the message with your public key, converting it back to the original plaintext challenge message.

All you have to do now is send the plaintext challenge message back to the original sender. The other party then knows that you have successfully decrypted the message and so must hold your private key. This works well, although if others learn your private key, they can pose as you.

Digital Signatures ■

Challenge and response systems work at the beginning of a session. Sometimes, however, there is no session. You merely send a single message.

Figure J.6 shows message-by-message authentication using a **digital signature**. Suppose you are using a public key encryption method. Then you encrypt outgoing messages with the other party's public key as usual.

However, you create a **digest** containing part of the message. You encrypt this digest with your own private key, which only you should hold. This digest becomes your **digital signature**.

FIGURE J.6
Authentication Through a Digital Signature

You encrypt an outgoing message in the other party's public key. However, you create a *digest* containing part of the message and encrypt it in your secret private key. The other party decrypts the digest with your public key. This encrypted digest is your *digital signature*. If the other party gets the correct result, he or she knows that you hold your private key. This establishes your identity.

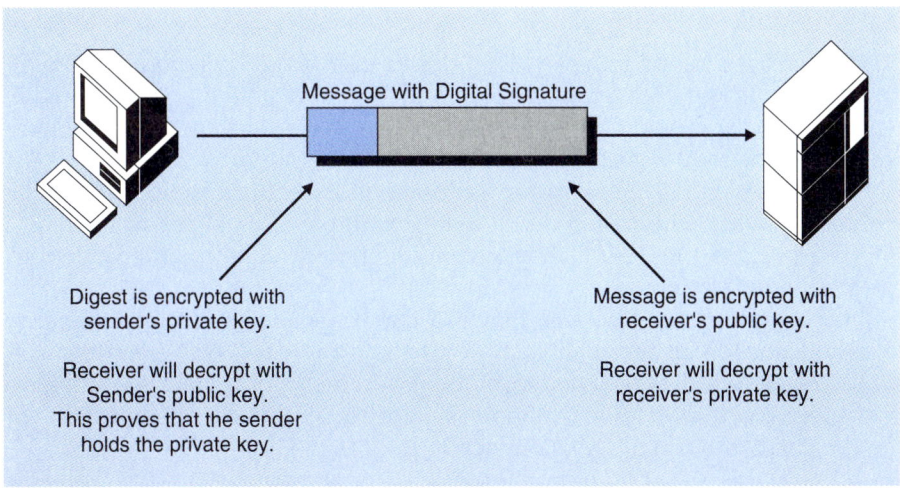

The other party decrypts the digest with *your public key*. If the decryption is successful, the other party knows that the sender holds your private key. Only you should have this key. Just as a normal signature establishes who wrote a letter, a digital signature assures the receiver that you sent the message.

Nonrepudiation ■

One important aspect of authentication is **nonrepudiation**. Repudiation means that you deny you have sent a message. For instance, you might claim that you never sent a purchase order for an expensive product. This allows you, in effect, to cancel an order without paying cancellation penalties. Authentication establishes that you *did* send in the order. You cannot **repudiate** the message unless you claim that your private key was compromised.

Data Integrity ■

Most forms of message digest authentication also provide **data integrity,** which is assurance that message contents have not been changed during transmission. This assures you that someone has not captured the packet, modified it, and then resubmitted it.

VIDEO COMPRESSION

As discussed in Chapter 10, videoconferencing codecs do more than convert the source analog signal format into a digital transmission format. They also **compress** the signal before sending it out.

Compression Standards ■

Television has a bandwidth of 6 MHz. Digitization should generate a bit stream of 96 Mbps. Using such a transmission line for a long period of time would be enormously expensive. We would like to use T1-speed lines or even lower-speed lines to reduce costs. *For more on digitization, see Chapter 4.*

The ITU-T H.320 standard for compression is **H.261**. It is also called **Px64** because it does compression down to any multiple of 64 kbps, up to about 2 Mbps. By supporting H.261, codecs from different vendors can exchange compressed video streams.

Of course, there is no free lunch in compression. As you increase compression, you lose image quality and voice quality as well. Voices become buzzy as voice bandwidth falls. In video, motions become jerky. At 64 kbps and other low speeds, you have more the impression of seeing the other people than you do the effect of normal television viewing.

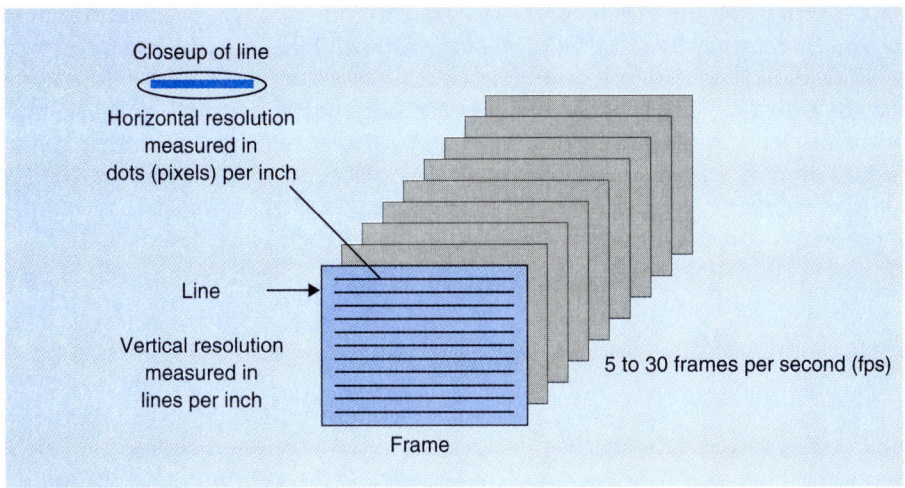

FIGURE J.7
Television Images

A television image consists of a succession of still frames. Broadcast television sends 25 to 30 frames per second. Reducing the frames per second (fps) before compression provides transmission savings at the cost of jerkiness. A frame's resolution is defined by the number of lines per frame and by the number of pixels on each line. Reducing resolution before compression is another way to reduce transmission requirements.

Reducing Frames per Second ■

Another way to reduce transmission requirements is to reduce video quality before compression. A frame is a full picture. A television picture is really a series of still frames, as shown in Figure J.7.[5] In the United States, for instance, television pictures are sent at 30 **frames per second (fps).** In Europe, the comparable frame rate is 25 frames per second. Many videoconferencing systems allow you to reduce the number of frames per second, down to 20, 15, or even 5 frames per second. This reduces transmission requirements, but it produces jerky movement.

Reducing Resolution ■

Another approach to reducing video quality is to reduce **screen resolution**. Figure J.7 shows that a frame consists of many scanning lines—typically 240 (in the NTSC standard used in the United States, Canada, Japan, and many other countries) or 288 (in the PAL standard used in Europe and many other parts of the world).

---

[5]Compression is also based on frames. Every five to twelve frames are *base frames*. Only difference information is transmitted for intermediate frames.

On each line, furthermore, there are pixels (dots) of lightness and (usually) color. There typically are about 350 pixels on each line. It is possible to cut vertical resolution (the number of lines) and the horizontal resolution (the number of pixels per line). This produces a more blurry picture, but it also cuts transmission requirements. If you halve both vertical and horizontal resolution, you only have to send one quarter as many bits. H.261 offers two standard resolutions—a high-resolution mode with 288 lines and 352 pixels per line, and a low resolution mode with 144 lines of 176 pixels [Rash, 1994].

Document Conferencing

In document conferencing the two parties can see a common document as well as see and hear the other person. There are several levels of document conferencing.

Remote Control Software

The simplest form of document conferencing uses **remote control software**, which slaves the two screens together. Both sides can see the same image. Often, however, only one side can modify the image.

Whiteboard Conferencing

In **whiteboard conferencing,** there is a common display area, similar to the blackboard or whiteboard in a conference room. One party copies a screen image onto the whiteboard, and both parties can see it.

Whiteboard systems normally have markup tools. With a mouse, either side can circle an area, draw arrows, write text notes, and mark up the image in other ways. Using one's cursor, furthermore, each can point to an area on the whiteboard.

Application Sharing

In application sharing, there is no need to copy images to a whiteboard. One of the partners brings up an application, such as a spreadsheet program. He or she can change cells and take other editing actions right in the program, without copying images to a whiteboard.

The other party does not simply watch. He or she can also work on the application, using his or her own keyboard and mouse. This allows each side to show the effects of edits in documents and of changed assumptions of spreadsheets.

In addition, the changes are actually made in data files. In document conferencing, in contrast, after changes are agreed to by the parties, someone has to print the image from the whiteboard and go back to the original program to make the changes after the meeting is over.

DATABASE MIDDLEWARE

All database middleware stands between the application program and the transport layer process. However, there are several types of database middleware software that have different benefits and costs. The categories that we will use were created by Dolgicer [1994].

Remote Procedure Calls (RPCs) ■

One type of middleware is called the **remote procedure call (RPC).** As shown in Figure J.8, the sending application program sends a call to the receiving application program, much as it would send a call to the operating system.

This call is in a special format called a remote procedure call. The RPC **stub** program on the sending machine transmits the call to the RPC stub on the receiving machine. The stub on the receiving machine passes the call to the receiving program.

RPCs are synchronous at the application layer. If a program sends an RPC, it waits until the receiver replies [Dolgicer, 1994]. This having to wait is very bad for many applications.

RPCs are general-purpose standards. They are not limited to database processing. They are used in many situations in which a program on one machine needs to call a program on another machine. There are several remote procedure call standards. The RPC standard built into the *Distributed Computing Environment* championed by the UNIX community is probably the most widely used.

FIGURE J.8
Remote Procedure Call (RPC)

In an RPC, the application program issues a call in the remote procedure call format. The RPC program (stub) passes the message to the RPC program on another machine. That RPC program delivers the message to the application program on the second machine.

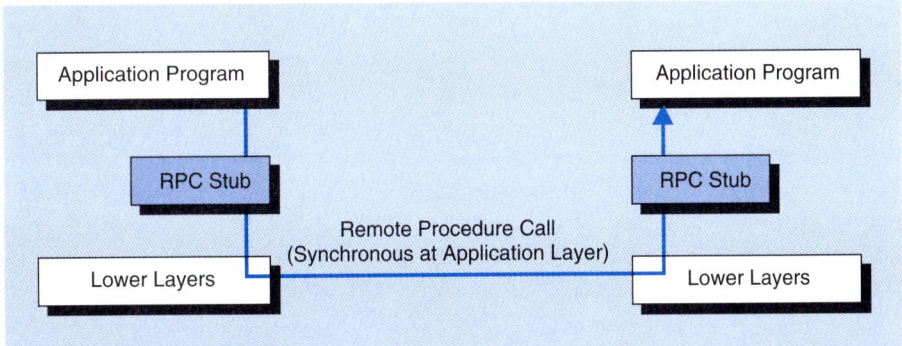

Message-Oriented Middleware (MOM)

Figure J.9 illustrates another approach called **message oriented middleware (MOM).** Obviously, MOM is similar to an RPC middleware. However, MOM is asynchronous at the application layer [Dolgicer, 1994]. Its message is like an electronic mail message. A reply is expected sometime and will be handled when it arrives, but in the meantime, the sending application program is free to do other work. MOMs have sophisticated functionality for database work.

On the downside, MOMs are expensive. They are also limited to working with only some database programs. There are as yet no standard APIs for MOMs to allow purchasing MOMs from different vendors. In general, the selection of a message oriented middleware product is a very complex choice.

Transaction Monitors

In many cases, application program interactions must be carefully synchronized and controlled. If something fails in the middle of a transaction, the client user might think he or she has updated the database this really did not happen. Or the user might not think that he or she has updated the database when it really has been updated.

MOM merely delivers messages. In contrast, **transaction monitor programs** manage the communication in detail. For instance, they will not allow the client program to believe that an update has occurred if it really has not. You might call transaction monitors "nagging MOMs."

Object Request Brokers (ORBs)

Database is moving toward object-oriented structure. If the objects are split across two or more machines, then there must be some way to request objects

FIGURE J.9
Message Oriented Middleware (MOM)

Message oriented middleware (MOM) is database specific. It is asynchronous in contrast to RPCs. MOMs have sophisticated functionality for database applications, but they are expensive, not standardized, and limited in what database products they support.

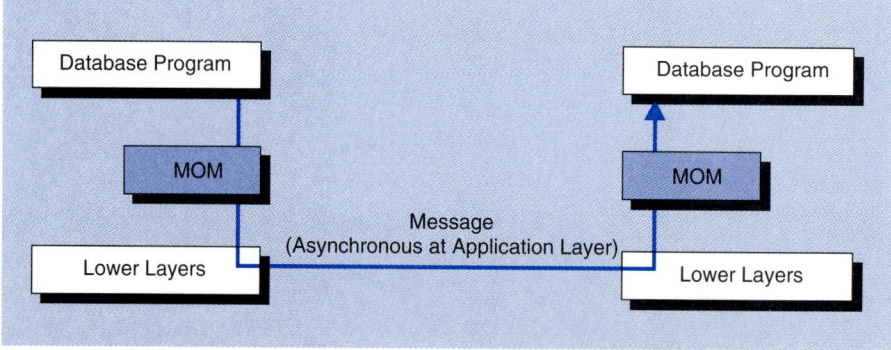

from the other machine. **Object request brokers (ORBs)** serve this function. Given the lack of market presence of object-oriented databases, ORBs are not likely to be important for some time.

CORE REVIEW QUESTIONS

1. Explain confidentiality and privacy.
2. Explain the encryption process. Be sure to distinguish between the encryption method and the encryption key.
3. Compare DES and RSA encryption in terms of processing requirements, the number of keys used, and the difficulty of distributing keys to communication partners.
4. Explain what public key encryption is and why it is good.
5. What is the key escrow problem?
6. What is authentication, and why is it important? What is nonrepudiation, and why is it important?
7. How does reducing the number of frames per second and the resolution of the picture aid in compression?
8. What do you gain when you move from remote control software to whiteboard conferencing software? From whiteboard conferencing software to application sharing?
9. Why do you need database middleware?

DETAILED REVIEW QUESTIONS

1. Describe the DES encryption process, commenting on its key length and the implications for security. How do organizations often get around this limited key length?
2. Describe challenge and response authentication. Describe message digest authentication.
3. What is the standard for video data compression? What is its other name?
4. What are the resolution modes in H.261?
5. Describe the three types of database middleware and the advantages and disadvantages of each.

THOUGHT QUESTIONS

1. What do you personally feel about law enforcement agencies' attempt to require mandatory key escrow?
2. When might you prefer to use challenge and response versus message digest authentication?

For online exercises, please visit this book's website at: http://www.prenhall.com/panko

REFERENCES

Davidson, E. (1995, December). University of Hawaii. Personal communication with the author.

Dolgicer, M. (1994, July). "Messaging Middleware: The Next Generation," *Data Communications*, pp. 77–84.

Kaufman, C., Perlman, R., and Speciner, M. (1995). *Network Security: Private Communication in a Public World*. Upper Saddle River, NJ: Prentice Hall.

Rash, W. (1994, January). "Virtual Meetings at the Desktop," *Windows Sources*, pp. 235–264.

Glossary

Legend
  [C4]   Chapter 4
  [MD]   Module D
  [F5-2] Figure 5-2
  [FD-4] Figure D-4
  [T5-3] Table 5-3

100Base-X. A set of three 100 Mbps versions of *Ethernet*. Created by the *802.3 Working Group* of the IEEE *802 Committee*. [F5-7]

100VG-AnyLAN. A 100 Mbps *shared media* LAN standard from the 802.12 Working Group of the *IEEE 802 Committee*. [F5-8]

10Base2. *Ethernet (802.3) physical layer* standard using thin *coaxial cable*. Stations are connected in *daisy chains*. [FF-3]

10Base5. A *physical layer* (OSI Layer 1) standard for *Ethernet*. [F4-12, FF-1]

10Base-F. *Ethernet (802.3) physical layer* standard using *optical fiber*. Used primarily in *interrepeater links*. [C4, FF-2]

10Base-T hub. Wiring concentrator in *Ethernet 10Base-T* LANs. [F4-3]

10Base-T. *Physical layer Ethernet* standard. Uses ordinary business telephone wiring. [C4]

23B+D. In *ISDN*, *Primary Rate Interface* with 23 *B channels* operating at 64 kbps and one *D channel* operating at 64 kbps. [F7-11, FG-8]

2B+D. In *ISDN*, two 64 kbps *B channels* plus one 16 kbps *D channel*. Used in the *BRI* in the United States. [F7-11, MG]

30B+D. In ISDN, *Primary Rate Interface* with 30 *B channels* operating at 64 kbps and one *D channel* operating at 64 kbps. [F7-11, FG-8]

3270 terminal. High-performance *terminal* preferred by IBM *mainframe* computers and other mainframe computers. [C3]

800/888 number portability. For a company to be able to keep *its 800 or 888 number* if it switches long-distance *carriers*. [MG]

800/888 number. Long-distance pricing system in which the called party (usually) pays for the call. [FG-5]

802 Committee. Another name for the *IEEE LAN MAN Standards Committee*. [C4]

802.3 MAC Layer Frame. *Media access control* layer *synchronous unit* for *Ethernet* LANs. [F4-8]

802.3 Working Group. The *802 Committee working group* responsible for *Ethernet* standards. [C4]

802.3. See *802.3 Working Group*.

802.5 frame. *MAC layer* frame in *802.5 Token-Ring Networks*. [FF-8]

802.5 Working Group. The working group that created the 802.5 *Token-Ring Network* standard. [MF]

802.5. *Token-Ring Network* LAN standards from *the 802.5 Working Group* of the IEEE *802 Committee*. [F4-2, C5]

802.11. Radio and infrared LAN standards

802.12. *100VG-AnyLAN* standards from the IEEE *802 Committee's 802.12 Working Group*. [F5-8]

900 numbers. Pricing system in which the calling party pays a price in excess of transmission charges. The excess goes to the calling party. [FG-5]

ABR. See *available bit rate*.

Abstract Syntax Notation Number 1 (ASN.1). In *OSI's presentation layer*, a computer-independent format for representing information. [FA-10]

access computer. A computer into which a user dials to receiver service from an *Internet service provider*. [F2-13]

access control. In Token-Ring Networks, a field in an *802.5 frame* that controls when a station may transmit. [FF-7, FF-8]. In security, preventing unauthorized people from accessing resources. [C9]

access line. A leased line to the point of presence of a *switched data network*. Required for service. [F7-10]

access unit. Wiring concentrator in 802.5 Token-Ring Networks. [F5-3]

accounting management. Keeping track of who is using which resources, so that departments can be charged for usage. [C9]

active monitor present. Supervisory *802.5 frame* that tells stations that an *active monitor* is present and functioning. [FF-8]

active monitor. In *802.5 Token-Ring Networks*, the station designated to ensure that there is one and only one *token* in circulation. [MF]

ad hoc. In standards, another name for *proprietary standards*. [C1]

adaption-layer software. On a client PC, a layer of software that *multiplexes* data streams using different *architectures* to the board driver for transmission and reception via the *NIC*. [FI-11]

Address Resolution Protocol (ARP). In *TCP/IP*, a protocol that allows a host to find the 48-bit IEEE 802 LAN address of another host or router. [FH-8]

address. In a network, each station has a unique address, similar to a telephone number. Other stations send it messages by addressing the messages to the station's address. [F1-2]

adhocracy. An organization characterized by constantly shifting project teams. These teams form quickly as needs arise, then disband to free people to work on the next project. [C1]

agenda system. In an *electronic meeting room*, an application that allows the moderator to move through items on the meeting agenda, using an appropriate application for each item. [C10]

agent. A hardware or software process that acts on behalf of a user or another process in dealing with a system. [C9 for network management agents, C10 for *electronic mail* agents, MC for *Internet* search agents]

alarm storm. When an element in a network fails, many *managed devices* may be affected and send out *alarms*. The *network management program* must sift through these alarms to deduce the root fault. [C9]

alarm. In a *client/server* system, a message initiated by the *server program* to *client programs* to warn of unusual conditions. Not prompted by a request message from the client program. [C9, MI]

alias. An easy-to-remember unofficial name that one uses to refer to a formal address. [C10]

alternative routing. The ability of switched LANs to choose among the multiple possible routes available to a packet at it travels between two *stations*. The network can select the *optimum route* for each packet, route around failures and congestion, and provide *load balancing* for the network's trunk lines and switches. [F5-10]

amplitude modulation. A form of *modulation* that varies the amplitude (intensity) of the transmitted signal. [F3-8].

analog. Type of signal characterized by smooth transitions between an infinite number of states (such as voltage levels). All older communications systems were analog. [F3-1, F3-6]

anchor tag. HTML *tag* for representing a *link*. [FC-6]

ANI. See *automatic number identification*.

anonymous FTP. Form of *File Transfer Protocol* in which a user can log into an FTP host without having a password. [FB-7, FC-14, FC-15]

anonymous input. Application that allows group members to generate ideas without giving their names. Allows the group to focus on the idea rather than the suggester. [C10]

ANS.1 See *Abstract Syntax Notation Number 1*.

ANSI terminal. Standard based on the *VT100 terminal* [C3]

API. See *application program interface*.

applet. A (small) application downloaded to a PC across a network, especially in the *Java* language. [FC-13]

application development tool. Tool for creating an application that a user will use to do corporate work. [F10-9]

application layer (OSI layer 7). In *OSI*, the layer that standardizes interactions between application programs. [MA]

application layer. In the *Basic Model of Communication*, the layer that standardizes exchanges between the application programs in the system. [F1-1]

application program interface (API). Standard governing interactions between an application program and a lower-layer process. [F10-1]

ARCHIE. *Internet* application to search for files available through *anonymous FTP*. [MB]

architecture. A standards agency's overall plan for setting *standards*. Individual standards are set within the context of the architecture. [F1-3]

ARP. See *Address Resolution Protocol*.

ARPA. See *Defense Advanced Research Projects Agency*.

ASCII. A code for representing English text. Each character is represented by a seven-bit sequence. [C3, TD-1]

asynchronous. In general, when two processes communicate by sending messages instead of through a real-time connection. In *data link layer* transmission, a specific way of packaging data in *asynchronous frames*.

asynchronous frame. A way of transmitting data in 10-bit *frames*. [F3-8]

asynchronous transfer mode (ATM). *Switched data network* technology supporting a broad range of speeds. Based on *broadband ISDN*. [C5]

AT&T. Formerly the monopoly long-distance telephone *carrier* in the United States. [MG]

AT. The command set used in *intelligent modems* following the *ad hoc* standard set by Hayes Computers. [TD-2]

attachment unit interface (AUI). [FF-1]

attachment. In *electronic mail*, a binary file that is delivered along with a text message. [F2-9]

attenuation. As a signal propagates, it weakens, making it more difficult to separate from *noise* on the line. [C4]

audit trail. Keeping track of the actions of individuals, so that malfeasance can be traced. [C9]

audit. Systematic analysis of whether a system or is functioning properly or whether a policy is being followed appropriately. [C9]

AUI. See *attachment unit interface*.

authentication. Ensuring that the person or process claiming to send a message actually sent the message. [C9, MH, MJ]

autoconfiguration host. A *TCP/IP* host that provides PCs and other hosts with temporary *IP numbers*. [C6, MH]

autoconfiguration. Protocol by which a host can request an IP number from an autoconfiguration server. [FH-9]

automatic number identification (ANI). Service in which the called party is told the telephone number of the calling party. [MG]

autonomous router. *Router* used within an *autonomous system*. [FH-13]

autonomous system. A network owned by a single organization. [FH-13]

availability. The proportion of time a network is available for use. [C9]

available bit rate (ABR). In *switched data networks*, the transmission speed at which a carrier allows a customer to transmit in bursts without guarantee of delivery. In excess of the CIR. The *access line* speed should reflect the ABR. [F7-10]

B channel. In *ISDN*, a 64 kbps channel for carrying user signals. [F3-17, C7, FG-8]

bandwidth. The range of frequencies in *hertz (Hz)* between the highest frequency that a channel can carry and its lowest frequency. The maximum transmission speed of a channel is proportional to its bandwidth. [C3]

baseband signaling. Type of physical signaling in which voltage changes are injected into the transmission line. Only one station may transmit at a time on a line. Used in most LANs. [C4]

Basic Model of Communication. The basic *layered architecture* used in this book. [F1-1]

Basic Rate Interface (BRI). The basic desktop interface in *narrowband ISDN*. Provides three *multiplexed* channels to desktop devices. [F3-17, C7, FG-8]

baud rate. The number of times a line changes per second. Almost always different than the *bit rate*. [F3-10, MD for modems, C4 for 10Base-T LANs]

BGP. See *Border Gateway Protocol*.

binary file transfer (BFT). In *facsimile modems*, a way of transferring binary data files from memory to memory instead of printing them. [TD-2]

binary. Representation of information in which there are abrupt changes between two states, such as voltage levels. One state is a 1 the other is a 0. Subcategory of *digital* [C1, C3]

bit. A single 1 or 0. [C1]

bit-mapped graphics. Image formed from many small dots (*pixels*). [FD-1]

bits per second (bps). Measure of transmission speed. The number of *bits* transmitted per second. [C1]

bit rate. The speed of a circuit in *bits per second*. [C1]

board driver. Software that mediates between the operating system and a *NIC*. [FI-9]

body. In *electronic mail*, the main portion of a message. [F2-9]

bookmark. In *browsers*, a way of recording a visited *uniform resource locator* location in a list for later retrieval. [MC]

Bootp. In *TCP/IP*, an *autoconfiguration* protocol. [MH]

Border Gateway Protocol (BGP). Router coordination protocol used by *border routers*. Also see *gateway*. [FH-13]

border router. Router that links an *autonomous system* to outside networks. [FH-13]

bps. See *bits per second*.

brainstorming. Application that helps groups to generate ideas. [C10]

BRI. See *Basic Rate Interface*.

bridge. The simplest and least expensive internetting device to link LANs into an *internet*. Bridges cannot be connected in loops, limiting them to simple internets and internets that do not require *alternative routing*. [F6-5]

broadband ISDN. Version of *ISDN* offering speeds in the range of hundreds of megabits per second to gigabits per second. [C5, MG]

broadband signaling. Type of physical signaling in which available transmission capacity is divided into channels. Different signals may propagate simultaneously in different channels. Used primarily in radio LANs. [C5]

browser view. View of a *webpage* as it will be seen by a *browser* user. [FC-5]

browser. A *client program* that sends requests to *client/server* server applications on hosts on the Internet and displays the returned information. Works with hosts providing *World Wide Web* service, *FTP* service, and other types of service. Common examples are Netscape,

MOSAIC, and Microsoft Internet Explorer. [F2-12, FC-3, FC-14]

buffer. Section of RAM used to transfer information from one process to another *asynchronously*. Allows for the two processes to work at somewhat different rates. [FI-10]

bulletin board system. Form of *computer conferencing*. [C10]

bus transmission. Physical transmission in which a signal is broadcast to all stations on the LAN. [C4]

bus. In a LAN, a *topology* in which a station broadcasts messages to all other stations. [C4] In a PC, the main communication path between the microprocessor and other circuits. [FI-5]

business information service. Among *commercial online services*, processing services oriented to organizations rather than to residential consumers. [C11]

business rule. In *groupware* and *database* system, a business policy that a system must implement. [C10, C11]

busy hour. The hour of the day having the highest traffic. [C9]

cable bundle. A bundle of many telephone wires. [FG-2]

call. A request issued by an application program to the operating system. [F8-13]

CAP. See *competitive access provider*.

capture. *Novell NetWare* command that maps a *virtual port* to a *print queue* on the *file server's* hard disk drive. [FI-12, FI-13]

Carrier Sense Multiple Access with Collision Detection (CSMA/CD). *Media access control* discipline. Stations may transmit if the LAN is unused but must wait if the LAN is already in use and must stop if another station starts transmitting when they do. [F4-6, F4-7]

carrier sensing. Listening for traffic on the LAN. See *Carrier Sense Multiple Access with Collision Detection*.

carrier. Regulated organization that provides transmission services for a fee. [C7, MG]

case worker. In *reengineering*, a single person who performs tasks formerly performed by several people. [C11]

Category 3, 4, and 5 UTP. Grades of *unshielded twisted pair* wiring suitable for data traffic. [F4-3, MD]

CDDI. Unofficial name for *FDDI over UTP*. [C5]

CDPD. See *cellular digital packet data*.

cell. In data networking, a fixed-length *packet*. This makes network transmission predictable, improving throughput. *ATM* uses cells of 53 octets. [F7-8] In cellular telephony, a limited region in *cellular telephony*. [FG-6]

cellphone. Mobile telephone used in *cellular telephony*. [FG-6]

cellular digital packet data (CDPD). *Digital cellular telephony* service. [FG-7]

cellular telephony. Mobile telephony in which the service region is divided into many areas called *cells*. [FG-6]

central network control center. In a distributed *network management system*, the central network control center that manages the overall network. [F9-6]

CGI. See *common gateway interface*.

challenge and response. *Authentication* implemented through exchanges of encrypted and decrypted messages. [C9, FJ-5]

channel service unit/data service unit (CSU/DSU). Combines a channel service unit to protect the carrier's network (see *termination equipment*) and a *data service unit*. [F7-4]

channel. A transmission facility that can carry a signal between two points. [C7]

chargeback. Charging work units for their usage of network resources. [C9]

CIR. See *committed information rate*.

circuit switching. Form of *switching* in which dedicated circuit capacity is assigned to the

calling parties for the duration of the call. If capacity is not used, it goes to waste. [F7-5]

circuit. An end-to-end connection between two (or more) parties. May extend across multiple transmission *media* and even multiple *carriers*. [T7-1]

claim token. Supervisory *802.5 frame* that tells other stations that the transmitting station has detected the absence of an *active monitor* and is claiming control over the management of the *token*. [MF]

Class 1 and Class 2. Command sets used in *intelligent modems* for *facsimile*. [TD-2]

Class A, B, C, and D IP addresses. Classes of *IP numbers* offering different trade-offs between the number of possible networks and the number of *hosts* per network. [FH-2]

clean data. In a *data warehouse*, extracted data from which inconsistencies and other data quality problems have been removed. [C11]

client PC. In a *PC network*, the PC on the desk of an ordinary user. Receives services from a *server*. [F2-2, C8]

client program. In *client/server processing*, the program on the user's desktop machine. [F2-8, C8]

client shell. For client operating systems that are not *network-capable*, added software to allow the PC to work with a *server*. [FI-11]

client/server processing. Processing is divided between two programs on two different computers. The *server program* on the server does the heavy processing work. The *client program* on the client machine handles user interface and other lighter chores. The client machine may be a PC, but there is no limit on the computer platform for either the client or the server computer. [F2-8, C8]

CMIP. See *Common Management Information Protocol*.

CMIS. See *Common Management Information Services*.

coaxial cable. Transmission *medium* consisting of a central wire and a concentric coaxial conductor. [F4-13]

codec. If you have an *analog* input signal and wish to transmit it over a *digital* transmission line, you need a codec to do the translation. [F3-17, F3-18] In *videoconferencing*, video codecs also *compress* the signal before transmission [C10, MJ]

collapsed backbone. In *local internets*, a transmission line runs point-to-point from each LAN to a central switch, router, or other internetting device. Also see *distributed backbone*. [F6-11]

collision domain. In *Ethernet*, a group of stations that hear the transmissions from other stations in the collision domain. [MF]

commercial online service. Network-based service that allows customers access to data and processing services for a fee. [F11-9, MB]

committed information rate (CIR). In *switched data networks*, a transmission speed that a carrier promises to provide except under unusual circumstances. [F7-10]

common gateway interface (CGI). On the *World Wide Web*, standard for passing user input to an external program, such as a *database* program, which generates a new *webpage* to be delivered to the *browser* user. [FC-12]

Common Management Information Protocol (CMIP). Part of the OSI *network management* standards. Defines protocols by which services are delivered. [F9-10]

Common Management Information Services (CMIP). Part of the OSI *network management* standards. Defines the services to be delivered. [F9-10]

communication server. On a PC network, a *server* that provides access to the world outside the PC network for input, output, or both. [C8]

communication software. See *Module E*.

competitive access provider (CAP). In the United States, a *carrier* that provides service within a *LATA*, other than the *LEC*. [F7-1, MG]

GLOSSARY

complex modulation. A form of *modulation* that combines two or more types of modulation. The most common forms combine amplitude and phase modulation. [F3-10]

compound document. Document combining text, graphics, and sometimes other information formats. [MD]

compress. To reduce the number of bits that will be needed to send a message. Possible in messages with *redundancy*. [F3-11, FD-2, MJ]

Compressed Serial Line Internet Protocol. Standard for telephone *modem* communication between a user PC and an *Internet service provider*. [F2-13]

computer conferencing. Application allowing group members to post messages to other members of the group. [F10-2]

computer layer. In the *Basic Model of Communication*, the layer that standardizes exchanges between the operating systems of the computers in the system. Synonym for *transport layer*. [F1-1]

confidentiality. Others cannot see a file or other resource, typically because of *encryption*. [C9, MJ]

congestion. When there are too many stations on a *shared media* LAN, there will be unacceptable delays in transmission. [F5-5]

connectionless service. *Switched data network* service in which each packet is routed individually. [F7-7]

connection-oriented service. *Switched data network* service in which all packets follow the same route after an initial connection negotiation. Dominates in switched data networking because it reduces the load on switches. [F7-7]

consolidate. In *reengineering*, to combine multiple tasks performed by different people or work units into a single task performed by a single *case worker* or work unit. [C11]

consumer-oriented service. Among *commercial online services*, services oriented to residential consumers rather than to organizations. [C11]

control console. In room-to-room *videoconferencing*, device that allows a participant to control the system. [C10].

core business function. In *reengineering*, a central business function that should nor be *outsourced*. [C11]

cover sheet. In *facsimile*, an initial sheet that specifies such information as who sent the fax, who should read it, and how many pages follow the cover sheet. [MD]

CSLIP. See *Compressed Serial Line Internet Protocol*.

CSMA/CD. See *Carrier Sense Multiple Access with Collision Detection*.

CSU/DSU. See *channel service unit/data service unit*.

customer premises. Property owned by a customer. Always plural. [C7, MG]

cyberself. The personality a person exhibits when communicating online in a fairly anonymous manner. [MB]

cyphertext. *Encrypted* information. [FJ-1]

D channel. In *ISDN*, a 16 kbps or 64 kbps channel for carrying control signals. [F3-17, C7, FG-8]

daisy chain. Interconnection of devices in which there is a line of devices without loops. [C4]

DARPA. See *Defense Advanced Research Projects Agency*.

Data communications. A broad term that embraces all forms of electronic communication involving computers. [C1]

Data Encryption Standard (DES). Commonly used *encryption method*. [FJ-2]

data field. In any *synchronous unit*, field containing data to be transmitted. Often the *PDU* of the next higher layer process. [e.g., F4-8]

datagram. Protocol data unit at the TCP/IP internet layer.

data link layer (OSI Layer 2). The second lowest layer in the OSI architecture. Deals with *media access control* and organizing data for transmission. [F4-1, F4-2, FA-3]

data modem. *Modem* designed to transmit data. [C3, MD]

data safety. In sophisticated database systems, a collection of features that reduces the risk of data loss. [C11]

data service unit (DSU). If you have a *digital* input signal and wish to transmit it over a digital transmission line, you will need a DSU to convert between the two digital signaling formats. [F3-17, C7]

data warehouse. System maintaining an extract of corporate production data and providing end user tools for retrieval and analysis. [F11-6]

database middleware. *Middleware* designed for *database* applications. [F11-5]

database. Structured collection of related information organized into records, fields, and relations or files. [C11]

datagram service. Older term for *connectionless service*. [F7-7]

de jure. In standards, de jure standards are set by officially recognized standards agencies, such as ISO and ITU-T. [C1]

dedicated database client development language. Tool for creating *client programs* in *client/server processing*. The tool is specific to *database* applications. [C11]

default router. Router to which IP packets will be sent if a *host* does not have specific information about which router it should use. [C6, MH]

default setting. Address setting set at the factory. If no action is taken, this address is used. [MI]

Defense Advanced Research Projects Agency (DARPA). The organization that created the *Internet*. [MB]

demand priority access method (DPAM). *Media access control* method in *100VG-AnyLAN*. Stations send high- or low-priority request to transmit to their hubs. Stations at each access level are served in order. [F5-8]

deregulation. The gradual movement from monopoly carriers to free-market competition among *carriers*. [C7, MG]

design rationale. Reason for making a design decision while developing a product. Should be retained to answer future questions. [C10]

desktop groupware. *Groupware* in which users work at desktop computers. [C10]

Desktop Management Interface (DMI). Standard for managing personal computers in a network environment. [C9]

desktop videoconferencing. *Videoconference* in which participants work at their desktop computers with added video hardware and software. [C10]

destination address. In any *synchronous unit*, specifies the address of the process that should receive the unit. [e.g., F4-8]

DHCP. See *Distributed Host Configuration Protocol*.

dial-back modem. *Modem* that dials a number and then, once connected, hangs up and waits to be dialed back. Used insecurity *access control*. [MD]

differential Manchester encoding. Electrical signaling format used in *802.5 Token-Ring Networks*. [FF-6]

digest. Portion of a packet providing *authentication* information. [FJ-6]

digital signature. *Authentication* system in which each *packet* carriers authentication information. [FJ-6]

digital switch. *Switch* that is essentially a computer. Many services can be added by software alone. [MG]

digital. Type of signal characterized by abrupt transitions between a limited number of states (such as voltage levels). Most data communications is digital. [F3-1, F3-6]

digitization. Converting an *analog* signal to *digital*, so that it may travel down a digital transmission line. [F3-18]

directory server. In a *synchronized server* network, a server that maintains of list of all resources on all servers on the network. Not just for PC networking. [F8-2, C9, F10-5]

directory. Menu-based *search engine* on the *World Wide Web*. [MC]

discovery. Legal process of searching through computer-based information in conjunction with a legal action. [C10]

disk cache. On a *file server*, a section of RAM that holds recently accessed files, important supervisory files, and files waiting to be written to disk. [FI-7]

distributed backbone. In *local internets*, a distributed backbone LAN runs between the individual LANs. Also see *collapsed backbones*. [F6-10]

Distributed Communication Service (DCS). *PCS* technology popular in Europe and many other locations. [FG-7]

distributed database. *Database* in which the data are distributed across multiple hosts or servers. [F11-7]

Distributed Host Configuration Protocol (DHCP). In TCP/IP, an *autoconfiguration* protocol. Native autoconfiguration protocol in Windows 95. [MH]

distributed network management system. *Network management* system with multiple *network management consoles*. [F9-6]

distribution list. In *electronic mail*, a list of addresses. Messages addressed to the distribution list are sent to everyone on the list. [F2-9]

DIX connector. Fifteen-pin connector in Ethernet *10Base5*. [F4-12, MF]

DMI. See *Desktop Management Interface*.

DNS. See *domain name system*.

docubase. Searchable collection of documents. [C10, MB]

document conferencing. Form of conferencing in which both participants can see the same information on-screen and can both manipulate the image. Often combined with *videoconferencing*. [C10]

domain name system (DNS). A *TCP/IP* service that allows a *host* to look up an *IP number* if it has an *IP name*. [C6, MB, FH-10]

domestic. In *carrier* transmission, transmission within a single country. [C7]

download. Transferring a file from a *host computer* to a PC doing *terminal emulation*. [F3-5]

downsizing. In information systems, moving applications from a large computer, such as a mainframe, to a smaller platform, such as a PC network server. In corporations in general, reducing the workforce. [C2]

DPAM. See *Demand Priority Access Method*.

drop cable. [FF-1]

DSU. See *data service unit*.

dual-homed. *Router* connected to two networks. It has a different *IP number* on each network. [FH-3]

E1. *Circuit* operating at 2.048 Mbps. [T7-1]

echo. *ICMP* protocol that sends a message to a host asking if it is available. [FH-11]

EDI clearinghouse. Vendor that collects *EDI* information and translates it into the format desired by other parties. [C11]

EDI. See *electronic data interchange*.

EDM. See *electronic document management*.

EFT. See *electronic funds transfer*.

EIA/TIA-232-E. Standard for PC *serial ports*.

electronic commerce. Buying and selling over a network [C11]

electronic data interchange (EDI). The electronic exchange of business documents (orders, invoices, and so on) between customers and vendors. [C11]

electronic document management (EDM). Application that manages documents and *docubases* over the live cycles of documents from initial research through composition, storage, retrieval, and eventual deletion. [C10]

electronic funds transfer (EFT). The exchange of money electronically. [C11]

electronic mail (email). A service for transmitting text messages and sometimes multimedia messages *asynchronously*. [F2-9, F2-10]

electronic mail host. Host computer that provides *email* service. [F10-3]

electronic meeting room (EMR). Meeting room in which each participant has a computer. [F10-14]

electronic software distribution (ESD). Distributing a new program or an upgrade to multiple servers and/or personal computers automatically. [C9]

eliminate. In *reengineering*, to delete a process or portion of a process entirely. [C11]

email. See *electronic mail*.

EMR. See *electronic meeting room*.

emulation (terminal). See *terminal emulation*.

encapsulation security. In *IPv6*, optional *encryption* procedure. [MH]

encode. In general, any process that converts one form of information into another; often analog information into digital information. In *Internet electronic mail*, a process that converts an *attachment* into a form that will pass over the Internet. The receiver must decode the attachment. [C10]

encryption method. Process for *encrypting* information. Uses a *key*. [FJ-1]

encryption. Transforming a message into *cyphertext* that people intercepting the transmission could not read. [C9, FJ-1]

end frame delimiter. Field that ends an 802.5 *token* frame and is the second-to-last frame in a full *802.5 frame*. [FF-7, FF-8]

error correction. If an error is detected, asking the original sender to retransmit the message. [C3, C7]

error detection and correction. Process in which a receiver checks a message for errors and, if it finds one, asks the sender for a retransmission. [MA]

error detection. Examining an incoming message for errors. [C3]

ESD. See *electronic software distribution*.

Ethernet switch. Switches that are compatible with *Ethernet NICs* and *10Base-T hubs*. [C5]

Ethernet. Name commonly given to LAN standards from *the 802.3 Working Group* of the *IEEE 802 Committee*. [C4]

existing application program. An application program that currently exists and must be maintained or converted. [C11]

expansion board. Printed circuit board placed in a PC to provide additional capabilities. The *NIC* often is an expansion board. [F4-3, FI-3]

expansion slot. In a personal computer, a connector on the mother board for adding expansion boards. [FI-3]

extended ASCII. Extended version of *ASCII* in which eight bits are used to represent each character, rather than seven. [C3]

external image. Graphic image not shown on a *webpage* but available by clicking on a *link*. Really, webpage in a graphic format, such as GIF or JPEG, rather than HTML format. [FC-9]

external modem. *Modem* outside the system unit. [FD-4]

facsimile modem (fax modem). Modem that transmits *facsimile* images. [MD]

facsimile software. Software on the PC that allows the PC to send *facsimile* images via *facsimile modems*. [MD]

facsimile. Service that sends *bit-mapped* pages to be printed from one machine to another.

FAQs. See *frequently asked questions*.

fault management. Identifying and correcting failures in a network. [C9]

fault repair. Fixing a *fault* (failure) in a network. [C9]

fault. Failure in the operation of a network. [C9]

fax modem. See *facsimile modem*.

FCC. See *Federal Communications Commission*.

FDDI over UTP. Version of *Fiber Distributed Data Interface* that can use *unshielded twisted pair* wiring. [C5]

FDDI. See *Fiber Distributed Data Interface*.

Federal Communications Commission (FCC). National telecommunications regulatory agency in the United States. [MG]

fiber. See *optical fiber*.

Fiber Distributed Data Interface (FDDI). A 100 Mbps token ring LAN technology created by the American National Standards Institute. [F5-6]

field. A section of a *synchronous frame*. [e.g., F4-8]

file server program access. In *PC networking*, a process of executing programs. The program is downloaded from a *file server* to the client PC, for execution on the client PC. Data files are also downloaded. Useful for smaller programs. [F2-3, F8-12]

file server. A server that stores data and program files for user retrieval and, in the case of program files, for execution on the client PC via *file server program access*. [C2, C8-6, MI]

file transfer program. A program that allows *file transfer*. [F3-5]

file transfer protocol (FTP). In general, any protocol (standard) to govern *file transfers* in *terminal emulation*. [F3-5, especially ME] If capitalized, File Transfer Protocol (FTP) is a *TCP/IP* standard used on the *Internet* for retrieving files from a host or sending files to the host. Also see *anonymous FTP*. [MB]

file transfer. In *terminal emulation*, when the PC and host transfer a file between them. [F3-5]

filtering rate. The rate at which an *internetting device* can handle *frames* or *packets*. In general, as sophistication increases, filtering rate falls for internetting devices of comparable cost. [C6]

filtering rule. In electronic mail, a rule for screening incoming mail for automatic processing. [C10]

firewall. A server that stands between the *Internet* and a corporate network for security *access control*. [F9-17, FB-10]

flaming. In electronic communication, sending a message that the receiver(s) will find insulting or inflammatory. [C10]

flat-rate local service. Telephone service in which there is no separate charge for individual local calls. [FG-5]

flow control. Process by which communicating parties can ask the others to reduce their rate of transmission or to pause. [MA]

flow. In *IPv6*, header field that identifies a packet as being part of a series of packets for which certain *quality of service* parameters have been reserved. [MH]

formal address. In electronic mail, a user's unique address on the mail system. May not be easy to remember. [C10]

forum. Form of *computer conferencing*. [C10]

fourth-generation programming language. Programming language that allows developers to work with larger chunks of code than individual statements. Hides details of underlying syntax and structure in the chunks. [C11]

fps. See *frames per second*.

fractional T1. A circuit operating at a fraction of T1 speed. [T7-1]

frame. Protocol data unit at the data link layer (OSI Layer 2). In LANs, this includes the *media access control* layer, the *MAC bridge layer*, and the *logical link control* layer. [C4]

frame check sequence field. The last field in the *802.3 MAC layer frame*. Used to identify errors in transmission. If an error is detected, the frame is discarded. Also used in other LAN standards. [F4-8]

frame control. Field in an *802.5 frame* that tells the receiver the purpose of the frame. [FF-8]

Frame Relay. *Switched data network* service offering low megabit speed at moderate prices. [T7-2]

frame status field. Field in an *802.5 frame* that ends the full data frame and tells the sending station of any errors that occurred during transmission. [FF-8]

frames per second (fps). The number of video frames transmitted each second. [FJ-7]

frequency modulation. A form of *modulation* that varies the frequency (pitch) of the transmitted signal. [F3-7]

frequency reuse. In *cellular telephony*, using the same frequency in different cells in order to increase the number of supportable customers. [FG-6]

frequency. The number of times a wave goes through a full cycle per second. [C3-14]

frequently asked questions (FAQs). Set of often-asked questions that new users should familiarize themselves with before participating actively in an *online community*. [MB]

FTP. See *file transfer protocol*.

full repeater. *Repeater* that connects directly to two or more LAN segments. Also see *half repeater*. [FF-2]

full-duplex. Transmission system that allows both sides to transmit simultaneously. Almost all modern transmission systems are full-duplex. [F3-14]

full-text searching. Form of string searching in documents that searches the entire text of documents for the string instead of limiting itself to *keywords*. [C10]

gateway server. On a PC network, a *communication server* that allows a user to access a network using a different standards architecture than the PC network. [C8]

gateway. In general data communications, a device that translates at all layers between two different standards *architectures*. [e.g., FB-10] In TCP/IP, an obsolete but still-used term for router [MH].

Gbps. See *gigabits per second*.

General System for Mobile telephony (GSM). Second-generation *cellular telephone* standard used in much of the world; digital. [FG-7]

GIF. Popular *bit-mapped graphics* format used in the *World Wide Web*. [MD]

gigabits per second (Gbps). Thousands of megabits per second. [F1-5]

Gopher client. Client program for the *Gopher* application on the *Internet*. [MB]

Gopher server. Client program for the *Gopher* application on the *Internet*. [MB]

Gopher. Menu-based retrieval service on the *Internet*. [FB-8]

Gopherspace. Collection of all *Gopher* servers. [MB]

group composition. *Groupware* application in which group members jointly create a document or set of documents. [C10]

groupware. Application aimed at groups rather than at individuals. [F10-6]

GSM. See *General System for Mobile telephony*.

H.261. Family of *ITU-T* standards for *videoconferencing*. [MJ]

H.320. Set of *OSI* standards governing *videoconferencing*. [C10]

hacker. Someone who accesses a computer system without permission. [C9]

half repeater. *Repeater* that connects to one LAN segment and, through an *interrepeater link*, to another half repeater. Also see *full repeater*. [FF-2]

half-duplex. Transmission system that allows both sides to transmit but requires them to take turns. Obsolete and rarely used. [F3-14]

hand-held cellphone. The common *cellphone* that can be carried in a hand. [MG]

handoff. In *cellular telephony*, being handled by a different cell's transceiver when you move from one *cell* to another. [FG-6]

hand-shaking. A process by which two systems exchange messages to establish the capabilities of the other system and to establish parameters that will be used in transmission. [MD]

hardware flow control. *Flow control* between a PC and a modem that uses hardware built into PC *serial ports*. [ME]

header. In electronic mail, a section at the top of the message that contains searchable fields, such as To: and From:. [F2-9]

help desk. Organizational unit that provides help for users when they have problems. [MI]

helper application. In *World Wide Web browsers*, auxiliary applications that allow users to view files in formats not supported by the *browser* itself. [MC]

hertz (Hz). The number of times a wave goes through a complete cycle per second. Formerly called cycles per second. [F3-7]

home page. A *webpage* that is designed to be a starting point for reading about a particular topic. [C2, MC]

host access gateway. On a *PC network*, a *communication server* that allows a user to access a *host* computer via *terminal emulation*. [F8-18]

host computer. In the context of the *Internet*, any computer attached to the Internet. [C1, MB, MH] In the context of *terminal–host* computer platforms, the central computer that governs terminals. [F2-1]

host ID bits. The bits in an *IP number* identifying a specific host on a network. [FH-2]

host name. See *IP name*.

htm. Common extension for *webpages* on MS-DOS and Windows computers. [MC]

HTML editor. Word processor for creating *HTML* documents. [MC]

html. Common extension for *webpages* on MS-DOS and Windows computers. [MC]

HTML. See *Hypertext Markup Language*. [FC-4]

HTTP. See *HyperText Transfer Protocol*.

hyperbase. Collection of linked *hypertext* pages. [MC]

Hypertext Markup Language (HTML). Standard for the formatting of *webpages*. [MC]

HyperText Transfer Protocol (HTTP). The protocol used in communication between *browser* clients and *World Wide Web* host computers. [C2, FC-2]

hypertext. A way of representing information in which pages contain *links* pointing to other pages. Users can jump between related pages by following the links. Invented in the 1960s by Engelbart and Nelson. Used in the World Wide Web. [F2-11, FC-1]

Hz. See *hertz*.

I/O address. The address of a register on a *NIC* or other *expansion board*. [FI-10]

ICC. See *international common carrier*.

ICMP. See *Internet Control Message Protocol*.

IEEE LAN MAN Standards Committee. *IEEE* standards committee that has created most LAN standards. Also called *802 Committee*. [C4]

IEEE. See *Institute of Electrical and Electronic Engineers*.

IETF. See *Internet Engineering Task Force*.

imaging. Capturing an image of a document or other information electronically, for storage and distribution. [C10]

implementation profile. Limited list of standards that an organization will implement internally. This allows the organization to build networks without having to deal with hundreds of different standards. Ensures that all stations will be able to communicate. [C1]

independent server operating system. *Server operating system* in which servers are unaware of one another. Users must know on which server each needed resource resides. Also see *synchronized server operating system*. [F8-2]

inheritance of rights. On a *file server*, the determination of *rights* in a directory on the basis of rights granted in the parent directory. [FI-2]

inline image. Graphic image shown on a *webpage* when viewed by a *browser*. [FC-8]

installation charge. When you purchase a *carrier* service, there may be an initial installation charge as well as ongoing use charges. [C3, C7]

Institute of Electrical and Electronic Engineers (IEEE). Professional society that has developed most LAN standards. [C4]

Integrated Services Digital Network (ISDN). [FG-8]

integrated services. In *ISDN*, services such as ANI that are standardized to operate worldwide instead of just locally. [MG]

integrity. Ensuring that a *frame* or *packet* has not been changed during transmission. [MH]

intelligent. A *modem* that can accept commands from the computer and then carry out actions autonomously. See *Module D*.

interexchange carrier (IXC). In the United States, a carrier that provides service between *LATAs*. [F7-1, MG]

interface. Vertical *standard* for communication between processes at adjacent layers a single *station* on a network. [C2]

interference. Unwanted signals on the transmission line. Produced by sources outside the line. [C4]

internal modem. *Modem* built onto an expansion board. [FD-3]

international common carrier (ICC). *Carrier* that provides transmission service between two countries. [F7-1]

international common carrier (ICC). See *Module G*.

International Organization for Standardization (ISO). A standards agency specializing in computer technology and other industrial standards. Joint creator of the *OSI architecture*. [F1-3, MA]

International Telecommunications Union–Telecommunications Standards Sector. A standards agency specializing in telecommunications. Joint creator of the *OSI* architecture. [F1-3, MA]

Internet Control Message Protocol (ICMP). Supervisory *TCP/IP* protocol that is used to deliver error messages and many other supervisory messages at the internet layer. [FH-11]

Internet Engineering Task Force (IETF). Standards arm of the *Internet Society*. Oversees the *TCP/IP architecture*. [F1-3, MB, MH]

internet layer. In our expanded *Basic Communication Model*, the layer that governs the *routing* across multiple networks in an internet. [F1-2]

Internet Protocol (IP). The protocol governing the transmission of data packets at the internet layer in *TCP/IP* networks, including the *Internet*. [F6-8, MH]

Internet Relay Chat (IRC). *Internet* application that allows users on different machines to type messages back and forth in real time. [MB]

Internet Service Provider (ISP). An organization providing access to the Internet. [F2-13, MB]

Internet Society (ISOC). The organization that oversees the Internet. [MB]

Internet. If spelled with a capital "I," a worldwide transmission system that connects *host computers* around the world. [C1, FB-1]

internet. If spelled with a lower case "i," an interconnected set of subnets (single networks) in which any station attached to any subnet can send a message to any other station on any of the internet's subnets simply by giving the internet address of the other station. In contrast, see Internet with a capital "I". [F1-1, F1-2]

internetting device. In *internets*, device that links *subnets*. *Bridges*, *LAN switches*, and *routers* are popular internetting devices [C6]

InterNIC. Internet Society organization that assigns IP names and IP numbers. [MB]

interrepeater link (IRL). Transmission link between two half repeaters. [FF-2]

interrupt request (IRQ). Address that a NIC uses to request service from the microprocessor. [FI-11]

intranet. Internal organizational *internet* based on Internet technology. Typically includes use of the *TCP/IP architecture* and services delivered to *browser* software on networked PCs and desktop *workstations*. [C1]

inventory management. Keeping information about devices and software on a network and keeping status information on each element. [C9]

IP address. The address of a host's or router's internet process on an internet. May be a 32-bit *IP number* or a longer *IP name*. [C6, MH]

IP host address. See *IP address*.

IP host name. See *IP name*.

IP host number. See *IP number*.

IP name. Easy-to-remember name for a host on an *internet* using *TCP/IP*, such as the Internet. [C6, MH]

IP number. 32-bit number that uniquely identifies a host on the *Internet* or on a corporate *internet*. Address at the *internet layer*. [C6, FH-2]

IP. See *internet protocol*.

IPv6. New version of the *Internet Protocol*. [MH]

IPX/SPX. *Proprietary* standards architecture created by *Novell* and used in Novell *PC networks*. [F1-3]

IRC. See *Internet Relay Chat*.

IRL. See *interrepeater link*.

ISO. See *International Organization for Standardization*.

ISOC. See *Internet Society*.

ISP. See *Internet Service Provider*.

ITU-T. See *International Telecommunications Union–Telecommunications Standards Sector*.

IXC. See *interexchange carrier*.

Java. A programming language whose programs (*applets*) are downloaded to client machines via *HTML* documents. The Java applets then execute on the user's machine. [FC-13]

Joint Photographic Experts Group (JPEG). Popular *bit-mapped graphics* format used on the *World Wide Web*. [MD]

JPEG. See *Joint Photographic Experts Group*.

JPG. Common extension for files in the *Joint Photographic Experts Group bit-mapped graphics* file format. [MD]

kbps. See *kilobits per second*.

Kermit. *File transfer protocol* used by PCs. [ME]

key distribution. In encryption, the secure distribution of secret *encryption keys* to communication partners. [FJ-4]

key escrow. Having a third party hold (and probably distribute) keys used by communicating partners. [FJ-4]

key. A binary value that an *encryption method* uses to *encrypt* files. [FJ-1]

keyword search engine. Keyboard-based *search engine* on the *World Wide Web*. [MC]

keyword. In text searching, one of a small set of words chosen to represent a document in searches. [C10, MC]

kilobits per second (kbps). Thousands of bits per second. Note that k is not capitalized in the metric system. [F1-5]

LAN switch. An internetting device of medium sophistication. Useful in *switched LANs* [F5-10] and *local internets* [F6-7]

LAN. See *Local Area Network*.

LATA. See *local access and transport area*.

latency. Latency is a synonym for delay. [F5-5]

layering. In standards *architectures*, dividing the responsibility of ensuring that two systems can communicate into a number of stacked tasks, each of which provides a foundation of services for the task at the next higher layer. [F1-4]

leased line. Point-to-point *circuit* for which an organization has exclusive use. [F7-3]

LEC. See *Local Exchange Carrier*.

legacy system. An system that currently exists and must be maintained or converted. [C11]

length field. Field present in most *synchronous units*. Specifies the overall length of the unit. [e.g., F4-8]

LEO. See *low earth orbit satellite*.

license fee. Fee paid to an application software vendor to use its product. The fee will depend on the number of possible simultaneous users. [C8]

link. In *hypertext*, a section of a page that points to another hypertext page. When the user hits a link, he or she is take to the other page. [F2-11, FC-1, FC-3]

LISTSERV. *Internet computer conferencing* system that delivers postings to users through their normal *email* system. [C10, MB].

LLC layer. See *logical link control layer*.

load balancing. In *switched data networks*, routing can be done to even the load on various switches and trunk lines, in order not to overload some components. [C5]

lobe. In *802.5 Token-Ring Networks*, the transmission line from the *access unit* to the station. [FF-11]

local access and transport area (LATA). In the United States, a region served by an *LEC* and several *CAPs*. The United States is divided into 161 LATAs. [F7-1, MG]

local area network (LAN). Network restricted to a small distance on a customer premises. Usually limited to a single office or a small building. [F1-6, C4, C5]

local bus. *Bus* in a PC that links the microprocessor to high-speed *expansion boards*. [MI]

local exchange carrier (LEC). In the United States, the traditional monopoly telephone *carrier* operating in a local region (see *LATA*). [F7-1, MG]

local internet. An *internet* at a single site. [C6]

local loop. The transmission line running from the *customer premises* to the first *switching office* of the telephone *carrier*. [F7-2, FG-1]

logical link control (LLC) layer. In the IEEE LAN standards framework, the highest sublayer of the *data link layer*. Concerned with providing error checking and other service quality benefits if desired. [F4-2, F4-9]

login script. In *terminal emulation*, a set of commands that can be stored and later executed to allow users to log into host computers more easily. [F3-4] In *Novell NetWare* networking, a set of commands that are executed when a user logs into the network. [FI-12, FI-13]

lossless compression. *Compression* that does not lose information from the original file. Also see *lossy compression*. See *Module D*.

lossy compression. *Compression* that permanently loses information from the original file. Usually provides greater compression than *lossless compression*. [MD]

low earth orbit satellite (LEO). Satellite system in which satellites orbit the earth at only 100 to 200 miles above the ground. [FG-7]

lurk. To be in a *computer conference* as an observer without making comments. [MB]

Ma Bell. Popular name for AT&T before its breakup in 1984. [MG]

GLOSSARY

MAC bridge layer. In the IEEE LAN standards framework, the middle sublayer of the *data link layer*. Concerned with bridging information. [F4-2]

MAC layer. See *media access control*.

mail server. On a *PC network*, a server that provides *electronic mail* service. [C10, C8]

main bus. *Bus* in a PC that links the microprocessor to normal-speed *expansion boards*. [MI]

MAN. See *metropolitan area network*.

mainframe. Large multiuser computer that serves many *terminal* users simultaneously. Normally employs the *SNA architecture*. [C2]

managed device. A PC, router, transmission circuit or other entity participating in a *network management system*. [F9-2]

managed modem. Modem with a *network management agent*. [MD]

management information base (MIB). Database that maintains data on the state of *managed devices* on the network. [F9-4]

Manchester encoding. See the electrical encoding of 1s and 0s used in *Ethernet 10Base*-T. [F4-5]

MAP INS. In *NetWare*, a *login script* command that specifies the order in which file server directories will be searched to find an application program. [FI-12]

MAP. In *NetWare*, a *login script* command that maps a *virtual drive* letter to a directory on the *file server*. [FI-12]

marketing channel. A system of wholesalers and retailers who link a vendor to its customers. [C11]

MAU. See *medium attachment unit*.

Mbps. See *megabits per second*.

MCU. See *multipoint control unit*.

mean time between failure (MTBF). Average time between failures in the network. Related to availability. [C9]

mean time to repair (MTTR). The average time between when a network fails and when it is repaired. [C9]

media. See *medium*.

media access control (MAC) layer. In the IEEE LAN standards framework, the lowest sublayer of the *data link layer*. Concerned with *media access control*. [F4-2]

media access control. Controlling when stations in a LAN may transmit (access the medium). [F4-2]

mediated. A process is mediated if the parties do not interact directly, but only through the mediation of software, computers, and transmission technologies. [C1]

medium. Physical matter through which electrical signals travel. Popular media are *unshielded twisted pair* wire, *coaxial cable*, and *optical fiber*. [C3, C4, C5]

medium attachment unit (MAU). Official name for *transceivers* in *10Base5*. [F4-12, FF-1]

megabits per second (Mbps). Millions of bits per second. [F1-5]

memory bus. *Bus* linking the microprocessor to RAM and ROM memory. [MI]

menu directory. Menu-based *search engine* on the *World Wide Web*. [MC]

message oriented middleware (MOM). *Middleware* standard that is *asynchronous* at the *application layer* and is *database*-specific. [FJ-9]

message units. Telephone service in which even local calls are charged, usually by distance and time. [FG-5]

metropolitan area network (MAN). Network restricted to a city and surrounding communities. [F1-6]

MIB. See *management information base*.

microcell. In *PCS*, a *cell* much smaller than a cell in traditional *cellular telephony*. [MG]

Microcom Network Protocol (MNP). Proprietary standards for both *error correction* and data *compression* in modems. [T3-1, TD-2]

middleware. Layer of software between an

application and the *transport layer* that performs services and hides details of the transport layer. [C10 for *email* middleware; F11-5, MJ for database middleware.]

MIME. See *Multipurpose Internet Mail Extensions*.

Ministry of Telecommunications. In many countries, the government agency that oversees telecommunications. [MG]

MNP. See *Microcom Network Protocol*.

MNP. See *Microcom Network Protocol*.

mobile cellphone. *Cellphone* larger than a hand-held cellphone. Has more power and a longer-lived battery. [MG]

mobile telephone switching office (MTSO). In *cellular telephony*, the office that manages the cellular network and interconnects it with the ordinary public switched telephone network. [FG-6]

modem. A device that *modulates* a *digital* computer signal into an *analog* signal that will travel down an analog transmission line, such as a telephone line. [F3-6, MD]

moderated conference. *Computer conference* in which a moderator screens incoming message and decides whether each message will be posted to the group. [C10]

moderator tools. Tools that allow a *computer conference* moderator maintain member lists, screen messages before postings, and do other administrative tasks. [C10]

modulation. The process of converting a *digital* computer signal into an *analog* signal that will travel down an analog transmission line. See *modem*. [F3-6]

MOM. See *message oriented middleware*.

monopoly. A carrier that enjoys exclusive rights to provide transmission service within a region or service category. [C7, MG]

monthly service charge. When your purchase a *carrier* service, there usually is a continuing charge for use. It may be a flat fee, or it may depend on usage. [C3, C7]

mother board. In a personal computer, the main printed circuit board, which holds the microprocessor and other core circuitry. [FI-3]

MTBF. See *mean time between failure*. [C9]

MTSO. See *mobile telephone switching office*.

MTTR. See *mean time to repair*.

multidimensional database. *Database* arranged in tabular form but allowing more than two dimensions. [C11]

multifamily network management program. *Network management program* that can work with *network management agents* communicating via different *network management protocols*. [F9-14]

multihomed. *Router* connected to multiple networks. It has a different *IP number* on each network. [MH]

multimedia database. *Database* whose records include not only text and numbers but also graphics and perhaps animation, voice and video as well [F11-8].

multiple access. Technique for ensuring that two station do not transmit simultaneously, jumbling their signals. [C4]

multiplex. To mix several signals over a single transmission channel to reduce cost. [F3-17, FA-4, C4]

multipoint control unit (MCU). Device that allows multiple sites to participate in a single *videoconference*. [F10-12].

multiport repeater. *Repeater* with multiple ports that can connect multiple *segments* of a LAN. [F4-3, F4-12, C6]

Multipurpose Internet Mail Extensions (MIME). *TCP/IP* standard used on the *Internet* to allow *electronic mail headers* and bodies to contain information other than plain *ASCII* text. [F10-4]

multivendor standard. A *standard* followed by multiple vendors, so that a purchaser is not restricted to buying compatible equipment from a single source. [C1]

narrowband ISDN. Form of *ISDN* in which the normal service brought to the customer is the 144 kbps *Basic Rate Interface*. Other low-megabit service rates are possible. [MG]

netiquette. Explicit or implicit set of rules that specify how one should behave while on network system, such as *email*. [C10]

NetView. Network management standards in *SNA*. Also the name of an IBM network management program. [F9-10]

NetWare Loadable Module (NLM). Application program that runs on the *NetWare* operating system on a *server*. [FI-8]

NetWare. *Server operating system* of *Novell*. The dominant server operating system. [C8]

network. An any-to-any transmission system in which any *station* can send a message to any other station simply by giving the network the *address* of the other station. [F1-2]

network capable. Client operating system that can connect to *servers* without additional software. [C8, MI]

network control center. Physical location from which a distributed network is managed. [F9-2]

network design program. Program for designing a network for optimal performance and for considering the impacts of modifications to existing networks. [9-15]

network drive. See *virtual drive*.

Network File System (NFS). *UNIX* file service component. [C8]

network part. The bits in an *IP number* identifying a specific network. [FH-2]

network information center (NIC) Administrative unit that *assigns IP* numbers and *IP names*. Also see *InterNIC*. [MH]

network interface card (NIC). In a LAN, an *expansion board* placed in a PC. Manages communication between the PC and the network at the *physical layer* and *data link layer* and, if applicable, at the *network layer*. [F4-3, MI]

network layer (OSI Layer 3). The *OSI* layer that governs *switched data networks* and OSI internets. Used in many switched data networks but not in many internets. [C5, FA-5]

network management. The processes involved in managing a large distributed network.

network management agent. Hardware, software, or both in a manag*ed device* that communicates with the *network management program*. [F9-5]

network management console. PC, workstation, or terminal from which the network administrator manages the network. [F9-2]

network management program. Software that manages a distributed network by collecting data, presenting reports to management, and changing the configurations of managed devices. [F9-2, F9-3]

network management system. The technology needed for network management. Includes the *network management console*, the *network management program*, *network management agents*, and *managed devices*.

network operating system (NOS). Older name for *server operating system*.

network-ready printer. A printer with a built-in *print server*. [C8]

network version. Version of an application program specially written to run from a *file server*. Can server multiple users simultaneously. [C8]

newbie. A new user in an *online community*. May be unfamiliar with norms of interaction (*netiquette*) [MB].

newsgroup. *Internet computer conferencing* system that requires users to check postings by using a newsgroup reader. [FB-6, C10]

NFS. See *Network File System*.

NIC. See *network interface card* in the context of local area networks [C4, MI] or *network information center* in the context of the Internet [MH].

NLM. See *NetWare Loadable Module*.

noise. Random stray signals in a transmission line that can cause errors. [C3, C4]

nonrepudiation. The inability to reject the claim that you sent a message. [MJ]

NOS. See *network operating system*.

Notes (Lotus). Popular *desktop groupware* application. [C10]

Novell. *Server operating system* vendor for PC networking. Its server operating system is *Net-Ware*. [F1-3, C8]

object request broker (ORB). *Middleware* process that uses objects. [MJ]

OC. Designations for *SONET circuits*. [T7-1]

octet. Synonym for byte. A group of eight bits. [C4]

ODBC. See *Open Database Connectivity*.

OLAP. See *online analytical processing*.

online analytical processing (OLAP). In a *data warehouse*, tool for end user analysis of the data in the warehouse. [C11]

online community. A group that meets online and forms social bonds. [MB]

Open Database Connectivity (ODBC). A standard, based on SQL for making *database* queries in *client/server* systems. [F11-4]

Open Shortest Path First (OSPF). Sophisticated *router coordination protocol*. [FH-13]

open. In standards, open standards are created by official (de jure) standards agencies. They are not under the control of any vendor but have an open development process. [C1]

optical fiber. Transmission *medium* consisting of a thin glass or plastic tube through which light passes. [F4-14]

optimal route. In a *switched data network*, this is the best *alternative route* for a packet, based on speed, security, reliability, or other grounds. [C5]

ORB. See *object request broker*.

ordered list. In *HTML*, a list of items. Items are numbered sequentially. Also see *unordered list*. [FC-11]

OSI. See *Reference Model of Open Systems Interconnection*.

OSPF. See *Open Shortest Path First*.

outsourcing. Having an outside organization manage a firm's network. [C9]

packet switched network. Switched data network that breaks messages into small packets and routes them from the source to the destination. Packets are multiplexed enroute. [C7, MA]

packet. *Synchronous protocol data unit* at the *network layer* (OSI layer 3). [C5, MA]

PAD field. In the *802.3 MAC layer frame*, field containing added bits if the frame would otherwise fall below the minimum size. [F4-8]

parallel port. Port at the back of a PC. Used to connect the PC with printers and other devices. Uses *parallel transmission*. [F3-15]

parallel transmission. Form of transmission in which multiple transmission lines are used for faster transmission. [F3-16]

parity bit. In an *asynchronous frame* using 7-bit *ASCII* character coding, the ninth bit in the frame. Used in error detection. [F3-12, F3-13]

password. A secret string of characters a user must type to get access to a system. [C9]

payload. In *ATM packets*, the *data field*. [F7-8]

PC network. Computer platform in which the machines on the desks of users are *client PCs* served by other computers called *servers*. [F2-2, C8, MI]

PCM. See *pulse code modulation*.

PCMCIA modem. *Modem* connected to a computer via a PCMCIA slot. Used mostly in mobile computers. [FD-5]

PCS. See *personal communication service*.

PCX. Popular *bit-mapped graphics* format. [MD]

PDF. See *Portable Document Format*.

PDU. See *protocol data unit*.

peak period. Time of maximum system usage. The network must have sufficient capacity to serve traffic during the peak period. [C9]

percentage of capacity. Throughput compared with possible throughput. [C9]

performance management. Keeping information about the performance of various elements in a network. Analyzing options for improving performance. [C9]

permanent virtual circuit (PVC). In *connection-oriented switched data networks*, a connection established days or months in advance of use. [F7-9]

personal communication service (PCS). Third generation of *cellular telephony*, characterized by small *cells*, *digital* operation, and large channel capacity. [FG-7]

personal computer network. See *PC network*.

personal telephone number. A telephone number assigned to a person rather than to a piece of equipment. [FG-5]

phase modulation. A form of *modulation* that varies the phase (start of each sign wave) of the transmitted signal. [F3-9]

physical control. Security *access control* by preventing physical access to terminals or other access devices. [C9]

physical layer (OSI Layer 1). The lowest layer in the *OSI architecture*. Deals with *media*, connectors, and electrical signaling. [F4-1, F4-2, FA-2]

piecemeal deployment. The ability to deploy *IPv6* in parts of the *Internet*, rather than in the entire Internet at a single time. [MH]

Ping. Application that implements the *echo* protocol. [MH]

pixel. Dot of intensity and color from which a *bit-mapped* picture is made. [FD-1]

plaintext. Information to be *encrypted* for transmission. [FJ-1]

platform. A type of computer system serving users. Typical platforms include *PC networks*, *workstation networks*, and *terminal–host systems*. [C2]

plenum wiring. *Wiring bundle* with a covering that gives off relatively few toxic fumes if burnt. Required in wiring run through false ceilings and other airways. [C4, MD]

point of presence (POP). A physical location on the premises of an *LEC switching office* at which the LEC is interconnected with other carriers. [F7-2, MG]

Point-to-Point Protocol. Standard for telephone *modem* communication between a user PC and an *Internet service provider*. [F2-13]

POP. For *carriers*, see *point of presence*. For *electronic mail*, see *Post Office Protocol*.

port number. In *TCP* and *UDP* addressing, the designation of a specific application process on a host computer. Also see *socket*. [MH]

Portable Document Format (PDF). Standard for representing information in *compound documents* in a form that can be viewed on a screen and printed. [MD]

Post Office Protocol (POP). In electronic mail, a *TCP/IP* protocol that allows users working at intelligent devices such as PCs to do a good deal of work on their local devices. Requires a POP-compliant *electronic mail host*. [F10-4]

PostScript. Popular *vector graphics* file format. [MD]

PPP. See *Point-to-Point Protocol*.

preamble. The first field in the *802.3 MAC layer frame*. Synchronizes the sender's and receiver's clocks. [F4-8]

presentation layer (OSI Layer 6). In *OSI*, the layer that standardizes how information will be presented compatibly across two different systems. [MA]

presentation tool. In an *electronic meeting room*, an application to help a participant make a structured presentation to the group. [C10]

PRI. See *Primary Rate Interface*.

Primary Rate Interface (PRI). The interface between an *ISDN* carrier and the *customer premises*. [F7-11, FG-8]

print queue. In *print service*, output first goes to the *file server*, where it is stored in a print queue until the *print server* is ready to accept it. [F8-15]

print server. A device standing between the printer and the *print queue* on the *file server* that accepts jobs to be printed and feeds them to the printer. The printer is not aware that the network even exists. [F8-15]

print service. When the user prints in an application program, the output is *redirected* to a printer on the PC network. [F8-15]

priority. *Media access control* method that gives preference in transmission to stations requiring very low delays. [C5]

privacy extension header. In *IPv6*, header that implements encapsulation security. [MH]

privacy. Others cannot read a file, either because of norms or because of *encryption*. See *discovery*. [C9, MJ]

private branch exchange (PBX). A telephone switching system located on the *customer premises*. [FG-2, FG-3]

private key. In *public key encryption*, a secret *key* known only to the receiver. [FJ-3]

private telephone network. Telephone network constructed by an organization for internal use from *leased lines* and *switches*. [F7-3, FG-3]

private virtual drive. A *virtual drive* pointing to a region on the *file server's* hard disk to which only the user should have access. [C8]

private. In general, information that only the owner should be able to see. Organizations and the legal system decide what information stored on user computers is private. [C10] This may require *encryption*.

production data processing. Data processing for operational transactions, such as order entry. [C11]

project management system. Form of *groupware* in which team members can manage a project involving multiple people and tasks. [F10-8]

propagation effects. Set of undesirable changes in a signal as it travels down a transmission line. [C4]

proprietary MIB. A *management information base* whose design is created by a vendor rather than by a standards committee. [C9]

proprietary. In standards, proprietary standards are created and controlled by a single company but are also used by other companies. [C1]

protocol analyzer. Network monitor that analyzes the contents of individual *frames* or *packets*. [C9, MI]

protocol data unit (PDU). *Synchronous* unit of information exchanged between peer processes at the same layer but on different *devices* on a network. [F2-15, F2-16, MA, FH-1]]

protocol. Horizontal *standard* for communication between peer processes at the same layer but on different *stations* on a network. [C2]

PSTN. See *Public Switched Telephone Network*.

PTT. See *public telephone and telegraph authority*.

public key encryption. *Encryption method* in which encryption is done with a non-secret *public key* but decryption is done with a secret *private key*. [F9-16, FJ-3]

public key. In *public key encryption*, a non-secret *key* known by many people. [FJ-3]

public screen. In an *electronic meeting room*, a large screen for presenting information to the whole group. [F10-14]

Public Switched Telephone Network (PSTN). Today's worldwide dialup telephone network.

public telephone and telegraph authority (PTT). In many countries (but not in the United

States) the traditional *domestic* and sometimes international monopoly *carrier*. [F7-1, MG]

public utilities commission (PUC). State-level telecommunications regulatory agency in the United States. [MG]

PUC. See *public utilities commission*.

pulse code modulation (PCM). A way of encoding an analog signal so that it may transmit down a digital transmission line. Used in Codecs. [F3-18]

PVC. See *permanent virtual circuit*.

Px64. Data *compression* standard in the *H.261 videoconferencing* family. [MJ]

QOS. See *quality of service*.

quality of service (QOS). In *IPv6*, header field specifies the quality of service (low latency, security, and so on) requested of the network. [MH]

radio LAN. LAN technology that uses radio to deliver frames among the stations on the LAN. [F5-11]

RAID. See *redundant array of inexpensive disks*.

rapid application development. In *client/server database* systems, the ability to develop many smaller applications usable immediately, instead of waiting for a single large complex application to be developed. [F11-3]

RARP. See *Reverse Address Resolution Protocol*.

RBOC. See *Regional Bell Operating Companies*.

red lining. Refusing to provide service to people in specific geographical areas. [MG]

redirect. *ICMP* protocol that tells a *host* that is should use a different *router* for communication with another host. [MH]

redirection. In *PC networks*, when an application program issues a *call* for service to the operating system, if the call is for a *virtual drive* or *virtual port*, it is redirected to the *file server* for handling. [F8-13, F8-14]

redirector. Software on a client PC that sends (redirects) *calls* to *virtual drives* and *virtual ports* to a *file server*. [FI-11]

redundant array of inexpensive disks. Disk drive system that uses multiple inexpensive disk drives. Reads and writes are done in *parallel* to increase speed. Data are stored *redundantly*, so that if one drive fails, no information is lost. [FI-6]

redundant. When more bits are in a file than are needed to represent the information in that file. Redundancy allows *compression*. [MD]

reengineering. The redesign of an organizational process for improved responsiveness, lower costs, or other benefits. [C11]

Reference Model of Open Systems Interconnection (OSI). The standards *architecture* created by *ISO* and *ITU-T*. [F1-3, MA]. Individual standards are set by ISO, ITU-T, or other standards agencies. For instance, most LAN standards are set by the *IEEE*. [C4].

regenerate. To recreate a signal, cleaning up the signal in the process. Regeneration removes distance limitations in transmission. [F6-1]

Regional Bell Operating Companies (RBOCs). After *AT&T* was broken up, one of the seven regional holding companies that owned local telephone companies after the breakup. [MG]

register. Temporary storage location on a *NIC* or other *expansion board*. [FI-10]

relational database. *Database* structured into data structures called relations. [C11]

reliable service. *Switched data network* service in which there is error correction at each hop between switches in the network. [F7-6]

remote access server. On a PC network, a *communication server* that allows users to log into the network from home or other outside locations. [F8-17]

remote control software. Software that connects to PCs so that one keyboard controls the other machine. [MJ]

Remote Monitoring (RMON). *Simple Network Management Protocol* standard for collecting data from remote LANs. [F9-11]

remote printing. In *PC networking*, when a user prints at a *client PC*, the printing is routed to a printer attached to the network. [C2, MI]

remote procedure call (RPC). *Middleware* standard that is synchronous at the application layer and is not database-specific. [FJ-8]

repeater. In a LAN, a device that retransmits an incoming signal. [F4-4, FF-1]

replication. In a *synchronized server* environment, copying changes in one server to other servers. [F10-10, F11-7]

repudiate. Rejecting the claim that you sent a message. [MJ]

request for comments (RFC). Any standard in *TCP/IP*. [C1]

request message. In *client/server processing*, the message sent by the *client program* to the *server program* to ask the server program to perform specific work. [F2-8, C8]

resegment. To divide an existing *shared media LAN* into multiple smaller LANs and then to interconnect them into a *local internet*. Resegmentation reduces *congestion*. [F6-2, F6-3]

resolution. In bit-mapped images, the number of *pixels* per inch horizontally and vertically. [MJ]

response message. In *client/server processing*, the message returned by the *server program* in response to a *client program's response message*. [F2-8, C8]

response time. The lag between when a request is issued and when a response comes back. [C9]

Reverse Address Resolution Protocol (RARP). In *TCP/IP*, an *autoconfiguration* protocol. [MH]

RFC. See *request for comment*.

rights. In *PC networking*, permission to see and do specific things in a directory or to a file. [TI-1]

ring. A LAN topology in which stations are connected in a loop. Used in *802.5 Token-Ring Networks* and *FDDI*. [F5-1, FF-9, FF-11].

RIP. See *Routing Information Protocol*.

RISC Workstation. Desktop machine that physically resembles a PC but has much higher processing power and a much higher price. [F2-4]

riser. In buildings, a vertical space through which telephone *cable bundles* can be pulled. [FG-2]

RJ-45. The wall jack standard used in business telephone wiring. Slightly larger than the RJ-11 wall jacks used in residential telephone wiring [F4-3]

RMON. See *Remote Monitoring*.

roll-about video conferencing system. Portable *videoconferencing* system that can be moved from room to room as needed. [F10-11]

root DNS host. A DNS host supposedly knowing all *IP names* and *IP numbers* on the Internet. See *Module H*.

route. An end-to-end path between *stations* that a *packet* takes as it moves across a *switched data network*. [F5-10, MA]

router coordination protocol. Protocol by which routers exchange information in their *routing tables* in order to provide peer coordination over routing. [FH-12, FH-13]

router. The most sophisticated internetting device. Also the most expensive. Used on the *Internet* and in many internal corporate *internets*. [F2-13 through F2-16, C6, C7, MH]

Routing Information Protocol (RIP). Simple *router coordination protocol* used within *autonomous systems*. [FH-13]

routing information. Field in an 802.5 frame that contains information on routing across source routing *bridges*. [FF-8]

routing table. A table in a *router* that lists information about other routers. This information is needed in networking. [FH-12]

RPC. See *remote procedure call*.

GLOSSARY

RS-232-C. Standard for *serial ports*. Replaced by *EIA/TIA-232-E*. [F3-15]

RSA. *Public key encryption* algorithm. [MJ]

sampling. A way of digitizing an analog signal so that it may travel down a digital transmission line. Used in codecs. [F3-18]

satellite network control center. In a distributed *network management system*, a *network control center* at a remote site. [F9-6]

scalable. In *switched data networks*, the ability of a switching protocol to handle a broad range of speeds. This avoids the need to change protocols as a network increases in size. [C5] In *servers*, the ability to switch hardware as demand grows without having to change the application or move to multiple servers. [C8, C10, C11, MI]

schema. The design of a *database*, including relations, attributes, and field characteristics. [e.g., F9-7]

scripting. [ME]

SDH. See *Synchronous Digital Hierarchy*.

search drive. In *NetWare*, virtual drive name associated with a directory in a *MAP INS* command. [FI-12]

search engine. Search tool for locating information on *World Wide Web* pages. [MC]

segment. In LANs, a section of LAN unbroken by repeaters. Several segments connected by repeaters form a LAN. [F4-12, C5, MF, FF-2] In TCP/IP, *TCP protocol data unit*. [MH]

self-test. A software routine in a device that allows it to test its functioning. [C9]

serial cable. Cable that connects the *serial port* on a PC to the serial port on a modem or other device. Serial port on a PC has 9 or 25 pins. [MD]

serial port. Port at the back of PCs for connecting the PC to *modems* and other devices requiring a serial port. Uses *serial transmission*. Follows the *RS-232-C* standard or the newer *EIA/TIA-232-E* standard. [F3-15]

serial transmission. Form of transmission in which a single transmission line carries signals in each direction. Slower than *parallel transmission* but useful at any distance. [F3-16]

server operating system (SOS). In PC networking, the operating system of the server. Popular SOSs include *Novell NetWare*, UNIX, and Microsoft *Windows NT Server*. [F8-1]

server program. In client/server processing, the program on the client/server server. [F2-8, C8]

server. In a PC network, a computer that provides services to *client PCs*. [F2-2, C8]

service customization. In ANI, sending the telephone number of the calling party to a service representative, in order to bring up specific information about the customer. [MG]

session layer (OSI Layer 5). In *OSI*, the layer that creates a session (connection) between two application programs on different computers. [FA-9]

SGML. See *Standardized General Markup Language*.

shared media LAN. LAN in which every station hears every transmitted message. [F5-5]

shielded twisted pair (STP). Transmission medium in which each wires are covered with a metal sheath shield. The entire *wiring bundle* is also covered with a metal sheath shield. [F5-2]

Signaling System 7 (SS7). Standard for supervisory communication between switches in ISDN. [MG]

signal-to-noise Ratio (SNR). The ratio of electrical power in a signal to electrical power in the noise in a medium. The better the SNR, the fewer errors there will be in the transmission. [MD]

signature. In electronic mail, an optional section at the bottom of the message that contains information about the sender. [F2-9]

Simple Mail Transfer Protocol (SMTP). *TCP/IP* protocol used on the *Internet* to deliver messages between *electronic mail hosts* and to specify message structure. [F10-4, MB]

Simple Network Management Protocol (SNMP). Network management standards created by the *Internet Engineering Task Force*. [F9-10]

single point of failure. A single device whose failure would have a strong negative impact on the network. A potential problem in *collapsed backbones*. [F6-11]

single point of service. A single location for handling multiple servicing problems. advantage of *collapsed backbones*. [F6-11]

single-key encryption. *Encryption method* that uses a single key for both encryption and decryption. [FJ-2]

SLIP. See *Compressed Serial Line Internet Protocol*.

smart card. Credit-card sized device with a microprocessor. Provides sophisticated security. [C9]

SMDS. See *switched megabit digital service*.

SMP. See *symmetric multiprocessing*.

SMTP. See *Simple Mail Transfer Protocol*.

SNA. See *Systems Network Architecture*.

SNMP. See *Simple Network Management Protocol*.

SNR. See *signal-to-noise ratio*.

socket. The combination of an *IP number* and a *port number*. Uniquely identifies an application process on a *TCP/IP internet*. [MH]

software inventory. Maintaining a list of software packages in use and the extent of their use. [C9]

software metering. Monitoring how much application programs are used and who uses them. [C9]

SONET. See *Synchronous Optical Network*.

SOS. See *server operating system*.

source address. In any *synchronous unit*, specifies the address of the process sending the unit. [e.g., F4-8]

source quench. *ICMP* protocol that tells a host to reduce its transmission rate, in order to achieve *flow control*. [FH-11]

spamming. In electronic mail, sending a message to many recipients who are unlikely to wish to see it. Often used in *Internet* advertising. [C10]

specialization. In *PC networking*, not having all applications run on a single host computer but instead assigning applications to multiple specialized *servers*. These servers can be optimized for their applications, improving economics and performance. [F2-2]

spoofing. Modifying the From: or source address field in a message or *protocol data* unit to indicate that the message or PDU is coming from someone other than the real sender. [C10]

SQL. See *Structured Query Language*.

SS7. See *Signaling System 7*.

Standardized General Markup Language (SGML). Predecessor of *HTML*. Way of placing formatting information in *ASCII* documents. [MD]

standards agency. An organization that produces *standards*. [F1-3]

standards. Expectations of how people (and technology) are to behave when they communicate. [C1]

start bit. In an *asynchronous frame*, a single 0 that tells that receiver that a new frame is beginning. [F3-8]

start of frame delimiter. For *Ethernet* LANs, the second field in the *802.3 MAC layer frame*. Completes the synchronization of the sender's and receiver's clocks and tells the receiver that a new field is starting. [F4-8] For *Token-Ring Networks*, field in an *802.5 frame* that begins a frame. [FF-7, FF-8]

station. Any computer or other intelligent device attached to a network that will communicate with other stations. [F1-2]

STP. See *shielded twisted pair*.

strategy support system. An information system designed to support a specific organizational strategy. [C2]

Structured Query Language (SQL). A standard for making database queries. [C11]

stub. Small program on a *client PC* needed to work with an application program on a *file server*. [F8-10, MI]

subnet access layer. In *OSI*, one of the three sublayers of the network layer required in OSI internetting. [FA-7]

subnet layer. In our expanded *Basic Communication Model*, the layer that governs the routing within a single *subnet* in an *internet*. [F1-2]

subnet mask. 32-bit number which, in conjunction with *IP numbers*, allows a network to be divided into *subnets* for *IP* addressing. [C6, FH-6]

subnet. In an *internet*, a single network. [F1-2, MA]. In IP addressing, the subdivision of a network using a subnet mask [FH-6]

subnet-dependent convergence layer. In *OSI*, one of the three sublayers of the network layer required in OSI internetting. [FA-7].

subnet-independent convergence layer. In *OSI*, one of the three sublayers of the network layer required in OSI internetting. [FA-7]

subscribe. To join a computer conference in LISTSERV or USENET. [FB-4 for LISTSERVs]

supervisory protocols. In TCP/IP, protocols used to control the operation of the network, as opposed to delivering messages. [MH]

switch. Device with several input and output ports. When a packet comes into an input port, the switch decides which output port to use to send the packet back out. This leads to complex *routing*. In addition, a switch can handle multiple messages at the same time, so congestion and latency need not grow as traffic increases. [F5-9]

switched data network. Data network that uses switches. Characterized by *alternative routing*. [F5-10, C7]

switched Ethernet. A LAN [C5] or local internet [C6] using *Ethernet switches*.

Switched Megabit Digital Service (SMDS). Medium-speed switched data network. [T7-2]

switched virtual circuit (SVC). In *connection-oriented switched data networks*, a connection is established at the beginning of a call, then disestablished at the end of the call. [C7]

switching decision. Whenever an incoming message arrives, a switch in a switched data network must decide which output port to use to send the message back out. This leads to *alternative routing*. [F5-9]

switching office. An *carrier* office containing a switch. [F7-2, FG-1]

symmetric multiprocessing (SMP). Processing that takes place on a host computer with multiple microprocessors. [C11, FI-4]

synchronized server operating system. *Server operating* system in which all of the resources on the servers in a network or part of a network are indexed on a directory server, so that users need not know on which server any particular resource resides. Also see *independent server operating system*. [F8-3, C10]

synchronous. In general, two processes communicating in real time, instead of *asynchronously* by sending messages to be read later. In *protocol data units*, a way of organizing messages into *fields*.

Synchronous Digital Hierarchy (SDH). Set of circuit standards extending into the gigabit range. Compatible with *SONET*. [T7-1]

Synchronous Optical Network (SONET). Set of circuit standards extending into the gigabit range. Compatible with *SDH*. [T7-1]

synchronous transmission. Form of transmission in which data is organized into complex *frames*, *packets*, or other *protocol data units* with multiple *fields*. [e.g., F4-8]

synchronous unit. *Protocol data unit* organized for *synchronous transmission.*

system login script. In *NetWare*, a *login script*. All users have the same system login script. Also see *user login script*. [FI-12]

systems management. Managing servers, PCs and other computers on a network, as opposed to the transmission components of the network. [F9-10]

Systems Network Architecture (SNA). Proprietary standards *architecture* created by IBM but used in many other mainframe networks. Dominant in mainframe networks. [F1-3]

T1. Circuit operating at 1.544 Mbps [T7-1]

tag view. View of a *webpage* showing its formatting *tags*. [FC-4]

tag. Way of representing formatting in *HTML*. [FC-4]

Tagged Image File Format (TIFF). Popular *bit-mapped graphics* format. [MD]

Talk. *Internet* application that allows users on different machines to type messages back and forth in real time. [MB]

tariff. A document filed by a *carrier* that specifies the details of a service and its pricing. [MG]

Tbps. See *terabits per second*.

TCP. See *Transmission Control Protocol*.

TCP/IP. Standards *architecture* managed by the *Internet Engineering Task Force* of the *Internet Society*. Used on the Internet. Also used in many other networks. [F1-3, MH]

Telecommunications Act of 1996. In the United States, an act that expanded national telecommunications *deregulation* and especially opened local telecommunications service to extensive competition. [MG]

telecommunications. Any form of communication using electrical or radio transmission [C1]

telecommuting. The ability to work from locations other than the normal office. [C11]

Telnet. Application that allows a user logged into one computer to log into another computer. [FB-2]

terabits per second (Tbps). Millions of megabits per second. [F1-5]

terminal adapter. In *narrowband ISDN*, a device that can connect computer *serial ports* and telephone lines to the ISDN. [F3-17]

terminal emulation. When a personal computer emulates (acts like) a *terminal* in order to communicate with a *host computer*. [F3-1]

terminal. A desktop device with little intelligence. The processing power resides on a *host computer*. [F2-1, C3]

terminal-host. Computer platform using an intelligent *host* computer and *terminals* with little or no local processing power [F2-1, C3, ME]

termination equipment. Equipment on the *customer premises* that protects the telephone system from improper voltages and signals. [FG-2]

thread. When a program is divided into small pieces called threads, which execute independently. [FI-4]

throughput. The amount of traffic flowing through a system per second. [C9]

TIFF. See *Tagged Image File Format*.

token passing. *Media access control* method in *802.5 Token-Ring Networks*. A station can only transmit when it has the token. [F5-4]

token. In *802.5 Token-Ring Networks*, a *MAC layer* supervisory frame used in *media access control*. A station may only transmit if it has the token. [FF-7]

Token-Ring Network. LAN standards from the 802.5 Working Group of the IEEE 802 Committee. [C5, MF]

toll call. Long-distance call for which there is a charge. [FG-5]

topology. Layout of the transmission medium in a LAN. Determines how frames or packets

GLOSSARY

get to stations. Popular topologies include the bus [C4] and ring [F5-1] topologies.

transaction automation. The automation of routine transactions, such as credit card checking in retail stores. [C1]

transaction monitor programs. *Middleware* process that provides extensive control over the communication. [MJ]

transceiver. In *10Base5*, a device that connects to the *trunk line* to transmit outgoing signals and receive incoming signals. Also called a *medium attachment unit*. [F4-12, FF-1] In *radio LANs*, a radio transmitter and receiver [C5]

Transmission Control Protocol (TCP). A *transport layer* standard in the *TCP/IP architecture*. Provides *reliable connection-oriented* service. See [C1, FH-7].

transmission layer. In the *Basic Model of Communication*, the layer that standardizes the transmission of bits across networks between the computers in the system. Divided into the *internet layer* and *subnet layer* [F1-1, F1-2]

transport layer (OSI Layer 4). In *OSI*, the layer that governs interactions between operating systems and ensures that two computers from different vendors can interact. [FA-8]

transport layer. In the *Basic Model of Communication*, the layer that standardizes exchanges between the operating systems of the computers in the system. Synonym for *computer layer*. [F1-1]

TrueType. Standard for representing type. Used especially on Windows computers. [MD]

trunk line. A transmission line connecting switches in a *switched data network* [F5-10, F7-2, FG-1]

trustee rights. See *rights*.

turnpike theorem. After an improved communication system implemented, traffic will be higher than it was before the implementation. [C9]

type of service (TOS). In *IP*, header field that specifies the quality of service requested of the network. [C6]

UDP. See *User Datagram Protocol*.

uniform resource locator (URL). A uniform way of specifying the addresses of resources on Internet hosts. [F2-12, especially MC]

UNIX. Operating system capable of acting both as a *server operating system* and as a client operating system. Popular server operating system for database and Internet servers. [C8]

unmoderated conference. *Computer conference* with no pre-screening of messages before they are posted to the group. [C10]

unordered list. In *HTML*, a list of items. Items are not numbered, as in *ordered lists*, but each item is proceeded by a graphical bullet. [MC]

unreliable service. *Switched data network* service in which there is no *error correction* within the network. Dominates in switched data networking because it reduces the load on switches. [F7-6]

unshielded twisted pair (UTP). A type of telephone wiring in which each pair of wires in the wiring bundle is twisted several times per foot to reduce interference. [F4-3, MD]

upload. Transferring a file from a PC doing *terminal emulation* to a *host computer*. [F3-5]

upper memory. Memory addresses between 640 KB and 1 MB. The addresses in the NIC buffer must be located in this range in MS-DOS machines. [MI]

URL. See *uniform resource locator*.

USENET newsgroup. Form of *computer conferencing*. [C10]

USENET. *Computer conferencing* service in which to see postings you must subscribe to *newsgroups* on particular topics. [FB-6]

User Datagram Protocol (UDP). *Unreliable, connectionless transport layer* protocol in the *TCP/IP architecture*, used to carry relatively unimportant messages efficiently. [FH-7]

user layer. In the *Basic Model of Communica-*

tion the layer that standardizes exchanges between the ultimate users of the system. [F1-1]

user login script. In *NetWare*, a *login script* executed when a specific person logs in. Executed after the *system login script*. [FI-13]

UTP. See *unshielded twisted pair*.

V.34. A standard for modem modulation at up to 28.8 kbps. [T3-1]

V.42 bis. ITU-T standard for *modem* data compression. [TD-2]

V.42. ITU-T standard for *modem error detection and correction*. [TD-2]

variable latency. Condition in which *latency* (delay) is sometimes small, sometimes excessive. [C5]

vector graphics. Graphics image formed by lines, circles, arcs, and other objects. [MD]

verbal communication. Communication involving words. Sometimes restricted to oral communication. [C10]

VERONICA. Search tool for locating information on multiple *Gopher* servers. [MB]

version. One in a series of refinements to a document or program. Must be controlled to manage document composition or program development. [C10]

videoconferencing. Communications tool in which participants can see and hear one another. Often combined with *document conferencing*. [C10]

virtual channel. Part of the *PVC* designation in *ATM*. [F7-8]

virtual circuit. Older name for *connection-oriented* service. The parties appear to have a dedicated circuit, but in reality their traffic is multiplexed with other traffic. [F7-9]

virtual corporation. Another name for *adhocracy*. [C1]

virtual drive. A drive letter, such as S:, that refers to part of a *file server's* hard disk drive, instead of to a real disk drive on the client PC. [C8-6]

virtual LAN. In a switched LAN, users with similar broadcast needs and access to resources are grouped into logical groups called virtual LANs. [C6]

virtual path. Part of the *PVC* designation in *ATM*. [F7-8]

virtual port. A port designation, such as LPT3:, that refers to a *printer* on the network, instead of to a real port on the client PC. [C8]

virtual private network. Service provided by carriers. Provides service similar to that of *private telephone networks* but without the need for the user organization to manage the network. [C7, MG]

virus control. Preventing viruses from attacking PCs on a network. [C9]

voice/data modem. *Modem* that can carry voice and data simultaneously. [FD-6]

voting tool. Application that allows a group to vote electronically on proposals. [C10]

VT100 Terminal. Low-performance *terminal* acceptable to many host computers. [F3-3]

WATS. See *wide area telephone service*.

wavelength. The distance between comparable points on succeeding cycles in a wave. [F3-7]

Web. See *World Wide Web*.

webcrawler. Search tools that finds *webpages* by following *links* in every webpage it finds. [MC]

webpage. A single page on the *World Wide Web*. [FC-3]

webserver. Internet host computer offering *World Wide Web* service. [FC-2, MC]

Webspace. Collection of all *World Wide Web* servers. [MC]

well-known port number. Standardized *port number* for a specific application. [MH]

what-if analysis. Anticipating the impacts of alternative assumptions. [C9]

whiteboard conferencing. Form of *document*

conferencing in which both parties can mark up a shared image. [MJ]

wide area telephone service (WATS). long-distance pricing scheme in which the calling party gets reduced rates on long-distance calls. [FG-5]

wideband. Strictly speaking, a transmission channel with a large bandwidth. Loosely speaking, any high-speed channel. [MD]

Windows NT Server. *Server operating system* from Microsoft. [C8]

Wiring Bundle. Business telephone wiring often comes in a bundle of two or four pairs. [F4-3, MD]

wiring closet. In buildings, a closet on a floor for holding telephone equipment and for breaking *cable bundles* into smaller bundles. [FG-2]

workflow. Form of *groupware* in which a task is passed from person to person (or person to process) several times before it is completed. [F10-7]

Working Group. In the *IEEE*, a subgroup of the *802 Committee*. Working groups propose actual *standards*. [T4-1]

World Wide Web. A service on the *Internet* in which users working at *browsers* can retrieve *webpages* from any *webserver* around the world. Each page is in *hypertext* format, containing *links* to other webpages. [F2-11, F2-12, MB, FC-1, FC-2]

wrapping. In a dual-ring network, a way of re-routing frames if there is a break in the main ring. [FF-9, FF-10]

WWW Client. Client program that calls on a *webserver* to receive *webpages*. Usually a *browser*. [FC-2]

WWW. See *World Wide Web*.

X.25. Low-speed *switched data network*. [T7-2]

X.400. Family of *OSI* protocols used to deliver messages between *electronic mail hosts* and to specify message structure. [C10]

X.500. *OSI directory server* standard. [C9]

XMODEM. *File transfer protocol* used by PCs. [ME]

YMODEM. *File transfer protocol* used by PCs. [ME]

ZMODEM. *File transfer protocol* used by PCs. [ME]

Index

A

Absolute addressing, 354
Abstract Syntax Notation 1 (ASN.1), 313–314
Access computer, 45
  reception of message, 49
Access control, media access control (MAC)
  layer frame, 407, 409
Access units, token-ring networks, 112–113
Accounting management, 245–246
Active monitor, 413
Active monitor present, 413
Adaptation-layer software, 477
Address
  of station in network, 9
  See also Internet Protocol (IP) number
Address Resolution Protocol (ARP), 453
Adhocracy, nature of, 1–2, 297
Ad hoc standards agencies, 11
Advanced Network Services, 322
Advanced Peer to Peer Networking (APPN), 17
Agenda system, 276
Agents, email, 253
Alarms
  alarm storms, 235
  LANalyzer, 488
  and network management, 225
  signal for fault, 234
Alternative routing
  rerouting around congestion, 123–124
  rerouting around failures, 123
  route optimization, 124
  switched internets, 147

America Online, 293, 335
Amplitude modulation, 66
Analog cellphones, 432–433
Analog information, 57
Analog transmission, nature of, 64
Analog voice-grade circuits, 158
Anchor tags, Hypertext Markup Language (HTML), 353, 355
Anonymous File Transfer Protocol (FTP), 329–330, 333–335
  ARCHIE, 333
  example of session, 329–330
  Gopher, 333–334
  VERONICA, 334
ANSI terminals, 62
  standard, 391
ANSNET, 322
Applets, 42
  Java, 362
Application development
  desktop groupware, 267–268
  Lotus Notes, 268–269
Application layer
  client/server processing, 34–37
  communication method, 47
  in communication process, 7–8
  electronic mail, 37–39
  OSI, 314
  PC networks, 193
  standards for, 34
  World Wide Web (WWW), 39–42
Application program interface (API), 50, 251
Application protocol data unit (APDU), 47, 209, 226
Apps directory
  inheritance in, 467
  PC network, 466
ARCHIE, 333, 363

Architectures
  meaning of, 12
  See also Standards architectures
ARPANET, development of, 319–322
ASCII
  character set, 377
  coding system, 57–58
  extended ASCII, 73
  limitations of, 376
Asynchronous ASCII transmission, 57–58, 72–74
  data bits, 73
  full-duplex, 74, 75
  half-duplex, 74, 75
  limitations of, 74
  parallel transmission, 76
  parity bits, 73–74
  serial transmission, 74, 76
  start and stop bits, 72
Asynchronous frame, 72
Asynchronous transfer mode (ATM), 165, 167, 169, 171, 177
  advantages of, 177
  cost factors, 177–180
  speeds and pricing, 179
Asynchronous transfer mode (ATM) switches, 124–125
  scalability of, 125
AT & T, break up of, 420
AT modems, 384
Attachment unit interface (AUI), 399
Attachments, electronic mail, 38, 252
Attenuation, 90
Audit trails, 244–245
AUI drop cable, 104
Authentication, 242–243, 500–502
  challenge and response, 500–501

INDEX 541

data integrity, 502
digital signatures, 501–502
and IP Version 6, 460
nonrepudiation, 502
Autoconfiguration hosts, routers, 143
Autoconfiguration protocols, 454–455
  Bootp, 455
  Distributed Host Configuration Protocol (DHCP), 455
  and IP Version 6, 461
  Reverse Address Resolution Protocol (RARP), 455
Automatic number identification (ANI), 429
  problems related to, 429
Autonomous routers, 458
Autonomous system, networks, 458
Available bit rate (ABR), 179

B

Backup
  systems for, 473
  on virtual drives, 196–197
Bandwidth
  and noise, 379
  and telephone transmission, 71–72
Baseband signaling
  meaning of, 94
  token-ring networks, 113
Base memory address, 479
Basic Rate Interface (BRI), 77
Baud rate, 381
  meaning of, 69–70, 93
B channels, 77, 176
Binary file transfer (BFT), 387
Binary notation, 18
Binhex, 259
Bit-mapped graphics, 374
BITNET, 322
Bits per second (bps), 18
Black-and-white images, 374
Board driver, 476
Body, electronic mail, 38, 252
Bookmarks, World Wide Web (WWW), 347
Bootp, 455
Border Gateway Protocol (BGP), 459
Border routers, 459
Brainstorming tool, 276

Bridges, 132, 133–138
  advantages of, 136–137
  connecting to digital leased lines, 162–163
  disadvantages of, 136
  evaluation of, 145–146
  forbidding loops, 136
  local bridges, 146
  operation of, 134–136
  standards for, 137–138
  transmission speed, 137
Bridging devices
  bridges, 132, 133–138
  LAN switch, 132–133, 138–141
  routers, 133, 141–145
Broadband ISDN, 124, 176
Broadband signaling, meaning of, 94
Broadcasting, meaning of, 91
Browsers
  operation of, 363–365
  and World Wide Web (WWW), 39, 363–365
Browser view, 349
Bulletin board systems, 253
Business information online services, 295
Business processing online services, 294
Business rules, 296–297
Busses, 470–471
  local bus, 471
  main bus, 470
  memory bus, 470
Bus transmission, meaning of, 91
Busy hour, 239
Bytes, 97

C

C++, 286, 362
Cable bundles, 423
Call, 203
  meaning of, 35
Call blocking, 429
Carrier sensing, meaning of, 95
Case worker, 297
Catalog entries, 265
Category 3/4/5 unshielded twisted pair, 90
CDDI (copper distributed data interface), 117

Cell switched network, 171
Cellular digital packet data (CDPD), 434
Cellular telephones, 429–436
  analog cellphones, 432–433
  cellphones, types of, 430
  digital cellphones, 434
  distributed communication service (DCS), 435
  frequency reuse, 430
  handoffs, 430
  LEO satellites, 436
  microcells, 434–435
  mobile telephone switching office (MTSO), 430, 432
  personal communication service (PCS), 435
  roaming, 432
  spectrum capacity, 430
Central network control center, 224
CERN, 343
Challenge and response identity verification, 243–244, 500–501
Channel service unit/data service unit, 162–163
Chargeback, 245
Circuits, 157–160
  analog voice-grade circuits, 158
  digital 64 kbps circuits, 159
  fractional T1 circuits, 159
  SONET/SDH circuits, 159
  T1 and E1 circuits, 159
  T3 and E3 circuits, 159
  transmission speeds, 158
  for video and data, 159–160
Circuit switching, 164
Claim token, 413
Class 1 standards, 387
Class 2 standards, 387
Classes, of networks, 443–444
Clean data, 290
Clearinghouses, 294
  electronic data interchange (EDI), 294–295
Client PCs, 28, 185–186
  client PC management, 230–232
  connectivity, 185–186
  limitations of, 186
Client program, 36, 205, 208, 282
  World Wide Web (WWW), 345

Index

Client/server databases, 282–289
　access to multiple databases, 284
　client software, 286
　data safety, 283
　load on client machine, 282–283
　middleware, 287–289
　Open Data Base Connectivity (ODBC), 287
　rapid application development, 284–285
　and reduced network load, 283
　relational databases as, 283
　server software, 285–286
　standards, 286–287
　Structured Query Language (SQL), 287
　symmetric multiprocessing (SMP), 286
Client/server processing, 34–37, 202, 205, 208–210
　advantages of, 209–210
　application-layer process, 209
　client program, 205, 208
　disadvantages of, 210
　location of processing, 205
　on network, 36
　request/response communication, 209
　server program, 205, 208–209
　and stand-alone PCs, 35
　World Wide Web (WWW), 40, 344–345
Client shell software, 477
Client software, 286
Co-axial cable, 104–106
　Ethernet 10Base5 standard, 104–105, 398
　10Base5T attachment, 104
Codec
　and ISDN, 80
　and videoconferencing, 271
Collapsed backbone internets, 148–149
　and single point of failure, 148
　and single point of service, 148
Collision detection, and CSMA/CD, 96
Collision domain, 401
Color images, 374–375

Commercial online services, 293–296
　business information services, 295
　business processing services, 294
　consumer-oriented services, 293–294
　electronic commerce on Internet, 295
　electronic data interchange (EDI), 294–295
　electronic funds transfer (EFT), 295
　elements of, 293
Committed information rate (CIR), 179
Common Gateway Interface (CGI), 361–362
Common Management Information Services (CMIS), 230
Common Management Information Protocol (CMIP), 230
Communication on networks. *See* Network communications
Communication process
　application layer, 7–8
　computer layer, 8
　transmission layer, 9–10
　user layer, 6–7
Communication program
　file transfer protocols (FTP), 394–395
　set up of, 392–394
Communication servers, 210–212
　gateway server, 210–212
　hardware for, 212
　remote access server, 210
Competing standards problem, 8
Competitive access providers (CAPs), 155, 418
Complex modulation, 68–70
Compound document, text/graphics integration, 378
Compression
　modems, 65, 70–71, 382–383
　pixel-mapped graphics, 375
　video compression, 502–504
　videoconferencing, 271
CompuServe, 293, 335
Computer layer, in communication process, 8
Computer platforms, 26–33
　multiple platforms, 33

　networked PCs, 28–31
　RISC workstations, 31–32
　terminal-host systems, 27–28
Computer protocol data unit (CPDU), 47
Conferencing
　by email, 253–254
　moderated conferencing, 254
　system elements for, 254
Confidentiality, 241
Configuration management, 235–237
　and changing states, 235–236
　current network map, 236
　directory servers, 237
　electronic software distribution (ESD), 236
　remote management, 236
Connectionless service, 167–168
Connection-oriented service, 168–169
Consolidation, in reengineering, 297
Consumer-oriented online services, 293–294
Control console, for videoconferencing, 272
Core business functions, 298
Credit checking online service, 295
CSLIP (Compressed Serial Line Internet Protocol), 44, 45, 52, 336
CSMA/CD (carrier sense multiple access with collision detection), 94–96
　carrier sensing in, 95
　collision detection in, 96
　limited throughput, 96
　multiple access in, 95
CSNET, 322
CUSEEME, 328
Customer premises, 20, 155
　telephone call, 417, 418
Cyberselves, 327–328
Cyphertext, 496

D

Daisy chain, connecting hubs, 103
Databases
　client/server databases, 282–289

commercial online services, 293–296
data warehouses, 289–290
distributed databases, 291–292
multidimensional database, 290
multimedia databases, 292–293
production data processing, 290–291
Data communications, meaning of, 1
Data Encryption Standard (DEC), 497
Data field, 99
Data integrity, and authentication, 502
Data link layer
LANs, 85
OSI, 305–306
Data safety, client/server databases, 283
Data services, circuits for, 159–160
Data warehouses, 289–290
online analytical processing (OLAP), 290
operation of, 289–290
D channel, 77, 176
Dedicated database client development language, 286
Default, meaning of, 480
Default router, 142, 445
Defense Advanced Research Projects Agency (DARPA), and Internet, 319–322
De jure standards agencies, 11
Delphi, 293, 335
Demand priority access method (DPAM), 120
Deregulation
telephone service, 419
of transmission carriers, 155–156
Design rationale systems, 265
Desktop conferencing, 273–274
document conferencing, 274
equipment for, 274
Desktop groupware
application development, 267–268
control aspects, 267
electronic document management (EDM), 265

group composition, 265
Lotus Notes, 264, 268–269
project management system, 266–267
verbal communication, 264
workflow systems, 265–266
Desktop Management Interface (DMI), 230–232
Destination address, 98
Dial-back modems, 388–389
Digest, 501
Digital 64 kbps circuits, 159
Digital cellphones, 434
Digital information, 57
Digital service unit (DSU), for ISDN, 78
Digital signature, 501–502
Digital transmission, nature of, 63
Direct memory access (DMA) request, 480
Directories, PC network, 465–466
Directory servers, 189, 237
email, 257
standard for, 237
standards, 258
Disk caches, 472, 483
Disk drives
slow, and networks, 483
types of, 471
Distortion, of signals, 90
Distributed backbone connections, 147–149
Distributed communication service (DCS), 435
Distributed databases, 291–292
Distributed Host Configuration Protocol (DHCP), 455
Distributed network management system, 223–224
Distribution lists, email, 39, 252
DLT (digital linear tape), 473
Docubases, 277, 335
Document conferencing, 274, 504
application sharing, 504
remote control software, 504
whiteboard conferencing, 504
Domain name system (DNS), 319, 455–457
hosts, 319, 455–456
routers, 143

Domestic carriers, 154–155
Downloading, 61
Downsizing, to small computers, 27
Drives, operations performed on, 195–196
Dual-homed router, 446

E

E1 circuits, 159
E3 circuits, 159
EBONE, 322
Echo message, 457
EDIFACT, 294
800/888 numbers, 427
802.3 100Base-X standards, 117–119
wiring/hubs, 118–119
802.5 Token-Ring Network standard, 111, 404–413
error handling, 410–413
Manchester encoding, 404–406
media access control (MAC) layer frame, 406–410
Electronic commerce, on Internet, 295
Electronic data interchange (EDI), 294–295
clearinghouses, 294–295
standards, 294
Electronic document management (EDM), 265, 277
Electronic funds transfer (EFT), 295
Electronic mail, 37–39, 251–262
attachment, 38
attachments, 252
body, 38, 252
for conferencing, 253–254
distribution lists, 39, 252
features of, 252
filtering rules, 253
header, 37, 252
hosts for, 255
mail server, 212
Post Office Protocol (POP), 258
receiving mail, 38–39
standard for, 251
Electronic mail etiquette, 259, 260–262
flaming, 260
forwarding, 261

Electronic mail etiquette (*continued*)
  privacy, 261–262
  spamming, 260
  spoofing, 261
Electronic mail standards, 254–257
  directory standards, 258
  endoding attachments, 259
  Internet email standards, 255–256, 258–259
  Mail Application Program Interface (MAPI), 251
  message handling system (MHS), 256
  for message structure, 258–259
  Multipurpose Internet Mail Extensions (MIME), 259
  Simple Mail Transfer Protocol (SMTP), 255
  X.400 standards, 255–256
Electronic meeting rooms, 274–276
  application software, 276
  hardware for, 275
Electronic software distribution (ESD), 236
Electronic switching services, 428–429
Encapsulation security, 461
Encryption, 241–242, 496–500
  Data Encryption Standard (DEC), 497
  elements of, 496–497
  key escrow, 499–500
  public key encryption, 242, 497–499
  single-key method, 497
End frame delimiter, media access control (MAC) layer frame, 408, 410
Enterprise internets
  circuits, 157–160
  leased lines, 160–163
  switched data networks, 171–180
  transmission carriers, 154–160
  transmission service, 156
  virtual private networks, 163, 170
Error correction, modems, 65, 381–382
Error detection

modems, 65
OSI, 315
and parity, 73–74
Error handling, token-ring network, 410–413
Error rates, measurement of, 239
Error reporting, routers, 142–143
Ethernet 10Base2 standard, 401–403
  cable, 401
  cabling method, 402
  distance factors, 402
  T-connector, 402
Ethernet 10Base5 standard, 398–401
  attachment unit interface (AUI), 399
  coaxial cable, 104–105, 398
  medium attachment unit (MAU), 398–399
  multisegment LANs, 401
  segments/repeaters, 399–400
Ethernet 10Base-F standard, 105–106
Ethernet 10Base-T hubs, 91, 100–103
  baseband signaling, 94
  bus operation, 91
  co-axial cable, 104–106
  Manchester encoding, 92–93
  receipt of signals, 91
  second hub for larger network, 100–103
  several hubs for larger network, 103
  signal transmission, 91
  transmission speed, 92
Ethernet 10Base-T standard, 88–91
  network interface cards (NICs), 89
  unshielded twisted pair, 89–91
Ethernet switches, elements of, 124
Existing application program, 286
Expansion board, network interface card, 89, 469
Expansion slots, 469
Explicit formatting, Hypertext Markup Language (HTML), 359
Extended ASCII, 73
External images, 357–358

F

Facsimile modems, 386–387
  binary file transfer (BFT), 387
  facsimile software for, 387
  intelligent modems, 387
  speed standards, 387
Fault, meaning of, 233–234
Fault management, 233–235
  diagnosis of faults, 234–235
  recognition of faults, 234
Fault repair, and changing states, 235
FDDI (Fiber Distributed Data Interface), 116–117
  FDDI over UTP, 117
  strengths/weaknesses of, 117
Federal Communication Commission (FCC), 419
File server program access, 31, 201–202
  advantage of, 201–202
  disadvantage of, 202, 205
File servers
  file server program access, 30–31
  non-PC servers, 31
  operation of, 30
  PC servers, 31
  use of term, 194
File service, 194–199
  for data files, 196–198
  for program files, 199–202
File transfer, 61
  downloading/uploading, 61
File transfer protocol (FTP), 61, 394–395
  Kermit, 395
  XMODEM, 394
  XMODEM/CRC, 394
  ZMODEM, 61, 395
Filtering rate, bridges, 137
Filtering rules, 253
Filters, email, 253
Financial information online service, 295
Firewalls, 244
  Internet, 337
Flaming, email, 260
Flat-rate local service, 427
Flow control, 74, 311
Fonts, 376–377
  PostScript fonts, 377, 378
  TrueType fonts, 377

INDEX 545

Formal address, 256
Forums, 253
Forwarding, email, 261
Fractional T1 circuits, 159
Frame, 58, 125, 308
Frame check sequence field, 99
    media access control (MAC) layer frame, 410
Frame control, media access control (MAC) layer frame, 409
Frame Relay, 165, 167, 169, 171, 176–177
    cost factors, 177–180
    permanent virtual circuits for, 179–180
    speeds and pricing, 179
Frames per second, and video compression, 503
Frame status field, media access control (MAC) layer frame, 410
Frequency modulation, 64, 66
Frequency reuse, cellular telephones, 430
Frequently asked questions (FAQ), 328
Full-duplex, meaning of, 74, 75
Full repeater, 400
Full-text searches, 265, 335

G

Gateway server, 210–212
    to Internet, 336
General fourth-generation programming language, 286
Gigabit per second (gbps), 18
Gopher, 333–334, 363, 365
Graphics. *See* Images/imaging
Group, network users, 466
Group composition, 276
    desktop groupware, 265
Groupware, 262–270
    definition of, 262
    different-place interactions, 264
    different-time work, 263
    Internet groupware, 270
    same-place interactions, 264
    same-time work, 263
    TCBworks, 270
    See also Desktop groupware
GSM, 434

H

H.261 standard, 502
H.320 standards, 271
Half-duplex, meaning of, 74, 75
Half repeater, 400
Hand-held cellphones, 430
Handoffs, cellular telephones, 430
Hand-shaking, modems, 381
Header, electronic mail, 37, 252
Help desk, network support, 490
Hertz (Hz), 66
High-end servers, 187
History list, World Wide Web (WWW), 347
Home pages, World Wide Web (WWW), 346
Home PCs, Internet connection, 52–53
Host access gateway, 210
Host addresses, Internet, 318–319
Hosts, for email, 255
Hyperbase, 343
HyperTerminal, 59
Hypertext
    nature of, 343
    and World Wide Web (WWW), 39
Hypertext Markup Language (HTML), 349–361
    browser image, 352
    compound documents, 378
    explicit formatting, 359
    first-level heading, 352
    horizontal rules, 352
    HTML editors, 350
    images, 356–359
    links to files in other directories, 354–355
    location of tags, 352
    logical formatting, 359–360
    ordered list, 360–361
    paired codes, 351
    paragraphs, 352
    referencing pages on other machines, 356
    referencing specific location in document, 355–356
    simple link, 353
    special characters, 360
    tags, 349, 350–351, 353
    text, adding properties to, 359–360
    text, listing of codes, 368–372
    title, 351
    word processors, use of, 349–350
Hypertext Transfer Protocol (HTTP), 47, 345–346

I

Images/imaging
    bit-mapped graphics, 374
    black-and-white, 374
    color, 374–375
    compression, 375
    external images, 357
    Hypertext Markup Language (HTML), 356–359
    inline images, 356
    pixel-mapped graphics, 374–375
    in production systems, 276–277
    text/graphics integration, 378
    vector graphics, 376
Implementation profiles, standards, 17
Incorrect packets per thousand packets, 239
Independent server operating systems (SOSs), 189
Information field, media access control (MAC) layer frame, 409
Informix, 285
Inherit rights, 466, 468
Inline images, 356
Installation charge, ISDN, 78
Institute for Electrical and Electronics Engineers (IEEE), 13
    LAN MAN Standards Committee (802 Committee), 85–87
Integrity, 460
Intelligent modems, 4, 384, 392
Interexchange carriers (IXCs), 155, 419
Interface standards, 50
Interference
    meaning of, 90–91
    and unshielded twisted pair, 89
International carriers, 155
International common carriers (ICCs), 419

Internet
  Anonymous File Transfer Protocol (FTP), 329–330, 333–335
  and business, 10
  connecting to, 44–45
  corporate access, 51
  CSLIPP or PPP connection, 336
  development of, 319–320
  electronic commerce on, 295
  firewalls, 337
  frequently asked questions (FAQ), 328
  future view, 337–338
  groupware, 270
  home connection, 52–53
  host addresses, 318–319
  Internet gateway server, 336
  Internet Relay Chat (IRC), 327
  legal issues, 337–338
  LISTSERV, 323–326
  management of, 322–323
  MUDs/MOOs, 327–328
  online services, 335–336
  real-time discussions, 327–328
  talk session, 327
  telephone service, 329
  USENET, 326
  videoconferencing, 328–329
Internet Control Message Protocol (ICMP), 457–458
Internet Engineering Task Force (IETE), 14, 255, 322
Internet gateway servers, 212
Internet PCs, 42
Internet Protocol (IP), 15, 321
  and routers, 143–144
  unreliability of, 451–452
  Version 6, 459–461
Internet Protocol (IP) names, 319, 448–449
  domain name system, 455–456
  parts of, 449
Internet Protocol (IP) number, 142, 318
  Address Resolution Protocol (ARP), 453–454
  assigning IP numbers, 442
  autoconfiguration protocol, 453–455
  Class A/B/C/D networks, 443–444
  format for, 449
  port number, 449
  of router, 446
  socket, 449
  subnet mask, 449–451
  32-bit addresses, 442, 448, 459
Internet Relay Chat (IRC), 327
Internets
  enterprise internets, 153–180
  local internets, 129–149
  use of term, 10
Internet Service Provider (ISP), 44
  access computer, 45
  and home use, 53
  routers, 45
  transmission method, 46
Internet Society, 322, 323
Internetting layer, PC networks, 193
Internetting protocol data unit (IPDU), 48
Internetting transmission layer, 46
InterNIC, 323
Interrupt request (IRQ), 479–480
Intranets
  advantages of, 42
  nature of, 42
Inventory management, 231
I/O address, 479
IPX/SPX, 17, 193
ISDN (Integrated Services Digital Network) service, 77–80, 175–176
  B and D channels, 176
  broadband ISDN, 124, 176
  and codec, 80
  cost factors, 78
  market factors, 176
  narrowband ISDN, 124, 175–176
  primary rate interface (PRI), 176
  sampling for pulse code modulation, 79–80
  speed of, 77
ISO (International Organization for Standardization), 12, 304
ITU-T (International Telecommunications Union-Telecommunications Standards Sector, 13, 304, 420

J

Jabbering, 235
Java, 42, 362
Joint Photographic Experts Group (JPEG), 375

K

Kermit, 61, 395
Key distribution, encryption, 497
Key escrow, encryption, 499–500
Key value, encryption, 497
Keyword search engine, World Wide Web (WWW), 348
Kilobits per second (kbps), 18

L

LANalyzer, 488–489
  alarms, 488
  frame decoding, 488–489
LAN layer, PC networks, 192
LAN MAN Standards Committee (802 Committee), 85–87
LAN switch, 132–133, 138–141
  multiple switches, 138–139
  operation of, 138
  switched Ethernet, 139–140
Latency, 130
  variable latency, 116
LATEX, 378
Leased lines, 160–163
  advantages/disadvantages of, 161
  plug-and-play service, 162
  private telephone networks, 162
  routers/bridges connected to, 162–163
  between sites, 161
  and switching, 161–162
  virtual private networks, 160, 163, 170
Length field, 99
LEO satellites, 436
License fees, 200
Lights out operation, 236
Links
  hypertext, 39
  Hypertext Markup Language (HTML), 353–356
Lists, Hypertext Markup Language (HTML), 360–361

INDEX 547

LISTSERV, 253–254, 323–326
    advantages of, 325–326
    moderated lists, 325
    submitting comments, 323–324
    subscribing to, 323
    unsubscribing, 324–325
Load balancing, 124
Lobe, 413
Local access and transport area (LATA), 155, 417, 419, 427
Local area networks (LANs), 19–20
    alternative routing, 123–124
    baseband signaling, 94
    broadband signaling, 94
    co-axial cable, 104–106
    connectivity, 19, 20
    802.5 Token-Ring Network standard, 111, 404–413
    Ethernet 10Base2 standard, 401–403
    Ethernet 10Base5 standard, 398–401
    Ethernet 10Base-T hubs, 91, 100–103
    Ethernet 10Base-T standard, 88–91
    FDDI (Fiber Distributed Data Interface), 116–117
    larger networks, construction of, 100–106
    logical link control (LLC) layer, 87
    logical link control (LLC) layer frame, 99–100
    media access control (MAC), 86–87, 94–96
    media access control (MAC) layer frame, 96–99
    network interface cards (NICs), 89
    100mbps LANs, 117–120
    optical fiber connection, 105–106
    radio LANs, 125–126
    switched LANs, 120–125
    10 Mbps LANs, limitations of, 115–116
    token-ring networks, 110–114
    transmission speed, 20
    transmission standards, 85–87
    unshielded twisted pair (UTP), 89–91

Local bridging, 146
Local bus, 471
Local carrier exchange (LCE), 418
Local internets
    collapsed backbone internets, 148–149
    connecting devices, 132–133
    distance spans, 130
    distributed backbone connections, 147–148
    need for, 129
    number of stations, 130–132
    compared to switched LANs, 132
Local loop, 156
    telephone call, 417, 418
Logical formatting, Hypertext Markup Language (HTML), 359–360
Logical link control (LLC) layer, 87
    functions of, 87
Logical link control (LLC) layer frame, 99–100
    802.2 layer frame, 99–100
LOGIN.EXE, 475, 482
Login scripts, 60–61, 484–487
    CAPTURE commands, 486
    MAP commands, 485–486
    system login script, 484–486
    user login script, 486–487
Lossless standards, 375
Lossy standards, 375
Lotus Notes, 264, 268–269
    application development, 268–269
    replication and multiple servers, 269

M

Mail Application Program Interface (MAPI), 251
Mail server, 212
Main bus, 470
Mainframes, 27
    as servers, 187
Managed devices, 220
Managed modems, 389
Management information base (MIB), 225
    analysis for faults, 234
    proprietary MIBs, 228–229
    schema for, 225–226

Manchester encoding, 92–93, 404–406
    differential encoding for a 0, 405
    differential encoding for a 1, 405
    differential encoding for a J and K, 405
    limitations of, 93
    line states in, 406
Marketing channels, 298
MBONE, 329
Mean time between failures (MTBF), 238
Mean time to repair (MTTR), 238
Media access control (MAC), 86–87, 94–96
    bridge layer, 87
    CSMA/CD (carrier sense multiple access with collision detection), 94–96
    functions of, 86–87
Media access control (MAC) layer frame, 96–99
    access control, 407, 409
    address fields, 98–99, 409
    data field, 99
    802.3 layer frame, 97
    802.5 layer frame, 406–410
    end frame delimiter, 408, 410
    frame check sequence, 410
    frame check sequence field, 99
    frame control, 409
    frame status field, 410
    full frame, 408–410
    information field, 409
    length field, 99
    PAD field, 99
    preamble, 98
    routing information, 409
    start of frame delimiter, 98, 407
Mediated communication
    meaning of, 5
    and standards, 5
Medium attachment unit (MAU), 398–399
Megabits per second (mbps), 18
Memory bus, 470
Menu directories, World Wide Web (WWW), 347–348
Message handling system (MHS), 258

Message-oriented middleware (MOM), 506
Message transmission, 47–48
    application protocol data unit (APDU), 47
    computer protocol data unit (CPDU), 47
    internetting protocol data unit (IPDU), 48
    reception of message, 49–50
    subnet transmission protocol data unit (SnPDU), 48
Message units, 427
Metropolitan area networks (MANs), 21
    transmission speed, 21
Microcells, cellular telephones, 434–435
Microcom Network Protcol (MNP), 383
Microprocessor, Pentium, 469
Microsoft Excel, 208
Microsoft Internet Explorer, 345
Microsoft Network, 293, 335
Microsoft Windows NT Server, 31, 32, 189
Middleware, 251, 505–507
    client/server databases, 287–289
    message-oriented middleware (MOM), 506
    object request brokers (ORBs), 506–507
    remote procedure calls (RPCs), 505
    transaction monitor programs, 506
MILNET, 321
Minicomputers, 27–28
Ministry of Telecommunications, 155, 420
MNP standards, 71
Mobile cellphones, 430
Mobile telephone switching office (MTSO), 430, 432
Modems, 52, 57, 64–71, 379–389
    baud rate, 69–70, 93, 381
    communication program setup, 392–394
    compression, 65, 70–71
    data compression, 382–383
    dial-back modems, 388–389
    error detection/error correction, 65, 381–382
    external, 384–385

facsimile modems, 386–387
file transfer protocols, 394–395
hand-shaking signals, 381
intelligent modems, 384
internal, 384
managed modems, 389
modulation, 64, 66–70
operation of, 63, 64
PCMCIA modems, 385–386
speed, 64–65, 380–381
standards, 64–65, 71, 380–381, 383
telephone transmission, 64–71, 384, 385, 392, 393
voice/data modems, 388
Moderated conferencing, 254
    moderator tools, 254
Modulation, 64, 66–70
    amplitude modulation, 66
    complex modulation, 68–70
    frequency modulation, 64, 66
    phase modulation, 66–68
MONITOR program, 482–483, 488
    disk information, 483, 488
    LAN information, 488
    processor information, 488
Monthly service charge, ISDN, 78
MOOs, 327–328
MOSAIC, 345
Mother board, 468–469
Motion Picture Experts Group (MPEG), 13
MS-DOS
    changing directories, 331
    viewing contents of file, 332
    viewing files in directory, 331
MUDs, 327–328
Multidimensional database, 290
Multifamily network management programs, 233
Multihomed router, 446
Multimedia databases, 292–293
Multiple access, meaning of, 95
Multiplexing, 164–165
    meaning of, 98
    OSI, 306, 315
Multipoint control unit (MCU), 272
Multiport repeater, 91, 403
Multiprotocol routers, 145

Multipurpose Internet Mail Extensions (MIME), 257
Multitasking, 29
Multivendor standards, 5–5

N

Name anchor, 355
Narrowband ISDN, 124, 175–176
Netscape Navigator, 42, 345
NetView, 230
Netware Core Protocol, 193
Netware LANalyzer for Windows, 488–489
NetWare loadable modules (NLMs), 474–475
Netware SOS, 470
Network capable operating system, 188
Network communications
    electronic mail, 251–262
    electronic meeting rooms, 274–276
    groupware, 262–270
    production systems, 276–277
    standards, 250–251
    videoconferencing, 270–274
Network control centers, 219
Network design programs, 237, 240
    and what-if analysis, 237, 240
Network File Service (NFS), 189, 190
Network interface cards (NICs), 89, 185, 476–480
    address settings, 478–480
    and board driver, 476
    expansion board NIC, 89
    manual installation, 477
    plug-and-play installation, 476–477
    32-bit cards, 473
Network layer, OSI, 306–308
Network management
    accounting management, 245–246
    areas of, 218
    configuration management, 235–237
    consoles for, 219, 220
    distributed network management system, 223–224
    elements of system, 219–220

fault management, 233–235
management agents, 222–223
management information base (MIB), 221
messages in, 224–225
multifamily network management programs, 233
NetView, 230
performance management, 237–241
performance monitoring, 232
security management, 241–245
software metering, 231–232
Network management program, functions of, 220–221
Network management server, 212
Network management standards
 areas for needed standards, 225–227
 carrier standards, 232
 Common Management Information Services (CMIS), 230
 Common Management Information Protocol (CMIP), 230
 Desktop Management Interface (DMI), 230–232
 limits of, 232–233
 products and multiple standards, 233
 server systems management standards, 232
 simple network management protocol (SNMP), 227–229
Network map, 236
Network-ready printers, 207
Networks
 client/server processing on, 36
 definition of, 9
 new, design of, 240–241
 See also specific types of networks
Network versions, 200
Newbies, 328
New networks, design of, 240–241

900 numbers, 428
Noise, 378–379
 meaning of, 90, 378
 signal-to-noise ratio (SNR), 379
Noise floor, 90
Nonrepudiation, in authentication, 502
Novell IPX/SPX architecture, 17
Novell NetWare, 31, 188–189, 193
 MONITOR utility, 482–483, 487–489
NSFNET, 322
Number of bits per second, 239
NuPools
 database applications, 296
 electronic mail, 262
 Internet access, 51
 local area network (LAN), 88–89
 overview of company, 3–4
 PC networks, 212–213
 videoconferencing, 273
 on WWW, 41, 44

O

Object request brokers (ORBs), 506–507
Octets, 97
Offline, 61
100mbps LANs, 117–120
 demand priority access method (DPAM), 120
 802.3 100Base-X standards, 117–118
 evaluation of, 120
 100VG-AnyLAN standards, 119–120
100VG-AnyLAN standards, 119–120
 hubs/wiring, 119–120
Online analytical processing (OLAP), 290
Online services, 335–336
 Internet connection, 335
 See also Commercial online services
Open Data Base Connectivity (ODBC), 287
Open Shortest Path First (OSPF), 458
Open standards, 11
Open system, meaning of, 12

Operating systems, 188–190
 network capability of, 188, 477
 server operating systems (SOSs), 188–190
Optical fiber, 10Base-F standard, 105–106
Optimization, in routing, 124
Oracle, 209, 285
Ordered lists, Hypertext Markup Language (HTML), 360–361
OSI architecture
 application layer, 314
 data link layer, 305–306
 error detection/correction, 315
 internetting subnet layers, 309–311
 multiplexing function, 306, 315
 network layer, 306–308
 physical layer, 304–305
 presentation layer, 313
 protocol data units (PDUs), 308–309
 security function, 315
 session layer, 312
 transport layer, 311
OSI Layer 7 standards, 314
OSI (Open System Interconnection), 12–14
 agencies producing standards, 13
 architecture. *See* OSI architecture
 layers of, 14–15
 market position of, 13–14
 standard agencies related to, 12–13
Outsourcing, 245–246, 298–299
 basic rules for, 246
 nature of, 298–299

P

Packets, 125, 164, 307, 308
 in switched networks, 121–122
Packet switching, 164–165
 OSI, 306–308
PAD field, 99
Parallel plugs, 75, 76
Parallel ports, 384
Parallel transmission, 76

Parity bits, 73, 393
Passwords, 60, 243
Payroll processing services, 294
PCMCIA modems, 385–386
PC networks, 28–31
    client PCs, 28, 185–186
    client/server processing, 205, 208–210
    communication layers, 192–193
    communication servers, 210–212
    corporate use of, 212–213
    electronic mail, 212
    elements of, 185
    file server program access, 199–202
    file service, 194–199
    LANalyzer, 488–489
    MONITOR program, 482–483, 488–489
    network management server, 212
    operating systems, 188–190
    print service, 205, 206–207
    redirection, 202–204
    servers, 28–29, 186–188
    size factors, 191
    user support, 489–490
PC networks installation
    application software on client PC, 475
    application software on server, 473–475
    backup systems, 473
    busses, 470–471
    client shell software, 477
    directories for file servers, 465–466
    disk caches, 472
    disk drive, 471
    inheritance in Apps, 467
    microprocessor, 469
    mother board, 468–469
    NetWare loadable modules (NLMs), 474–475
    network interface card (NIC), 476–480
    PC versus RISC server, 468
    RAM, 472
    symmetric multiprocessing (SMP) machines, 469–470
    user account installation on file server, 481–482
    user rights, 466–467

Peak periods, 239
Percentage of capacity, 239
Percent of packets with errors, 239
Performance management, 237–241
    client/server performance management, 241
    new network design, 240–241
    what-if analysis, 237, 239, 241
Performance monitoring, 232
Permanent virtual circuits, 170–171
    cost factors, 179–180
Personal communication service (PCS), cellular telephones, 435
Personal computers. *See* PC networks
Personal telephone number, 428
Phase, 66–67
Phase modulation, 66–68
Physical access control, 243
Physical layer
    LANs, 85
    OSI, 304–305
Piecemeal deployment, and IP Version 6, 461
Ping program, 457
Pixel-mapped graphics
    black-and-white images, 374
    color images, 374–375
    compression, 375
    pixel-mapped graphics codes, 374–375
    standards, 375
Pixels, 374
Plaintext, 496
Plug-and-play installation, network interface cards (NICs), 476–477
Point of presence (POP), 156–157, 419
Point-to-point protocol (PPP), 44, 45, 52, 336
Point-to-point transmission
    bandwidth, 379
    modems, 379–389
    noise, 378–379
    pixel-mapped graphics codes, 374–375
    text/graphics documents, 376–378
    vector graphics, 376

Portable Document Format (PDF), 378
Port address, 479
Port number
    Internet Protocol (IP) number, 449
    well-known port number, 449
Ports, 91
    virtual port, 205
Post Office Protocol (POP), 256
PostScript fonts, 377, 378
PowerBuilder, 286
PPP (Point-to-Point Protocol), 44, 45, 52, 336
Preamble, media access control (MAC) layer frame, 98
Presentation layer, OSI, 313
Presentation tool, 276
Primary rate interface (PRI), 176
Print service, 205, 206–207
    network-ready printers, 207
    print queues, 207
    and virtual ports, 206–207
Priority, of frame in token-ring networks, 113–114
Privacy, 241
    email, 261–262
Privacy extension header, 461
Private branch exchanges (PBX), 421–423
    network, 423–424
    wiring, 421, 423
Private telephone networks, 162, 424
Private virtual drive, 198
Prodigy, 293, 335
Production data processing, 290–291
    three-tier processing, 291
Production systems, 262, 276–277
    electronic document management (EDM), 277
    imaging systems, 276–277
    workflow systems, 277
Program file service
    file server program access, 201–202
    file storage, 199–200
    program execution, 200–201
Programming languages, fourth-generation, 286
Project management system, 266–267, 276

INDEX 551

Propagation effects, 91
Proprietary MIBs, 228–229
Proprietary standards, nature of, 11
ProShare, 274
Protocol analyzers
　error rate measurement, 239
　LANalyzer, 488–489
Protocol data units (PDUs), 441
　OSI, 308–309
　reception of, 49–50
　transmission of, 47–48
Protocols, 50
　message exchanges, 47
PSNT (Public Switched Telephone Network), 416
Public key encryption, 242, 497–499
　limitations of, 499
Public screen, 275
Public Telephone and Telegraph (PTT), 155, 419
Public utilities commission (PUCs), 419
Pulse code modulation (PCM), sampling for, 79–80
Px64 standard, 502

Q

Quality of service field, and IP Version 6, 461

R

Radio LANs, 125–126
　benefits of, 125–126
　limitations of, 126
RAID drives, 471
RAM, 472
　insufficient and networks, 483
Rapid application development, 284–285
Real-time discussions, Internet, 327–328
Redirection
　in client shells with older operating systems, 204
　in network-capable operating systems, 203–204
　PC networks, 202–204
Redirect message, 458
Redirector program, 477
Red lining, 429

Reengineering, 296–299
　consolidation in, 297
　goals of, 296–297
　interorganizational systems, 298
　outsourcing, 298–299
　and teamwork, 297–298
　telecommuting, 299
Reference model, meaning of, 12
Reference Model of Open Systems Interconnection. *See* OSI architecture
Regulation, telephone service, 419–421
Relational databases, 283
Relative addressing, 354, 355
Reliable packet switched networks, 166
Remote access client software, 210
Remote access server, 210
Remote control software, 504
Remote management, 236
Remote printing, 30
Remote procedure calls (RPCs), 505
Repeaters
　full repeater, 400
　half repeater, 400
　LAN connection, 132, 400
　multiport repeater, 403
Replication, and Lotus Notes, 269
Request, 209
　and network management, 224–225
Request message, 36
Requests for comments (RFCs), 14
Resegmentation, of LAN, 131
Response messages, 36
　and network management, 225
Response time, 238
Reverse Address Resolution Protocol (RARP), 455
Rights of user, PC network, 466–467, 481
RISC workstations, 31–32, 468
　operating system of, 32
　operation of, 31–32
RJ-45 plugs/jacks, 90, 101
RMON (remote monitoring), 229

Roll-about system, videoconferencing, 271–272
Route, 308
　alternative routing, 123–124
　remote routing, 147
　switched LANs, 123
Router coordination protocols, 458–459
Routers, 133, 141–145, 318
　address, 446
　autoconfiguration hosts, 143
　autonomous routers, 458
　border routers, 459
　connecting to digital leased lines, 162–163
　coordination protocols, 458–459
　default router, 142, 445
　domain name system (DNS), 143
　dual-homed router, 446
　error reporting, 142
　evaluation of, 145
　functions of, 133
　and Internet Protocol (IP) packets, 143–144
　Internet Service Provider (ISP), 45, 46
　multihomed router, 446
　multiprotocol routers, 145
　operation of, 141, 142–144
　reception of message, 49
　role in internetting, 446–447
　TCP/IP routing standards, 141, 142–144
Routing information, media access control (MAC) layer frame, 409
Routing Information Protocol (RIP), 458
RSA encryption method, 497–499
Rules, email, 253

S

Sampling, for pulse code modulation (PCM), 79–80
Satellite network control centers, 223–224
Scalability
　ATM switches, 125
　meaning of, 269
　symmetric multiprocessing (SMP), 286

Schema of database, elements of, 225–226
Screen resolution, and video compression, 503–504
Scripts, login script, 60–61
SDH (synchronous digital hierarchy), 159
Security, OSI, 315
Security management, 241–245
    audit trails, 244–245
    authentication, 242–243
    challenge and response identity verification, 243–244
    encryption, 241–242
    firewalls, 244
    passwords, 243
    physical access control, 243
    virus control, 245
Segment, LAN links, 100–101, 398, 399–401
Self-test, to diagnose faults, 235
Sequenced packet exchange (SPX) protocol, 193
Serial cable, 384
Serial plug, 74–75
Serial ports, COM 1/2/3/4, 392
Serial transmission, 74, 76
Server operating systems (SOSs), 188–190
    independent SOSs, 189
    Microsoft Windows Server NT, 189
    Novell NetWare, 188–189
    synchronized SOSs, 189–190
    UNIX, 189
Server program, 36, 205, 208–209, 282
Servers, 28–29, 186–188
    application software, 188
    file servers, 30
    high-end servers, 187
    compared to hosts, 28–29
    mainframes, 187
    operating systems, 31
    specialization of, 29, 186–187
    workstation servers, 187
Server software, 285–286
Service customization, automatic number identification (ANI), 429
Services, in network management, 226
Session layer, OSI, 312

Share directory, PC network, 466
Shared virtual drive, 198
Shielded twisted pair (STP) wiring, 112
Simple Mail Transfer Protocol (SMTP), 255, 321
Simple network management protocol (SNMP), 227–229
    and proprietary MIBs, 228–229
    RMON (remote monitoring), 229
    SNMPv2, 228
Single-key encryption, 497
Single point of failure, 148
Smart cards, 244
SMDS, 165
Socket, Internet Protocol (IP) number, 449
Software, terminal emulation, 59–60
Software inventory tools, 236
    limitation of, 236
Software metering, 231–232, 236
SONET (synchronous optical network), 159
Source address, 98
Source quench, 458
Spamming, email, 260
Specialization
    and servers, 29
    of servers, 186–187
Speed of transmission, 18–19
    error rates, 239
    modems, 64–65, 380–381
    and network availability, 238
    outage measures, 238
    peak periods, 239
    response time, 238
    throughput, 239
    types of, 18–19
Spoof, Internet address, 337
SQL Server, 285
Standardized General Markup Language (SGML), 378
Standards
    and application layer, 34
    and client/server databases, 286–287
    communication methods, 251
    competing standards problem, 8

    electronic data interchange (EDI), 294
    electronic mail, 254–257
    facsimile modems, 387
    implementation profiles, 17
    interface standards, 50
    local area networks, 85–89, 91, 100–103, 398–403
    meaning of, 5
    and mediated communication, 5
    for modems, 64–65, 71, 380–381, 383
    multivendor standards, 5–5
    network management standards, 225–233
    open standards, 11
    organizational standard setting, 17
    pixel-mapped graphics, 375
    proprietary standards, 11
    TCP/IP (Transmission Control Protocol/Internet Protocol), 14–16
    token-ring networks, 111, 119–120, 404–413
    video compression, 502
    videoconferencing, 270–271
    See also specific standards
Standards agencies, 10–11
    ad hoc agencies, 11
    de jure agencies, 11
    major agencies, 11
Standards architecture
    IPX/SPX, 17
    OSI (Open System Interconnection), 12–14
    Systems Network Architecture (SNA), 17
Start bits, 72
Start of frame delimiter, media access control (MAC) layer frame, 98, 407
State
    changing states, 235–236
    meaning of, 235
Stations, 9
Stop bits, 72
Strategy support systems, 42–43
Structured Query Language (SQL), client/server databases, 287
Stub, 200, 473
Subnet, meaning of, 310

Index

Subnet-dependent convergence layer, OSI, 310–311
Subnet-independent convergence layer, OSI, 310
Subnet layer, 9, 13
    example of, 46
    subnet access layer OSI, 310
Subnet mask, 142, 449–451
Subnet transmission protocol data unit (SnPDU), 48
Supervisory protocols
    Address Resolution Protocol (ARP), 453
    autoconfiguration protocols, 453–455
Switched data networks, 164–180
    asynchronous transfer mode (ATM), 177
    cell switched network, 171
    circuit switching, 164
    connectionless service, 167–168
    connection-oriented service, 168–169
    cost factors, 173
    elements of, 172–173
    Frame Relay, 176–177
    ISDN service, 175–176
    load reduction on, 171
    packet switching, 164–165
    reliable packet switched networks, 166
    unreliable packet switched networks, 167
    virtual circuits, 169–171
    X.25 data networks, 174–175
Switched Ethernet, 139–140
    operation of, 140
    simple switch, 139–140
    virtual LANs, 141
Switched LANs, 120–125
    alternative routing, 123–124
    asynchronous transfer mode (ATM) switches, 124–125
    Ethernet switches, 124
    and OSI architecture, 125
    route, 123
    switched data network, 122–123
    switching decision, 123
    switching mechanism, 121–122
Switched virtual circuits, 171, 180

Switches
    evaluation of, 147
    LAN switch, 132–133
Switching
    elements of, 121–122
    and leased lines, 161–162
Switching office, 156
    telephone call, 417, 418
Sybase, 285
Symmetric multiprocessing (SMP), 286
    application threads, 469–470
    machines for, 469–470
    operating systems for, 470
    scalability, 286
Synchronized server operating systems, 189–190
System directory, PC network, 466
Systems management, server systems management standards, 232
Systems Network Architecture (SNA), 17
    market share, 17

T

T1 circuits, 159
    fractional, 159
T3 circuits, 159
Tags, Hypertext Markup Language (HTML), 349, 350–351, 353
Tag view, 349
Talk session, Internet, 327
Tape backup, 473
Tariffs, telephone service, 426–427
TCBworks, 270
TCP/IP (Transmission Control Protocol/Internet Protocol), 14–16, 193, 321
    Address Resolution Protocol (ARP), 453
    autoconfiguration protocols, 453–455
    domain name system (DNS), 455–456
    host, role in delivery, 444–446
    Internet Control Message Protocol (ICMP), 457–458
    Internet Protocol (IP) number, 448–451

    layers of, 16, 17, 440–441
    market acceptance, 14–16
    router address, 446
    router coordination protocols, 458–459
    router role, 446–447
    and routers, 141, 142–144
    standard setting, 14
    TCP (Transmission Control Protocol), 15–16, 47, 452–453
    User Datagram Protocol (UDP), 453
TCP (Transmission Control Protocol), 15–16, 47, 452–453
Telecommunications Act of 1966, 156, 420
Telecommuting, 299
Telenet, 293
Telephone lines
    CSLIP (Compressed Serial Line Internet Protocol), 45
    and home use, 52
    PPP (Point-to-Point Protocol), 45
Telephone service
    automatic number identification (ANI), 429
    carrier services/pricing, 426
    cellular telephones, 429–436
    competitive access providers (CAPs), 418
    deregulation, 419
    800/888 numbers, 427
    electronic switching services, 428–429
    interexchange carriers (IXCs), 419
    international common carriers (ICCs), 419
    international service, 419, 420
    Internet, 329
    local access and transport area (LATA), 417, 419
    local calling, 427
    local carrier exchange (LCE), 418
    900 numbers, 428
    personal telephone number, 428
    placing call, events of, 417–418
    private branch exchanges (PBX), 421–423

Telephone service (*continued*)
  private branch exchanges (PBX) network, 423–424
  regulation, 419–421
  tariffs, 426–427
  toll calls, 427
  user services, 425
  wide area telephone services (WATS), 428
Telephone transmission, 62–72
  analog transmission, 64
  codec, 78, 80
  digital transmission, 63
  versus ISDN transmission, 77–80
  limitations of, 71–72
  and modems, 64–71, 384, 385, 392, 393
  terminal-host communication connection, 57
Telephony, 274
Telnet, 319–321
10 Mbps LANs, limitations of, 115–116
10Base-T hubs. *See* Ethernet 10Base-T hubs
Terabit per second (tbps), 19
Terminal adapter, for ISDN, 78
Terminal emulation, 56–57, 58–62
  ANSI terminals, 62
  file transfer, 61
  login scripts, 60–61
  meaning of, 58
  software, 59–60
  3270 terminals, 62
  TTY terminals, 62
  virtual terminals, 62
  VT100 terminals, 58–60, 391–395
Terminal-host communication
  asynchronous ASCII transmission, 57–58, 72–74
  telephone transmission, 57, 62–72
  terminal emulation, 56–57, 58–62
Terminal-host systems, 27–28
  mainframes, 27
  minicomputers, 27–28
Termination equipment, 421
Text
  ASCII text, 376–377
  fonts, 376–377
  formatting, 376–377

Hypertext Markup Language (HTML), 359–360, 368–372
LATEX format, 378
Portable Document Format (PDF), 378
text/graphics integration, 378
Thinnet, 401–402
3270 terminals, 62
TIFF (Tagged Image File Format), 375
Token, in token-ring networks, 113
Token-ring networks, 110–114
  access units, 112–113
  baseband signaling, 113
  802.5 Token-Ring Network standard, 111, 404–413
  market status of, 114
  100VG-AnyLAN standards, 119–120
  ring topology, 111
  shielded twisted pair (STP) wiring, 112
  speed of transmission, 113
  token passing, 113–114
Toll calls, 427
Transaction automation, 2
Transaction monitor programs, 506
Transceivers
  medium attachment unit, 104, 399
  radio LAN, 125
Transmission carriers, 154–160
  circuits, 157–160
  competitive access providers (CAPs), 155
  deregulation of, 155–156
  domestic carriers, 154–155
  interexchange carriers (IXCs), 155
  international carriers, 155
  local access and transport areas (LATAs), 155
  local transmission service, 156–157
  network management, 232
  public telephone and telegraph authority (PTT), 155
  tiers of carriers, 154
Transmission layer
  communication method, 47

  in communication process, 9–10
  internetting transmission layer, 46
  local area networks (LANs), 19–20
  message transmission, 47–48
  metropolitan area networks (MANs), 21
  protocols, 47
  speed of transmission, 18–19
  subdivisions of, 9
  wide area networks (WANs), 20
Transmission service, 156
  local loop, 156
  point of presence (POP), 156–157
Transport layer
  in communication process, 8
  computer platforms, 26–33
  OSI, 311, 311–312
  PC networks, 193
  TCP (Transmission Control Program), 47, 452–453
  User Datagram Protocol (UDP), 453
Transport-layer protocol data unit (TPDU), 311, 312
TrueType fonts, 377
Trunk lines, 122, 156
  telephone call, 417, 418
TTY terminals, 62
Turnpike theorem, 240–241
TYMNET, 293
Type of service field, and IP Version 4, 461

U

Uniform resource locator (URL), 346, 363
UNIX, 31, 189, 190, 285–286, 470
  changing directories, 332
  and RISC workstations, 31
  viewing contents of file, 332
  viewing files in directory, 332
Unreliable packet switched networks, 167
Unshielded twisted pair (UTP), 89–91
  Category 3/4/5, 90

distance limitations, 90–91
quality standards, 90
RJ-45 plugs/jacks, 90
twisting and interference, 89
wire bundles, 90
Uploading, 61
Upper memory, 479
URL (uniform resource locator), 39
USENET, 326
newsgroups, 253, 326
reader program, 326
User Datagram Protocol (UDP), 453
User directory, PC network, 465
User layer, in communication process, 6–7

V

V.42 *bis*, 383, 392
Variable latency, 116
Vector graphics, 376
advantages of, 376
Verbal communication, desktop groupware, 264
VeriFone
database applications, 296
electronic mail, 259, 262
Internet access, 51
local internet, 148–149
overview of company, 2–3
videoconferencing, 273
on WWW, 41
VERONICA, 334, 363
Video compression, 502–504
frames per second, reduction of, 503
screen resolution, reduction of, 503–504
standards, 502
Videoconferencing, 270–274
benefits of, 272–273
compression, 271
control consoles, 272
corporate uses, 273
cost factors, 270
desktop conferencing, 273–274
document conferencing, 274
graphics, 272
Internet, 328–329
multipoint control unit (MCU), 272
roll-about system, 271–272

room-to-room conferencing, 270–271
standards, 270–271
telephony, 274
Video services, circuits for, 159–160
Virsus, virus control software, 245
Virtual circuits, 169–171
permanent virtual circuits, 170–171, 179–180
switched virtual circuits, 171, 180
Virtual corporations, 2, 297
Virtual drives, 195–196
operations performed on, 195–198
private virtual drive, 198
shared virtual drive, 198
Virtual LANs, elements of, 141
Virtual ports, 205, 206–207
Virtual private networks, 160, 163, 170, 424
Virtual terminals, 62
Visual Basic, 286, 362
Voice/data modems, 388
Voting tool, 276
VT100 terminals, 58–60, 391–395
communication program setup, 392–393
emulation software, 59–60
file transfer protocols, 394–395
host setup, 393–394
saving setup information, 394

W

WAIS (Wide-Area Information Server), 335
Wavelength, 66
Webcrawlers, 348
Webpages, 344
Webserver, 45, 47, 344
reception of message, 49–50
switching between, 50
Well-known port number, 449
What-if analysis, 237, 239, 241
and network design programs, 237, 240
Whiteboard conferencing, 504
Wide area networks (WANs), 20
transmission speed, 20
Wide area telephone services (WATS), 428

Wideband lines, 379
Windows 95, 59
Windows NT Server, 286, 470
Wiring closet, 423
Workflow systems
desktop groupware, 265–266
in production systems, 277
Workstation servers, 187
World Wide Web (WWW), 10, 39–42
bookmarks, 347
browsers, 39, 363–365
client program, 345
client/server processing, 40, 344–345
commercial transactions, 348
Common Gateway Interface (CGI), 361–362
corporate uses of, 41
history list, 347
home pages, 346
and hypertext, 39
Hypertext Markup Language (HTML), 349–361
Hypertext Transfer Protocol (HTTP), 345–346
interactive functions, 361–362
Java, 362
jumping links, 347
keyword search engine, 348
menu directories, 347–348
and platform independence, 40
speed, 349
uniform resource locator (URL), 346
Wrapped, token-ring network, 411

X

X.12 standard, 294
X.25 data networks, 165, 171, 174–175
limitations of, 174
X.400 standards, electronic mail standards, 255–256
X.500 standard, 237
XMODEM, 61, 394
XMODEM/CRC, 394

Z

ZMODEM, 61, 395